AF763786

Ionic Liquids in Separation Technology

Ionic Liquids in Separation Technology

Edited by

ANTONIA PÉREZ DE LOS RÍOS

Chemical Engineering Department, University of Murcia

FRANCISCO JOSÉ HERNÁNDEZ FERNÁNDEZ

Chemical & Environmental Engineering Department, Technical University of Cartagena

ELSEVIER

Amsterdam • Boston • Heidelberg • London • New York • Oxford
Paris • San Diego • San Francisco • Singapore • Sydney • Tokyo

Elsevier
Radarweg 29, PO Box 211, 1000 AE Amsterdam, Netherlands
The Boulevard, Langford Lane, Kidlington, Oxford OX5 1GB, UK
225 Wyman Street, Waltham, MA 02451, USA

Copyright © 2014 Elsevier B.V. All rights reserved.

No part of this publication may be reproduced or transmitted in any form or by any means, electronic or mechanical, including photocopying, recording, or any information storage and retrieval system, without permission in writing from the publisher. Details on how to seek permission, further information about the Publisher's permissions policies and our arrangements with organizations such as the Copyright Clearance Center and the Copyright Licensing Agency, can be found at our website: www.elsevier.com/permissions.

This book and the individual contributions contained in it are protected under copyright by the Publisher (other than as may be noted herein).

Notices
Knowledge and best practice in this field are constantly changing. As new research and experience broaden our understanding, changes in research methods, professional practices, or medical treatment may become necessary.

Practitioners and researchers must always rely on their own experience and knowledge in evaluating and using any information, methods, compounds, or experiments described herein. In using such information or methods they should be mindful of their own safety and the safety of others, including parties for whom they have a professional responsibility.

To the fullest extent of the law, neither the Publisher nor the authors, contributors, or editors, assume any liability for any injury and/or damage to persons or property as a matter of products liability, negligence or otherwise, or from any use or operation of any methods, products, instructions, or ideas contained in the material herein.

ISBN: 978-0-444-63257-9

British Library Cataloguing in Publication Data
A catalogue record for this book is available from the British Library

Library of Congress Cataloging-in-Publication Data
A catalog record for this book is available from the Library of Congress

For information on all Elsevier publications visit
our web site at store.elsevier.com

This book has been manufactured using Print On Demand technology. Each copy is produced to order and is limited to black ink. The online version of this book will show color figures where appropriate.

ELSEVIER
Book Aid International
Working together to grow libraries in developing countries
www.elsevier.com • www.bookaid.org

This Book is Dedicated—

To our parents, Paco & Lola, and Mateo & Toñi.
&
To our brothers and sisters, Juanjo, Mª Luisa, Mentxu, Jesús, Mª Carmen, and Laura.
&
Specially to our nieces and nephews, Carmen, Víctor, Pedro, the two babies coming, and to our first baby who is also coming to our lives.

CONTENTS

LIST OF CONTRIBUTORS

Enrique Alvarez-Guerra
Departamento de Ingenierías Química y Biomolecular, Universidad de Cantabria, Santander, Cantabria, Spain

Juan Arturo Santaballa
Chemical Reactivity and Photoreactivity Group, Department of Physical Chemistry and Chemical Engineering I, University of A Coruña, A Coruña, Spain

Péter Bakonyi
Research Institute on Bioengineering, Membrane Technology and Energetics, University of Pannonia, Veszprém, Hungary

Katalin Bélafi-Bakó
Research Institute on Bioengineering, Membrane Technology and Energetics, University of Pannonia, Veszprém, Hungary

Oscar Cabeza
Departamento de Física, Facultade de Ciencias, Universidade da Coruña, A Coruña, Spain

Elena Fernández
Departamento de Química Analítica, Nutrición y Bromatología e Instituto Universitario de Materiales, Universidad de Alicante, Alicante, Spain

Francisco José Hernández Fernández
Department of Chemical and Environmental Engineering, Regional Campus of International Excellence "Campus Mare Nostrum," Technical University of Cartagena, Campus La Muralla, Cartagena, Murcia, Spain

Josefa Fernández
Departamento de Física Aplicada, Universidade de Santiago de Compostela, Santiago de Compostela, Spain

Jesús Hernández Fernández
Department of Chemical Engineering, Regional Campus of International Excellence "Campus Mare Nostrum," University of Murcia, Campus de Espinardo, Murcia, Spain

Zvjezdana Findrik
Faculty of Chemical Engineering and Technology, University of Zagreb, Croatia

Félix M. Gaciño
Departamento de Física Aplicada, Universidade de Santiago de Compostela, Santiago de Compostela, Spain

Laszlo Gubicza
Research Institute on Bioengineering, Membrane Technology and Energetics, University of Pannonia, Veszprem, Hungary

Sofiane Hassab
Department of Applied Physics, Technical University of Cartagena (UPCT), Cartagena, Spain

Angel Irabien
Departamento de Ingenierías Química y Biomolecular, Universidad de Cantabria, Santander, Cantabria, Spain

María Isabel Fernández Pérez
Chemical Reactivity and Photoreactivity Group, Department of Physical Chemistry and Chemical Engineering I, University of A Coruña, A Coruña, Spain

Elena López Lago
Grupo de Microóptica y Sensores de Frente de Onda, Departamento de Física Aplicada, Universidade de Santiago de Compostela, Santiago de Compostela, Spain

Katarzyna Materna
Poznan University of Technology, Institute of Chemical Technology and Engineering, Poznan, Poland

Michiaki Matsumoto
Department of Chemical Engineering and Materials Science, Doshisha University, Japan

Trinidad Méndez-Morales
Grupo de Nanomateriais e Materia Branda, Departamento de Física da Materia Condensada, Universidade de Santiago de Compostela, Santiago de Compostela, Spain

Humberto Michinel
Area de Óptica, Facultade de Ciencias de Ourense, Universidade de Vigo, Ourense, Spain

M. Canle L.
Chemical Reactivity and Photoreactivity Group, Department of Physical Chemistry and Chemical Engineering I, University of A Coruña, A Coruña, Spain

Nándor Nemestóthy
Research Institute on Bioengineering, Membrane Technology and Energetics, University of Pannonia, Veszprém, Hungary

José A. Nóvoa López
Area de Óptica, Facultade de Ciencias de Ourense, Universidade de Vigo, Ourense, Spain

Joaquín A. Ortuño
Department of Analytical Chemistry, University of Murcia, Murcia, Spain

Javier Padilla
Department of Applied Physics, Technical University of Cartagena (UPCT), Cartagena, Spain

Antonia Pérez de los Ríos
Department of Chemical Engineering, Regional Campus of International Excellence "Campus Mare Nostrum," University of Murcia, Campus de Espinardo, Murcia, Spain

Aulia I. Pratiwi
Department of Chemical Engineering and Materials Science, Doshisha University, Japan

Magdalena Regel-Rosocka
Poznan University of Technology, Institute of Chemical Technology and Engineering, Poznan, Poland

Aurora M. Rubio
Department of Chemical Engineering, Regional Campus of International Excellence "Campus Mare Nostrum," University of Murcia, Campus de Espinardo, Murcia, Spain

Josefa Salgado
Applied Physics Department, University of Santiago de Compostela, Santiago de Compostela, Spain

Luisa Segade
Departamento de Física, Facultade de Ciencias, Universidade da Coruña, A Coruña, Spain

Francisca Tomás-Alonso
Department of Chemical Engineering, Regional Campus of International Excellence "Campus Mare Nostrum," University of Murcia, Campus de Espinardo, Murcia, Spain

Luis M. Varela
Grupo de Nanomateriais e Materia Branda, Departamento de Física da Materia Condensada, Universidade de Santiago de Compostela, Santiago de Compostela, Spain

Lorena Vidal
Departamento de Química Analítica, Nutrición y Bromatología e Instituto Universitario de Materiales, Universidad de Alicante, Alicante, Spain

María Villanueva
Applied Physics Department, University of Santiago de Compostela, Santiago de Compostela, Spain

ABOUT THE EDITORS

Antonia Pérez de los Ríos is a professor at the Department of Chemical Engineering and Coordinator of Knowledge Transfer & Entrepreneurship at the University of Murcia, Spain. She graduated in Chemistry (2000) and in Chemical Engineering (2004) at the University of Murcia, being awarded with the First National Award in Chemical Engineering by the Spanish Ministry of Education and Science in 2005. She received her PhD degree in Chemical Engineering at the Faculty of Chemistry of the University of Murcia, being awarded with the distinction of Extraordinary Award of Doctorate for the best PhD doctoral work conducted at the University of Murcia in her area of knowledge. She did postdoctoral stays at the Technical University of Delft (The Netherlands), at the University of Nottingham (United Kingdom), and the Technical University of Cartagena (Spain).

She has vast experience in applying innovative chemical process engineering tools to design sustainable and safe chemical processes based on the use of neoteric solvents (ionic liquids and supercritical fluids), membrane technology, and biocatalysis. She has over hundred publications in scientific journals and several patents in the fields of Engineering, Chemical Engineering, Biotechnology, and Applied Chemistry. She has been member of the Scientific Committees of different congresses and she has been invited as speaker in many scientific events. She is also editor of many international journals.

Francisco José Hernández Fernández is a professor at Department of Chemical and Environmental Engineering, Technical University of Cartagena, Spain. He received his degree in Chemistry and degree in Chemical Engineering from the University of Murcia (Spain), being awarded with the Extraordinary Award in both degrees and with Special Mention in the National Award in Chemical Engineering by the Spanish Ministry of Education and Science. He received his PhD degree (European Doctorate) in Chemical Engineering at the Faculty of Chemistry of the University of Murcia, being awarded with the distinction of Extraordinary Award of Doctorate. He has realized research stays in Delft University of Technology (The Netherlands) with the Professor Roger Sheldon and Professor Fred van Rantwijk in The University of Nottingham (United Kingdom) with the Professor Gillian Stephens and Professor Peter Licence.

His current research lines include the development of green reaction and separation processes using ionic liquids and of sustainable processes to produce energy using microbial fuel cells. Derived from his research, he has published over hundred papers in the fields of Engineering, Chemical; Engineering, Environmental; Biotechnology & Applied Microbiology and Chemistry and several patents. He has contributed with many invited lectures in several congresses and he has being member of several Scientific Committees of different congresses. He is also editor of several international journals.

PREFACE

This book will focus on applications of ionic liquids in separation technology. Ionic liquids are organic salts, usually consisting of an organic cation and a polyatomic inorganic anion, which are liquid at room temperature. The most relevant property of ionic liquids is their very low vapor pressure. Furthermore, their physical and chemical properties can be fine-tuned by the adequate selection of the cation and anion constituents, for that reason being described as *design solvents*. Application of ionic liquids in chemical processes has blossomed only within the last decade. Indeed, these media have been used in replacement of volatile organic solvents in a wide variety of chemical processes, such as separation and purification, reaction media in biochemical, and chemical catalysis. Several books have covered their applications as reaction media in chemical and biochemical processes and also in catalysis. However, there is not any specific book in the field of separation processes, in spite of being one of the most active topics in ionic liquids research.

This work reports on the most important and latest fundamental and technological advances in separation processes using ionic liquids and it reaches a multidisciplinary audience. This book is aimed at scholars and scientists in the fields of chemistry, chemical engineering, physics, biochemistry, and biology interested in the application of these new solvents, as well as industrial researchers interested in the implementation of this new technology. It brings together the latest developments in this fascinating field, supplemented by numerous practical tips, and thus provides those working in both research and industry with an indispensable source of information. The specialist audience can find in this book a recompilation of the most important knowledgments is this field and it can be also of great help for audience who are interested in starting researching in ionic liquid separation technology. This book can be also useful for university students and audience interested in the study of new tendencies and technologies within the chemical engineering field.

The volume begins with an opening chapter that introduces the reader to the knowledge of the main properties of ionic liquids, which are interesting in the context of separation technology, and the green properties of these new solvents. A discussion of the physical–chemical basis of ionic liquids effects on separation and transformation processes is included in the second chapter. The third chapter highlights some of the recent advances and provides an overview at the current status of ionic liquids and extraction techniques, such as liquid–liquid extraction, single-drop microextraction, hollow-fiber liquid-phase microextraction, and dispersive liquid–liquid microextraction. The fourth chapter reviews the use of ionic liquids as solvents for efficient separation of the metal ions and organic compounds from aqueous solutions. The fifth chapter includes the recent progress on liquid separations using ionic liquid membranes (such as bulk, emulsion, and supported liquid membranes containing ionic liquids). Ionic liquid-based three phase partitioning is presented in the sixth chapter as a novel technique to recover proteins from waste streams or complex media that overcomes the main drawbacks of conventional methods. In the seventh chapter, the role of ionic liquids in enzyme-membrane integrated systems, which makes possible the development of processes that are environmentally friendly, consume less amount of chemicals, energy, have higher yields, and generate more profit, is discussed. The eighth chapter summarizes the benefits, features of supported ionic liquid membranes, and their recent progress and application possibilities for various gas separation purposes. The ninth chapter seeks to provide an overview of a special type of ion-selective electrode based on ionic liquids and their role in modern potentiometry. The different roles that ionic liquids can play in several aspects of electrochromic technology are reviewed in the tenth chapter. The last chapter includes an outlook that summarizes the advantages of using ionic liquids in separation technology and describes the latest progress and challenges remaining in this field.

The authors are working scientists, experts in the topics they cover. The Editors edited the content. At Elsevier, the project was coordinated by Anita Koch (Acquisition Editor), and we are grateful to her for her leadership and advice throughout the Project.

The Editors hope that you find the subject matter of Ionic Liquids in Separation Technology as interesting and exciting as they do.

ACKNOWLEDGMENTS

This book could not have been completed without the assistance of a series of valuable partners. We would like to heartily thank all them for all their efforts and patient along the publication process.

The authors are also grateful to Anita Koch (Acquisition Editor at Elsevier) for her leadership and advice throughout the Project and to Sharmila Vadivelan (Project Manager at Elsevier) for her kind help in the production process of the book.

For outstanding collaboration over several years, were are particularly indebted to our colleagues Francisca Tomás Alonso, Luis Javier Lozano Blanco, Joaquín Quesada Medina, Carlos Godínez Seoane and Sergio Sánchez Segado. Special thanks are due to all the students and collaborators who have worked in the ionic liquid field with us for all their efforts.

Last, but not least, it is our pleasure to thank our families for their constant encouragement and support. We love you.

LIST OF ABBREVIATIONS OF CATIONS AND ANIONS CONSTITUENTS OF IONIC LIQUIDS

Cations

Short Name	Long Name
$[C_nC_1Im]^+$	1-alkyl-3-methyl imidazolium
$[C_nC_mC_pIm]^+$	1,2,3 trialkyl imidazolium
$[(C_2CN)C_nC_1Im]^+$	1-propyronitrile-3-alkylmethyl imidazolium
$[C_4C_{3A}Im]^+$	1-allyl-3-butylimidazolium
$[SC_3C_1Im]$ or [MIMPS]	1-(3-sulfonic acid) propyl-3-methyl imidazolium
$[C_{2V}C_4Im]^+$	1-vinyl-3-butyl imidazolium
$[BzC_1Im]^+$	1-benzyl-3-methylimidazolium
$[C_4tmsIm]^+$	1-butyl-3-trimethylsilylimidazolium
$[C_{16}POHIm]^+$	1-(1,2-dihydroxypropyl)-3-hexadecylimidazolium
$[C_2OHC_1C_1Im]^+$	1-(2-hydroxyethyl)-2,3-dimethylimidazolium
$[C_2OHC_1Im]^+$	1-(2-hydroxyethyl)-3-methylimidazolium
$[(C_nSO_3)C_1Im]^+$	1-alkylsulfonate-3-methylimidazolium
$[(C_nCOOH)C_1Im]^+$	1-alkylcarboxy-3-methylimidazolium
$[(C_nCOOH)_2Im]^+$	1,3-dialkylcarboxy-3-methylimidazolium
$[PMVIm]^+$	poly(methyl 2-(3-vinylimidazolidin-1-yl)acetate
$[PMMEIm]^+$	poly-1-methyl-3-[2-(methacryloyloxy)ethyl] imidazolium
$(EO_2\text{-}TMSC_3Im]^+$	1-(2-(2-methoxyethoxy)ethoxy)ethyl)-3-(3-(trimethoxysilyl)propylimidazolium
$[C_nPy]^+$	alkyl-pyridinium
$[C_nC_mPy]^+$	dialkyl-pyridinium
$[C_4C_1C_1NPy]$	1-n-butyl-4-dimethylaminopyridinium
$[C_nPyrr]^+$	alkyl-pyrrolidinium
$[C_nC_mPyrr]^+$	1,1-dialkyl pyrrolidinium
$[C_1OC_2C_1Pyrr]^+$	1-(2-methoxyethyl)-1-methyl-pyrrolidinium
$[HNH_2C_1Pyrr]^+$	1-(6-amino-hexyl)-1-methylpyrrolidinium
$[C_nC_mPip]^+$	dialkyl-piperidinium
$[C_nC_mOxa]$	dialkyl-oxazolidinium
$[C_nC_mMorp]$	dialkyl-morphonium
$[C_nIsoq]^+$	N-alkyl isoquinolinium
[Chol]	cholinium
$[C_nN]^+$	alkyl ammonium
$[C_nC_mN]^+$	1,3-dialkylammonium
$[C_nC_mC_pC_qN]^+$ or N_{nmqp}	1-4 tetra-alkyl ammonium
$[C_{16}C_1C_1C_1N]^+$ or $[CTMA]^+$	cetyltrimethylammonium

(Continued)

—cont'd

Cations

Short Name	Long Name
[TMEDAPS] $^+$	N, N, N ′, N′-tetramethyl-N, N′-disulfonate propylacetamide diammonium
$[C_nC_mC_pC_qP]^+$ or $[P_{n,m,p,q}]^+$	tetraalkyl phosphonium
$[C_{12}Phe_2C_2P]^+$ or $[DEDPP]^+$	dodecylethyl-diphenyl phosphonium
$[Phe_2C_4C_2P]^+$ or[DPBEP-]$^+$	diphenylbutylethylphosphonium
$[(C_nSO_3)Phe_3P]$	alkylsulfonatetriphenylphosphonium
$[TMG]^+$	1,1,3,3-tetramethylguanidine
$[Gly]^+$	glycine

Anions

Short Name	Long Name
$[Cl]^-$	chloride
$[Br]^-$	bromide
$[I]^-$	iodide
$[I_7]^-$	pentaiodide
$[I_9]^-$	eneaiodide
$[BrI_2]^-$	diiodobromate
$[Br_2I]^-$	dibromoiodate
$[ClI_2]^-$	diodochlorate
$[BF_4]^-$	tetrafluoroborate
$[FeCl_4]$	tetrachloroferrate(III)
$[PF_6]^-$	hexafluorophosphate
$[C(CN)_3]^-$	tricyano methanide
$[N(CN)_2]^-$ or $[DCA]^-$	dicyanamide
$[B(CN)_4]^-$	tetracyanoborate
$[Au(CN)_2]^-$	dicyanoaurate
[CNS]	thiocyanate
$[ClO_4]^-$	perchlorate
$[NO_3]^-$	nitrate
$[HCOO]^-$	formate
$[CH_3COO]^-$	acetate
$[CF_3COO]^-$	trifluoroacetate
[Met]	methionate
[BA]	benzoate
[Hex]	hexanoate
$[OF]^-$	pentadeca fluorooctanoate

—cont'd

Anions

Short Name	Long Name
$[C_nSO_4]^-$	alkyl sulfate
$[C_1(OC_2)_2SO_4]^-$	diethyleneglycol monomethylethersulfate
[TfO] or $[CF_3SO_3]$	trifluoromethanesulfonate
$[C_{12}BzSO_3]^-$	dodecylbenzenesulfonate
$[NTf_2]^-$	bis(trifluoromethylsulfonyl)imide
$[PFSI]^-$	bis(perfluoroethyl(sulfonyl))imide
$[C_4C_4N]^-$	bis(nonafluorobutylsulfonyl)imide
$[FSI]^-$	bis(fluorosulfonyl)amide
$[NfO]^-$	nonafluoromethylsulfonate
$[PTA]^-$	(phenylthio)acetate
$[MTBA]^-$	2-(methylthio)benzoate
$[TOS]^-$	tosylate
$[TA]^-$	thioacetamide
[DHSS]	dihexylsulfosuccinate
$[C_nC_nPO_4]^-$	dialkylphosphate
$[DEHP]^-$	bis(2-ethylhexyl)phosphate
$[FAP]^-$	tris(pentafluoroethyl)trifluorophosphate
[Bis] or $[C_8C_8PO_2]^-$	bis(2,4,4-trimethylpentyl)phosphinate
$[DEHP]^-$	bis(2-ethylhexyl)phosphoric acid
$[sacch]^-$	saccharinate
$[Van]^-$	vanillate
$[LS]^-$	lauroyl sarcosinate
$[Sal]^-$	salycilate
$[SSal]^-$ or $[TS]^-$	thiosalycilate
$[BSB]^-$	bis(salicylat[2-])borate
$[BPhe_4]^-$	tetraphenylborate
$[B_{12}H_{11}N(C_2H_5)_3]^-$	triethylammonium-*closo*-dodecaborate
$[F_2NO_4S_2]^-$	bis(fluorosulfonyl)azanide
$[CF_3BF_3]^-$	(trifluoromethyl)trifluoroborate
$[C_2F_5BF_3]^-$	perfluoroethyltrifluoroborate
$[C_3F_7BF_3]^-$	(heptafluoro- n-propyl)trifluoroborate
$[C_4F_9BF_3]^-$	(nonafluoro- n-butyl)trifluoroborate
$[DGA]^-$	bis(2-ethylhexyl)diglycolamate
$[CA\text{-}12]^-$	*sec*-octylphenoxy acetate
$[MTBA]^-$	2-(methylthio) benzoate
$[PW_{12}O_{40}]^-$	phosphotungstate

CHAPTER 1

Properties and Green Aspects of Ionic Liquids

Oscar Cabeza
Departamento de Física, Facultade de Ciencias, Universidade da Coruña, A Coruña, Spain

The first chapter of this book will introduce the reader to the knowledge of the main properties that this huge group of compounds present, which gives its importance as green materials in many chemical and technology applications. Let us remember that ionic liquid (IL) is a common name given to very different substances that had the particularity of being fused salts below 100 °C (and usually at room temperature). Up to date, more than 1000 ILs have been synthetized, but millions of them can be prepared. This is because there are many combinations of organic cations (with alkyl chains of variable length) with anions (usually inorganic but also organic with their own alkyl chains).

The great applied interest of ILs comes mainly for its ability to transport electrical charge, broad temperature range in liquid state, very small vapor pressure (in many of them negligible), no inflammability, and a big capacity to dissolve specific compounds in a selective form depending on the given IL. In addition, the physic-chemical properties of the ILs can be tailored by mixing two of them, or with different solvents, therefore adapting the resulting mixture to the given application. Although many potential applications of ILs have been proposed in the literature, and for many of them the ILs improve the studied process, only some have reached the industrial scale up to now due mainly to lack of experimental data but also to the high prize that ILs have by now to be used in huge scales.

This chapter has been separated into six subchapters giving the majority of properties of pure ILs (physical, thermal, and optical), the published measurements about properties of IL mixtures, one subchapter about theory and computer simulations, and the last one exploring the "green" character of ILs. A different Spanish research group wrote each subchapter; each of them specialists in their subject:

- Physical properties of pure liquids by J. Fernández and F. Gaciño from the University of Santiago de Compostela.
- Thermal properties of pure ILs by J. Salgado and M. Villanueva from the University of Santiago de Compostela.
- Optical properties of ILs by E. López, J.A. Novoa, and H. Michinel from the Universities of Santiago de Compostela and Vigo.

Ionic Liquids in Separation Technology
http://dx.doi.org/10.1016/B978-0-444-63257-9.00001-8
Copyright © 2014 Elsevier B.V.
All rights reserved.

- Physical properties of mixtures by O. Cabeza and L.M. Segade from the University of A Coruña.
- ILs: theory and simulations by L.M. Varela and T. Méndez-Morales from the University of Santiago de Compostela.
- Green aspects of ILs by A.M. Rubio et al. from the Universities of Murcia and Cartagena.

The first five groups belong to the Galician Network of Ionic Liquids (REGALIs, CN2012/210) and acknowledge the Xunta de Galicia funds to create it.

SUBCHAPTER 1.1

Physical Properties of Pure Liquids

Josefa Fernández, Félix M. Gaciño
Departamento de Física Aplicada, Universidade de Santiago de Compostela, Santiago de Compostela, Spain

1. INTRODUCTION

This chapter summarizes the experimental measurements performed up to date in pure ionic liquid (IL) for four physical magnitudes: density (ρ), viscosity (η), surface tension (σ), and electrical conductivity (κ). A good understanding of these physical properties of ILs is required before using them in industrial applications.

Accurate values of *liquid density* are required in the design of equipment, such as condensers and *reboilers*, to perform material and energy balances involving liquids.

Viscosity is needed in the scale-up of IL applications. In general, a low viscosity is desired for solvent applications to minimize pumping costs, whereas higher viscosities may be favorable in lubrication or in supported membrane separation processes [1].

Electrical conductivity is a crucial property of ionic liquids to check their suitability in electrochemical processes, solar power applications, or in lithium batteries.

Surface tension may affect the manner in which these electrolytes are adsorbed into the structure of porous electrodes during capacitor manufacture. Surface tensions are also important in storage techniques or in the use of ILs in the synthesis of nanoparticles. Interfacial tensions are important in separation technology. We should remark that for ILs all these properties strongly depend on the impurities, especially with the water content.

2. DENSITY

Apart from the necessity to know the density of the ILs in most of their applications, densities are also needed to determine phase equilibrium, heat capacity, viscosity, and the sensible heat and phase change storage using some experimental techniques. For ILs, typical densities range from 0.96−1.65 g/cm^3 at 293 K, which are, in general, larger than most of the current molecular liquids. Figure 1 shows the temperature dependence of the densities of different ILs, where it can be seen that those based on tris(pentafluoroethyl)trifluorophosphate (FAP^-) anion are the densest, whereas the ILs 1-ethyl-3-methylimidazolium hexylsulfate ($[C_2C_1Im][C_6SO_4]$) and 1-butyl-1-methylpyrrolidinium tetracyanoborate ($[C_4C_1Pyrr][B(CN)_4]$) are the lightest. The liquids containing several halogen atoms are the densest. Thus, perfluoroalkanes are

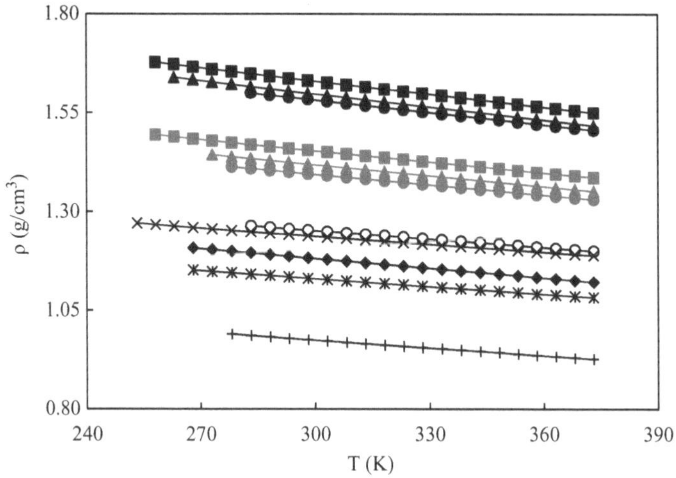

Figure 1 Experimental density of several ionic liquids as a function of temperature at atmospheric pressure: (■) $[C_1OC_2C_1Pyrr][FAP]$, (▲) $[C_4C_1C_1Im][FAP]$, (●) $[C_4C_1Pyrr][FAP]$, (■) $[C_1OC_2C_1Pyrr][NTf_2]$, (▲) $[C_4C_1C_1Im][NTf_2]$, (●) $[C_4C_1Pyrr][NTf_2]$, (○) $[C_4C_1Pyrr][OTf]$, (X) $[C_2C_1Im][C_2SO_4]$, (◆) $[C_{14}C_6C_6C_6P]$ [FAP], (⁎) $[C_2C_1Im][C_6SO_4]$, (+) $[C_4C_1Pyrr][B(CN)_4]$ [4].

denser than alkanes because the fluorine atoms are heavier than the hydrogen atoms, as well as strongly electronegative. Recently, Regueira et al., among other authors, have analyzed the effect of the anion structure in the ILs density [2–4]. These authors found for densities of $[C_4C_1Im]^+$ ILs the following trend: $[N(C_2F_5SO_2)_2]^- > [NTf_2]^- > [PF_6]^- > [TfO]^- > [ClO_4]^- > [CF_3COO]^- > [C_1SO_4]^- > [BF_4]^- > [C_1(OC_2)_2SO_4]^- > [NO_3]^- > [C_8SO_4]^- > [N(CN)_2]^- > [CH_3COO]^- > [C_4C_4PO_4]^- > [C(CN)_3]^-$. From Figure 1, it can be also concluded that the $[FAP]^- > [NTf_2]^- > [OTf]^- > [B(CN)_4]^-$ [4].

The effect of the cation on the density has been analyzed by several authors [5–7]. Regueira et al. [2] have found, for ILs with the $[NTf_2]^-$ anion and $[C_4C_1Py]^+$, $[C_4C_1Pyrr]^+$ or $[C_4C_1Im]^+$ cations, the following density trend: imidazolium > pyridinium > pyrrolidinium. This sequence can be explained by the planarization of the cation because of the formation of conjugated structures (as imidazolium and pyridinium). However, this trend seems to depend strongly on the anion. Thus, Sánchez et al. [5] have reported that for dicyanamide ILs, pyrrolidinium ILs are denser than imidazolium and pyridinium ILs. Furthermore, Chiappe et al. [7] concluded that density trend for imidazolium ILs are denser than pyrazolium ILs.

The effect on the densities [8–10] of the alkyl chains of ILs with alkyl sulfate anions and with the alkyl methyl imidazolium or *N*-alkylisoquinolinium ($[C_nIsoq]^+$) cations are shown in Figure 2. It can be seen that the densities decrease with the number of methylene

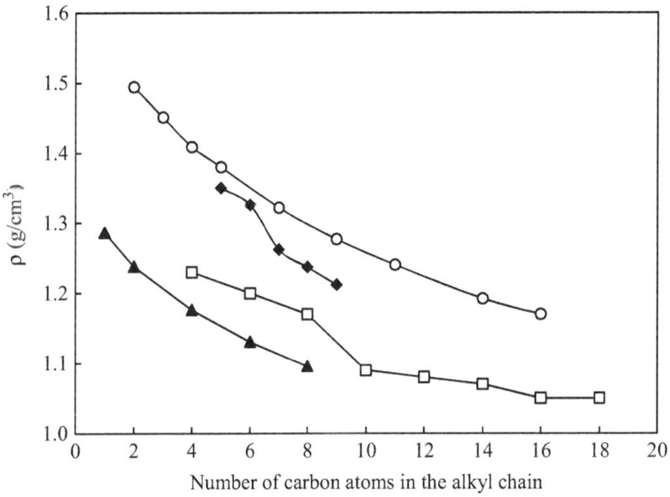

Figure 2 Experimental density of several ionic liquids as a function of the alkyl chain length: (○) $[C_nC_1Im][NTf_2]$ [8], (◆) $[C_nC_1Im][PF_6]$ [9], (□) $[C_nIsoq][N(C_2F_5SO_2)_2]$ [9], (▲) $[C_2C_1Im][C_nSO_4]$ [10]. $[C_nC_1Im][PF_6]$ and $[C_2C_1Im][C_nSO_4]$ values at T = 298.15 K. Values for $[C_nC_1Im][NTf_2]$ at T = 323.15 K.

chains. The same effect is found for polar molecular compounds. In all these types of liquids, increasing chain length of molecules leads to an increasing of the dispersive interactions between the aliphatic carbon chains (but also a decreasing of the polar or H-bond interactions) leading to a lower dense packing. Thus, the increase of the molecular mass produced by the increase of methylene groups is less important than the increment produced in the molar volume [11]. Similarly for quaternary ammonium ILs, the higher their number of carbons the lower the density [9]. For alkyltrioctylphosphonium $[C_nC_8C_8C_8P][Cl]$ ILs at 30 °C, Adamova et al. [12] have found two different but parallel trends for odd (4,6,8) and even (3,5,7,9) n values.

Most of the studied imidazolium-derived ILs are based on dialkylimidazolium $[C_nC_mIm]$ cations. The scarce studies on the density of ILs containing trialkylimidazoliums $[C_nC_mC_pIm]$ show that the introduction of a third alkyl substituent on the imidazolium ring at the C_2 position reduces the density [13].

In regard to the temperature dependence on the density, it is interesting to point out that, depending of the IL and its pressure and temperature conditions, its isobaric thermal expansivity may increase or decrease with the temperature. Thus, non-monotonic dependence of isobaric thermal expansivity was found for several ILs [2–4]. Besides, some ILs, such as $[C_2C_1Im][C_2SO_4]$, have very small thermal compressibility, which means that they are excellent as hydraulic fluids [2,14]. It is interesting to point out that several group contribution models have been developed for the density of ILs and its temperature and pressure dependence [15,16].

3. VISCOSITY

The viscosity of ILs vary widely depending on the type of cation and anion, ranging at least from 6−7600 mPa at 20 °C and atmospheric pressure. This property depends strongly on the molecular structure and is highly dependent on the interactions between the ions: electrostatic, van der Waals interactions, and hydrogen bonding [4,17]. Alkyl chain lengthening makes the salt more viscous because of an increase in van der Waals interactions. Moreover, delocalization of the charge on the anion, such as through fluorination, decreases the viscosity by weakening hydrogen bonding [17,18]. It seems that the viscosity of ILs is more dependent on the anion structure than on the cation. Several authors have analyzed the effect of the structure on the viscosity and its temperature and pressure dependence [17,19−21]. Recently, Yu et al. [22] have reported an extensive analysis of the viscosity of ILs according their structure.

Because of the high variety of anions it is not possible to have all the viscosity values for ILs with a common cation in order to obtain a general trend for viscosity. Figure 3 shows the viscosity of several ILs based on $[C_2C_1Im]^+$ and $[C_4C_1Pyrr]^+$ cations. For the $[C_2C_1Im]^+$ ILs, the following increasing trend with the anions is found: $[F_2NO_4S_2]^- < [CF_3BF_3]^- < [C_2F_5BF_3]^- < [C_3F_7BF_3]^- < [NTf_2]^- < [C_4F_9BF_3]^- < [OTf]^- < [N(C_2F_5SO_2)_2]^- < [C_2SO_4]^- < [C_6SO_4]^-$, whereas for $[C_4C_1Pyrr]^+$ ILs the trend is slightly different: $[C_3F_7BF_3]^- < [N(FSO_2)_2]^- < [C_2F_5BF_3]^- < [NTf_2]^- < [C_4F_9BF_3]^- < [CF_3BF_3]^- < [OTf]^- < [FAP]^-$. It can be seen that there are slight differences between these last trends. Taking into account these sequences and the partial trends found by Gardas and Coutinho [21], Pensado et al. [17], and Yu et al. [22], as well as Figure 4, we can conclude that:

1. The ILs with lowest viscosities are based in the following anions $[N(CN)_2]^-$, $[N(FSO_2)_2]^-$, $[C_nF_{2n+1}BF_3]^-$ with $n < 4$ and $[NTf_2]^-$ ILs. The low viscosity is due

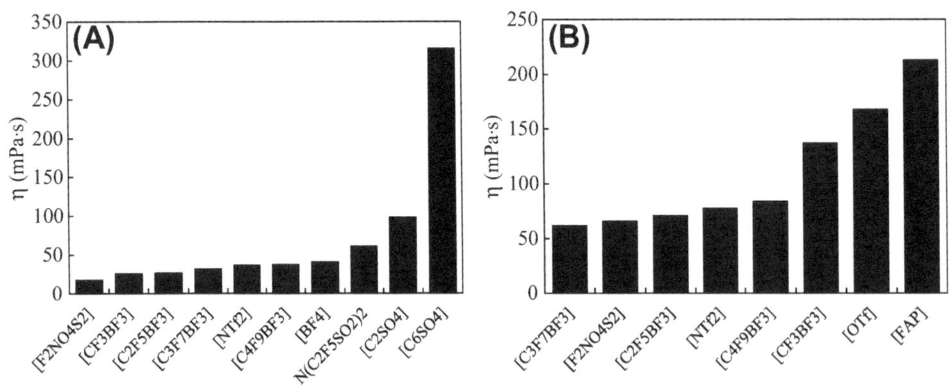

Figure 3 Viscosity trend of several ionic liquids according their anion at 298.15 K and atmospheric pressure. (A) $[C_2C_1Im]^+$ based ionic liquids, (B) $[C_4C_1Pyrr]^+$. *(Data taken from Yu et al. [22] and Gaciño et al. [4].)*

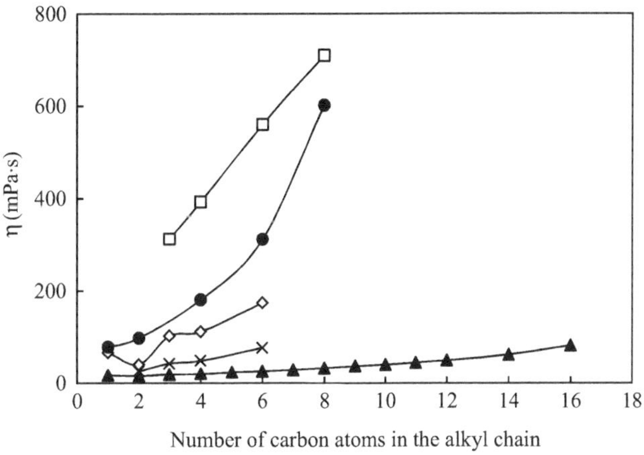

Figure 4 Experimental viscosity of several ionic liquids as a function of the alkyl chain length: (□) $[C_nC_1Im][PF_6]$ [9], (●) $[C_2C_1Im]$ $[C_nSO_4]$ [16] (◆) $[C_nC_1Im][BF_4]$ [9], (X) $[C_nC_1Im][OTf]$ [9], (▲) $[C_nC_1Im]$ $[NTf_2]$ [8,23]. All values at T = 298.15 K except for $[C_nC_1Im][NTf_2]$ which values are at T = 323.15 K.

to the high flexibility (e.g., for $[NTf_2]^-$) and to the electronic delocalization (e.g., $[N(CN)_2]^-$). Besides, the delocalization through fluorination diminishes the viscosity by weakening hydrogen bonding.

2. More viscous ILs follow the increasing general viscosity trend $[FAP]^- < [OTf]^- < [BF_4]^- < [C_2SO_4]^- < [C_1SO_4]^- < [C_6SO_4]^- < [PF_6]^- < [CH_3COO]^- < [Cl]^- < [Br]^-$. The high viscosity is due to the lack of conformational degrees of freedom of the anion (rigid structures), the localization of negative charge, and the increasing of the van der Waals forces because of the increasing of methylene groups (e.g., $[C_6SO_4]^-$). In Figure 4, it can been seen how the viscosity changes with the number of methylene groups for $[C_2C_1Im][C_nSO_4]$.

Viscosity values of the ILs containing a cation with a ring present the following trend: imidazolium < pyridinium < pyrrolidinium < oxazolidinium < piperidinium < morphonium. Those ILs formed by a cation with a six-member ring are more viscous than those with a five-member ring, thus, it can be concluded these increasing viscous trends: pyridinium > imidazolium, piperidinium > pyrrolidinium, and morphonium > oxazolidinium. Furthermore, ILs based of cations with saturated rings are more viscous than those based in aromatic cations (i.e., piperidinium > pyridinium, morphonium > pyridinium, pyrrolidinium > imidazolium, oxazolidinium > imidazolium) [22].

As it happens with the effect of the number of methylene groups of the anions, there is a pronounced increase in viscosity as the alkyl chain in the cation grows, as was reported for imidazolium, pyridinium, alkylammonium, and pyrrolidinium based ILs [5,18,24]. As for $[C_nSO_4]^-$ ILs, this effect can be explained because of the increase in the van der Waals interactions. However, some ILs with ethyl chains are less viscous with the corresponding ones with methyl chains (see Figure 4). This fact may be due to more flexibility of the ethyl chain than that of the methyl group (more conformational degrees of freedom, which may partly compensate the increase of van der Waals interactions) [22].

On the other hand, the higher the number of alkyl chains, the higher the viscosity. Fumino et al. [25,26] showed that when methylation of the C_2 position in $[C_2C_1Im]^+$ cation occurs, the long-range van der Waals interaction decrease and Coulomb interactions increase [27]. Thereby rearranging the charge network and making it more ionic-like with a more regular and tightly packed structure, an increase can be observed in viscosity and also in the melting point [20]. This increasing on viscosity from the addition of an alkyl chain was also reported by Gaciño et al. [20]. These authors compared viscosity data at high pressure of $[C_4C_1C_1Im][NTf_2]$ and $[C_4C_1Im]$ $[NTf_2]$ [28], obtaining that the addition of another methyl group on the $[C_4C_1Im]^+$ cation leads to a viscosity increase of up to 75%.

For some applications of the ILs, such as lubricants or hydraulic fluids, it is envisaged that the viscosity decreases less as possible with the temperature. ILs with $[NTf_2]^-$ anion and $[C_1OC_2C_1Pyrr]^+$, $[C_2C_1Im]^+$, and $[C_4C_1Pyrr]^+$ cations have low viscosity temperature dependences [4]. For the last applications, the pressure-viscosity coefficient is an important characteristic parameter. Fernández et al. [29] have found that $[C_4C_1C_1Im][FAP]$ has strong pressure viscosity dependence similar to that of some synthetic oils with similar kinematic viscosity, whereas $[C_2C_1Im][C_2SO_4]$ present a very low viscosity pressure dependence.

4. ELECTRICAL CONDUCTIVITY

Because ILs are entirely composed by ions, a high electrical conductivity, σ, should be expected. However, their σ values are quite low in comparison with conventional aqueous electrolyte solutions [30] used in electrochemical applications (40–75 S/m). The low conductivities can be attributed to the available charge carriers because of ion paring and/or ion aggregation, as well as the reduced mobility resulting from the large ion size. Their tunability allows the possibility of the design of a liquid with a specific conductivity value. Thus, at 25 °C the electrical conductivity is 0.0017 S/m for $[C_8C_4C_4C_4N][OTf]$ and 11.874 S/m for $[C_1C_1Im][Cl]$ [9].

Electrical conductivity of an IL depends on the mobility of its ions, which is influenced by the viscosity, ion size, and the ion association. Small ions with little interionic interactions usually result in high conductivities [31]. In regard to the cation, σ values increase as its molecular mass and size decrease, or when it becomes less symmetrical. Thus, for imidazolium and pyrrolidinium ILs, σ decreases as the alkyl chain length increases, whereas it increases when ether groups are added in the alkyl chain [18,32]. Conductivity generally follows the decreasing trend: imidazolium > pyrrolidinium > ammonium. Liu et al. [33] and MacFarlane et al. [34]

explain this fact through the different cation planarity, thus, the flatness of the imidazolium ring seems to confer a higher conductivity than the tetrahedral arrangement of alkyl groups displayed by the ammonium ILs.

As for concern to the anion effect on the conductivity, several trends were reported. For $[C_2C_1Im]^+$ ILs, Vila et al. [35] found the trend: $[TOS]^- < [Cl]^- < [Br]^- < [C_2SO_4]^- < [PF_6]^- < [BF_4]^-$ and Ignat'ev et al. [36] this sequence: $[FAP]^- < [NTf_2]^- \approx [OTf]^- < [CF_3COO]^-$. These last authors indicate that although $[NTf_2]^-$, $[OTf]^-$, $[CF_3COO]^-$ are less bulky than $[FAP]^-$, the triflate and trifluoroacetate anions coordinate more strongly to the cation and hinders their mobility. Besides, Leys et al. [37] found for the σ values of $[C_4C_1Im]^+$ ILs the following trend: $[Cl]^- < [Br]^- < [I]^- < [BF_4]^- < [NCS]^- < [N(CN)_2]^-$.

Electrical conductivity is approximately inversely proportional to the viscosity. Both properties are related by the empirical Walden law:

$$\Lambda\eta^{\alpha} = C$$

where Λ is the electrical molar conductivity, C is a constant, and α is an adjustable parameter. For ILs, α generally is slightly smaller than unity, which indicates that the dependence of Λ with pressure and temperature is a little lower in strength than that of viscosity. López et al. [38] have found that the roles of the density and temperature in the electrical conductivity and in the viscosity are very similar.

5. SURFACE TENSION

Surface tension, γ, is an important property in the study of physics and chemistry at free surfaces because of its influence on transfer rates of vapor absorption at the vapor–liquid interfaces. For this reason, the relationships between the chemical structure and the surface tension are essential in many fields, such as chemical process and reactor engineering, flow and transport in porous media, materials selection and engineering, biomedical and biochemical engineering, electronic and electrical engineering, as well as in environmental science and biology [1]. Tariq et al. [39] have recently published a critical review on surface tension of ILs.

In general, the liquid/air surface tension values of ILs are higher than those of conventional solvents (hexane 18 mN/m at 298 K) and organic compounds, but not so high as water (71.97 mN/m at 298 K) [9]. In general, the surface tension decreased when the ions had a higher packing efficiency or increased when the cohesiveness of the ILs also increased, such as through more hydrogen bonding [17]. The decrease of the surface tension of most of the ILs with increasing temperature is smaller than that of water, and similar to those of *n*-octane, *n*-perfluorobutane, or NaCl [39].

Concerning the alkyl chains of the imidazolium, the surface tension of the $[C_nC_1Im][NTf_2]$, $[C_nC_1Im][PF_6]$, and $[C_nC_1Im][BF_4]$ decreases with the increase of n. This decrease is more intense in the last two families. Besides, the same behavior was found for $[C_nSO_4]^-$, $[C_nC_1Pyrr]^+$, and tetra-alkylammonium ILs. Thus, the long alkyl side chains tend to be segregated toward the surface, forming an alkane-like layer [39]. Furthermore, the alkyl substitution at the C_2 position of the imidazolium seems to lead to slightly higher γ values.

For $[C_2C_1Im]^+$ ILs, the following increasing trend was found for the anion type: $[NTf_2]^- < [N(CN)_2]^- < [BF_4]^- < [I_7]^- < [I_9]^-$, whereas for $[C_4C_1Im]^+$ ILs it is verified $[OTf]^- < [BF_4]^- < [Cl]^- < [I_7]^-$. These trends indicate that larger, flexible, and/or asymmetrical anions where the charge is delocalized between several atoms tend to yield ionic liquids with lower surface tension values.

ACKNOWLEDGMENTS

This work was supported by Spanish Ministry of Science and Innovation and EU FEDER Program through CTQ2011-23925 project.

REFERENCES

[1] D. Rooney, J. Jacquemin, R. Gardas, Thermophysical properties of ionic liquids, Top. Curr. Chem. 290 (2009) 185–212.

[2] T. Regueira, L. Lugo, J. Fernández, Influence of the pressure, temperature, cation and anion on the volumetric properties of ionic liquids: new experimental values for two salts, J. Chem. Thermodyn. 58 (2013) 440–448.

[3] T. Regueira, L. Lugo, J. Fernández, High pressure volumetric properties of 1-ethyl-3-methylimidazolium ethylsulfate and 1-(2-methoxyethyl)-1-methyl-pyrrolidinium bis(trifluoromethylsulfonyl)imide, J. Chem. Thermodyn. 48 (2012) 213–220.

[4] F.M. Gaciño, T. Regueira, L. Lugo, M.J.P. Comuñas, J. Fernández, Influence of molecular structure on densities and viscosities of several ionic liquids, J. Chem. Eng. Data 56 (2011) 4984–4999.

[5] L.G. Sánchez, J.R. Espel, F. Onink, G.W. Meindersma, A.B.D. Haan, Density, viscosity, and surface tension of synthesis grade imidazolium, pyridinium, and pyrrolidinium based room temperature ionic liquids, J. Chem. Eng. Data 54 (2009) 2803–2812.

[6] S. Seki, T. Kobayashi, Y. Kobayashi, K. Takei, H. Miyashiro, K. Hayamizu, S. Tsuzuki, T. Mitsugi, Y. Umebayashi, Effects of cation and anion on physical properties of room-temperature ionic liquids, J. Mol. Liq. 152 (2010) 9–13.

[7] C. Chiappe, A. Sanzone, D. Mendola, F. Castiglione, A. Famulari, G. Raos, A. Mele, Pyrazolium-versus imidazolium-based ionic liquids: structure, dynamics and physico-chemical properties, J. Phys. Chem. B 117 (2012) 668–676.

[8] M.A.A. Rocha, C.M.S.S. Neves, M.G. Freire, O. Russina, A. Triolo, J.A.P. Coutinho, L.M.N.B.F. Santos, Alkylimidazolium based ionic liquids: impact of cation symmetry on their nanoscale structural organization, J. Phys. Chem. B 117 (2013) 10889–10897.

[9] S. Zhang, N. Sun, X. He, X. Lu, X. Zhang, Physical properties of ionic liquids: database and evaluation, J. Phys. Chem. Ref. Data 35 (2006) 1475–1517.

[10] A.J.L. Costa, J.M.S.S. Esperança, I.M. Marrucho, L.P.N. Rebelo, Densities and viscosities of 1-ethyl-3-methylimidazolium n-alkyl sulfates, J. Chem. Eng. Data 56 (2011) 3433–3441.

[11] J. Palomar, V.R. Ferro, J.S. Torrecilla, F. Rodríguez, Density and molar volume predictions using COSMO-RS for ionic liquids. An approach to solvents design, Ind. Eng. Chem. Res. 46 (2007) 6041–6048.

[12] G. Adamova, R.L. Gardas, L.P.N. Rebelo, A.J. Robertson, K.R. Seddon, Alkyltrioctylphosphonium chloride ionic liquids: synthesis and physicochemical properties, Dalton Trans. 40 (2011) 12750−12764.

[13] C.P. Fredlake, J.M. Crosthwaite, D.G. Hert, S.N.V.K. Aki, J.F. Brennecke, Thermophysical properties of imidazolium-based ionic liquids, J. Chem. Eng. Data 49 (2004) 954−964.

[14] T. Regueira, L. Lugo, J. Fernández, Ionic liquids as hydraulic fluids: comparison of several properties with those of conventional oils, Lubr. Sci. (2013), http://onlinelibrary.wiley.com/doi/10.1002/ls.1235/abstract.

[15] R.L. Gardas, J.A.P. Coutinho, Extension of the Ye and Shreeve group contribution method for density estimation of ionic liquids in a wide range of temperatures and pressures, Fluid Phase Equilibr. 263 (2008) 26−32.

[16] J. Jacquemin, P. Husson, V. Mayer, I. Cibulka, High-pressure volumetric properties of imidazolium-based ionic liquids: effect of the anion, J. Chem. Eng. Data 52 (2007) 2204−2211.

[17] A.S. Pensado, M.J.P. Comuñas, J. Fernández, The pressure−viscosity coefficient of several ionic liquids, Tribol. Lett. 31 (2008) 107−118.

[18] T.L. Greaves, C.J. Drummond, Protic ionic liquids: properties and applications, Chem. Rev. 108 (2007) 206−237.

[19] F.M. Gaciño, X. Paredes, M.J.P. Comuñas, J. Fernández, Effect of the pressure on the viscosities of ionic liquids: experimental values for 1-ethyl-3-methylimidazolium ethylsulfate and two bis (trifluoromethyl-sulfonyl) imide salts, J. Chem. Thermodyn. 54 (2012) 302−309.

[20] F.M. Gaciño, X. Paredes, M.J.P. Comuñas, J. Fernández, Pressure dependence on the viscosities of 1-butyl-2,3-dimethylimidazolium bis(trifluoromethylsulfonyl)imide and two tris(pentafluoroethyl) trifluorophosphate based ionic liquids: new measurements and modelling, J. Chem. Thermodyn. 62 (2013) 162−169.

[21] R.L. Gardas, J.A.P. Coutinho, A group contribution method for viscosity estimation of ionic liquids, Fluid Phase Equilibr. 266 (2008) 195−201.

[22] G. Yu, D. Zhao, L. Wen, S. Yang, X. Chen, Viscosity of ionic liquids: database, observation, and quantitative structure-property relationship analysis, AIChE J. 58 (2012) 2885−2899.

[23] M. Tariq, P.J. Carvalho, J.A.P. Coutinho, I.M. Marrucho, J.N.C. Lopes, L.P.N. Rebelo, Viscosity of (C2−C14) 1-alkyl-3-methylimidazolium bis(trifluoromethyl-sulfonyl)amide ionic liquids in an extended temperature range, Fluid Phase Equilibr. 301 (2011) 22−32.

[24] Y. Yoshida, O. Baba, G. Saito, Ionic liquids based on dicyanamide anion: influence of structural variations in cationic structures on ionic conductivity, J. Phys. Chem. B 111 (2007) 4742−4749.

[25] K. Fumino, A. Wulf, R. Ludwig, Strong, localized, and directional hydrogen bonds fluidize ionic liquids, Angew. Chem. Int. Ed. 47 (2008) 8731−8734.

[26] K. Fumino, A. Wulf, R. Ludwig, The cation-anion interaction in ionic-liquids probed by far-infrared spectroscopy, Angew. Chem. Int. Ed. 47 (2008) 3830−3834.

[27] K. Noack, P.S. Schulz, N. Paape, J. Kiefer, P. Wasserscheid, A. Leipertz, The role of the C2 position in interionic interactions of imidazolium based ionic liquids: a vibrational and NMR spectroscopic study, Phys. Chem. Chem. Phys. 12 (2010) 14153−14161.

[28] K.R. Harris, M. Kanakubo, L.A. Woolf, Temperature and pressure dependence of the viscosity of the ionic liquids 1-hexyl-3-methylimidazolium hexafluorophosphate and 1-butyl-3-methylimidazolium bis(trifluorosulfonyl)imide, J. Chem. Eng. Data 52 (2007) 1080−1085.

[29] J. Fernández, X. Paredes, F.M. Gaciño, M.J.P. Comuñas, A.S. Pensado, Pressure-viscosity behaviour and film thickness in elastohydrodynamic regime of lubrication of ionic liquids and other base oils, Lubr. Sci. (2013), http://onlinelibrary.wiley.com/doi/10.1002/ls.1236/abstract.

[30] M. Galiński, A. Lewandowski, I. Stepniak, Ionic liquids as electrolytes, Electrochim. Acta 51 (2006) 5567−5580.

[31] A. Pinkert, K.L. Keng, L. Ang, K.N. Marsh, S. Pang, Density, viscosity and electrical conductivity of protic alkanolammonium ionic liquids, Phys. Chem. Chem. Phys. 13 (2011) 5136−5143.

[32] T. Makino, M. Kanakubo, T. Umecky, A. Suzuki, T. Nishida, J. Takano, Electrical conductivities, viscosities, and densities of N-methoxymethyl- and N-butyl-N-methylpyrrolidinium ionic liquids with the bis(fluorosulfonyl)amide anion, J. Chem. Eng. Data 57 (2012) 751−755.

[33] H. Liu, Y. Liu, J. Li, Ionic liquids in surface electrochemistry, Phys. Chem. Chem. Phys. 12 (2010) 1685−1697.
[34] D.R. MacFarlane, P. Meakin, J. Sun, N. Amini, M. Forsyth, Pyrrolidinium imides: A new family of molten salts and conductive plastic crystal phases, J. Phys. Chem. B 103 (1999) 4164−4170.
[35] J. Vila, L.M. Varela, O. Cabeza, Cation and anion sizes influence in the temperature dependence of the electrical conductivity in nine imidazolium based ionic liquids, Electrochim. Acta 52 (2007) 7413−7417.
[36] N.V. Ignat'ev, U. Welz-Biermann, A. Kucheryna, G. Bissky, H. Willner, New ionic liquids with tris(perfluoroalkyl)trifluorophosphate (FAP) anions, J. Fluorine Chem. 126 (2005) 1150−1159.
[37] J. Leys, R.N. Rajesh, P.C. Menon, C. Glorieux, S. Longuemart, P. Nockemann, M. Pellens, K. Binnemans, Influence of the anion on the electrical conductivity and glass formation of 1-butyl-3-methylimidazolium ionic liquids, J. Chem. Phys. 133 (2010) 034503.
[38] E.R. López, A.S. Pensado, M.J.P. Comuñas, A.A.H. Pádua, J. Fernández, K.R. Harris, Density scaling of the transport properties of molecular and ionic liquids, J. Chem. Phys. 134 (2011) 144507.
[39] M. Tariq, M.G. Freire, B. Saramago, J.A.P. Coutinho, J.N.C. Lopes, L.P.N. Rebelo, Surface tension of ionic liquids and ionic liquid solutions, Chem. Soc. Rev. 41 (2012) 829−868.

SUBCHAPTER 1.2

Thermal Properties of Pure Ionic Liquids

Josefa Salgado, María Villanueva
Applied Physics Department, University of Santiago de Compostela, Santiago de Compostela, Spain

1. MELTING AND FREEZING POINTS, GLASS TRANSITIONS, AND CRYSTALLIZATION TEMPERATURES

Pure crystalline solids have a characteristic temperature at which the solid melts to become liquid named the "melting point" and, reciprocally, liquids have a characteristic temperature at which they turn into solids, known as the "freezing point." Both temperatures are commonly determined using differential scanning calorimetry (DSC); melting point is measured under heating experiences and freezing point under cooling the sample. In theory, the melting point of a solid should be the same as the freezing point of the liquid. In practice, small differences between these quantities can be observed in the most common liquids, and important differences, even higher than 100 K, have been reported in literature of thermal properties of ILs [1]. Additionally, the glass transition is an atypical phase transition that does not involve a discontinuous change in structure and it does not have a sharp transition temperature; there is rather a transformation range that extends over several Kelvin. This transition is associated with a smooth step in the thermal expansion coefficient and in the specific heat when cooling or heating through this temperature range. The exact temperature where these effects are observed depends, however, on the temperature variation rate and, more generally, on the preparation protocol. This transition takes place on amorphous polymers, although some crystalline (or semicrystalline) polymers have an amorphous portion, showing this transition, together the melting and freezing points. The determination can be also done using DSC, however, a useful and extended approach is determined from the relation between the viscosity and temperature as the temperature at which the viscosity is 10^{12} Pa·s. The results obtained for both methods are in good agreement [2].

Fredlake et al. (2004) [3] suggested three types of thermal behavior for the ILs. The first group of ILs has a different freezing point on cooling than melting point on heating. These ILs readily crystallize and do not form glasses, as an example [C_4C_1Im][OTf]. The second type of behavior is characterized by no true phase transitions but only the formation of an amorphous glass on cooling and reformation of the liquid on heating, showing only glass-transition and no melting or freezing points. The IL [C_4C_1Im][BF_4] shows this behavior. The third group of ILs is characterized by the liquid subcooling to a glass state, as is observed for the previous group. However, upon heating, the compound passes from the glass to a subcooled liquid phase, and then a cold crystallization occurs at temperatures higher than T_g. After that, as the sample is heated further, it melts at T_m. An example of this behavior is shown by $[C_4C_1Im]$ $[NTf_2]$.

The cold crystallization temperature is an exothermic transition observed in DSC curves on heating from a subcooled liquid state to a crystalline solid state. This behavior is well known in polymers and other amorphous materials [4].

It is important to know that, in some occasions and in order to be able to do comparative studies, it is suitable to perform cyclical essays of heating/cooling/heating in DSC technique. The first heating allows "erasing" the thermal history of the sample and it is important that the final temperature be high enough to ensure melting but low enough to avoid decomposition of the sample. The cooling generates a known thermal history and then the last heating allows the comparison between materials. For the interpretation of the thermal scans, the melting temperature (T_m) is usually taken as the onset of an endothermic peak on heating, the freezing temperature (T_f) as the onset of an exothermic peak on cooling, the cold crystallization temperature (T_{cc}) as the onset of an exothermic peak on heating from a subcooled liquid state to a crystalline solid state, and the glass transition temperature (T_g) as the midpoint of a small heat capacity change on heating from the amorphous glass state to a liquid state [5]. It is must be taken into account that an increase in the heating (or cooling) rate provokes an increase in glass transition and crystallization temperature and a decrease in the freezing temperature, whereas the melting process is less sensitive to the changes in the rate of the DSC experiences [6,7].

Figure 1 shows the variation with the alkyl chain of some of the most studied ILs: $[C_nC_1Im][BF_4]$, $[C_nC_1Im][PF_6]$, and $[C_nC_1Im][NTf_2]$. Results from the literature [8,9] show a tendency to reducing the melting temperature when alkylic chain lengthens until up to a certain length, from which it grows, and usually the imidazolium based ILs presented the lowest values of melting temperature when comparing ILs with similar structural conformations.

Available data showed, in some cases, for example $[C_4C_1Im][NTf_2]$, remarkable differences for the same IL between different literature sources. Values presented in Figure 1 correspond to the mean of the reported values in the database of Zhang et al. [8]. To avoid these discrepancies, $[C_6C_1Im][NTf_2]$ was recommended by the International Union of Pure and Applied Chemistry [10] as a reference IL, with a glass transition between 183−192 K and a melting point between 266−272 K [11].

Dissolved gasses have remarkable effects on the melting point of ILs, thus, the addition of CO_2, especially in ILs with fluorinated anions, can provoke falls around 120 K. Furthermore, increases in the melting point values with the increasing pressure are reported in the literature [11].

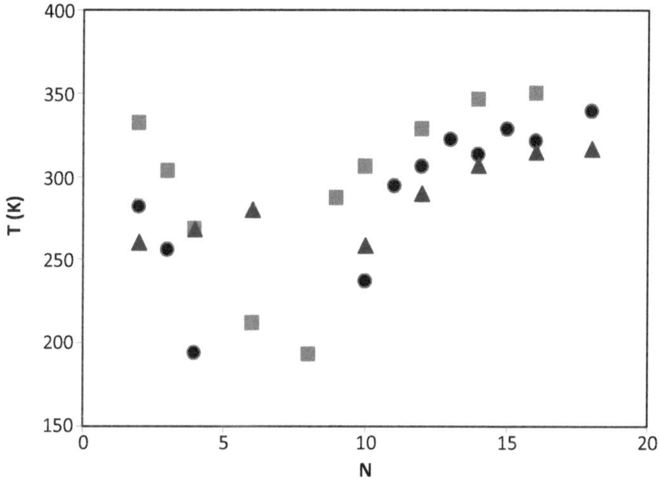

Figure 1 Melting temperature of the ionic liquids (ILs) with the common cation [C_nC_1Im] and with the anions [BF_4] (●), [PF_6] (■), and [NTf_2] (▲) [8].

2. THERMAL STABILITY

In the last several years, the definitions of stability and of the maximum operation temperature for ILs are an open question. Thermogravimetric analysis is the most used technique to determine the thermal stability of a substance. The faster and most common method to do this is performing a single linear heating rate in a wide interval of temperature in a controlled atmosphere. Although different criteria are used in the literature [12], the study of the thermal stability, in this condition, is mainly characterized through the onset temperature, T_{onset}. The selection of experimental conditions, such as the atmosphere, the heating rate, and the mass sample when the thermal stability is analyzed, is fundamental to obtain reproducible results and comparable with the literature [13].

Moreover, because of this scanning nature of the experiment, T_{onset} often overestimates the long thermal stability of the ILs. This fact indicates that the degradation of the IL starts at lower temperatures than T_{onset} [13,14]. Nevertheless, this value can be used as a relative parameter of thermal stability.

Thermal stability of ILs is affected by many factors (e.g., the cation and anion type), structural modifications of the cation (alkyl chain length, different functionalities in the alkyl chain), and impurities (water, chlorides, etc.). Although the anion is the most relevant moiety in the ILs thermal stability [15,16], having influence not only in the degradation temperature but also in the pathway of this process [17]. The thermal stability increases with the size of the anion, whereas the cation seems to have lower influence on this property [18,19].

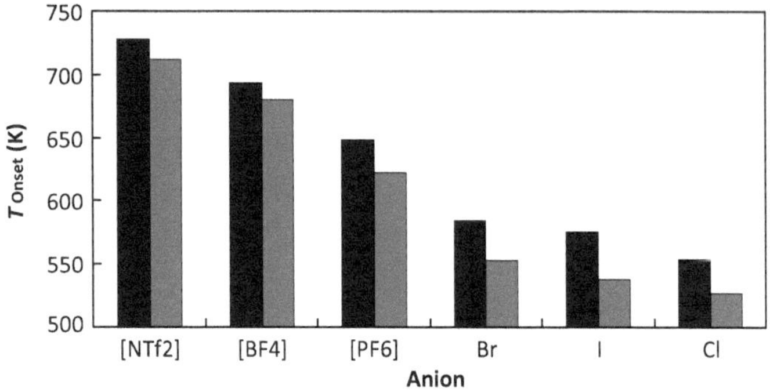

Figure 2 Dependence of T_{onset} on the anion: $[C_2C_1Im]X$ (black), $[C_4C_1Im]X$ (gray).

From the database IL Thermo [9], the sequence of thermal stability when changing the anion is: $[NTf_2] \approx [OTf] > [BF_4] > [PF_6] > [FAP] > [I] \approx [Cl] \approx [Br]$. Figure 2 presents the values of T_{onset} for two imidazolium cations and different anions and Figure 3 presents the values of this parameter for ILs with the same anion $[NTf_2]$ and different cations.

Based on experimental results, imidazolium ILs are more stable than pyridinium, phosphonium, ammonium, pyrrolidinium, and piperidinium ILs [20]. Additionally, as it can be seen in the figures, the decomposition temperature values slightly decreased with growth of the alkyl chain length [13,21]. Recent papers suggest that dicationic ILs showed higher thermal stability than the corresponding monocationic ones, finding differences up to 100 K in studies of geminal dicationic ILs using flame ionization technique [22].

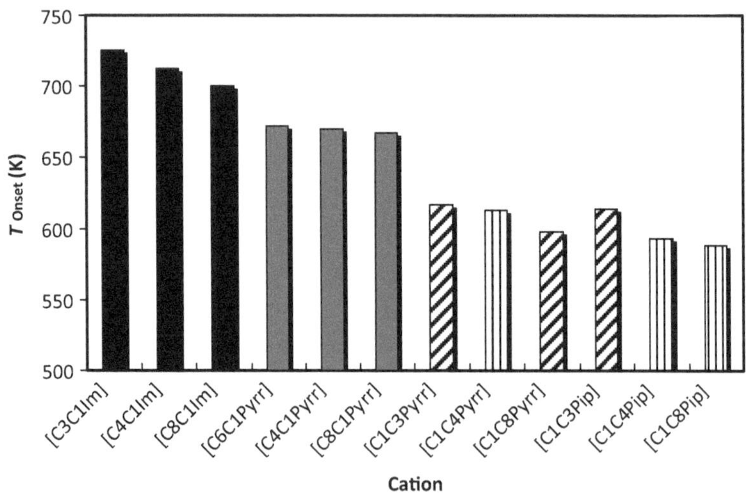

Figure 3 Dependence of T_{onset} on the cation for different ionic liquids (ILs) with anion $[NTf_2]$.

As it was pointed out above, the analytical conditions have strong influence on the thermal analysis results. In particular, decomposition temperature increases with the heating rate in the dynamic experiments, and several authors [23] suggest the use of low heating rates to obtain more realistic values. Although this is completely true, a recent paper [13] demonstrates that even the lowest values overrated the true stability temperatures and the long-term thermal stability should only be evaluated by isothermal scans [14,24,25].

Additionally, the isothermal analysis allows us to investigate thermal degradation kinetics. At a given temperature, the mass loss has a linear behavior, although at higher temperatures this linear regime can be abandoned because of the overlapping of different processes that take place in the degradation (evaporation, combustion, pyrolysis, and even intermediate products degradation). Weight loss rate increases with temperature and it can be assumed that this is a pseudo-zero order process, presupposing that evaporation is too small with regard to degradation, the slope in the mass versus time plot is the zero-order rate constant, k, measured in % loss per minute. This degradation rate corresponds to Arrhenius law [18,24] and it is pressure independent.

$$k = Ae^{\frac{-E_a}{RT}}$$

A is a pre-exponential factor, E_a the activation energy, R the constant of ideal gas, and T the absolute temperature.

The activation energy and the pre-exponential factor can be obtained from the Arrhenius plot (log k versus T^{-1}). Although isothermal studies are frequent in literature, activation energy determinations are scarce and tendencies are not still found. The E_a values are between 100 and 200 kJ/mol for the most common imidazolium based ILs [14,21].

On the other hand, the knowledge of these parameters allows predicting the lifetime at different temperatures [14,21].

3. SPECIFIC HEAT AND MOLAR HEAT CAPACITY

Before the analysis of this property in ILs, a clarification is needed because of the fact that it is frequently found in literature as "heat capacity" instead "molar heat capacity" or "specific heat."

Heat capacity is the ratio of the amount of energy absorbed to the associated temperature rise, measured in J/K, whereas, specific heat is the heat capacity of a substance per mass unit measured in J/kg K.

From these definitions, the volumetric heat capacity (heat per volume unit and Kelvin) and the molar heat capacity (heat per mole and Kelvin) can be deduced.

Specific heat is usually determined by calorimetric methods. Commercial DSCs and modulated DSCs are the most used [26–29], although references using adiabatic and isoperibol calorimeters can also be found [30].

The dataset of this property in the references presents a wide variation, up to 20% in some cases [31], because of differences in the purity samples, especially water content, and others factors of the technique used, such as the sample preparation and handling.

This property is not the most determined in ILs, the imidazolium is the most studied. In Figure 4, the specific heat (in J/g K) of several ILs at 298.15 K is shown and compared with the corresponding value of sulfolane (organosulphur compound commonly used in the chemical industry as a solvent for extractive distillation and chemical reactions). On the other hand, Figure 5 presents the molar heat capacity (in J/mol K) of the same fluids as Figure 4 at the same temperature (298.15 K), as well as the values for water. As it can be seen comparing these two figures, the behavior is completely different; the ILs present higher values than sulfolane and water, and this property increases with the molecular mass of the IL.

This property increases with temperature and with pressure, and polynomial equations can be used to express the dependence between these properties among the temperature and/or pressure [6,19,28]. It is important to remark that this behavior with pressure is the opposite of that obtained in the most common organic solvents [28,32].

The pyridinium based ILs present higher heat capacities than their analogous imidazolium based ILs. These values are clearly affected by the anion, being the general trend that the higher the molar mass of the anion, the higher the heat capacity, and

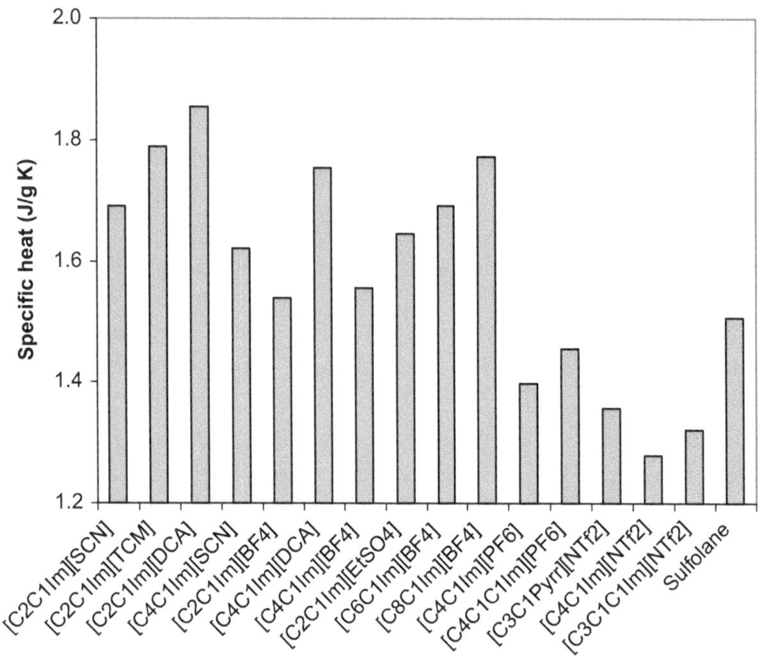

Figure 4 Specific heat of several ionic liquids (ILs) [8,9].

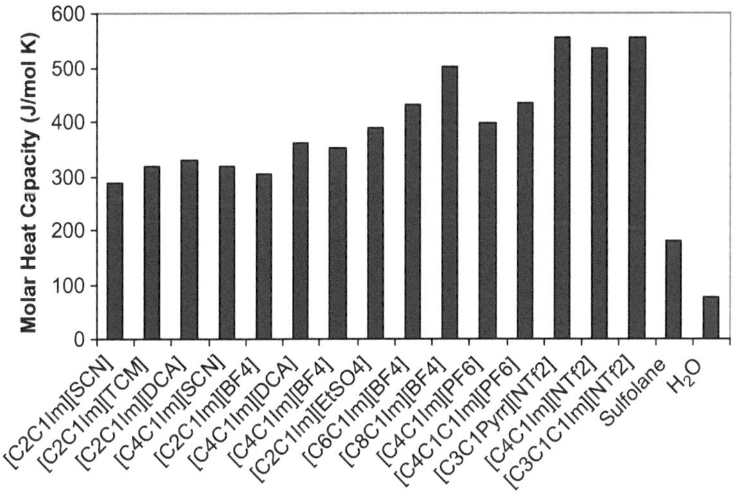

Figure 5 Molar heat capacity of several ionic liquids (ILs).

probably related to the number of atoms of the anion ($[NTf_2] > [OTf] > [DCA]$). Additionally, the enlargement of the alkyl side chain leads the molar heat capacities to higher values, whether in imidazolium, as in pyridinium-based ILs, this property is independent of the anion [2,6].

As it was already discussed, the alkyl side chain length has a clear influence on the molar heat capacity values, but not on the volumetric heat capacity. However, the anion does present its influence on volumetric heat capacities, obtaining the higher values for the ILs with [DCA] and [OTf] anions and the lower for the ILs with $[NTf_2]$ anion, whereas in molar heat capacities, the ILs with this latter anion presented the higher values [6].

4. THERMAL CONDUCTIVITY

The thermal conductivity of a substance, λ, also known as heat conductivity, is the heat flow across a surface per unit area per unit time, divided by the rate of change of temperature with distance in a direction perpendicular to the surface [33].

One of the most used methods to measure the thermal conductivity of ILs is the transient hot wire method [34—36] The advantage of this method is that it is a direct and absolute method of determination of the thermal conductivity, minimizing the experimental error [37].

Thermal conductivity at 323 K of several ILs is presented in Figure 6. As it can be observed for all the ILs presented, λ ranges between 0.1 and 0.2 W/m K, whereas the thermal conductivity of water at the same temperature is around 0.643 W/m K. There is a large difference between the values for pure ILs and that for water or even for toluene and ethylene glycol. Then, ILs are poor thermal conductors, with λ ranging

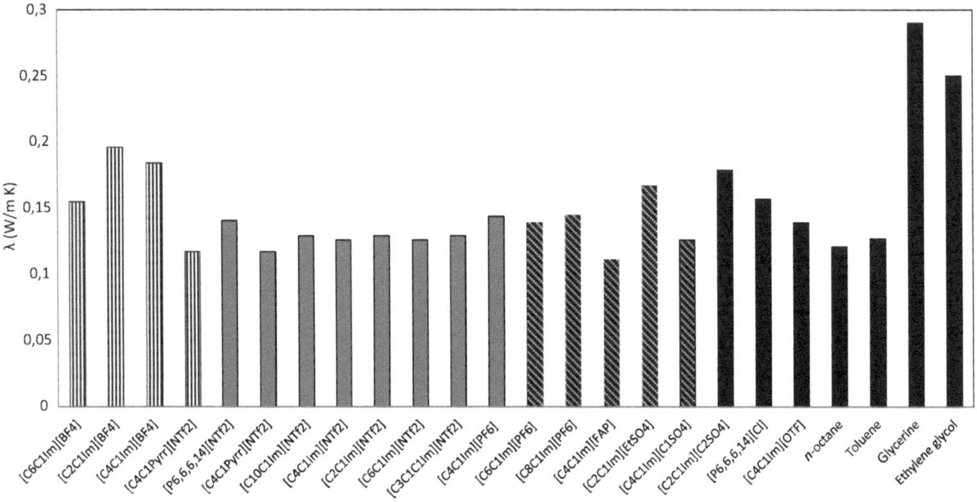

Figure 6 Thermal conductivity of several ILs and of some commonly used molecular solvents or heat transfer liquids at 323 K [9].

approximately between 15 and 31% of that of water at this temperature. Additionally, the variation of this property is lower than other properties (i.e., the experimental determination must be done by extreme rigor) [31]. Owing that many ILs are hygroscopic, caution must be taken when measuring thermal conductivity because water contained in the IL can increase notably in this property [37].

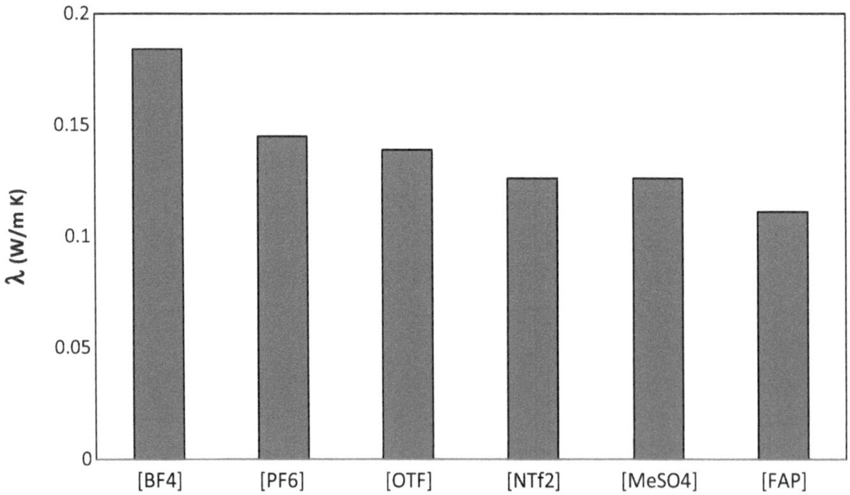

Figure 7 Thermal conductivity of several ionic liquids (ILs) with cation $[C_4C_1Im]$ at 323 K [8].

In Figure 7, thermal conductivity of different ILs with common cation [C_4C_1Im] is shown. As some authors have indicated [37] and as is pictured in this figure, the influence of the anion on the thermal conductivity is high; for example, replacing the anion [BF_4] to [FAP] in the ILs from Figure 7, causes a decrease from approximately 0.18 to 0.11 W/m K.

Some authors [34,37,38] have detected slight linear decrease of thermal conductivity with temperature (in the range between approximately 293 and 393 K, depending on the author and also on the IL) or even almost independent on the temperature [39].

ACKNOWLEDGMENTS

This work was supported by Spanish Ministry of Science and Innovation and EU FEDER Program through CTQ2011-23925 project and by the Xunta de Galicia through EM2013/031 project.

REFERENCES

[1] H.L. Ngo, K. LeCompte, L. Hargens, A.B. McEwen, Thermal properties of imidazolium ionic liquids, Thermochim. Acta 357–358 (2000) 97–102.

[2] F.M. Gaciño, T. Regueira, L. Lugo, M.J.P. Comunas, J. Fernandez, Influence of molecular structure on densities and viscosities of several ionic liquids, J. Chem. Eng. Data 56 (2011) 4984–4999.

[3] C.P. Fredlake, J.M. Crosthwaite, D.G. Hert, S.N.V.K. Aki, J.F. Brennecke, Thermophysical properties of imidazolium-based ionic liquids, J. Chem. Eng. Data 49 (2004) 954–964.

[4] Y. Wang, M. Rodriguez-Perez, R. Reis, J. Mano, Thermal and Thermomechanical behaviour of polycaprolactone and starch/polycaprolactone blends for biomedical applications, Macromol. Mat. Eng. 290 (2005) 792–801.

[5] E. Gomez, N. Calvar, A. Dominguez, E.A. Macedo, Thermal analysis and heat capacities of 1-alkyl-3-methylimidazolium ionic liquids with NTf_2^-, TFO^-, and DCA^- anions, Ind. Eng. Chem. Res. 52 (2013) 2103–2110.

[6] N. Calvar, E. Gomez, E.A. Macedo, A. Dominguez, Thermal analysis and heat capacities of pyridinium and imidazolium ionic liquids, Thermochim. Acta 565 (2013) 178–182.

[7] L.F.O. Faria, J.R. Matos, M.C.C. Ribeiro, Thermal analysis and Raman spectra of different phases of the ionic liquid butyltrimethylammonium bis(trifluoromethylsulfonyl)imide, J. Phys. Chem. B 116 (2012) 9238–9245.

[8] S. Zhang, X. Lu, Q. Zhou, X. Li, X. Zhang, S. Li, Ionic Liquids. Thermophysical Properties, first ed., Elsevier, Amsterdam, 2009.

[9] Ionic Liquids Database- (ILThermo), The National Institute of Standards and Technology, 2006.

[10] R.D. Chirico, V. Diky, J.W. Magee, M. Frenkel, K.N. Marsh, Thermodynamic and thermophysical properties of the reference ionic liquid: 1-hexyl-3-methylimidazolium bis[(trifluoromethyl)sulfonyl] amide (including mixtures). Part 2. Critical evaluation and recommended property values (IUPAC Technical Report), Pure Appl. Chem. 81 (2009) 791–828.

[11] S. Aparicio, M. Atilhan, F. Karadas, Thermophysical properties of pure ionic liquids: review of present situation, Ind. Eng. Chem. Res. 49 (2010) 9580–9595.

[12] T.J. Wooster, K.M. Johanson, K.J. Fraser, D.R. MacFarlane, J.L. Scott, Thermal degradation of cyano containing ionic liquids, Green Chem. 8 (2006) 691–696.

[13] M. Villanueva, A. Coronas, J. García, J. Salgado, Thermal stability of ionic liquids for their application as new absorbents, Ind. Eng. Chem. Res. 52 (2013) 15718–15727.

[14] J. Salgado, M. Villanueva, J.J. Parajo, J. Fernandez, Long-term thermal stability of five imidazolium ionic liquids, J. Chem. Thermodyn. 65 (2013) 184–190.

[15] J.M. Crosthwaite, M.J. Muldoon, J.K. Dixon, J.L. Anderson, J.F. Brennecke, Phase transition and decomposition temperatures, heat capacities and viscosities of pyridinium ionic liquids, J. Chem. Thermodyn. 37 (2005) 559–568.
[16] J. Huddleston, A. Visser, W. Reichert, H. Willauer, G. Broker, R. Rogers, Characterization and comparison of hydrophilic and hydrophobic room temperature ionic liquids incorporating the imidazolium cation, Green Chem. 3 (2001) 156–164.
[17] F. D'Anna, H.Q.N. Gunaratne, G. Lazzara, R. Noto, C. Rizzo, K.R. Seddon, Solution and thermal behaviour of novel dicationic imidazolium ionic liquids, Org. Biomol. Chem. 11 (2013) 5836–5846.
[18] V. Kamavaram, R.G. Reddy, Thermal stabilities of di-alkylimidazolium chloride ionic liquids, Int. J. Therm. Sci. 47 (2008) 773–777.
[19] P. Navarro, M. Larriba, E. Rojo, J. Garcia, F. Rodriguez, Thermal properties of cyano-based ionic liquids, J. Chem. Eng. Data 58 (2013) 2187–2193.
[20] E.M. Siedlecka, M. Czerwicka, S. Stolte, P. Stepnowski, Stability of ionic liquids in application conditions, Curr. Org. Chem. 15 (2011) 1974–1991.
[21] C. Maton, N. De Vos, C.V. Stevens, Ionic liquid thermal stabilities: decomposition mechanisms and analysis tools, Chem. Soc. Rev. 42 (2013) 5963–5977.
[22] J. Anderson, R. Ding, A. Ellern, D. Armstrong, Structure and properties of high stability geminal dicationic ionic liquids, J. Am. Chem. Soc. 127 (2005) 593–604.
[23] U. Domanska, Thermophysical properties and thermodynamic phase behavior of ionic liquids, Thermochim. Acta 448 (2006) 19–30.
[24] M. Kosmulski, J. Gustafsson, J.B. Rosenholm, Thermal stability of low temperature ionic liquids revisited, Thermochim. Acta 412 (2004) 47–53.
[25] P. Verdía, M. Hernaiz, E.J. González, E.A. Macedo, J. Salgado, E. Tojo, Effect of the number, position and length of alkyl chains on the physical properties of polysubstituted pyridinium ionic liquids, J. Chem. Thermodyn. 69 (2014) 19–26.
[26] M. Zhang, V. Kamavaram, R. Reddy, Thermodynamic properties of 1-nutyl-3-methylimidazolium chloride (C_4mim[Cl]) ionic liquid, J. Phase Equilib. Diffus 26 (2005) 124–130.
[27] R.L. Gardas, R. Ge, P. Goodrich, C. Hardacre, A. Hussain, D.W. Rooney, Thermophysical properties of amino acid-based ionic liquids, J. Chem. Eng. Data 55 (2010) 1505–1515.
[28] Y.A. Sanmamed, P. Navia, D. Gonzalez-Salgado, J. Troncoso, L. Romani, Pressure and temperature dependence of isobaric heat capacity for [Emim][BF_4], [Bmim][BF_4], [Hmim][BF_4], and [Omim][BF_4], J. Chem. Eng. Data 55 (2010) 600–604.
[29] U. Domanska, R. Bogel-Lukasik, Physicochemical properties and solubility of alkyl-(2-hydroxyethyl)-dimethylammonium bromide, J. Phys. Chem. B 109 (2005) 12124–12132.
[30] Y.U. Paulechka, Heat capacity of room-temperature ionic liquids: a critical review, J. Phys. Chem. Ref. Data 39 (2010) 033108.
[31] C.A. Nieto de Castro, Thermophysical properties of ionic liquids: do we know how to measure them accurately? J. Mol. Liquids 156 (2010) 10–17.
[32] J.L. Valencia, D. Gonzalez-Salgado, J. Troncoso, J. Peleteiro, E. Carballo, L. Romani, Thermophysical characterization of liquids using precise density and isobaric heat capacity measurements as a function of pressure, J. Chem. Eng. Data 54 (2009) 904–915.
[33] M. Sorai, Comprehensive Handbook of Calorimetry and Thermal Analysis, John Wiley and Sons, United Kingdom, 2004.
[34] M. Van Valkenburg, R. Vaughn, M. Williams, J. Wilkes, Thermochemistry of ionic liquid heat-transfer fluids, Thermochim. Acta 425 (2005) 181–188.
[35] J. Healy, J. Degroot, J. Kestin, The theory of the transient hot-wire method for measuring thermal conductivity, Physica B & C 82 (1976) 392–408.
[36] J.M.P. Franca, S.I.C. Vieira, M.J.V. Lourenco, S.M.S. Murshed, C.A. Nieto de Castro, Thermal conductivity of [C_4mim][$(CF_3SO_2)_2N$] and [C_2mim][$EtSO_4$] and their ionanofluids with carbon nanotubes: experiment and theory, J. Chem. Eng. Data 58 (2013) 467–476.
[37] R. Ge, C. Hardacre, P. Nancarrow, D.W. Rooney, Thermal conductivities of ionic liquids over the temperature range from 293 K to 353 K, J. Chem. Eng. Data 52 (2007) 1819–1823.

[38] C.A. Nieto de Castro, M.J.V. Lourenco, A.P.C. Ribeiro, E. Langa, S.I.C. Vieira, P. Goodrich, C. Hardacre, Thermal properties of ionic liquids and ionanofluids of imidazolium and pyrrolidinium liquids, J. Chem. Eng. Data 55 (2010) 653–661.
[39] H. Chen, Y. He, J. Zhu, H. Alias, Y. Ding, P. Nancarrow, C. Hardacre, D. Rooney, C. Tan, Rheological and heat transfer behaviour of the ionic liquid, [C_4mim][NTf_2], Int. J. Heat Fluid Flow 29 (2008) 149–155.

SUBCHAPTER 1.3

Optical Properties of Ionic Liquids

José A. Nóvoa López[1], Humberto Michinel[1], Elena López Lago[2]
[1]Area de Óptica, Facultade de Ciencias de Ourense, Universidade de Vigo, Ourense, Spain
[2]Grupo de Microóptica y Sensores de Frente de Onda, Departamento de Física Aplicada, Universidade de Santiago de Compostela, Santiago de Compostela, Spain

1. INTRODUCTION

Ionic liquids (ILs) have been the subject of great scientific and technological interest because of their singular properties that can be tuned by an adequate combination of the cation and anion or by mixing two or more ILs. This fact allows designing ILs adapted to specific applications [1−3]. ILs are considered emergent materials in fields, such as nanotechnology, biotechnology, spectroscopy, or mineralogy, and even as photonic materials, although not so much systematic research has been done in this direction. The existing publications refer to interesting photonic applications involving the best optically known ILs: operating in a Moon mirror telescope [4], in optical diffusers [5], in variable focus liquid lens [6,7], in optical thermometers [3], or as matching index liquids [8], for example.

The use of an IL in a given photonic device requires previous knowledge of its optical behavior. Depending on the application, it will be needed for the previous characterization of one or more optical properties of the material [9]. Among them, the refractive index plays a crucial role and it is well known that it varies with the frequency of the optical radiation, as in all optical properties. The dependence of the refractive index on frequency (or wavelength) is called chromatic dispersion. Although some applications require minimizing dispersion [10] (for example, imaging systems, optical communications systems, etc.), others benefit from it [11] (for example, dispersive prisms in laser cavities, for compensating dispersion introduced by other optical components, or in optical spectrometers). In both situations, an accurate characterization of the refractive index dispersion is required to ensure an optimal performance of the optical device.

There are many works, involving a great variety of ILs, that determine the refractive index at a wavelength of $\lambda_D = 589.3$ nm [2,12−14]. Systematic studies explore the relationship between the refractive index values and the structural parameters of the ILs (that means the cationic and the anionic part) and/or analyze its dependence on temperature [15−17], but only a few works deal with the wavelength dependence [18,19].

In this chapter, we revise the influence of the structural parameters of ILs on the refractive index, taking into account not only the results published during these last years but also new results measured specifically for this work. However, because of space limitations, we restrict our discussion to a reduced but meaningful set of ILs. We also focus our attention on those papers that refer to the wavelength dependence and we will complete the discussion with original data that have also been measured namely for this work. Finally, we center on the characterization of the thermo-optic coefficient, which gives information about the variation of the refractive index with temperature.

2. CATIONIC AND ANIONIC INFLUENCE ON THE REFRACTIVE INDEX

The goal of this section is to give a representative overview about the influence of the anionic and cationic parts of the ILs on the refractive index value. It is well known that the number of carbon atoms in the alkyl chain, n, is a parameter that serves to tune the refractive index in a given family of ILs, as can be seen in Table 1. Imidazolium based ILs are the best characterized family. In series of ILs, such us $[C_nC_1Im][NTf_2]$, $[C_nC_1Im][PF_6]$, $[C_nN][HCOO]$ [30], or $[C_nC_1Im][BF_4]$ (being n an odd number), the reported data show that the refractive index increases as n does. The same behavior has been observed for the $[C_nPy][NTf_2]$ series [31] in which the refractive index varies from 1.4437 ($n = 4$) to 1.4496 ($n = 12$). However, in other families, such us $[C_nC_1Im][Br_2I]$, $[C_nC_1Im][BrI_2]$, $[C_nC_1Im][Br]$, or $[C_nC_1Im][Cl]$, the behavior is just the contrary: the refractive index decreases as n does. A similar tendency was detected in $[(C_2CN)C_nC_1Im][Cl]$ or $[(C_2CN)C_nC_1Im][Br]$ series of ILs. In some families, the monotonic behavior is broken if ILs with even values of n are considered. This is the case of the $[C_nC_1Im][BF_4]$ series, because the reported refractive index value for $[C_3C_1Im][BF_4]$ is higher than reported for $[C_6C_1Im][BF_4]$. A similar behavior has been detected in the $[C_nN][NO_3]$ family because $[C_3N][NO_3]$ has a higher refractive index than the other isotropic ILs of the series with two, four, and six carbon atoms in the alkyl chain [17]. However, the lack of systematic refractive index data relative to ILs series with even values of n ($n = 5, 7\ldots$) prevents us to draw definitive conclusions.

The choice of the cationic family has a similar effect, the refractive index change may alter up to a second decimal place. For example, we can cite the work [32] where authors report the refractive index of several ILs resulting from the combination of the anion $[NTf_2]^-$ different cations (*N*-alkyl-*N*-methylpyrrolidinium series, dimethyl[isopropyl] alkylammonium series, and cations related to the latter by the substitution or the addition of a hydroxyl functionality). The measured refractive index values vary from 1.415−1.434. Froba et al. [18] reported a value of 1.4475 with the cation $[C_8C_1N]^+$, Yunus et al. [31] reported 1.4472 with the $[C_8Py]^+$, and Huddleston et al. [22] gave a value of 1.4665 with the cation $[C_4Py]^+$.

It is well known that the choice of the anionic part has a strong influence on the refractive index value. For example, in Table 2, the refractive index values of several 1-ethyl-3-methylimidazolium-based ILs are shown. The refractive index varies from 1.4 (if it is combined with $[TA]^-$) to higher values than 2 (if it is combined with $[I_9]^-$).

Table 1 Refractive Index Data for Different Methyl-Imidazolium Based Ionic Liquids

Cation	$[BF_4]^-$	$[NTf_2]^-$	$[PF_6]^-$	$[Br]^-$	$[Cl]^-$	$[BrI_2]^-$	$[Br_2I]^-$
$[C_2C_1Im]^+$	1.4109	1.4229 [20]				1.833 [8]	1.715 [8]
	1.4352 [6]	1.4225					
		1.42 [19]					
$[C_3C_1Im]^+$	1.4285 [6]	1.4525 [21]					
$[C_4C_1Im]^+$	1.4195 [6]	1.4263	1.4084	1.54 [12]		1.810 [8]	1.701 [8]
	1.4217	1.4265 [6]	1.4093 [14]				
	1.416 [8]	1.4271 [22]	1.4089 [16]				
$[C_6\ C_1Im]^+$	1.4282	1.4295 [16]	1.4178 [14]	1.5329	1.515 [23]	1.768 [8]	1.685 [8]
	1.4237 [24]				1.517 [13]		
$[C_8C_1Im]^+$	1.4342 [25]	1.4325 [16]	1.4235 [16]	1.5247	1.505 [23]		
	1.4322 [26]	1.4331 [27]	1.423 [22]		1.506 [13]		
$[C_{10}C_1Im]^+$	1.4380	1.4356 [16]					
$[C_{12}C_1Im]^+$		1.4376 [16]					
$[(C_2CN)C_4C_1Im]^+$				1.5454 [28]	1.5193 [29]		
$[(C_2CN)C_6C_1Im]^+$				1.5287 [28]	1.5162 [29]		
$[(C_2CN)C_8C_1Im]^+$				1.51473 [28]	1.5081 [29]		

Table 2 Refractive Index, n_D, Data for 1-Ethyl-3-Methylimidazolium Based Ionic Liquids at 298 K

$[C_2C_1Im][X]$	n_D	References	$[Y][NTf_2]^-$	n_D	References
$[TA]^-$	1.4009	[19]	$[C_1C_4Pyrr]^+$	1.423	[32]
$[CF_3COO]^-$	1.4056	[6]	$[C_4C_1IM]^+$	1.4275	This work
$[BF_4]^-$	1.4109	This work	$[C_8C_8C_8C_1N]$	1.4475	[18]
$[Tf_2N]^-$	1.4225	[6]	$[C_{14}C_6C_6C_6P]$	1.4496	[16]
$[C_2SO_4]^-$	1.4788	[25]	$[C_4Py]^+$	1.4637	[31]
$[N(CN)_2]^-$	1.5101	[18]			
$[Br_2I]^-$	1.715	[8]			
$[ClI_2]^-$	1.796	[8]			
$[BrI_2]^-$	1.833	[8]			
$[I_7]^-$	2.01	[33]			
$[I_9]^-$	2.08	[33]			

These tuning capabilities make them interesting candidates as sensors [34], matching index liquids [8], as fluids in liquid lens [6,8] or in microfluid channels for waveguide optics [35], among others. However, few studies try to explain the influence of the anions on the refractive index values. One approach has been recently published by Seki et al. [36]. By using computational ab-initio methods and experimental data, they found a linear relationship between the refractive index of ionic liquid and the molecular polarizability of ion pairs. This fact suggests that it is possible to design ILs with a given refractive index using the polarizabilities calculated for ions.

Finally, a reference must be done about reproducibility of the measurements. Several factors affect the refractive index measurement. Some of them are inherent to the chosen technique but others are related to the synthesis process of the ILs or to their handle and aging [8]. For optical applications, the refractive index must be known to an accuracy of $\pm 10^{-3}$. Most of the reported data satisfy this condition but differences up to ± 0.02 can be found in $[C_2C_1Im][BF_4]$ and differences up to ± 0.003 have been encountered in other ILs (see Table 1). Although these differences are irrelevant for other interesting technological applications, they are not desirable in photonic applications.

3. CHROMATIC DISPERSION

The chromatic dispersion in the transparent region of the materials is accurately described by the Sellmeier equation [37]:

$$n^2(x) = 1 + \frac{B_1\lambda^2}{\lambda^2 - C_1} + \frac{B_2\lambda^2}{\lambda^2 - C_2} + \frac{B_3\lambda^2}{\lambda^2 - C_3} \tag{1}$$

where Bi and Ci ($i = 1$, 2, and 3) stand for the Sellmeier coefficients of the material. They are determined empirically starting from, at least, six values of the refractive index measured at different wavelengths. Dispersion is also accurately described in the VIS spectral region by the Cauchy equation [38]:

$$n(x) = A + \frac{B}{\lambda^2} + \frac{C}{\lambda^4} + \frac{D}{\lambda^6} + \cdots \quad (2)$$

Usually, the first two terms are enough to describe the refractive index change.

Optical materials can be represented in an Abbe diagram [19] whose axes are the refractive index and the Abbe number, usually given at $\lambda_d = 587.6$ nm or $\lambda_D = 589.3$ nm. The Abbe number is defined as:

$$\nu_i = \frac{n_i - 1}{n_F - n_C}, \quad i = \mathrm{d}, \mathrm{D} \quad (3)$$

where n_F and n_C are the refractive indices evaluated at $\lambda_F = 486.1$ nm and $\lambda_C = 656.3$ nm, respectively. The $n_F - n_C$ is called mean dispersion and $(1 - n_i)$ is the refractivity of the medium at λ_i. As the material is more dispersive, the Abbe number (ν_i) takes lower values.

Although a great amount of papers report data about the refractive index at 589.3 nm, only few studies concern the dispersive effects on ILs. One of them is cited in Reference [19] where the authors characterize the chromatic dispersion of several ILs ([C_2C_1Im][NTf_2], [C_2C_1Im][TA], [C_4C_1Im][PF_6], [C_4C_1Im][NO_3], [C_4Py] [NTf_2], and [C_1Pyrr][HSO_4]) in order to fabricate variable focus liquid lenses. They show that the dispersive behavior (Abbe number) is higher for the [C_1Pyrr][HSO_4] and lower for the [C_4C_1Im]-based ILs. In Reference [18], they characterize the mean dispersion of [C_2C_1Im][NTf_2], [C_2C_1Im][C_2SO_4], [C_2C_1Im][$N(CN)_2$], and [$C_8C_8C_8C_1$N][NTf_2] and they introduce it in a linear model to calculate the refractive index at given wavelengths. This linear model is not formally correct because the dispersion follows a more complicated function (Eqn (1) or Eqn (2)), although the experimental data corresponding to a narrow band of the visible spectrum actually seems to follow a linear behavior. Once the mean dispersion data of these four ILs are known, the calculation of the corresponding Abbe number is immediate: 52, 50, 36, and 89, respectively.

In this work, we present the characterization results of other sets of ILs, which are listed in Table 3. The measurement has been done by using the angle of minimum deviation technique and by multiwavelength Abbe refractometry. The refractive index at 589 nm has also been measured with a Zuzi refractometer and an Atago DR-M2 refractometer to ensure reproducibility of the data. The measurement was taken at 20 °C ± 0.3 °C.

In Table 3, we show the refractive index at λ_D, λ_F, and λ_C of five ammonium based ILs and of five 1-butyl-3-methylimidazolium based ILs together with the corresponding

Table 3 Refractive Index at λ_d, λ_F and λ_C, Abbe Number ν_d at 293 K. Absorption Coefficient at 800 nm (m^{-1}), Thermal Conductivity (W/mK), Normalized Thermal Lens Strength, and Thermo-Optic Coefficient (K^{-1}) of the Listed Ionic Liquids at 298 K

Ionic Liquids	n_F	n_d	n_C	ν_d	α_0	κ	θ/P	d_n/d_T
$[C_2N][NO_3]$	1.4576	1.4521	1.4506	49	14.4	0.244	−0.83	−1
$[C_3N][NO_3]$	1.4594	1.4537	1.4528	50	4.4	0.218	−0.97	−3.6
$[C_4N][NO_3]$	1.4573	1.4527	1.4519	57	1.6	0.192	−1.21	−11
$[C_6N][NO_3]$	1.4567	1.4519	1.4511	53	–	–	–	–
$[C_2C_2C_1N][C_1SO_3]$	1.4583	1.4535	1.4527	63	–	–	–	–
$[C_2C_2C_1N][OTf]$	1.4029	1.3993	1.3987	66	–	–	–	–
$[C_2N][OF]$	1.4610	1.4569	1.4556	63	8.2		−1.16	–
$[C_3C_1Im][NO_3]$	1.4584	1.4527	1.4506	41	5.1		−0.78	–
$[C_2C_1Im][NTf_2]$	1.4264	1.4240	1.4212	81	17.3	0.130[b]	−0.64	−0.4
$[C_4C_1Im][NTf_2]$	1.4307	1.4269	1.4236	59	5[a]	0.128[b]	–	−6[a]
$[C_6C_1Im][NTf_2]$	1.4357	1.4294	1.4281	62	1.2[a]	0.127[b]	–	−3[a]
$[C_4C_1Im][BF_4]$	1.4250	1.4220	1.4194	66	1.4[a]	0.176	–	−20[a]
$[C_4C_1Im][C_1SO_4]$	1.4830	1.4775	1.4743	55	–	–	–	–

[a]Ref. [54].
[b]Ref. [55].

Abbe number. The n_F and n_C have been estimated by fitting Eqn (1) to the refractive index values taken at six wavelengths of the 475–680 nm spectral band.

It must be noted that the measured refractive index of PAN at λ_D is rather lower than that reported in Reference [17] and [30], which are, respectively, 1.4561 and 1.4565. The same discrepancy has been found with the measured refractive index of $[C_4C_1Im][C_1SO_4]$ [14,15,39–41]. The origin of these differences remains unclear and is under study.

4. EFFECT OF TEMPERATURE

The variation of the refractive index with temperature can be characterized in terms of the thermo-optic coefficient, dn/dT [37]. There are a great variety of papers involving a wide set of ILs that study directly or indirectly this property at $\lambda_d = 589.3$ nm [15,26,41,42]. Although the thermo-optic coefficient uses are less dispersive than the refractive index, at least in the visible spectral region, it is interesting to study this behavior in other spectral regions, such as the near infrared region. The Z-scan technique [43] in combination with the thermal lens effect (TLE) [44] is a good option. It is well known that dn/dT plays an important role in the development of the TLE in liquid and solid samples, together with the optical absorbance (α_0) and the thermal conductivity (κ). The TLE obeys an inhomogeneous heating of the medium induced by an optical beam. The heating is originated by absorption processes developed in the medium. The TLE can be generated with continuous wave lasers and even with laser pulses. In this last case, it is needed that the temporal width of the pulse or the temporal interval between the pulses be less than the thermal characteristic time of the ILs, t_c ($t_c = w_0^2/4D$, being w_0 the width of the beam at the waist and D the thermal diffusivity). The characterization of TLE is not only interesting for fluid photonics devices, but also it has important applications in spectroscopy [45], microscopy [46], and calorimetry [47].

Several models have been developed to describe the TLE [48–50] considering linear and/or nonlinear absorption processes as triggering mechanisms. In this work, we refer to the TLE induced by linear absorption processes. Following the corresponding Falconieri's model, the normalized transmittance when the irradiation time is high enough that t_c is given by:

$$T(z) = 1 + \theta \tan^{-1}\left[\frac{2\dfrac{z}{z_0}}{\left[3 + \left(\dfrac{z}{z_0}\right)^2\right]}\right] \tag{4}$$

where z_0 is the Rayleigh distance given by $\pi w_0^2/\lambda_0$, and θ is the strength of the TLE, which is related to the thermo-optic coefficient by:

$$\theta = \frac{P\alpha_0}{\lambda_0 \kappa} L_{eff} \frac{dn}{dT} \tag{5}$$

P represents the input power, α_0 the linear absorption coefficient, and L_{eff} the effective length of the medium:

$$L_{eff} = \frac{1 - e^{-\alpha_0 L}}{\alpha_0} \tag{6}$$

Z-scan has already been used to characterize the TLE in some ILs as $[C_4C_1Im][BF_4]$, $[C_4C_1Im][PF_6]$, $[C_nC_1Im][NTf_2]$, $[C_2C_1Im][CF_3CO_2]$, $[C_4C_1Im][CF_3CO_2]$, and IL(s) of the 1-methylpirrolidine family. The experimental conditions (wavelength, pulse duration, repetition rate, etc.) are given in the corresponding References [51–53]. A systematic study [54] concludes that the influence of the anion on the development of the TLE is higher than the influence of the alkyl chain. We have characterized the TLE strength in $[C_2N][NO_3]$, $[C_3N][NO_3]$, $[C_4N][NO_3]$, and $[C_2N][OF]$ in order to evaluate the influence of the alkyl chain length and of the anion in this family of protic ILs. We have also studied the $[C_3C_1Im][NO_3]$ IL to analyze the influence of the cationic part. Because we could not measure the thermal conductivity of $[C_2N][OF]$ and $[C_3C_1Im][NO_3]$, the thermo-optic coefficient of these two ILs could not been calculated. We use 80 fs laser pulses delivered by a Ti: Sapphire oscillator (repetition rate of

80.75 MHz and $\lambda_0 = 800$ nm) and irradiation times long enough to reach the thermal steady state.

In Table 3, we show the power-normalized values of θ (i.e., θ/P) together with the absorption coefficient the thermal conductivity and thermo-optic coefficient when available. The thermal lens strength, θ/P, in alkylammonium nitrate based ILs increases with the alkyl chain length. A similar behavior is observed in the thermo-optic coefficient but the contrary occurs in the optical absorption and thermal conductivity. θ/P is higher in $[C_2N][OF]$ than in $[C_2N][NO_3]$, although the optical absorption is lower. The thermal lens strength is lower in $[C_3C_1Im][NO_3]$ than in $[C_3N][NO_3]$, although the linear absorption coefficient is similar. The explanation of this behavior and the developing of a model that describes it are still under research.

REFERENCES

[1] M. Freemantle, An Introduction to Ionic Liquids, RCS Publishing, 2010.
[2] R.D. Rogers, K.R. Seddon (Eds.), Ionic Liquids IIIA: Fundamentals, Progress, Challenges and Opportunities. Properties and Structure, ACS Symposium Series 901, Washington DC, 2005.
[3] R.D. Rogers, K.R. Seddon (Eds.), Ionic Liquids IIIB Fundamentals, Progress, Challenges and Opportunities. Transformations and Processes, ACS Symposium Series 902, Washington DC, 2005.
[4] E.F. Borra, O. Seddiki, D. Eisenstein, P. Hickson, K.R. Seddon, Deposition of metal films on an ionic liquid as a basis for a lunar telescope, Nature 447 (2007) 979–981.
[5] H. Lin, P.W. Oliveira, M. Veith, M. Gros, I. Grobelsek, Optic diffusers based on photopolimerizable hologram material with an ionic liquid as additive, Opt. Lett. 34 (2009) 1150–1153.
[6] S. Calixto, M.E. Sánchez-Morales, F.J. Sánchez-Marín, M. Rosete-Aguilar, A. Martínez Richa, K.A. Barrera-Rivera, Optofluidic focus variable lenses, Appl. Opt. 48 (2009) 2308–2310.
[7] X. Hu, S. Zhang, C. Qu, Q. Zhang, L. Lu, X. Ma, Y. Deng, Ionic liquid based variable focus lenses, Soft Matter 7 (2011) 5941–5943.
[8] M. Deetlefs, K.R. Seddon, M. Shara, Neoteric optical media for refractive index determination of gems and minerals, New J. Chem. 30 (2006) 317–326.
[9] B.E.A. Saleh, M.C. Teich, Fundamentals of Photonics (Wiley Series in Pure and Applied Optics), August 15, 1991.
[10] H. Tsuchida, T. Nagaoka, K. Yamamoto, Design of imaging lens systems that use low dispersive radial gradient-index rod, Jpn J. Appl. Phys.1 37 (1998) 3633–3637.
[11] D. Kopf, G.J. Spühler, K.J. Weingarten, U. Keller, Mode-locked laser cavities with a single prism for dispersion compensation, Appl. Opt. 35 (1996) 912–915.
[12] K.S. Kim, B.K. Shin, H. Lee, F. Ziegler, Refractive index and heat capacity of 1-butyl-3-methylimidazolium bromide and 1-butyl-3-methylimidazolium tetrafluoroborate, and vapor pressure of binary systems for 1-butyl-3-methylimidazolium bromide + trifluoroethanol and 1-butyl-3-methylimidazolium tetrafluoroborate + trifluoroethanol, Fluid Phase Equilibr. 218 (2004) 215–220.
[13] E. Gómez, B. González, A. Domínguez, E. Tojo, J. Tojo, Dynamic viscosities of a series of 1-alkyl-3-methylimidazolium chloride ionic liquids and their binary mixtures with water at several temperatures, J. Chem. Eng. Data 51 (2006) 696–701.
[14] A.B. Pereiro, J.L. Legido, A. Rodríguez, Physical properties of ionic liquids based on 1-alkyl-3-methylimidazolium cation and hexafluorophosphate as anion and temperature dependence, J. Chem. Thermodyn. 39 (2007) 1168–1175.
[15] B. Pereiro, P. Verdía, E. Tojo, A. Rodríguez, Physical properties of 1-butyl-3-methylimidazolium methyl sulfate as a function of temperature, J. Chem. Eng. Data 52 (2007) 377–380.

[16] M. Tariq, P.A.S. Forte, M.F. Costa Gomes, J.N. Canongia Lopes, L.P.N. Rebelo, Densities and refractive indices of imidazolium-based and phosphonium-based ionic liquids: effect of temperature, alkyl chain length and anion, J. Chem. Thermodyn. 41 (2009) 790–798.
[17] S.B. Capelo, T. Mendez-Morales, J. Carrete, E. López Lago, J. Vila, O. Cabeza, J.R. Roddríguez, M. Turmine, L.M. Varela, Effect of temperature and cationic chain length on the physical properties of ammonium nitrate-based protic ionic liquids, J. Phys. Chem. B 116 (2012) 11302–11312.
[18] A.P. Fröba, H. Kremer, A. Leipertz, Density, refractive index, interfacial tension, and viscosity of ionic liquids [EMIM][$EtSO_4$], [EMIM][NTf_2], [EMIM][$N(CN)_2$], and [OMA][NTf_2] in dependence on temperature at atmospheric pressure, J. Phys. Chem. B 112 (2008) 12420–12430.
[19] S. Calixto, N. Roscte-Aguilar, F.J. Sánchez-Marín, O. Torres-Rocha, E.M. Martínez Prado, M. Calixto Solano, Optofluidic compound lenses made with ionic liquids, in: S. Handy (Ed.), Applications of Ionic Liquids in Science and Technology, InTech, 2011, pp. 498–516.
[20] S. Lago, H. Rodriguez, A. Soto, A. Arce, Deterpenation of citrus essential oil by liquid– liquid extraction with 1-alkyl-3-methylimidazolium bis (trifluoromethylsulfonyl) amide ionic liquids, J. Chem. Eng. Data 56 (2011) 1273–1281.
[21] E. Gómez, N. Calvar, E.A. Macedo, A. Domínguez, Effect of the temperature on the physical properties of pure 1-propyl 3-methylimidazolium bis (trifluoromethylsulfonyl) imide and characterization of its binary mixtures with alcohols, J. Chem. Thermodyn. 45 (2012) 9–15.
[22] J.G. Huddleston, A.E. Visser, W.M. Reichert, H.D. Willauer, G.A. Broker, R.D. Rogers, Characterization and comparison of hydrophilic and hydrophobic room temperature ionic liquids incorporating the imidazolium cation, Green Chem. 3 (2001) 156–164.
[23] C. Chiappe, D. Pieraccini, Ionic liquids: solvent properties and organic reactivity, J. Phys. Org. Chem. 18 (2005) 275–297.
[24] M. Wagner, O. Stanga, W. Schroer, The liquid liquid coexistence of binary mixtures of the room temperature ionic liquid 1-methyl-3-hexylimidazolium tetrafluoroborate with alcohols, Phys. Chem. Chem. Phys. 6 (2010) 4421–4431.
[25] A. Arce, E. Rodil, A. Soto, Volumetric and viscosity study for the mixtures of 2-ethoxy-2-methylpropane, ethanol, and 1-ethyl-3-methylimidazolium ethyl sulfate ionic liquid, J. Chem. Eng. Data 51 (2006) 1453–1457.
[26] A. Mokhtarani, M.M. Mojtahedi, H.R. Mortaheb, M. Mafi, F. Yazdani, F. Sadeghian, Densities, refractive indices, and viscosities of the ionic liquids 1-Methyl-3-octylimidazolium tetrafluoroborate and 1-methyl-3-butylimidazolium perchlorate and their binary mixtures with ethanol at several temperatures, J. Chem. Eng. Data 53 (2009) 677–682.
[27] L. Alonso, A. Arce, M. Francisco, O. Rodríguez, A. Soto, Liquid-liquid equilibria for systems composed by 1-methyl-3-octylimidazolium tetrafluoroborate ionic liquid, thiophene, and n-hexane or cyclohexane, J. Chem. Eng. Data 52 (2007) 1729–1732.
[28] A.K. Ziyada, C.D. Wilfred, M.A. Bustam, Z. Man, T. Murugesan, Thermophysical properties of 1-propyronitrile-3-alkylimidazolium bromide ionic liquids at temperatures from (293.15 to 353.15) K, J. Chem. Eng. Data 55 (2010) 3886–3890.
[29] A.K. Ziyada, C.D. Wilfred, T. Murugesan, Densities, viscosities and refractive indices of 1-alkyl-3-propanenitrile imidazolium chloride ionic liquids, Phys. Chem. Liq 50 (2012) 152–160.
[30] T.L. Greaves, C.J. Drummond, Protic ionic liquids: properties and applications, Chem. Rev. 108 (2008) 206–237.
[31] N.M. Yunus, M.I. Abdul Mutalib, Z. Man, M.A. Bustam, T. Murugesan, Thermophysical properties of 1-alkylpyridinum bis (trifluoromethylsulfonyl) imide ionic liquids, J. Chem. Thermodyn. 42 (2010) 491–495.
[32] H. Jin, B. O'Hare, J. Dong, S. Arzhantsev, G.A. Baker, J.F. Wishart, M. Maroncelli, Physical properties of ionic liquids consisting of the 1-butyl-3-methylimidazolium cation with various anions and the bis(trifluoromethylsulfonyl) imide anion with various cations, J. Phys. Chem. B 112 (2008) 81–92.
[33] M. Deetlefs, K.R. Seddon, M. Shara, Predicting physical properties of ionic liquids, Phys. Chem. Chem. Phys. 8 (2006) 642–649.
[34] X. Hu, J. Huang, W. Zhang, M. Li, T. Chengan, G. Li, Photonic ionic liquid polymer for naked eye detection of anions, Adv. Mater. 20 (2008) 4074–4078.

[35] L. Pang, H.M. Chen, L.M. Freeman, Y. Fainman, Optofluidic devices and applications in photonics, sensing and imaging, Lab Chip 12 (2012) 3543–3551.
[36] S. Seki, S. Tsuzuki, K. Hayamizu, Y. Umebayashi, N. Serizawa, K. Takei, H. Miyashiro, Comprehensive refractive Index property for room-temperature ionic liquids, J. Chem. Eng. Data 57 (2012) 2211–2216.
[37] G. Ghosh, Sellmeier coefficients and dispersion of thermo-optic coefficients for some optical glasses, Appl. Opt. 36 (1997) 1540–1546.
[38] M.F. Al-Kuhaili, Optical properties of hafnium oxide thin films and their application in energy-efficient windows, Opt. Mater. 27 (2004) 383–387.
[39] T. Singh, A. Kumar, Temperature dependence of physical properties of imidazolium based ionic liquids: internal pressure and molar refraction, J. Solution Chem. 38 (2009) 1043–1053.
[40] M.A. Iglesias-Otero, J. Troncoso, E. Carballo, L. Romaní, Density and refractive index in mixtures of ionic liquids and organic solvents: correlations and predictions, J. Chem. Thermodyn. 40 (2008) 949–956.
[41] A.N. Soriano, B.T. Doma Jr., M.H. Li, Measurements of the density and refractive index for 1-n-butyl-3-methylimidazolium-based ionic liquids, J. Chem. Thermodyn. 41 (2009) 301–307.
[42] E. Vercher, F.J. Llopis, M.V. Gonzalez-Alfaro, A. Martinez-Andreu, Density, speed of sound, and refractive index of 1-ethyl-3-methylimidazolium trifluoromethanesulfonate with acetone, methyl acetate, and ethyl acetate at temperatures from (278.15 to 328.15), J. Chem. Eng. Data 55 (2010) 1377–1388.
[43] M. Sheik-Bahae, A.A. Said, T.H. Wei, D.J. Hagan, E.W. Van Stryland, Sensitive measurement of optical nonlinearities using a single beam, IEEE J. Quantum Elect. 26 (1990) 760–769.
[44] R.C.C. Leite, R.S. Moore, J.R. Whinnery, Low absorption measurements by means of the thermal lens effect using an He:Ne laser, Appl. Phys. Lett. 5 (1964) 141–143.
[45] M. Franko, C.D. Tran, Thermal Lens Spectroscopy, Encyclopedia of Analytical Chemistry, 2010.
[46] T. Kitamori, A. Hibara, A.M. Tokeshi, Thermal Lens Microscope, European Patent No. EP 1324024 M, European Patent Office, Munich, Germany, 2008.
[47] J.M. Harris, N.J. Dovichi, Thermal lens calorimetry, Anal. Chem. 52 (1980) 695A–706A.
[48] S.J. Sheldon, L.V. Knight, J.M. Thorne, Laser-induced thermal lens effect: a new theoretical model, Appl. Optics 21 (1982) 1663–1669.
[49] A. Carter, J.M. Harris, Comparison of models describing the thermal lens effect, Appl. Opt. 23 (1984) 476–481.
[50] M. Falconieri, Thermo-optical effects in Z-scan measurements using high-repetition-rate lasers, J. Opt. A-pure Appl. Opt. 1 (1999) 662–667.
[51] R.F. Souza, M.A.R.C. Alencar, M.R. Meneghetti, J. Dupont, J.M. Hickmann, Nonlocal optical nonlinearity of ionic liquids, J. Phys. Condens. Mat 20 (2008) paper number 155102.
[52] E. Valencia-Loredo, M. Barrera-Rivera, M. Trejo-Duran, E. Alvarado-Mendez, A. Martínez Richa, J.A. Andrade-Lucio, Nonlinear optical characterization of ionic liquids, Photonics North, Proc. SPIE 7386 (August 04, 2009) 738610, http://dx.doi.org/10.1117/12.839505.
[53] M. Trejo-Duran, E. Alvarado-Mendez, E. Vargas-Rodriguez, J.M. Estudillo-Ayala, R.I. Mata-Chavez, Nonlinear optical characterization of ionics liquids of 1-methylpyrrolidine family, Photonics North, Proc. SPIE 8412 (October 23, 2012) 84121X, http://dx.doi.org/10.1117/12.2001408.
[54] E. Santos, M.A. Alencar, P. Migowski, J. Dupont, J.M. Hickmann, Anionic and cationic influence on the nonlocal nonlinear optical response of ionic liquids, Chem. Phys. 403 (2012) 33–36.
[55] R. Ge, C. Hardacre, P. Nancarrow, D.W. Rooney, Thermal conductivities of ionic liquids over the temperature range from 293 K to 353 K, J. Chem. Eng. Data 52 (2007) 1819–1823.

SUBCHAPTER 1.4

Physical Properties of Mixtures

Luisa Segade, Oscar Cabeza
Departamento de Física, Facultade de Ciencias, Universidade da Coruña, A Coruña, Spain

The physical properties of the binary mixtures already measured were obtained from SciFinder from 2010 up to date because the IL Thermo database has all published papers up to 2009 (the last time this database was updated was the middle of 2010). The words used in the search were "ionic liquid" (as entered and as concept), "2010," "mixture," and "the desired magnitude". Before 2005, very low data were published devoted to mixtures containing ILs, but from that year many works have been published on this subject, which reveals its interest from the theoretical and applied point of views. The values behavior of the systems with the concentration depends a lot on the physical magnitude studied but that behavior followed by each property is similar for the majority of mixtures of the ILs with the different solvents. Thus, density can be deduced easily taking into account that molar volume has a near ideal behavior (i.e., it changes linearly between the values of both pure components). In contrast, viscosity varies exponentially between the pure values with the molar fraction, and so a small quantity of solvent concentration substantially changes the viscosity value of the mixture. Electrical conductivity also changes substantially with solvent, but in a different manner than viscosity. Its value usually presents a maximum at a given concentration that could be up to 100 times higher than the value of the pure IL (the solvent is usually isolant). The concentration and value of that maximum depends on the given pair IL + solvent. Finally, surface tension of mixtures depends greatly with the solvent nature. For some mixtures, IL acts as a surfactant and, therefore, a critical micelle concentration-like point can be defined that happens, for example, in aqueous mixtures of imidazolium based ILs. For other solvents, the surface tension value varies linearly with solvent concentration, as for mixtures with ethanol.

The measured four magnitudes in the different binary mixtures are included in Table 1. The files are ordered following the cation family name, later the solvent (water, alkanols, and others), and then the anion type in alphabetic order. References for each mixture can be easily obtained in different databases and are not included because of the lack of space.

As observed, only density have been measured in many binary mixtures of ILs with a solvent, there is still a lack of experimental data for many other magnitudes and possible mixtures, and so experimental work is necessary to properly understand the influence of the different solvents mixed with ILs.

Table 1 Binary Mixtures with an Ionic Liquid where Density, ρ, Viscosity, η, Surface Tension, σ, or Electrical Conductivity, κ, were Measured

	Ionic Liquid			Property Studied			
Solvent	**Cation Family**	**Anion Family**	**Name**	ρ	η	σ	κ
Water	Ammonium	$[C_nCOO]$	Ethylammonium acetate	×			
Water	Ammonium	$[C_nCOO]$	Propylammonium acetate	×			
Water	Ammonium	$[C_nCOO]$	Butylammonium acetate	×	×		
Water	Ammonium	$[C_nCOO]$	2-hydroxy Ethylammonium acetate	×			
Water	Ammonium	$[C_nCOO]$	bis(2-hydroxyethyl)ammonium acetate	×			
Water	Ammonium	$[C_nCOO]$	Diisopropyl-ethylammonium heptanoate	×			
Water	Ammonium	$[C_nSO_3]$	Diethylmethylammonium methanesulfonate	×	×		
Water	Ammonium	$[C_nSO_4]$	Tris-(2-hydroxylethyl) methylammonium methylsulfate		×		
Water	Ammonium	$[NO_3]$	Ethylammonium nitrate	×			
Water	Ammonium	$[NO_3]$	Propylammonium nitrate	×			
Water	Ammonium	$[NO_3]$	Butylammonium nitrate	×	×		
Water	Ammonium	$[NO_3]$	Tetradecyltrimethylammonium nitrate	×			
Water	Ammonium	$[NTf_2]$	Butyltrimethylammonium bis(trifluoromethylsulfonyl)imide	×	×		
Methanol	Ammonium	$[C_nCOO]$	bis(2-hydroxyethyl)ammonium propanoate	×			
Methanol	Ammonium	$[C_nCOO]$	Butylammonium acetate	×	×		
Methanol	Ammonium	$[C_nCOO]$	2-hydroxy ethylammonium acetate	×			
Methanol	Ammonium	$[C_nCOO]$	bis(2-hydroxyethyl)ammonium acetate	×			
Methanol	Ammonium	$[C_nCOO]$	bis(2-hydroxyethyl) methylammonium formate	×	×		
Ethanol	Ammonium	$[C_nCOO]$	bis(2-hydroxyethyl)ammonium propanoate	×			
Ethanol	Ammonium	$[C_nCOO]$	Butylammonium acetate	×	×		

(Continued)

Table 1 Binary Mixtures with an Ionic Liquid where Density, ρ, Viscosity, η, Surface Tension, σ, or Electrical Conductivity, κ, were Measured—cont'd

	Ionic Liquid			Property Studied			
Solvent	**Cation Family**	**Anion Family**	**Name**	ρ	η	σ	κ
Ethanol	Ammonium	$[C_nCOO]$	2-hydroxy ethylammonium acetate	×			
Ethanol	Ammonium	$[C_nCOO]$	bis(2-hydroxyethyl)ammonium acetate	×			
Ethanol	Ammonium	$[C_nCOO]$	bis(2-hydroxyethyl) methylammonium formate	×	×		
1-propanol	Ammonium	$[C_nCOO]$	bis(2-hydroxyethyl)ammonium propanoate	×			
1-propanol	Ammonium	$[C_nCOO]$	Butylammonium acetate	×	×		
1-propanol	Ammonium	$[C_nCOO]$	bis(2-hydroxyethyl)ammonium acetate	×			
1-propanol	Ammonium	$[C_nCOO]$	bis(2-hydroxyethyl) methylammonium formate	×	×		
1-butanol	Ammonium	$[C_nCOO]$	Butylammonium acetate	×	×		
Methanol	Ammonium	$[NO_3]$	Butylammonium nitrate	×	×		
Ethanol	Ammonium	$[NO_3]$	Butylammonium nitrate	×	×		
1-propanol	Ammonium	$[NO_3]$	Butylammonium nitrate	×	×		
1-butanol	Ammonium	$[NO_3]$	Butylammonium nitrate	×	×		
Methanol	Ammonium	$[NTf_2]$	Methyltrioctylammonium bis(trifluoromethylsulfonyl)imide	×			
Ethanol	Ammonium	$[NTf_2]$	Methyltrioctylammonium bis(trifluoromethylsulfonyl)imide	×			
1-propanol	Ammonium	$[NTf_2]$	*N*,*N*,*N*-trimethyl-*N*-propylammonium bis(trifluoromethanesulfonyl)imide		×		
1-propanol	Ammonium	$[NTf_2]$	*N*,*N*,*N*-trimethyl-*N*-propylammonium bis(trifluoromethanesulfonyl)imide		×		
N-Methyl-2-pyrrolidone	Ammonium	$[C_nC_nPO_4]$		×	×		

			Trimethylammonium dihydrogenphosphate				
N-Methyl-2-pyrrolidone	Ammonium	[$C_nC_nPO_4$]	Triethylammonium dihydrogenphosphate	×	×		
Acetonitrile	Ammonium	[C_nCOO]	Diisopropyl-ethylammonium octanoate	×			
Acetonitrile	Ammonium	[NTf_2]	Triethylammonium bis(trifluoromethanesulfonyl)imide	×			
N-Methyl-2-pyrrolidone	Ammonium	[C_nCOO]	Trimethylammonium acetate	×	×		
Monoethanolamine	Ammonium	[C_nCOO]	bis(2-hydroxyethyl)ammonium acetate	×			
N-Methyl-2-pyrrolidone	Ammonium	[C_nSO_4]	Trimethylammonium hydrogensulfate	×	×		
Dimethyl carbonate	Ammonium	[Halide]	Tricaprylmethylammonium chloride	×			×
γ-butyrolactone	Ammonium	[NO_3]	Ethylammonium nitrate	×			
Acetonitrile	Ammonium	[NTf_2]	*N*,*N*,*N*-trimethyl-*N*-propylammonium bis(trifluoromethanesulfonyl)imide		×		
Acetonitrile	Ammonium	[NTf_2]	Trimethylammonium bis(trifluoromethylsulfonyl)imide		×		×
Methyl acetate	Ammonium	[NTf_2]	Methyltrioctylammonium bis(trifluoromethylsulfonyl)imide	×			
Ethyl acetate	Ammonium	[NTf_2]	Methyltrioctylammonium bis(trifluoromethylsulfonyl)imide	×			
Propylene carbonate	Ammonium	[NTf_2]	Trimethylammonium bis(trifluoromethylsulfonyl)imide		×		
γ-butyrolactone	Ammonium	[NTf_2]	Trimethylammonium bis(trifluoromethylsulfonyl)imide		×		
Poly(ethylene glycol)	Ammonium	[C_nSO_4]	2-Ethoxy-1-ethyl-1,1-dimethyl-2-oxoethanaminium ethyl sulfate				×
Water	Imidazolium	[BF_4]	1,2-dimethyl-3-propyl imidazolium tetrafluoroborate	×	×		
Water	Imidazolium	[BF_4]	1-propyl-2,3-dimethyl imidazolium tetrafluoroborate	×			

(*Continued*)

Table 1 Binary Mixtures with an Ionic Liquid where Density, ρ, Viscosity, η, Surface Tension, σ, or Electrical Conductivity, κ, were Measured—cont'd

	Ionic Liquid			Property Studied			
Solvent	**Cation Family**	**Anion Family**	**Name**	ρ	η	σ	κ
Water	Imidazolium	$[BF_4]$	1-butyl-2,3-dimethyl imidazolium tetrafluoroborate	×	×		
Water	Imidazolium	$[BF_4]$	1,3-dimethyl imidazolium tetrafluorobrorate	×			
Water	Imidazolium	$[BF_4]$	1-ethyl-3-methyl imidazolium tetrafluoroborate	×	×	×	×
Water	Imidazolium	$[BF_4]$	1-butyl-3-methyl imidazolium tetrafluoroborate	×	×	×	×
Water	Imidazolium	$[BF_4]$	1-hexyl-3-methyl imidazolium tetrafluoroborate	×	×	×	×
Water	Imidazolium	$[C(CN)_3]$	1-butyl-3-methyl imidazolium tricyanomethane	×	×		
Water	Imidazolium	$[C_nC_nPO_4]$	1,3-dimethyl imidazolium dimethylphosphate	×	×		
Water	Imidazolium	$[C_nC_nPO_4]$	1-ethyl-3-methyl imidazolium dimethylphosphate	×	×		
Water	Imidazolium	$[C_nC_nPO_4]$	1-butyl-3-methyl imidazolium dimethylphosphate	×	×		
Water	Imidazolium	$[C_nCOO]$	Imidazolium octanoate	×	×		
Water	Imidazolium	$[C_nCOO]$	1-ethyl-3-methyl imidazolium acetate	×	×		
Water	Imidazolium	$[C_nCOO]$	1-ethyl-3-methyl imidazolium L-lactate	×		×	
Water	Imidazolium	$[C_nCOO]$	1-butyl-3-methyl imidazolium L-lactate	×		×	
Water	Imidazolium	[CNS]	1-ethyl-3-methyl imidazolium thiocyanate	×		×	
Water	Imidazolium	[CNS]	1-butyl-3-methyl imidazolium thiocyanate	×	×		

Water	Imidazolium	$[C_nSO_3]$	1-ethyl-3-methyl imidazolium methanesulfonate			×	
Water	Imidazolium	$[C_nSO_3]$	1-butyl-3-methyl imidazolium methanesulfonate	×		×	×
Water	Imidazolium	$[C_nSO_4]$	1,3-dimethyl imidazolium methylsulfate	×	×	×	
Water	Imidazolium	$[C_nSO_4]$	1-ethyl-3-methyl imidazolium methylsulfate	×	×		
Water	Imidazolium	$[C_nSO_4]$	1-butyl-3-methyl imidazolium methylsulfate	×	×	×	
Water	Imidazolium	$[C_nSO_4]$	1-ethyl-3-methyl imidazolium ethylsulfate	×	×	×	×
Water	Imidazolium	$[C_nSO_4]$	1-ethyl-3-methyl imidazolium butylsulfate	×		×	
Water	Imidazolium	$[C_nSO_4]$	1-ethyl-3-methyl imidazolium hexylsulfate	×		×	
Water	Imidazolium	$[C_nSO_4]$	1-ethyl-3-methyl imidazolium octylsulfate	×		×	
Water	Imidazolium	$[C_nSO_4]$	1-butyl-3-methyl imidazolium octylsulfate	×		×	
Water	Imidazolium	$[C_nSO_4]$	1-butyl-3-methyl imidazolium hydrogensulfate	×	×		
Water	Imidazolium	[Halide]	1-methyl-3-octyl imidazolium chloride		×		
Water	Imidazolium	[Halide]	1,3-dimethyl imidazolium chloride	×			
Water	Imidazolium	[Halide]	1-butyl-3-methyl imidazolium chloride	×	×		
Water	Imidazolium	[Halide]	1-butyl-3-methyl imidazolium chloride	×		×	×
Water	Imidazolium	[Halide]	1-ethyl-3-methyl imidazolium chloride		×		
Water	Imidazolium	[Halide]	1-hexyl-3-methyl imidazolium chloride	×	×	×	

(*Continued*)

Table 1 Binary Mixtures with an Ionic Liquid where Density, ρ, Viscosity, η, Surface Tension, σ, or Electrical Conductivity, κ, were Measured—cont'd

	Ionic Liquid			Property Studied			
Solvent	**Cation Family**	**Anion Family**	**Name**	ρ	η	σ	κ
Water	Imidazolium	[Halide]	1-octyl-3-methyl imidazolium chloride	×			
Water	Imidazolium	[Halide]	1-decyl-3-methyl imidazolium bromide	×			
Water	Imidazolium	[Halide]	1-methyl-3-octyl imidazolium bromide	×			
Water	Imidazolium	[Halide]	1-ethyl-3-methyl imidazolium bromide	×	×	×	×
Water	Imidazolium	[Halide]	1-butyl-3-methyl imidazolium bromide	×	×	×	×
Water	Imidazolium	[Halide]	1-hexyl-3-methyl imidazolium bromide	×			
Water	Imidazolium	[Halide]	1-octyl-3-methyl imidazolium bromide	×			
Water	Imidazolium	[Halide]	1-butyl-3-methyl imidazolium iodide	×			
Water	Imidazolium	[Halide]	1-hexyl-3-methyl imidazolium iodide	×			
Water	Imidazolium	[Halide]	1-octyl-3-methyl imidazolium iodide	×			
Water	Imidazolium	$[N(CN)_2]$	1-ethyl-3-methyl imidazolium dicyanamide	×	×		
Water	Imidazolium	$[N(CN)_2]$	1-butyl-3-methyl imidazolium dicyanamide	×	×		×
Water	Imidazolium	$[N(CN)_2]$	1-hexyl-3-methyl imidazolium dicyanamide	×			
Water	Imidazolium	$[NTf_2]$	1-ethyl-3-methyl imidazolium bis(trifluoromethylsulfonyl)imide	×	×		

Water	Imidazolium	[NTf_2]	1-butyl-3-methyl imidazolium bis(trifluoromethylsulfonyl)imide	×	×	×	×
Water	Imidazolium	[NTf_2]	1-hexyl-3-methyl imidazolium bis(trifluoromethylsulfonyl)imide	×	×		×
Water	Imidazolium	[NTf_2]	1-octyl-3-methyl imidazolium bis(trifluoromethylsulfonyl)imide	×	×		×
Water	Imidazolium	[NTf_2]	1-decyl-3-methyl imidazolium bis(trifluoromethylsulfonyl)imide	×	×		×
Water	Imidazolium	[OTf]	1-ethyl-3-methyl imidazolium trifluoromethanesulfonate	×	×		×
Water	Imidazolium	[OTf]	1-ethyl-3-butyl imidazolium trifluoromethanesulfonate	×			
Water	Imidazolium	[OTf]	1-butyl-3-methyl imidazolium trifluoromethanesulfonate	×	×		
Water	Imidazolium	[PF_6]	1-butyl-3-methyl imidazolium hexafluorophosphate	×	×	×	×
Water	Imidazolium	[PF_6]	1-hexyl-3-methyl imidazolium hexafluorophosphate	×	×	×	×
Water	Imidazolium	[PF_6]	1-octyl-3-methyl imidazolium hexafluorophosphate	×	×	×	×
Methanol	Imidazolium	[AA]	1-butyl-3-methyl imidazolium alanine acid salt	×	×		
Methanol	Imidazolium	[AA]	1-butyl-3-methyl imidazolium aspartate	×			×
Methanol	Imidazolium	[AA]	1-butyl-3-methyl imidazolium glutamic acid salt	×	×		
Methanol	Imidazolium	[AA]	1-butyl-3-methyl imidazolium glycine acid salt	×	×		
1-butanol	Imidazolium	[AA]	1-butyl-3-methyl imidazolium glycine acid salt	×			
Benzyl alcohol	Imidazolium	[AA]	1-butyl-3-methyl imidazolium alanine acid salt	×	×		
Benzyl alcohol	Imidazolium	[AA]		×			

(Continued)

Table 1 Binary Mixtures with an Ionic Liquid where Density, ρ, Viscosity, η, Surface Tension, σ, or Electrical Conductivity, κ, were Measured—cont'd

	Ionic Liquid			Property Studied			
Solvent	**Cation Family**	**Anion Family**	**Name**	ρ	η	σ	κ
			1-butyl-3-methyl imidazolium glutamic acid salt				
Benzyl alcohol	Imidazolium	[AA]	1-butyl-3-methyl imidazolium glycine acid salt	×			
Isopropanol	Imidazolium	[AA]	1-butyl-3-methyl imidazolium glycine acid salt	×			
Methanol	Imidazolium	$[BF_4]$	1-methyl-3-octyl imidazolium tetrafluoroborate	×	×		
Methanol	Imidazolium	$[BF_4]$	1-butyl-3-methyl imidazolium tetrafluoroborate	×			×
Ethanol	Imidazolium	$[BF_4]$	1-methyl-3-octyl imidazolium tetrafluoroborate	×	×	×	
Ethanol	Imidazolium	$[BF_4]$	1-ethyl-3-methyl imidazolium tetrafluoroborate	×		×	
Ethanol	Imidazolium	$[BF_4]$	1-butyl-3-methyl imidazolium tetrafluoroborate	×		×	
Ethanol	Imidazolium	$[BF_4]$	1-hexyl-3-methyl imidazolium tetrafluoroborate	×		×	
1-propanol	Imidazolium	$[BF_4]$	1-methyl-3-octyl imidazolium tetrafluoroborate	×	×		
1-propanol	Imidazolium	$[BF_4]$	1-butyl-3-methyl imidazolium tetrafluoroborate	×			
1-butanol	Imidazolium	$[BF_4]$	1-methyl-3-octyl imidazolium tetrafluoroborate	×			
1-pentanol	Imidazolium	$[BF_4]$	1-methyl-3-octyl imidazolium tetrafluoroborate	×			
1-pentanol	Imidazolium	$[BF_4]$	1-hexyl-3-methyl imidazolium tetrafluoroborate	×	×		

2-propanol	Imidazolium	$[BF_4]$	1-methyl-3-octyl imidazolium tetrafluoroborate	×	×
2-propanol	Imidazolium	$[BF_4]$	1-butyl-3-methyl imidazolium tetrafluoroborate	×	
1,2-ethanediol	Imidazolium	$[BF_4]$	1-methyl-3-octyl imidazolium tetrafluoroborate	×	
1,2-ethanediol	Imidazolium	$[BF_4]$	1-butyl-3-methyl imidazolium tetrafluoroborate	×	
2-(2-methoxyethoxy) ethanol	Imidazolium	$[BF_4]$	1-methyl-3-octyl imidazolium tetrafluoroborate	×	
2-[2-(2-methoxyethoxy) ethoxy]ethanol	Imidazolium	$[BF_4]$	1-methyl-3-octyl imidazolium tetrafluoroborate	×	
2-methoxyethanol	Imidazolium	$[BF_4]$	1-methyl-3-octyl imidazolium tetrafluoroborate	×	
2,2,2-trifluoroethanol	Imidazolium	$[BF_4]$	1-ethyl-3-methyl imidazolium tetrafluoroborate	×	
2,2,2-trifluoroethanol	Imidazolium	$[BF_4]$	1-butyl-3-methyl imidazolium tetrafluoroborate	×	
3,6-dioxa-1-octanol	Imidazolium	$[BF_4]$	1-butyl-3-methyl imidazolium tetrafluoroborate	×	
Diethylene glycol monomethyl ether	Imidazolium	$[BF_4]$	1-butyl-3-methyl imidazolium tetrafluoroborate	×	
Ethylene glycol	Imidazolium	$[BF_4]$	1-butyl-3-methyl imidazolium tetrafluoroborate	×	×
Ethylene glycol monomethyl ether	Imidazolium	$[BF_4]$	1-butyl-3-methyl imidazolium tetrafluoroborate	×	
Triethylene glycol monomethyl ether	Imidazolium	$[BF_4]$	1-butyl-3-methyl imidazolium tetrafluoroborate	×	
Ethylene glycol monomethyl ether	Imidazolium	$[BF_4]$	1-hexyl-3-methyl imidazolium tetrafluoroborate	×	
Ethanol	Imidazolium	$[ClO_4]$	1-butyl-3-methyl imidazolium perchlorate	×	×

(*Continued*)

Table 1 Binary Mixtures with an Ionic Liquid where Density, ρ, Viscosity, η, Surface Tension, σ, or Electrical Conductivity, κ, were Measured—cont'd

	Ionic Liquid			Property Studied			
Solvent	**Cation Family**	**Anion Family**	**Name**	ρ	η	σ	κ
Methanol	Imidazolium	[$C_nC_nPO_4$]	1,3-dimethyl imidazolium dimethylphosphate	×	×		
Methanol	Imidazolium	[$C_nC_nPO_4$]	1-ethyl-3-methyl imidazolium dimethylphosphate	×	×		
Methanol	Imidazolium	[$C_nC_nPO_4$]	1-butyl-3-methyl imidazolium dimethylphosphate	×	×		
Ethanol	Imidazolium	[$C_nC_nPO_4$]	1,3-dimethyl imidazolium dimethylphosphate	×	×		
Ethanol	Imidazolium	[$C_nC_nPO_4$]	1-ethyl-3-methyl imidazolium dimethylphosphate	×	×		
Ethanol	Imidazolium	[$C_nC_nPO_4$]	1-butyl-3-methyl imidazolium dimethylphosphate	×	×		
Methanol	Imidazolium	[C_nCOO]	1-methyl imidazolium acetate	×	×		
Methanol	Imidazolium	[C_nCOO]	1-butyl-3-methyl imidazolium L-lactate	×		×	
Ethanol	Imidazolium	[C_nCOO]	Imidazolium octanoate	×	×		
Ethanol	Imidazolium	[C_nCOO]	1-ethyl-3-methyl imidazolium acetate	×	×		
Ethanol	Imidazolium	[C_nCOO]	1-methyl imidazolium acetate	×	×		
Ethanol	Imidazolium	[C_nCOO]	1-butyl-3-methyl imidazolium L-lactate	×		×	
1-propanol	Imidazolium	[C_nCOO]	1-methyl imidazolium acetate	×	×		
1-butanol	Imidazolium	[C_nCOO]	1-methyl imidazolium acetate	×	×		
1-butanol	Imidazolium	[C_nCOO]	1-butyl-3-methyl imidazolium L-lactate	×		×	
1-octanol	Imidazolium	[C_nCOO]	Imidazolium octanoate	×	×		
Methanol	Imidazolium	[CNS]	1-butyl-3-methyl imidazolium thiocyanate	×	×		
Ethanol	Imidazolium	[CNS]		×	×		

			1-butyl-3-methyl imidazolium thiocyanate			
1-propanol	Imidazolium	[CNS]	1-butyl-3-methyl imidazolium thiocyanate	×	×	
1-heptanol	Imidazolium	[CNS]	1-butyl-3-methyl imidazolium thiocyanate	×		
1-octanol	Imidazolium	[CNS]	1-butyl-3-methyl imidazolium thiocyanate	×		
1-nonanol	Imidazolium	[CNS]	1-butyl-3-methyl imidazolium thiocyanate	×		
1-decanol	Imidazolium	[CNS]	1-butyl-3-methyl imidazolium thiocyanate	×		
Methanol	Imidazolium	[C_nSO_4]	1,3-dimethyl imidazolium methylsulfate	×		
Methanol	Imidazolium	[C_nSO_4]	1-ethyl-3-methyl imidazolium methylsulfate	×		×
Methanol	Imidazolium	[C_nSO_4]	1-butyl-3-methyl imidazolium methylsulfate	×		
Methanol	Imidazolium	[C_nSO_4]	1-hexyl-3-methyl imidazolium methylsulfate	×		
Methanol	Imidazolium	[C_nSO_4]	1-ethyl-3-methyl imidazolium ethylsulfate	×	×	
Methanol	Imidazolium	[C_nSO_4]	1-butyl-3-methyl imidazolium ethylsulfate	×		
Methanol	Imidazolium	[C_nSO_4]	1-hexyl-3-methyl imidazolium ethylsulfate	×		
Methanol	Imidazolium	[C_nSO_4]	1-butyl-3-methyl imidazolium octylsulfate	×		
Ethanol	Imidazolium	[C_nSO_4]	1-methyl-3-methyl imidazolium methylsulfate	×		
Ethanol	Imidazolium	[C_nSO_4]	1,3-dimethyl imidazolium methylsulfate	×	×	

(*Continued*)

Table 1 Binary Mixtures with an Ionic Liquid where Density, ρ, Viscosity, η, Surface Tension, σ, or Electrical Conductivity, κ, were Measured—cont'd

	Ionic Liquid			Property Studied			
Solvent	**Cation Family**	**Anion Family**	**Name**	ρ	η	σ	κ
Ethanol	Imidazolium	$[C_nSO_4]$	1-ethyl-3-methyl imidazolium methylsulfate	×		×	
Ethanol	Imidazolium	$[C_nSO_4]$	1-butyl-3-methyl imidazolium methylsulfate	×	×		
Ethanol	Imidazolium	$[C_nSO_4]$	1-hexyl-3-methyl imidazolium methylsulfate	×			
Ethanol	Imidazolium	$[C_nSO_4]$	1-ethyl-3-methyl imidazolium ethylsulfate	×	×	×	
Ethanol	Imidazolium	$[C_nSO_4]$	1-butyl-3-methyl imidazolium ethylsulfate	×			
Ethanol	Imidazolium	$[C_nSO_4]$	1-hexyl-3-methyl imidazolium ethylsulfate	×			
Ethanol	Imidazolium	$[C_nSO_4]$	1-ethyl-3-methyl imidazolium butylsulfate	×		×	
Ethanol	Imidazolium	$[C_nSO_4]$	1-ethyl-3-methyl imidazolium hexylsulfate	×		×	
Ethanol	Imidazolium	$[C_nSO_4]$	1-ethyl-3-methyl imidazolium octylsulfate	×		×	
1-propanol	Imidazolium	$[C_nSO_4]$	1-butyl-3-methyl imidazolium methylsulfate	×			
1-propanol	Imidazolium	$[C_nSO_4]$	1-ethyl-3-methyl imidazolium ethylsulfate	×	×		
1-propanol	Imidazolium	$[C_nSO_4]$	1-butyl-3-methyl imidazolium octylsulfate	×			
1-butanol	Imidazolium	$[C_nSO_4]$	1,3-dimethyl imidazolium methylsulfate	×			
1-butanol	Imidazolium	$[C_nSO_4]$	1-ethyl-3-methyl imidazolium methylsulfate	×		×	

1-butanol	Imidazolium	[C_nSO_4]	1-butyl-3-methyl imidazolium methylsulfate	×
1-butanol	Imidazolium	[C_nSO_4]	1-ethyl-3-methyl imidazolium ethylsulfate	×
1-butanol	Imidazolium	[C_nSO_4]	1-butyl-3-methyl imidazolium octylsulfate	×
1-pentanol	Imidazolium	[C_nSO_4]	1-ethyl-3-methyl imidazolium ethylsulfate	×
1-hexanol	Imidazolium	[C_nSO_4]	1-butyl-3-methyl imidazolium methylsulfate	×
1-hexanol	Imidazolium	[C_nSO_4]	1-ethyl-3-methyl imidazolium ethylsulfate	×
1-hexanol	Imidazolium	[C_nSO_4]	1-butyl-3-methyl imidazolium octylsulfate	×
1-heptanol	Imidazolium	[C_nSO_4]	1-ethyl-3-methyl imidazolium ethylsulfate	×
1-octanol	Imidazolium	[C_nSO_4]	1-butyl-3-methyl imidazolium methylsulfate	×
1-octanol	Imidazolium	[C_nSO_4]	1-ethyl-3-methyl imidazolium ethylsulfate	×
1-octanol	Imidazolium	[C_nSO_4]	1-butyl-3-methyl imidazolium octylsulfate	×
1-nonanol	Imidazolium	[C_nSO_4]	1-ethyl-3-methyl imidazolium ethylsulfate	×
1-decanol	Imidazolium	[C_nSO_4]	1-butyl-3-methyl imidazolium methylsulfate	×
1-decanol	Imidazolium	[C_nSO_4]	1-ethyl-3-methyl imidazolium ethylsulfate	×
1-decanol	Imidazolium	[C_nSO_4]	1-butyl-3-methyl imidazolium octylsulfate	×
2-propanol	Imidazolium	[C_nSO_4]	1,3-dimethyl imidazolium methylsulfate	×

(Continued)

Table 1 Binary Mixtures with an Ionic Liquid where Density, ρ, Viscosity, η, Surface Tension, σ, or Electrical Conductivity, κ, were Measured—cont'd

	Ionic Liquid			Property Studied			
Solvent	**Cation Family**	**Anion Family**	**Name**	**ρ**	**η**	**σ**	**κ**
2-propanol	Imidazolium	[C_nSO_4]	1-butyl-3-methyl imidazolium methylsulfate	×			
2-propanol	Imidazolium	[C_nSO_4]	1-ethyl-3-methyl imidazolium ethylsulfate	×	×		
1,2-ethanediol	Imidazolium	[C_nSO_4]	1-butyl-3-methyl imidazolium methylsulfate	×			
1,2-ethanediol	Imidazolium	[C_nSO_4]	1-butyl-3-methyl imidazolium octylsulfate	×			
3,6-dioxa-1-octanol	Imidazolium	[C_nSO_4]	1-butyl-3-methyl imidazolium methylsulfate	×			
Ethylene glycol	Imidazolium	[C_nSO_4]	1-butyl-3-methyl imidazolium methylsulfate	×			
Methanol	Imidazolium	[Halide]	1-methyl-3-octyl imidazolium chloride		×		
Methanol	Imidazolium	[Halide]	1-methyl-3-octyl imidazolium bromide	×			
Methanol	Imidazolium	[Halide]	1-butyl-3-methyl imidazolium bromide	×			
Methanol	Imidazolium	[Halide]	1-butyl-3-methyl imidazolium chloride	×			
Ethanol	Imidazolium	[Halide]	1-propyronitrile-3-hexyl imidazolium bromide	×			
Ethanol	Imidazolium	[Halide]	1-methyl-3-octyl imidazolium chloride		×		
Ethanol	Imidazolium	[Halide]	1-methyl-3-octyl imidazolium bromide	×			
Ethanol	Imidazolium	[Halide]	1-butyl-3-methyl imidazolium bromide	×			

Ethanol	Imidazolium	[Halide]	1-butyl-3-methyl imidazolium chloride	×	
Ethanol	Imidazolium	[Halide]	1-hexyl-3-methyl imidazolium chloride	×	
1-propanol	Imidazolium	[Halide]	1-methyl-3-octyl imidazolium chloride		×
1-propanol	Imidazolium	[Halide]	1-methyl-3-octyl imidazolium bromide	×	
1,2-ethanediol	Imidazolium	[Halide]	1-decyl-3-methyl imidazolium bromide	×	
1,2-ethanediol	Imidazolium	[Halide]	1-methyl-3-octyl imidazolium bromide	×	
1,2-ethanediol	Imidazolium	[Halide]	1-butyl-3-methyl imidazolium bromide	×	
1,2-propanediol	Imidazolium	[Halide]	1-butyl-3-methyl imidazolium bromide	×	
1,3-propanediol	Imidazolium	[Halide]	1-butyl-3-methyl imidazolium bromide	×	
1,2-butanediol	Imidazolium	[Halide]	1-butyl-3-methyl imidazolium bromide	×	
Ethylene glycol monomethyl ether	Imidazolium	[Halide]	1-hexyl-3-methyl imidazolium bromide	×	
Ethylene glycol	Imidazolium	[Halide]	1-butyl-3-methyl imidazolium chloride	×	
Ethylene glycol	Imidazolium	[Halide]	1-octyl-3-methyl imidazolium chloride	×	
Ethanol	Imidazolium	$[N(CN)_2]$	1-ethyl-3-methyl imidazolium dicyanamide	×	×
1-propanol	Imidazolium	$[N(CN)_2]$	1-butyl-3-methyl imidazolium dicyanamide	×	
1-butanol	Imidazolium	$[N(CN)_2]$	1-butyl-3-methyl imidazolium dicyanamide	×	

(*Continued*)

Table 1 Binary Mixtures with an Ionic Liquid where Density, ρ, Viscosity, η, Surface Tension, σ, or Electrical Conductivity, κ, were Measured—cont'd

	Ionic Liquid			Property Studied			
Solvent	Cation Family	Anion Family	Name	ρ	η	σ	κ
1-pentanol	Imidazolium	$[N(CN)_2]$	1-butyl-3-methyl imidazolium dicyanamide	×			
2-propanol	Imidazolium	$[N(CN)_2]$	1-butyl-3-methyl imidazolium dicyanamide	×			
2-butanol	Imidazolium	$[N(CN)_2]$	1-butyl-3-methyl imidazolium dicyanamide	×			
Methanol	Imidazolium	$[NO_3]$	1-ethyl-3-methyl imidazolium nitrate	×		×	
Ethanol	Imidazolium	$[NO_3]$	1-ethyl-3-methyl imidazolium nitrate	×		×	
Ethanol	Imidazolium	$[NO_3]$	1-butyl-3-methyl imidazolium nitrate	×	×		
1-propanol	Imidazolium	$[NO_3]$	1-butyl-3-methyl imidazolium nitrate	×	×		
1-butanol	Imidazolium	$[NO_3]$	1-butyl-3-methyl imidazolium nitrate	×	×		
Methanol	Imidazolium	$[NTf_2]$	1-butyl-3-methyl imidazolium bis(trifluoromethylsulfonyl)imide				×
Methanol	Imidazolium	$[NTf_2]$	1-propyl-3-methyl imidazolium bis(trifluoromethylsulfonyl)imide	×	×		
Methanol	Imidazolium	$[NTf_2]$	1-hexyl-3-methyl imidazolium bis(trifluoromethylsulfonyl)imide	×			
Methanol	Imidazolium	$[NTf_2]$	1-octyl-3-methyl imidazolium bis(trifluoromethylsulfonyl)imide	×	×		
Ethanol	Imidazolium	$[NTf_2]$	1-propyl-3-methyl imidazolium bis(trifluoromethylsulfonyl)imide	×	×		
Ethanol	Imidazolium	$[NTf_2]$	1-butyl-3-methyl imidazolium bis(trifluoromethylsulfonyl)imide	×	×		

Ethanol	Imidazolium	[NTf_2]	1-hexyl-3-methyl imidazolium bis(trifluoromethylsulfonyl)imide	×			
Ethanol	Imidazolium	[NTf_2]	1-octyl-3-methyl imidazolium bis(trifluoromethylsulfonyl)imide	×	×		
1-propanol	Imidazolium	[NTf_2]	1-propyl-3-methyl imidazolium bis(trifluoromethylsulfonyl)imide	×	×		
1-propanol	Imidazolium	[NTf_2]	1-butyl-3-methyl imidazolium bis(trifluoromethylsulfonyl)imide	×			
1-propanol	Imidazolium	[NTf_2]	1-butyl-3-methyl imidazolium bis(trifluoromethylsulfonyl)imide	×		×	
1-propanol	Imidazolium	[NTf_2]	1-hexyl-3-methyl imidazolium bis(trifluoromethylsulfonyl)imide	×			
1-butanol	Imidazolium	[NTf_2]	1-butyl-3-methyl imidazolium bis(trifluoromethylsulfonyl)imide	×		×	×
2-propanol	Imidazolium	[NTf_2]	1-propyl-3-methyl imidazolium bis(trifluoromethylsulfonyl)imide	×	×		
2-propanol	Imidazolium	[NTf_2]	1-butyl-3-methyl imidazolium bis(trifluoromethylsulfonyl)imide	×			
2-propanol	Imidazolium	[NTf_2]	1-hexyl-3-methyl imidazolium bis(trifluoromethylsulfonyl)imide	×			
2-propanol	Imidazolium	[NTf_2]	1-octyl-3-methyl imidazolium bis(trifluoromethylsulfonyl)imide	×	×		
1,2-hexanediol	Imidazolium	[NTf_2]	1-butyl-3-methyl imidazolium bis(trifluoromethylsulfonyl)imide	×			
Isopropanol	Imidazolium	[NTf_2]	1-butyl-3-methyl imidazolium bis(trifluoromethylsulfonyl)imide	×	×		
Isopropanol	Imidazolium	[NTf_2]	1-octyl-3-methyl imidazolium bis(trifluoromethylsulfonyl)imide	×	×		
Methanol	Imidazolium	[OTf]	1-ethyl-3-methyl imidazolium trifluoromethanesulfonate	×			
Ethanol	Imidazolium	[OTf]	1-ethyl-3-methyl imidazolium trifluoromethanesulfonate	×			
Ethanol	Imidazolium	[OTf]		×			

(Continued)

Table 1 Binary Mixtures with an Ionic Liquid where Density, ρ, Viscosity, η, Surface Tension, σ, or Electrical Conductivity, κ, were Measured—cont'd

	Ionic Liquid			Property Studied			
Solvent	Cation Family	Anion Family	Name	ρ	η	σ	κ
			1-butyl-3-methyl imidazolium trifluoromethanesulfonate				
1-propanol	Imidazolium	[OTf]	1-ethyl-3-methyl imidazolium trifluoromethanesulfonate	×			
1-propanol	Imidazolium	[OTf]	1-butyl-3-methyl imidazolium trifluoromethanesulfonate	×			
1-butanol	Imidazolium	[OTf]	1-butyl-3-methyl imidazolium trifluoromethanesulfonate	×			
1-pentanol	Imidazolium	[OTf]	1-butyl-3-methyl imidazolium trifluoromethanesulfonate	×			
2-propanol	Imidazolium	[OTf]	1-ethyl-3-methyl imidazolium trifluoromethanesulfonate	×			
2-propanol	Imidazolium	[OTf]	1-butyl-3-methyl imidazolium trifluoromethanesulfonate	×			
2-butanol	Imidazolium	[OTf]	1-butyl-3-methyl imidazolium trifluoromethanesulfonate	×			
Methanol	Imidazolium	$[PF_6]$	1-butyl-3-methyl imidazolium hexafluorophosphate	×	×		
Ethanol	Imidazolium	$[PF_6]$	1-butyl-3-methyl imidazolium hexafluorophosphate	×	×		×
Ethanol	Imidazolium	$[PF_6]$	1-hexyl-3-methyl imidazolium hexafluorophosphate	×			
Ethanol	Imidazolium	$[PF_6]$	1-octyl-3-methyl imidazolium hexafluorophosphate	×			
2-propanol	Imidazolium	$[PF_6]$	1-butyl-3-methyl imidazolium hexafluorophosphate	×			
2-propanol	Imidazolium	$[PF_6]$	1-hexyl-3-methyl imidazolium hexafluorophosphate	×			
Triethylene glycol monoethyl ether	Imidazolium	$[PF_6]$	1-butyl-3-methyl imidazolium hexafluorophosphate	×			

2-(2-methoxyethoxy) ethanol	Imidazolium	[PF_6]	1-butyl-3-methyl imidazolium hexafluorophosphate	×			
2-(*N*,*N*-dimethylamino) ethanol	Imidazolium	[PF_6]	1-butyl-3-methyl imidazolium hexafluorophosphate	×	×		×
2-[2-(2-methoxyethoxy) ethoxy]ethanol	Imidazolium	[PF_6]	1-butyl-3-methyl imidazolium hexafluorophosphate	×			
2-aminoethanol	Imidazolium	[PF_6]	1-butyl-3-methyl imidazolium hexafluorophosphate	×	×		×
2-methoxyethanol	Imidazolium	[PF_6]	1-butyl-3-methyl imidazolium hexafluorophosphate	×			
Diethylene glycol monoethyl ether	Imidazolium	[PF_6]	1-butyl-3-methyl imidazolium hexafluorophosphate	×			
Diethylene glycol monomethyl ether	Imidazolium	[PF_6]	1-butyl-3-methyl imidazolium hexafluorophosphate	×			
Ethylene glycol monoethyl ether	Imidazolium	[PF_6]	1-butyl-3-methyl imidazolium hexafluorophosphate	×			
Propylene glycol monoethyl ether	Imidazolium	[PF_6]	1-butyl-3-methyl imidazolium hexafluorophosphate	×			
Propylene glycol monomethyl ether	Imidazolium	[PF_6]	1-butyl-3-methyl imidazolium hexafluorophosphate	×			
Ethylene glycol monomethyl ether	Imidazolium	[PF_6]	1-hexyl-3-methyl imidazolium hexafluorophosphate	×			
Triton X-45	Imidazolium	[PF_6]	1-butyl-3-methyl imidazolium hexafluorophosphate				×
Triton X-100	Imidazolium	[PF_6]	1-butyl-3-methyl imidazolium hexafluorophosphate				×
Methanol	Imidazolium	Other	1-ethyl-3-methyl imidazolium 1,1,2,2,2-pentafluoro-*N*-(pentafluoroethyl)sulfonyl] ethanesulfonamide	×			
Methanol	Imidazolium	Other	1-methyl-3-octyl imidazolium 2-(2-methoxyethoxy)ethyl sulfate	×			
Methanol	Imidazolium	Other	1-butyl-3-methyl imidazolium 2-(2-methoxyethoxy)ethyl sulfate	×			

(*Continued*)

Table 1 Binary Mixtures with an Ionic Liquid where Density, ρ, Viscosity, η, Surface Tension, σ, or Electrical Conductivity, κ, were Measured—cont'd

	Ionic Liquid			Property Studied			
Solvent	**Cation Family**	**Anion Family**	**Name**	ρ	η	σ	κ
Acetonitrile	Imidazolium	$[BF_4]$	1-butyl-2,3-dimethyl imidazolium tetrafluoroborate	×			
Dimethyl sulfoxide	Imidazolium	$[BF_4]$	1-butyl-2,3-dimethyl imidazolium tetrafluoroborate	×			
2-methylaniline	Imidazolium	$[BF_4]$	1-ethyl-3-methyl imidazolium tetrafluoroborate	×			
Aniline	Imidazolium	$[BF_4]$	1-ethyl-3-methyl imidazolium tetrafluoroborate	×			
N-methylaniline	Imidazolium	$[BF_4]$	1-ethyl-3-methyl imidazolium tetrafluoroborate	×			
1-methyl-2-pyrrolidinone	Imidazolium	$[BF_4]$	1-ethyl-3-methyl imidazolium tetrafluoroborate	×			
2-pyrrolidinone	Imidazolium	$[BF_4]$	1-ethyl-3-methyl imidazolium tetrafluoroborate	×			
Acetone	Imidazolium	$[BF_4]$	1-ethyl-3-methyl imidazolium tetrafluoroborate	×			
Acetonitrile	Imidazolium	$[BF_4]$	1-ethyl-3-methyl imidazolium tetrafluoroborate	×			×
Dimethyl sulfoxide	Imidazolium	$[BF_4]$	1-ethyl-3-methyl imidazolium tetrafluoroborate	×			
Pyridine	Imidazolium	$[BF_4]$	1-ethyl-3-methyl imidazolium tetrafluoroborate	×			
α-picoline	Imidazolium	$[BF_4]$	1-ethyl-3-methyl imidazolium tetrafluoroborate	×			
β-picoline	Imidazolium	$[BF_4]$	1-ethyl-3-methyl imidazolium tetrafluoroborate	×			
γ-picoline	Imidazolium	$[BF_4]$	1-ethyl-3-methyl imidazolium tetrafluoroborate	×			

1,2-propanediyl carbonate	Imidazolium	[BF_4]	1-butyl-3-methyl imidazolium tetrafluoroborate	×			×
1,3-dichloropropane	Imidazolium	[BF_4]	1-butyl-3-methyl imidazolium tetrafluoroborate	×			
1,4-dioxane	Imidazolium	[BF_4]	1-butyl-3-methyl imidazolium tetrafluoroborate		×		
Acetonitrile	Imidazolium	[BF_4]	1-butyl-3-methyl imidazolium tetrafluoroborate	×	×		×
Aniline	Imidazolium	[BF_4]	1-butyl-3-methyl imidazolium tetrafluoroborate	×			
Benzaldehyde	Imidazolium	[BF_4]	1-butyl-3-methyl imidazolium tetrafluoroborate	×			
Benzene	Imidazolium	[BF_4]	1-butyl-3-methyl imidazolium tetrafluoroborate	×			
Benzeneamine	Imidazolium	[BF_4]	1-butyl-3-methyl imidazolium tetrafluoroborate	×			
Butanone	Imidazolium	[BF_4]	1-butyl-3-methyl imidazolium tetrafluoroborate	×	×		
γ-butyrolactone	Imidazolium	[BF_4]	1-butyl-3-methyl imidazolium tetrafluoroborate				×
Dichloromethane	Imidazolium	[BF_4]	1-butyl-3-methyl imidazolium tetrafluoroborate	×	×		×
Dimethyl carbonate	Imidazolium	[BF_4]	1-butyl-3-methyl imidazolium tetrafluoroborate	×			
Dimethyl sulfoxide	Imidazolium	[BF_4]	1-butyl-3-methyl imidazolium tetrafluoroborate	×	×		×
Ethyl methanoate	Imidazolium	[BF_4]	1-butyl-3-methyl imidazolium tetrafluoroborate	×	×		
Methyl acetate	Imidazolium	[BF_4]	1-butyl-3-methyl imidazolium tetrafluoroborate	×	×		
Methyl methanoate	Imidazolium	[BF_4]	1-butyl-3-methyl imidazolium tetrafluoroborate	×	×		
N,N-dimethylacetamide	Imidazolium	[BF_4]	1-butyl-3-methyl imidazolium tetrafluoroborate	×			

(Continued)

Table 1 Binary Mixtures with an Ionic Liquid where Density, ρ, Viscosity, η, Surface Tension, σ, or Electrical Conductivity, κ, were Measured—cont'd

	Ionic Liquid			Property Studied			
Solvent	**Cation Family**	**Anion Family**	**Name**	ρ	η	σ	κ
N,N-dimethylformamide	Imidazolium	$[BF_4]$	1-butyl-3-methyl imidazolium tetrafluoroborate	×	×		
Nitromethane	Imidazolium	$[BF_4]$	1-butyl-3-methyl imidazolium tetrafluoroborate	×			
N-methyl-2-pyrrolidone	Imidazolium	$[BF_4]$	1-butyl-3-methyl imidazolium tetrafluoroborate	×			
N-methyldiethanolamine	Imidazolium	$[BF_4]$	1-butyl-3-methyl imidazolium tetrafluoroborate	×			
Propanone	Imidazolium	$[BF_4]$	1-butyl-3-methyl imidazolium tetrafluoroborate	×	×		
Tetrahydrofuran	Imidazolium	$[BF_4]$	1-butyl-3-methyl imidazolium tetrafluoroborate	×			
Butanone	Imidazolium	$[BF_4]$	1-hexyl-3-methyl imidazolium tetrafluoroborate	×			
Butylamine	Imidazolium	$[BF_4]$	1-hexyl-3-methyl imidazolium tetrafluoroborate	×			
Ethyl acetate	Imidazolium	$[BF_4]$	1-hexyl-3-methyl imidazolium tetrafluoroborate	×			
Tetrahydrofuran	Imidazolium	$[BF_4]$	1-hexyl-3-methyl imidazolium tetrafluoroborate	×			
Acetonitrile	Imidazolium	$[BF_4]$	1-hexyl-3-methyl imidazolium tetrafluoroborate	×			×
Nitromethane	Imidazolium	$[BF_4]$	1-hexyl-3-methyl imidazolium tetrafluoroborate	×			
Benzene	Imidazolium	$[BF_4]$	1-methyl-3-octyl imidazolium tetrafluoroborate	×		×	
Butanone	Imidazolium	$[BF_4]$	1-octyl-3-methyl imidazolium tetrafluoroborate	×			

Butyl acetate	Imidazolium	$[BF_4]$	1-octyl-3-methyl imidazolium tetrafluoroborate	×		
Ethyl acetate	Imidazolium	$[BF_4]$	1-octyl-3-methyl imidazolium tetrafluoroborate	×		
Methyl acetate	Imidazolium	$[BF_4]$	1-octyl-3-methyl imidazolium tetrafluoroborate	×		
Propyl acetate	Imidazolium	$[BF_4]$	1-octyl-3-methyl imidazolium tetrafluoroborate	×		
Indoline	Imidazolium	[CNS]	1-ethyl-3-methyl imidazolium thiocyanate	×		×
Quinoline	Imidazolium	[CNS]	1-ethyl-3-methyl imidazolium thiocyanate	×		×
Pyridine	Imidazolium	[CNS]	1-ethyl-3-methyl imidazolium thiocyanate	×		×
Pyrrole	Imidazolium	[CNS]	1-ethyl-3-methyl imidazolium thiocyanate	×		×
Thiophene	Imidazolium	[CNS]	1-ethyl-3-methyl imidazolium thiocyanate	×		×
Thiophene	Imidazolium	$[C(CN)_3]$	1-ethyl-3-methyl imidazolium tricyanomethanide	×		
2-amino-2-methyl-1-propanol	Imidazolium	$[C_nCOO]$	Imidazolium octanoate	×		
Acetonitrile	Imidazolium	$[C_nCOO]$	Imidazolium octanoate	×	×	
Diethanolamine	Imidazolium	$[C_nCOO]$	1-butyl-3-methyl imidazolium acetate	×	×	
Diisopropanolamine	Imidazolium	$[C_nCOO]$	1-butyl-3-methyl imidazolium acetate	×		
N-methyldiethanolamine	Imidazolium	$[C_nCOO]$	1-butyl-3-methyl imidazolium acetate	×	×	
Acetonitrile	Imidazolium	$[C_nSO_4]$	1,3-dimethyl imidazolium methyl sulfate	×		
Dimethyl formamide	Imidazolium	$[C_nSO_4]$	1,3-dimethyl imidazolium methyl sulfate	×		

(*Continued*)

Table 1 Binary Mixtures with an Ionic Liquid where Density, ρ, Viscosity, η, Surface Tension, σ, or Electrical Conductivity, κ, were Measured—cont'd

	Ionic Liquid			Property Studied			
Solvent	**Cation Family**	**Anion Family**	**Name**	ρ	η	σ	κ
Dimethyl sulfoxide	Imidazolium	$[C_nSO_4]$	1,3-dimethyl imidazolium methyl sulfate	×			
Butanone	Imidazolium	$[C_nSO_4]$	1,3-dimethyl imidazolium methylsulfate	×			
Ethyl acetate	Imidazolium	$[C_nSO_4]$	1,3-dimethyl imidazolium methylsulfate	×			
1,3-dichloropropane	Imidazolium	$[C_nSO_4]$	1-butyl-3-methyl imidazolium methylsulfate	×			
Nitromethane	Imidazolium	$[C_nSO_4]$	1-butyl-3-methyl imidazolium methylsulfate	×			
Acetone	Imidazolium	$[C_nSO_4]$	1-ethyl-3-methyl imidazolium ethylsulfate	×			
Acetonitrile	Imidazolium	$[C_nSO_4]$	1-ethyl-3-methyl imidazolium ethylsulfate	×			
Dichloromethane	Imidazolium	$[C_nSO_4]$	1-ethyl-3-methyl imidazolium ethylsulfate	×			
Indoline	Imidazolium	$[C_nSO_4]$	1-ethyl-3-methyl imidazolium ethylsulfate	×		×	
Nitromethane	Imidazolium	$[C_nSO_4]$	1-ethyl-3-methyl imidazolium ethylsulfate	×			
Quinoline	Imidazolium	$[C_nSO_4]$	1-ethyl-3-methyl imidazolium ethylsulfate	×		×	
Propylene carbonate	Imidazolium	$[C_nSO_4]$	1-ethyl-3-methyl imidazolium ethylsulfate	×			
Pyridine	Imidazolium	$[C_nSO_4]$	1-ethyl-3-methyl imidazolium ethylsulfate	×		×	
Pyrrole	Imidazolium	$[C_nSO_4]$	1-ethyl-3-methyl imidazolium ethylsulfate	×		×	
Thiophene	Imidazolium	$[C_nSO_4]$		×		×	

			1-ethyl-3-methyl imidazolium ethylsulfate				
Dimethyl sulfoxide	Imidazolium	[Halide]	1-decyl-3-methyl imidazolium bromide	×			
N,*N*-dimethylmethanamide	Imidazolium	[Halide]	1-decyl-3-methyl imidazolium bromide	×			
Dimethyl sulfoxide	Imidazolium	[Halide]	1-methyl-3-octyl imidazolium bromide	×			
N,*N*-dimethylmethanamide	Imidazolium	[Halide]	1-methyl-3-octyl imidazolium bromide	×			
N,*N*-dimethylacetamide	Imidazolium	[Halide]	1-propyl-3-methyl imidazolium bromide	×			
Acetonitrile	Imidazolium	[Halide]	1-butyl-3-methyl imidazolium chloride	×			
N,*N*-dimethylmethanamide	Imidazolium	[Halide]	1-butyl-3-methyl imidazolium chloride	×			
Acetonitrile	Imidazolium	[Halide]	1-ethyl-3-methyl imidazolium chloride		×		
N-methyldiethanolamine	Imidazolium	$[N(CN)_2]$	1-butyl-3-methyl imidazolium dicyanamide	×			
Dimethyl carbonate	Imidazolium	$[NTf_2]$	1,2-dimethyl-3-hexyl imidazolium bis(trifluoromethylsulfonyl)imide	×			×
1,3-cyclohexadiene	Imidazolium	$[NTf_2]$	1-butyl-2,3-dimethyl imidazolium bis(trifluoromethylsulfonyl)imide				×
Acetonitrile	Imidazolium	$[NTf_2]$	1-ethyl-3-methyl imidazolium bis(trifluoromethylsulfonyl)imide	×		×	
Dimethyl sulfoxide	Imidazolium	$[NTf_2]$	1-ethyl-3-methyl imidazolium bis(trifluoromethylsulfonyl)imide	×		×	
Chloroform	Imidazolium	$[NTf_2]$	1-ethyl-3-methyl imidazolium bis(trifluoromethylsulfonyl)imide				×
Tetrahydrofuran	Imidazolium	$[NTf_2]$	1-ethyl-3-methyl imidazolium bis(trifluoromethylsulfonyl)imide	×		×	
Acetonitrile	Imidazolium	$[NTf_2]$	1-butyl-3-methyl imidazolium bis(trifluoromethylsulfonyl)imide	×		×	×

(Continued)

Table 1 Binary Mixtures with an Ionic Liquid where Density, ρ, Viscosity, η, Surface Tension, σ, or Electrical Conductivity, κ, were Measured—cont'd

	Ionic Liquid			Property Studied			
Solvent	**Cation Family**	**Anion Family**	**Name**	**ρ**	**η**	**σ**	**κ**
Ethyl acetate	Imidazolium	$[NTf_2]$	1-butyl-3-methyl imidazolium bis(trifluoromethylsulfonyl)imide	×	×		
Isopropyl acetate	Imidazolium	$[NTf_2]$	1-butyl-3-methyl imidazolium bis(trifluoromethylsulfonyl)imide	×	×		
Propylene carbonate	Imidazolium	$[NTf_2]$	1-butyl-3-methyl imidazolium bis(trifluoromethylsulfonyl)imide	×	×		
Tetrahydrofuran	Imidazolium	$[NTf_2]$	1-butyl-3-methyl imidazolium bis(trifluoromethylsulfonyl)imide	×		×	
Dimethyl sulfoxide	Imidazolium	$[NTf_2]$	1-butyl-3-methyl imidazolium bis(trifluoromethylsulfonyl)imide	×		×	
Acetophenone	Imidazolium	$[NTf_2]$	1-butyl-3-methyl imidazolium bis(trifluoromethylsulfonyl)imide	×			
Anisole	Imidazolium	$[NTf_2]$	1-butyl-3-methyl imidazolium bis(trifluoromethylsulfonyl)imide	×			
1,3-cyclohexadiene	Imidazolium	$[NTf_2]$	1-butyl-3-methyl imidazolium bis(trifluoromethylsulfonyl)imide				×
Dichloromethane	Imidazolium	$[NTf_2]$	1-butyl-3-methyl imidazolium bis(trifluoromethylsulfonyl)imide				×
1-octene	Imidazolium	$[NTf_2]$	1-hexyl-3-methyl imidazolium bis(trifluoromethylsulfonyl)imide	×		×	
Acetone	Imidazolium	$[NTf_2]$	1-hexyl-3-methyl imidazolium bis(trifluoromethylsulfonyl)imide	×			
Acetonitrile	Imidazolium	$[NTf_2]$	1-hexyl-3-methyl imidazolium bis(trifluoromethylsulfonyl)imide	×			
Dichloromethane	Imidazolium	$[NTf_2]$	1-hexyl-3-methyl imidazolium bis(trifluoromethylsulfonyl)imide	×			
Isopropyl acetate	Imidazolium	$[NTf_2]$	1-octyl-3-methyl imidazolium bis(trifluoromethylsulfonyl)imide	×	×		

1-methylethyl ethanoate	Imidazolium	[NTf_2]	1-octyl-3-methyl imidazolium bis(trifluoromethylsulfonyl)imide	×	×		
Ethyl acetate	Imidazolium	[NTf_2]	1-octyl-3-methyl imidazolium bis(trifluoromethylsulfonyl)imide	×	×		
Methyl acetate	Imidazolium	[NTf_2]	1-octyl-3-methyl imidazolium bis(trifluoromethylsulfonyl)imide	×	×		
Tetrahydrofuran	Imidazolium	[OTf]	1-ethyl-3-methyl imidazolium trifluoromethanesulfonate	×			
Ethyl acetate	Imidazolium	[OTf]	1-ethyl-3-methyl imidazolium trifluoromethanesulfonate	×			
Methyl acetate	Imidazolium	[OTf]	1-ethyl-3-methyl imidazolium trifluoromethanesulfonate	×			
Nitromethane	Imidazolium	[OTf]	1-ethyl-3-methyl imidazolium trifluoromethanesulfonate	×			
N-methyldiethanolamine	Imidazolium	[OTf]	1-ethyl-3-methyl imidazolium trifluoromethanesulfonate	×			
Propanone	Imidazolium	[OTf]	1-ethyl-3-methyl imidazolium trifluoromethanesulfonate	×			
Nitromethane	Imidazolium	[OTf]	1-butyl-3-methyl imidazolium trifluoromethanesulfonate	×			
1,2-dimethoxyethane	Imidazolium	[PF_6]	1-butyl-3-methyl imidazolium hexafluorophosphate	×			
2,5,8,11,14-pentaoxapentadecane	Imidazolium	[PF_6]	1-butyl-3-methyl imidazolium hexafluorophosphate	×			
2,5,8,11-tetraoxadodecane	Imidazolium	[PF_6]	1-butyl-3-methyl imidazolium hexafluorophosphate	×			
2,5,8-trioxanonane	Imidazolium	[PF_6]	1-butyl-3-methyl imidazolium hexafluorophosphate	×			
3-pentanone	Imidazolium	[PF_6]	1-butyl-3-methyl imidazolium hexafluorophosphate	×	×		
Acetonitrile	Imidazolium	[PF_6]	1-butyl-3-methyl imidazolium hexafluorophosphate	×	×		×
Benzene	Imidazolium	[PF_6]	1-butyl-3-methyl imidazolium hexafluorophosphate	×			

(Continued)

Table 1 Binary Mixtures with an Ionic Liquid where Density, ρ, Viscosity, η, Surface Tension, σ, or Electrical Conductivity, κ, were Measured—cont'd

	Ionic Liquid			Property Studied			
Solvent	**Cation Family**	**Anion Family**	**Name**	ρ	η	σ	κ
Butanone	Imidazolium	[PF_6]	1-butyl-3-methyl imidazolium hexafluorophosphate	×	×		
Cyclopentanone	Imidazolium	[PF_6]	1-butyl-3-methyl imidazolium hexafluorophosphate	×	×		
Dimethyl sulfoxide	Imidazolium	[PF_6]	1-butyl-3-methyl imidazolium hexafluorophosphate	×	×		
Ethyl acetate	Imidazolium	[PF_6]	1-butyl-3-methyl imidazolium hexafluorophosphate	×	×		
N,N-dimethylacetamide	Imidazolium	[PF_6]	1-butyl-3-methyl imidazolium hexafluorophosphate	×			
N,N-dimethylformamide	Imidazolium	[PF_6]	1-butyl-3-methyl imidazolium hexafluorophosphate	×	×		
N-methyl-2-pyrrolidone	Imidazolium	[PF_6]	1-butyl-3-methyl imidazolium hexafluorophosphate	×			
Propanone	Imidazolium	[PF_6]	1-butyl-3-methyl imidazolium hexafluorophosphate	×	×		×
Propenoic acid, 2-methyl-methyl ester	Imidazolium	[PF_6]	1-butyl-3-methyl imidazolium hexafluorophosphate	×	×		
Tetrahydrofuran	Imidazolium	[PF_6]	1-butyl-3-methyl imidazolium hexafluorophosphate	×	×		
Poly(ethylene glycol)	Imidazolium	[PF_6]	1-methyl-3-pentyl imidazolium hexafluorophosphate				×
2-pentanone	Imidazolium	[PF_6]	1-hexyl-3-methyl imidazolium hexafluorophosphate	×			
Butanone	Imidazolium	[PF_6]	1-hexyl-3-methyl imidazolium hexafluorophosphate	×			
Butyl acetate	Imidazolium	[PF_6]	1-hexyl-3-methyl imidazolium hexafluorophosphate	×			
Diethyl carbonate	Imidazolium	[PF_6]	1-hexyl-3-methyl imidazolium hexafluorophosphate	×			

Dimethyl carbonate	Imidazolium	$[PF_6]$	1-hexyl-3-methyl imidazolium hexafluorophosphate	×			
Ethyl acetate	Imidazolium	$[PF_6]$	1-hexyl-3-methyl imidazolium hexafluorophosphate	×			
Methyl acetate	Imidazolium	$[PF_6]$	1-hexyl-3-methyl imidazolium hexafluorophosphate	×			
Propanone	Imidazolium	$[PF_6]$	1-hexyl-3-methyl imidazolium hexafluorophosphate	×			
Butanone	Imidazolium	$[PF_6]$	1-octyl-3-methyl imidazolium hexafluorophosphate	×			
Ethyl acetate	Imidazolium	$[PF_6]$	1-octyl-3-methyl imidazolium hexafluorophosphate	×			
Propanone	Imidazolium	Other	1-ethyl-3-methyl imidazolium 1,1,2,2,2-pentafluoro-*N*-(pentafluoroethyl)sulfonyl] ethanesulfonamide	×			
Water	Other	$[C_nCOO]$	1,1,3,3-tetramethylguanidinium lactate	×	×		
Water	Other	Other	(3-aminopropyl) tributyl phosphonium L-α-amino-4-methylvaleric acid salt	×			
Water	Other	Other	(3-aminopropyl) tributyl phosphonium L-α-aminoisovaleric acid salt	×			
Water	Other	Other	(3-aminopropyl) tributyl phosphonium L-α-aminopropionic acid salt	×			
Acetonitrile	Other	$[BF_4]$	Tributyl phosphonium tetrafluoroborate				×
Acetonitrile	Other	$[BF_4]$	Tretrabutyl phosphonium tetrafluoroborate				×
Tetrahydrofurane	Other	$[BF_4]$	Tretrabutyl phosphonium tetrafluoroborate				×

(*Continued*)

Table 1 Binary Mixtures with an Ionic Liquid where Density, ρ, Viscosity, η, Surface Tension, σ, or Electrical Conductivity, κ, were Measured—cont'd

	Ionic Liquid			Property Studied			
Solvent	**Cation Family**	**Anion Family**	**Name**	ρ	η	σ	κ
1,3-dioxolane	Other	$[BF_4]$	Tretrabutyl phosphonium tetrafluoroborate				×
Water	Other	$[C_nCOO]$	Choline lactate	×	×		
Methanol	Other	$[C_nCOO]$	Ethylpiperazinium propanoate	×			
Methanol	Other	$[C_nCOO]$	2-hydroxyethanaminium propanoate	×			
Methanol	Other	$[C_nCOO]$	2-hydroxyethanaminium acetate	×			
Methanol	Other	$[C_nCOO]$	2-hydroxyethanaminium formate	×			
Methanol	Other	$[C_nCOO]$	2-hydroxyethanaminium lactate	×			
Ethanol	Other	$[C_nCOO]$	Choline lactate	×	×		
Ethanol	Other	$[C_nCOO]$	Ethylpiperazinium propanoate	×			
1-propanol	Other	$[C_nCOO]$	Ethylpiperazinium propanoate	×			
1-butanol	Other	$[C_nCOO]$	Ethylpiperazinium propanoate	×			
1-butanol	Other	$[NTf_2]$	*N*-octylisoquinolinium bis(trifluoromethylsulfonyl)imide	×			
1-hexanol	Other	$[NTf_2]$	*N*-octylisoquinolinium bis(trifluoromethylsulfonyl)imide	×			
2-phenylethanol	Other	$[NTf_2]$	*N*-octylisoquinolinium bis(trifluoromethylsulfonyl)imide	×			
Ethanol	Other	Other	Choline cyclohexane carboxylate	×			
Ethanol	Other	Other	Choline cyclopentane carboxylate	×			
Benzene	Other	$[NTf_2]$	*N*-octylisoquinolinium bis(trifluoromethylsulfonyl)imide	×	×		
Pyridine	Other	$[NTf_2]$	*N*-octylisoquinolinium bis(trifluoromethylsulfonyl)imide	×	×		
Thiophene	Other	$[NTf_2]$	*N*-octylisoquinolinium bis(trifluoromethylsulfonyl)imide	×	×		
Toluene	Other	$[NTf_2]$	*N*-octylisoquinolinium bis(trifluoromethylsulfonyl)imide	×	×		

Acetonitrile	Other	[NTf_2]	Trimethylsulfonium bis(trifluoromethylsulfonyl)imide		×	
Propylene carbonate	Other	[NTf_2]	Trimethylsulfonium bis(trifluoromethylsulfonyl)imide		×	
γ-butyrolactone	Other	[NTf_2]	Trimethylsulfonium bis(trifluoromethylsulfonyl)imide		×	
Water	Piperidinium	[CNS]	1-butyl-1-methylpiperidinium thiocyanate	×	×	
Acetophenone	Piperidinium	[NTf_2]	1-methyl-1-propylpiperidinium bis(trifluoromethylsulfonyl)imide	×		
Anisole	Piperidinium	[NTf_2]	1-methyl-1-propylpiperidinium bis(trifluoromethylsulfonyl)imide	×		
Water	Pyridinium	[BF_4]	1-butyl-4-methylpyridinium tetrafluoroborate	×		
Water	Pyridinium	[BF_4]	1-butylpyridinium tetrafluoroborate	×	×	
Water	Pyridinium	[CNS]	1-butyl-4-methylpyridinium thiocyanate	×	×	
Water	Pyridinium	[C_nSO_4]	1-methylpyridinium methylsulfate	×		
Water	Pyridinium	[C_nSO_4]	1,2-diethylpyridinium ethylsulfate	×		
Water	Pyridinium	[NO_3]	1-butylpyridinium nitrate	×		×
Methanol	Pyridinium	[BF_4]	1-butyl-3-methylpyridinium tetrafluoroborate	×	×	
Methanol	Pyridinium	[BF_4]	1-butyl-4-methylpyridinium tetrafluoroborate	×	×	
Methanol	Pyridinium	[BF_4]	1-butylpyridinium tetrafluoroborate	×	×	
Ethanol	Pyridinium	[BF_4]	1-butyl-3-methylpyridinium tetrafluoroborate	×	×	
Ethanol	Pyridinium	[BF_4]	1-butyl-4-methylpyridinium tetrafluoroborate	×	×	
Ethanol	Pyridinium	[BF_4]	1-butylpyridinium tetrafluoroborate	×	×	
1-propanol	Pyridinium	[BF_4]	1-butyl-4-methylpyridinium tetrafluoroborate	×		
1-butanol	Pyridinium	[BF_4]		×		

(Continued)

Table 1 Binary Mixtures with an Ionic Liquid where Density, ρ, Viscosity, η, Surface Tension, σ, or Electrical Conductivity, κ, were Measured—cont'd

	Ionic Liquid			Property Studied			
Solvent	Cation Family	Anion Family	Name	ρ	η	σ	κ
			1-butyl-4-methylpyridinium tetrafluoroborate				
1-pentanol	Pyridinium	$[BF_4]$	1-butyl-4-methylpyridinium tetrafluoroborate	×			
Ethanol	Pyridinium	$[C_nSO_4]$	1-ethylpyridinium ethylsulfate	×	×		
1-propanol	Pyridinium	$[C_nSO_4]$	1-ethylpyridinium ethylsulfate	×	×		
Methanol	Pyridinium	$[NO_3]$	octylpyridinium nitrate	×		×	
Methanol	Pyridinium	$[NO_3]$	1-butylpyridinium nitrate	×			
Ethanol	Pyridinium	$[NO_3]$	octylpyridinium nitrate	×		×	
Ethanol	Pyridinium	$[NO_3]$	1-butylpyridinium nitrate	×			
1-butanol	Pyridinium	$[NO_3]$	octylpyridinium nitrate	×		×	
1-butanol	Pyridinium	$[NO_3]$	1-butylpyridinium nitrate	×			
Methanol	Pyridinium	$[NTf_2]$	1-ethyl-pyridinium bis(trifluoromethylsulfonyl)imide	×			
Methanol	Pyridinium	$[NTf_2]$	1-butylpyridinium bis(trifluoromethylsulfonyl)imide	×			
Methanol	Pyridinium	$[NTf_2]$	*N*-pentyl-pyridinium bis(trifluoromethylsulfonyl)imide	×			
Water	Pyrrolidinium	$[C_nCOO]$	Pyrrolidinium octanoate	×	×		
Water	Pyrrolidinium	$[C_nCOO]$	Pyrrolidinium trifluoroacetate		×		×
Water	Pyrrolidinium	$[C_nSO_4]$	Pyrrolidinium hydrogensulfate		×		×
Water	Pyrrolidinium	[CNS]	1-butyl-1-methylpyrrolidinium thiocyanate		×		
Water	Pyrrolidinium	$[N(CN)_2]$	1-butyl-1-methylpyrrolidinium dicyanamide	×			
Water	Pyrrolidinium	[OTf]	1-butyl-1-methylpyrrolidinium trifluoromethanesulfonate	×			
Water	Pyrrolidinium	[OTf]	1-butyl-1-methylpyrrolidinium trifluoromethanesulfonate	×			
Methanol	Pyrrolidinium	$[C_nCOO]$	Pyrrolidinium octanoate	×	×		

Ethanol	Pyrrolidinium	[C_nCOO]	Pyrrolidinium octanoate	×	×	
1-butanol	Pyrrolidinium	[C_nCOO]	Pyrrolidinium octanoate	×	×	
1-propanol	Pyrrolidinium	[$N(CN)_2$]	1-butyl-1-methylpyrrolidinium dicyanamide	×		
1-butanol	Pyrrolidinium	[$N(CN)_2$]	1-butyl-1-methylpyrrolidinium dicyanamide	×		
2-propanol	Pyrrolidinium	[$N(CN)_2$]	1-butyl-1-methylpyrrolidinium dicyanamide	×		
Methanol	Pyrrolidinium	[NTf_2]	*N*-butyl-*N*-methylpyrrolidinium bis(trifluoromethylsulfonyl)imide	×	×	
Ethanol	Pyrrolidinium	[NTf_2]	1-butyl-1-methylpyrrolidinium bis(trifluoromethylsulfonyl)imide	×		
1-propanol	Pyrrolidinium	[NTf_2]	1-butyl-1-methylpyrrolidinium bis(trifluoromethylsulfonyl)imide	×		
2-propanol	Pyrrolidinium	[NTf_2]	1-butyl-1-methylpyrrolidinium bis(trifluoromethylsulfonyl)imide	×		
N,*N*-dimethylformamide	Pyrrolidinium	[NTf_2]	1-butyl-1-methylpyrrolidinium bis(trifluoromethylsulfonyl)imide	×	×	
N-methylformamide	Pyrrolidinium	[NTf_2]	1-butyl-1-methylpyrrolidinium bis(trifluoromethylsulfonyl)imide	×		
Acetonitrile	Pyrrolidinium	[NTf_2]	1-butyl-1-methylpyrrolidinium bis(trifluoromethane sulfonyl)imide			×
Butyronitrile	Pyrrolidinium	[NTf_2]	*N*-butyl-*n*-methylpyrrolidinium bis(trifluoromethane sulfonyl)imide			×
Benzyl cyanide	Pyrrolidinium	[NTf_2]	*N*-butyl-*n*-methylpyrrolidinium bis(trifluoromethane sulfonyl)imide			×
Propylene carbonate	Pyrrolidinium	[NTf_2]	*N*-butyl-*n*-methylpyrrolidinium bis(trifluoromethane sulfonyl)imide			×
Benzonitrile	Pyrrolidinium	[NTf_2]	*N*-butyl-*n*-methylpyrrolidinium bis(trifluoromethane sulfonyl)imide			×
Acetonitrile	Pyrrolidinium	[C_nCOO]	Pyrrolidinium octanoate	×	×	
Propylene carbonate	Pyrrolidinium	[NO_3]	Pyrrolidinium nitrate	×	×	×
Acetonitrile	Pyrrolidinium	[NTf_2]		×		×

(*Continued*)

Table 1 Binary Mixtures with an Ionic Liquid where Density, ρ, Viscosity, η, Surface Tension, σ, or Electrical Conductivity, κ, were Measured—cont'd

	Ionic Liquid			Property Studied			
Solvent	**Cation Family**	**Anion Family**	**Name**	ρ	η	σ	κ
			N-methyl-*N*-pentylpyrrolidinium bis(trifluoromethylsulfonyl)imide				
Diethyl carbonate	Pyrrolidinium	$[NTf_2]$	*N*-methyl-*N*-pentylpyrrolidinium bis(trifluoromethylsulfonyl)imide	×			
Dimethyl carbonate	Pyrrolidinium	$[NTf_2]$	*N*-methyl-*N*-pentylpyrrolidinium bis(trifluoromethylsulfonyl)imide	×			×
Ethyl acetate	Pyrrolidinium	$[NTf_2]$	*N*-methyl-*N*-pentylpyrrolidinium bis(trifluoromethylsulfonyl)imide	×			×
Ethyl butyrate	Pyrrolidinium	$[NTf_2]$	*N*-methyl-*N*-pentylpyrrolidinium bis(trifluoromethylsulfonyl)imide	×			
Ethyl propanoate	Pyrrolidinium	$[NTf_2]$	*N*-methyl-*N*-pentylpyrrolidinium bis(trifluoromethylsulfonyl)imide	×			×
Ethylene carbonate	Pyrrolidinium	$[NTf_2]$	*N*-methyl-*N*-pentylpyrrolidinium bis(trifluoromethylsulfonyl)imide	×			
Methyl acetate	Pyrrolidinium	$[NTf_2]$	*N*-methyl-*N*-pentylpyrrolidinium bis(trifluoromethylsulfonyl)imide	×			
Methyl butyrate	Pyrrolidinium	$[NTf_2]$	*N*-methyl-*N*-pentylpyrrolidinium bis(trifluoromethylsulfonyl)imide	×			
Methyl isobutyrate	Pyrrolidinium	$[NTf_2]$	*N*-methyl-*N*-pentylpyrrolidinium bis(trifluoromethylsulfonyl)imide	×			
Methyl propanoate	Pyrrolidinium	$[NTf_2]$	*N*-methyl-*N*-pentylpyrrolidinium bis(trifluoromethylsulfonyl)imide	×			
Propylene carbonate	Pyrrolidinium	$[NTf_2]$	*N*-methyl-*N*-pentylpyrrolidinium bis(trifluoromethylsulfonyl)imide	×			×
γ-butyrolactone	Pyrrolidinium	$[NTf_2]$	*N*-methyl-*N*-pentylpyrrolidinium bis(trifluoromethylsulfonyl)imide	×			
γ-valerolactone	Pyrrolidinium	$[NTf_2]$	*N*-methyl-*N*-pentylpyrrolidinium bis(trifluoromethylsulfonyl)imide	×			×

There has also been some published work about measurement of physical properties in binary mixtures of two ILs. About this subject, a review paper has been recently published in which the majority of the experimental work made up to date is reported [1].

ACKNOWLEDGMENT

This work was supported by the Directorate General for R + D + i of the Xunta de Galicia (Grants N° 10-PXIB-103-294 PR).

REFERENCE

[1] H. Niedermeyer, J.P. Hallett, I.J. Villar-Garcia, P.A. Hunt, T. Welton, Mixtures of ionic liquids, Chem. Soc. Rev. 41 (2012) 7780–7802.

SUBCHAPTER 1.5

Ionic Liquids: Theory and Simulations

Trinidad Méndez-Morales, Luis M. Varela
Grupo de Nanomateriais e Materia Branda, Departamento de Física da Materia Condensada, Universidade de Santiago de Compostela, Santiago de Compostela, Spain

1. INTRODUCTION

As described in other chapters of this book, ionic liquids (ILs) are highly asymmetric organic salts formed solely by ions, which melt below 100 °C. They form a completely new class of "designer solvents" with peculiar properties and can be seen as infinitely concentrated ionic solutions or as room temperature molten salts. Hence, their properties are essentially dominated by electrostatic long-range interactions, although their well-known nanostructuring [1,2] can only be interpreted including short-range dispersive interactions. Consequently, they fall entirely within the category of charged complex fluids, which comprise ionic solutions, molten salts, and liquid metals, so covering most of soft matter. Electrostatic long-ranged interactions are ubiquitous in this type of condensed matter, where they are essential for describing, among other features, solubilization, concentration gradients, and mesoscopic structures. Accordingly, the efforts of all kind dedicated to the description of the structure and dynamics they give rise to in homogeneous and heterogeneous condensed phases have been truly gargantuan.

On the theoretical side, starting with early continuous, mean-field theories of electrolyte solutions and up to current pseudo-lattice formalism, the amount of theoretical developments for these systems is truly immense, including integral equation techniques and even field theoretic descriptions (see Reference [3] and references contained therein for a review). However, only a small selection of theories are currently being applied to the understanding of ILs, including the previously mentioned pseudo-lattice theory [4—7], hole theory [8], interstice model [9], and self-consistent mean-field formalism [10], and so our attention here will be focused on them, especially in the first two.

On the other hand, computer simulations are currently used as a general tool essential for getting insight of the structure and dynamics of physical systems at a microscopic level. It was in Los Alamos in the 1950s that Metropolis et al. [11] performed the first computer simulation of a liquid, and, since then, the number of applications to this state of matter has grown exponentially. As for ILs, the much slower dynamics of these compounds make them more difficult to be investigated by means of classical and, even more, quantum computer simulations; and this major problem can only be avoided using very high temperatures of simulation or very long simulation times. However, in spite of these limitations, the use of computational techniques for shedding some light on the fundamental nature of this new class of solvents, and also for predicting their properties at a level that cannot be achieved by means of experimental studies, has been continuously increasing during the last 2 decades.

In this chapter, we will briefly review the most relevant works reported up to now in the fields of theory and computer simulations of ILs, a research-intensive area that has been reviewed several times in the last few years [12–14].

2. THEORETICAL DEVELOPMENTS

The theoretical understanding of charged complex fluids has been one of the most outstanding problems of condensed matter and chemical physics since the seminal work of Debye and Hückel (DH) [15] concerning the structure and thermodynamics of dilute electrolyte solutions. Throughout the rest of the twentieth century, the theory of ionic fluids underwent an intense evolution since that early mean-field Poisson–Boltzmann-based theory and the corresponding linear response Debye–Hückel–Onsager transport theory [16–19], the first extension being that of Gronwall, Lamer, and Sandved [20]. The correction of finite radius of the ions, ion pairing [21], and short-range interactions [22] followed in the next years. On the other hand, it was in the 1940s that integral equation techniques started being used for the calculation of pair correlation functions from the Ornstein–Zernike equation using different closure relations (for a review of the evolution of the application of these techniques to ionic fluids see Reference [3]). Finally, the last developments of continuous formalisms to date are the formally exact version of the classical mean-field theory, introduced in the 1990s with the formulation of the dressed-ion theory [23,24] and dressed-ion transport theory [25,26], where renormalized charges interacting through potentials with renormalized screening lengths are used as kinetic entities. Theories of heterogeneous systems developed mostly in parallel since the Gouy–Chapman theory of the electric double layer (edl) [27,28] up to the latest developments for the Stern layer [29–31].

Apart from the formalisms cited above, all based on a continuous picture of the IL, it soon became clear that the picture of ionic solutions behind the DH theory, based on the structural unit of the ionic atmosphere, breaks at quite low concentrations. Thus, at concentrations above 0.01 mol/l, this image breaks down because, as pointed out by Bockris [32], at this concentration "only one ion produces the 50% of the effect of the ionic atmosphere on the central ion." Frank and Thompson [33,34] and later Robinson and Stokes [35] introduced the hypothesis and some experimental evidence of a concentration-dependent structure of ionic solutions, where an infinite-dilution, continuous ionic cloud model progressively evolves into a disordered lattice model as concentration increases, leading to a cube root law in salt concentration for the logarithm of the activity coefficient. However, it was Bahe [4–6] in the 1970s, and later Varela et al. in the 1990s [7] who theoretically founded the pseudo-lattice formalism on solid grounds. This theory, together with the hole theory [8,36]—where cavities are supposed to form in the bulk liquid controlled by thermal energy and surface tension that act as kinetic, charge-transport entities—and interstice model [9], are the major formalisms currently applied to the description of thermodynamics and transport properties of ILs, and so are the only ones to be considered here.

According to Bockris [32], there are lattice oriented and gas oriented theories of ILs (although he was referring mainly to molten salts at the time). Upon melting, empty space is introduced in the liquid, and the various models (vacancy model, hole model, cell theory, and liquid free-volume) differ basically in the way of treating this empty space [32]. According to the hole theory of ILs, melting gives rise to empty spaces of random size undergoing a random walk in the bulk material associated to thermally generated fluctuations in local density [8,32]. The energy of creating a hole, W, is essentially dominated by surface tension, γ, and was proved to be $W = 4\pi r^2\gamma$ for a hole of radius r by Fürth [36], treating holes as bubbles in the liquid. Thus, the radius of the average sized void is given by $4\pi\langle r^2\rangle = 3.5k_B T/\gamma$ where k_B is the Boltzmann constant and T the absolute temperature. This theory allowed the rationalizing of the empirical relation of activation energies and thermal energy at melting $E_D = E\eta = 3.5k_B T$and for calculating transport coefficients in fused salts. The essential feature of transport calculations is the consideration that holes in the pure ILs can be considered as molecules in ideal gases and, therefore, described by means of conventional kinetic theory. The momentum and charge transport phenomena would then be associated with (and restricted by) the availability of holes of random size and place in the bulk liquid. This scheme has been successfully applied by Abbott and coworkers for calculating viscosity and electrical conductivity coefficients in room temperature ILs [8,37,38], reporting the following expression for viscosity:

$$\begin{aligned} \eta &= \frac{m\langle v\rangle}{2\sigma P(r > R)} \\ P(r > R)\mathrm{d}r &= \frac{16}{15\sqrt{\pi}} a^{7/2} r^6 e^{-ar^2}\,\mathrm{d}r \end{aligned} \tag{1}$$

where $k_B T$ is the average molecular velocity $\sigma = 4\pi R^2$, is the hard-sphere collision probability of the molecule, R is the radius of the solvent molecule, and $a = 4\pi\gamma/k_B T$.

However, as successful as the hole theory is, its structural image is not fully consistent with a view of ILs as the maximally concentrated solutions of the corresponding salt in some molecular solvent. As previously mentioned, it has long been well known that as ionic concentration increases, a statistical lattice (or pseudo-lattice) develops in bulk ionic solutions. This statistical order is progressively reinforced up to the limit of pure ILs. This theory is lattice oriented and considers that empty space corresponds to the vacancies left by ions in the crystal structure upon migration of the free surface of the IL. Moreover, this long-range ordered structure is more compatible with the nano-structuring of low-temperature fused salts than the hole model. This theoretical formalism has been used for describing thermodynamic properties by Bahe [4–6], who successfully calculated the activity coefficients of 1:1 and 1:2 electrolyte solutions considering electrostatic and ion solvent interactions. Bahe's theory was generalized in the late 1990s by Varela et al. [7] to include short-range dispersive interionic interactions, and the combination of both results (the so-called Bahe–Varela theory) produced the basic expression for the rational mean activity coefficient of an ionic solution [7]:

$$\ln \gamma_{\pm} = -Ac^{1/3} + Bc + Dc^{2} \tag{2}$$

where c is the concentration of the solutions in mol l^{-1} and the expressions of coefficients have been reported elsewhere [7]. This theory has been successfully applied to the description of volumetric properties of pure ILs [39,40], surface tension of IL cosolvent mixtures using a Bragg–Williams-like distribution for the ions in the bulk IL, and to explain the essential features of the universal mechanism of charge transport in IL-insulator mixtures, specifically the conducting dome [41]. In this model, ionic motion is assumed to take place through hops between cells of two different types separated by non-random-energy barriers of different heights depending on the cell type. Assuming non-correlated ion transport and concentration-independent hopping probabilities, the authors reported a universal conducting dome and tested it with several electrolyte solutions and IL cosolvent mixtures [41,42]. Other authors reported pseudo-lattice approaches to thermodynamic properties [43,44] and transport properties replacing Debye's length by the average interionic distance (see Reference [45] and references therein).

Concerning the edl model, since the formulation of Gouy–Chapman theory in the early twentieth century, many theories have been formulated. This history is very well summarized in Reference [29]. Currently, the classical theory of the edl has been generalized in order to describe its structure in such dense ionic systems as ILs. In 2007, Kornyshev reported a possible paradigm change in the analysis of the double layer, calling for the usage of modern statistical mechanics of dense Coulomb systems or density functional theory, rather than conventional mean-field-like dilute solution frameworks. Moreover, Bazant et al. reported a simple Landau–Ginzburg-type continuum theory and analyzed the structure of the edl in these systems [31], specifically the interplay between crowding and overscreening in the inner structure of the edl in ILs and their effect of the subtleties of the screening of charges in ILs. In this sense, it is also worth citing a very recent and controversial paper by Gebbie et al. that claims to provide evidence that screening of charged surfaces in ILs would take place as in dilute electrolyte solutions (i.e., a monotonic exponential screening of the electrostatic potential would take place controlled by a very small amount of dissociated ions in a sea of associated ions [46]), something that must be definitely proved in the years to come.

3. COMPUTATIONAL DEVELOPMENTS

As we mentioned in the introduction to this chapter, computational methods have been widely used recently by the scientific community to investigate various aspects of ILs, and it is virtually impossible to review all the contributions reported to date. Because of this fact, even though several problems remain unsolved, an enormous amount of progress in this field has been made in the past decade. Using simulations in the study of the structural and dynamic properties not only of pure ILs [1,47—63], but also of their mixtures with several organic solvents [64—80], their interactions with other compounds [81—89], and even their behavior at electrochemical interfaces [90—95].

The choice of the simulation method depends on the scale of the question at hand and the different problems to face; for example, ab initio quantum chemical (AIQC) techniques include polarization and charge transfer and can provide us with a full understanding of the electronic structure, but their computational costs are considerably high and this makes them not suitable for studying a great number of molecules. On the other hand, molecular dynamics (MD) or Monte Carlo methods are appropriate for describing with accuracy the liquid environment, because they can consider a large number of molecules interacting; but they cannot be used for analyzing the movement of the electrons within the system. Halfway between the classical and quantum methods, as they contain both electronic and dynamics information, we can find ab initio molecular dynamic (AIMD) simulations, which cannot investigate a number of molecules as large as in the case of classical simulations but include information on the electronic structure. Thus, the description of these compounds should usually be obtained by combining more than one of these techniques.

The first report for an atomistic simulation of an IL dates back to 2001, when Hanke et al. [69] performed MD simulations of ILs based on imidazolium cations. Particularly, they analyzed some solvation properties of a series of small molecules in dilute solution in 1,3-dimethylimidazolium chloride ($[C_1C_1Im][Cl]$) and found that the strongest solute IL interactions were hydrogen bonds between the solute and the anions and that this was dominant for the solvation properties of solutes with hydroxyl groups. However, in the case of quantum computer simulations, it would not be until a few years later that several results were first published. For example, in 2003, Turner et al. [96] applied, for the first time, AIQC techniques in a detailed study of ILs. They searched for a correlation between the melting points of 1-alkyl-3-methylimidazolium halides and their interaction energies between cations and anions and found that there is more than one factor contributing to the melting point behavior of ILs based on 1-alkyl-3-methylimidazolium halides. On the other hand, the first AIMD simulations for an IL were performed in 2005 by Del Pópolo et al. [60], when the comparison of the liquid structure for $[C_1C_1Im][Cl]$ with that obtained from two classical methods and neutron scattering was carried out; and also by Bühl et al. [97], who described the local structure of $[C_1C_1Im][Cl]$ in terms of radial distribution functions and spatial distributions, revealing a considerable extent of hydrogen bonding.

After these first steps, many research groups were devoted to the development of new accurate force fields, more evolved computational tools, and suitable approximations for the simulation of ILs [48,50,52,53,87,98–107]. Because of this effort, the characterization of ILs and the comprehension of their behavior have advanced significantly and the number of publications focused on computational studies of these materials has increased. A remarkable analysis of the structure of 1-alkyl-3-methylimidazolium hexafluorophosphate ($[C_nC_1Im][PF_6]$) and 1-alkyl-3-methylimidazolium bis(trifluoromethanesulfonyl)amide ($[C_nC_1Im][NTf_2]$) using MD simulations was made by Canongia-Lopes and Pádua [1], in which they reported a structuring of their liquid phases in a similar way to microphase separation between polar and nonpolar domains, in agreement with previous publications [47,48].

In addition, because of their importance in several applications, computer simulations have been a fundamental tool for making clear the solvation process of mixtures of ILs with various solvents. One of the most important is water [64–67,69,72,73,75–80] and, very recently, Varela and coworkers [67] performed extensive MD simulations to analyze the influence of cation and anion natures, and of water concentration, on the structure and dynamics of aqueous mixtures of ILs of the $[C_nC_1Im]$ family. Although the studies of mixtures with other molecular solvents, such as alcohols, are much scarcer, some are the publications that can be currently found in literature [68–72,74,75,79]. As an example, Jahangiri et al. [68] reported a computational study of the mixture of 1-ethyl-3-methylimidazolium chloride ($[C_2C_1Im][Cl]$) and 1-ethyl-3-methylimidazolium hexafluorophosphate ($[C_2C_1Im][PF_6]$) with both methanol and ethanol in order to investigate excess properties and some physical and structural properties of the mixture.

A very important application of ILs is as electrolytes in electrochemical devices. In this context, many works have recently not only focused on their bulk mixtures with alkali salts trying to get a further understanding of the solvation process and the resultant aggregates [83–85,89,108], but also on their behavior at the proximities of an interface and the analysis of the well-known layered structure of the IL [90–95]. To name a few among the former, Niu et al. [108] provided a detailed picture of the structure and transport properties of mixtures of 1-ethyl-2,3-dimethylimidazolium hexafluorophosphate ($[C_2C_1C_1Im][PF_6]$) and lithium hexafluorophosphate ($LiPF_6$), and they reported a strong coordination of lithium cations with the anions in their first solvation shell. Additionally, Varela and coworkers analyzed the formation of stable and long-lived $[Li(Anion)_n]^{n-1}$ anionic clusters in both protic [85] and aprotic ILs [84] doped with lithium salts with a common anion. Taking a step further, Lynden-Bell et al. [91,92] studied the mechanisms of the interfacial layer formation on [C_1C_1Im][Cl] near charged and uncharged walls. They reported a significant enrichment of IL cations at the surface that leads to the formation of several distinct IL layers at the surface, as well as that the orientation of IL cations in the first layer depends on the charge of the wall and that ILs provide excellent electrostatic screening at distances above 1−2 nm. Moreover, Merlet et al. [109] provided the first quantitative picture of the structure of an IL adsorbed inside realistically modeled microporous carbon electrodes by means of MD simulations.

4. CONCLUSIONS AND FUTURE WORK

We have briefly reviewed the most important theoretical and computational contributions recently reported in the field of ILs. One can conclude that the theoretical understanding of these systems is still in its infancy and relies mostly on theoretical developments made for concentrated ionic solutions. A theory of screening in pure ILs and the actual role of ionic pairing is probably the most urgently needed for the development of the field. On the other hand, despite the great efforts attributed to computer simulations, there is still a large need for results in several areas, especially screening at interfaces. The very fast evolution of this field is expected to go on in the years to come.

ACKNOWLEDGMENTS

The authors wish to thank the financial support of Xunta de Galicia through the research projects of references 10-PXIB-103-294 PR, 10-PXIB-206-294 PR, both partially supported by FEDER. T. Méndez-Morales thanks the Spanish ministry of Education for her FPU grant.

REFERENCES

[1] J.N. Canongia-Lopes, A.A.H. Pádua, Nanostructural organization in ionic liquids, J. Phys. Chem. B 110 (7) (2006) 3330−3335.
[2] A. Triolo, O. Russina, H.-J. Bleif, E.D. Cola, Nanoscale segregation in room temperature ionic liquids, J. Phys. Chem. B 111 (2007) 4641−4644.
[3] L.M. Varela, M. García, V. Mosquera, Exact mean-field theory of ionic solutions: non-Debye screening, Phys. Rep. 382 (2003) 1−111.
[4] L.W. Bahe, Structure in concentrated solutions of electrolytes. Field-dielectric-gradient forces and energies, J. Phys. Chem. 76 (7) (1972) 1062−1071.
[5] L.W. Bahe, Relative partial molar enthalpies and heats of dilution of electrolytes in water, J. Phys. Chem. 76 (11) (1972) 1608−1611.
[6] L.W. Bahe, D.J. Parker, Activity coefficients of 2:1 electrolytes in structured electrolyte solutions, J. Am. Chem. Soc. 92 (20) (1975) 5664−5670.
[7] L.M. Varela, M. García, F. Sarmiento, D. Attwood, V. Mosquera, Pseudolattice theory of strong electrolyte solutions, J. Chem. Phys. 107 (16) (1997) 6415−6419.

[8] A.P. Abbott, Application of hole theory to the viscosity of ionic and molecular liquids, Chem. Phys. Chem. 5 (8) (2004) 1242–1246.
[9] J.-Z. Yang, X.-M. Lu, J.-S. Guic, W.-G. Xu, A new theory for ionic liquids – the interstice model part 1. The density and surface tension of ionic liquid EMISE, Green Chem. 6 (2004) 541–543.
[10] Y. Lauw, F. Leermakers, Self-consistent Mean-field Theory for Room-Temperature Ionic Liquids, InTech, 2011 (Chapter 14), pp. 329–346.
[11] N. Metropolis, A.W. Rosenbluth, M.N. Rosenbluth, A.H. Teller, E. Teller, Equation of state calculations by fast computing machines, J. Chem. Phys. 21 (1953) 1087–1092.
[12] P.A. Hunt, The simulation of imidazolium-based ionic liquids, Mol. Sim. 32 (1) (2006) 1–10.
[13] B.L. Bhargava, S. Balasubramanian, M.L. Klein, Modeling room temperature ionic liquids, Chem. Commun. 29 (2008) 3339–3351.
[14] E.J. Maggin, Molecular simulation of ionic liquids: current status and future opportunities, J. Phys. Condens. Matter 21 (37) (2009) 373101(1)–373101(17).
[15] P. Debye, E. Hückel, Zur theorie der elektrolyte. I. Gefrierpunktserniedrigung und verwandte erscheinungen (The theory of electrolytes. I. Lowering of freezing point and related phenomena), Phys. Z. 98 (24) (1923) 185–206.
[16] L. Onsager, Zur theorie der elektrolyte I, Phys. Z. 27 (1926) 388–392.
[17] L. Onsager, Zur theorie der elektrolyte II, Phys. Z. 28 (1927) 277–298.
[18] L. Onsager, Theories and problems of liquid diffusion, Ann. N. Y. Acad. Sci. 46 (1945) 241–265.
[19] L. Onsager, S.K. Kim, Wien effect in simple strong electrolytes, J. Phys. Chem. 61 (1957) 198–215.
[20] T.H. Gronwall, V.K.L. Mer, K. Sandved, Uber den einfluss der sogenannten höheren glieder in der Debye-Hückelschen theorie der lösungen starker elektrolyte, Phys. Z. 29 (1928) 358–393.
[21] N. Bjerrum, Untersuchungen über ionenassoziation 1. Der einflüss der ionenassoriation auf die aktivität der ionen bei mittleren assoziationsgraden, K. Danske. Vidensk. Selsk. 7 (9) (1926) 38013.
[22] E.A. Guggenheim, The specific thermodynamic properties of aqueous solutions of strong electrolytes, Philos. Mag. 19 (1935) 588–643.
[23] R. Kjellander, D.J. Mitchell, Dressed ion theory for electrolyte solutions: a Debye-Hückel-like reformulation of the exact theory for the primitive model, J. Chem. Phys. 101 (1995) 603–626.
[24] L.M. Varela, M. Pérez-Rodríguez, M. Garcia, F. Sarmiento, V. Mosquera, Static structure of electrolyte systems and the linear response function on the basis of a dressed-ion theory, J. Chem. Phys. 109 (1998) 1930–1938.
[25] L.M. Varela, C. Rega, M. Pérez-Rodríguez, M. García, V. Mosquera, F. Sarmiento, Relaxation of the ionic cloud on the basis of a dressed-ion theory, J. Chem. Phys. 110 (1999) 4483–4493.
[26] L.M. Varela, M. Pérez-Rodríguez, M. García, F. Sarmiento, V. Mosquera, Conductance of symmetric electrolyte solutions: formulation of the dressed-ion transport theory, J. Chem. Phys. 111 (1999) 10986–10997.
[27] G. Gouy, Sur la constitution de la charge électrique à la surface d'un électrolyte, J. Phys. 9 (1) (1910) 457–468.
[28] D.L. Chapman, A contribution to the theory of electrocapillarity, Philos. Mag. 25 (1913) 475–481.
[29] A. Kornyshev, Double-layer in ionic liquids: paradigm change? J. Phys. Chem. B 111 (2007) 5545–5557.
[30] M.V. Fedorov, A.A. Kornyshev, Ionic liquid near a charged wall: structure and capacitance of electrical double layer, J. Phys. Chem. B 112 (2008) 11868–11872.
[31] M.Z. Bazant, B.D. Storey, A.A. Kornyshev, Double layer in ionic liquids: overscreening versus crowding, Phys. Rev. Lett. 106 (2011) 046102.
[32] J.O. Bockris, A.K.N. Reddy, Modern Electrochemistry. Vol. 1 Ionics, second ed., Plenum Press, New York, 2000.
[33] H.S. Frank, P.T. Thompson, Fluctuations and the limit of validity of the Debye-Hückel theory, J. Chem. Phys. 31 (4) (1959) 1086–1095.
[34] H.S. Frank, P.T. Thompson, The Structure of Electrolyte Solutions, John Wiley and Sons, Inc., New York, 1959.
[35] R.A. Robinson, R.H. Stokes, Electrolyte Solutions, Butterworths, London, 1959.

[36] R. Fürth, On the theory of the liquid state, Cambridge Philos. Soc. (London) 37 (17) (1941), 252, 276, 281.
[37] A.P. Abbott, G. Capper, S. Gray, Design of improved deep eutectic solvents using hole theory, Chem. Phys. Chem. 7 (4) (2006) 803–806.
[38] A.P. Abbott, R.C. Harris, K.S. Ryder, Application of hole theory to define ionic liquids by their transport properties, J. Phys. Chem. B 111 (18) (2007) 4910–4913.
[39] S. Bouguerra, I.B. Malham, P. Letellier, A. Mayaffre, M. Turmine, Part 2: limiting apparent molar volume of organic and inorganic 1:1 electrolytes in (water + ethylammonium nitrate) mixtures at 298 K. Thermodynamic approach using Bahe–Varela pseudo-lattice theory, J. Chem. Thermodyn. 40 (2) (2008) 146–154.
[40] E. Rilo, J. Pico, S. García-Garabal, L.M. Varela, O. Cabeza, Density and surface tension in binary mixtures of CnMIM-BF4 ionic liquids with water and ethanol, Fluid Phase Equilib. 285 (1–2) (2009) 83–89.
[41] L. Varela, J. Carrete, M. García, L. Gallego, M. Turmine, E. Rilo, O. Cabeza, Pseudolattice theory of charge transport in ionic liquid mixtures: corresponding states law for the electric conductivity, Fluid Phase Equilib. 298 (2010) 280–286.
[42] E. Rilo, J. Vila, S. García-Garabal, L.M. Varela, O. Cabeza, Electrical conductivity of seven binary systems containing 1-ethyl-3-methyl imidazolium alkyl sulfate ionic liquids with water or ethanol at four temperatures, J. Phys. Chem. B 117 (2013) 1411–1418.
[43] E. Moggia, B. Bianco, Mean activity coefficients of electrolyte solutions, J. Phys. Chem. B 111 (2007) 3183–3191.
[44] E. Moggia, Osmotic coefficients of electrolyte solutions, J. Phys. Chem. B 112 (2008) 1212–1217.
[45] A. Chagnes, J. Swiatowska, Electrolyte and Solid-Electrolyte Interphase Layer in Lithium-Ion Batteries, InTech, 2011, pp. 145–172.
[46] M.A. Gebbie, M. Valtiner, X. Banquy, E.T. Fox, W.A. Henderson, J.N. Israelachvili, Ionic liquids behave as dilute electrolyte solutions, Proc. Natl. Acad. Sci. U.S.A. 110 (24) (2013) 9674–9679.
[47] Y. Wang, G.A. Voth, Unique spatial heterogeneity in ionic liquids, J. Am. Chem. Soc. 127 (35) (2005) 12192–12193.
[48] S.M. Urahata, M.C.C. Ribeiro, Structure of ionic liquids of 1-alkyl-3-methylimidazolium cations: a systematic computer simulation study, J. Chem. Phys. 120 (4) (2004) 1855–1863.
[49] S.M. Urahata, M.C.C. Ribeiro, Single particle dynamics in ionic liquids of 1-alkyl-3-methylimidazolium cations, J. Chem. Phys. 122 (2) (2005) 024511(1)–024511(9).
[50] T.I. Morrow, E.J. Maginn, Molecular dynamics study of the ionic liquid 1-n-butyl-3-methylimidazolium hexafluorophosphate, J. Phys. Chem. B 106 (49) (2002) 12807–12813.
[51] N.M. Micaelo, A.M. Baptista, C.M. Soares, Parametrization of 1-butyl-3-methylimidazolium hexafluorophosphate/nitrate ionic liquid for the GROMOS force field, J. Phys. Chem. B 110 (29) (2006) 14444–14451.
[52] C.J. Margulis, H.A. Stern, B.J. Berne, Computer simulation of a "green chem-istry" room-temperature ionic solvent, J. Phys. Chem. B 106 (46) (2002) 12017–12021.
[53] C.J. Margulis, Computational study of imidazolium-based ionic solvents with alkyl substituents of different lengths, Mol. Phys. 102 (9–10) (2004) 829–838.
[54] M.G.D. Pópolo, G.A. Voth, On the structure and dynamics of ionic liquids, J. Phys. Chem. B 108 (5) (2004) 1744–1752.
[55] C.E.R. Prado, L.C.G. Freitas, Molecular dynamics simulation of the room-temperature ionic liquid 1-butyl-3-methylimidazolium tetrafluoroborate, J. Mol. Struct. (THEOCHEM) 847 (1–3) (2007) 93–100.
[56] O. Borodin, G.D. Smith, Structure and dynamics of n-methyl-n-propylpyrrolidinium bis(tri-fluoromethanesulfonyl)imide ionic liquid from molecular dynamics simulations, J. Phys. Chem. B 110 (23) (2006) 11481–11490.
[57] S. Alavi, D.L. Thompson, Molecular dynamics studies of melting and some liquidstate properties of 1-ethyl-3-methylimidazolium hexafluorophosphate [emim][PF6], J. Chem. Phys. 122 (15) (2005) 154704(1)–154704(12).

[58] M.H. Kowsari, S. Alavi, M. Ashrafizaadeh, B. Najafi, Molecular dynamics simulation of imidazolium-based ionic liquids. I. Dynamics and diffusion coefficient, J. Chem. Phys. 129 (22) (2008) 224508(1)−224508(13).
[59] M.H. Kowsari, S. Alavi, M. Ashrafizaadeh, B. Najafi, Molecular dynamics simulation of imidazolium-based ionic liquids. II. Transport coefficients, J. Chem. Phys. 130 (1) (2009) 014703(1)−014703(10).
[60] M.G. Del Pópolo, R.M. Lynden-Bell, J. Kohanoff, Ab initio molecular dynamics simulation of a room temperature ionic liquid, J. Phys. Chem. B 109 (12) (2005) 5895−5902.
[61] C. Rey-Castro, L.F. Vega, Transport properties of the ionic liquid 1-ethyl-3-methylimidazolium chloride from equilibrium molecular dynamics simulation. The effect of temperature, J. Phys. Chem. B 110 (29) (2006) 14426−14435.
[62] E.J. Maginn, Atomistic simulation of the thermodynamic and transport properties of ionic liquids, Acc. Chem. Res. 40 (11) (2007) 1200−1207.
[63] M.H. Ghatee, Y. Ansari, Ab initio molecular dynamics simulation of ionic liquids, J. Chem. Phys. 126 (15) (2007) 154502(1)−154502(5).
[64] S. Feng, G.A. Voth, Molecular dynamics simulations of imidazolium-based ionic liquid/water mixtures: alkyl side chain length and anion effects, Fluid Phase Equilib. 294 (1−2) (2010) 148−156.
[65] C.G. Hanke, R.M. Lynden-Bell, A simulation study of water-dialkylimidazolium ionic liquid mixtures, J. Phys. Chem. B 107 (39) (2003) 10873−10878.
[66] W. Jiang, Y. Wang, G.A. Voth, Molecular dynamics simulation of nanostructural organization in ionic liquid/water mixtures, J. Phys. Chem. B 111 (18) (2007) 4812−4818.
[67] T. Méndez-Morales, J. Carrete, O. Cabeza, L.J. Gallego, L.M. Varela, Molecular dynamics simulation of the structure and dynamics of water-1-alkyl-3-methylimidazolium ionic liquid mixtures, J. Phys. Chem. B 115 (21) (2011) 6995−7008.
[68] S. Jahangiri, M. Taghikhani, H. Behnejad, S.J. Ahmadi, Theoretical investigation of imidazolium based ionic liquid/alcohol mixture: a molecular dynamic simulation, Mol. Phys. 106 (8) (2008) 1015−1023.
[69] C.G. Hanke, N.A. Atamas, R.M. Lynden-Bell, Solvation of small molecules in imidazolium ionic liquids: a simulation study, Green Chem. 4 (2) (2002) 107−111.
[70] T. Méndez-Morales, J. Carrete, O. Cabeza, L.J. Gallego, L.M. Varela, Molecular dynamics simulations of the structural and thermodynamic properties of imidazolium-based ionic liquid mixtures, J. Phys. Chem. B 115 (38) (2011) 11170−11182.
[71] T. Méndez-Morales, J. Carrete, M. García, O. Cabeza, L.J. Gallego, L.M. Varela, Dynamical properties of alcohol + 1-hexyl-3-methylimidazolium ionic liquid mixtures: a computer simulation study, J. Phys. Chem. B 115 (51) (2011) 15313−15322.
[72] J.N. Canongia-Lopes, M.F. Costa-Gomes, A.A.H. Pádua, Nonpolar, polar, and associating solutes in ionic liquids, J. Phys. Chem. B 110 (34) (2006) 16816−16818.
[73] C.E.S. Bernardes, M.E.M. da Piedade, J.N. Canongia-Lopes, The structure of aqueous solutions of a hydrophilic ionic liquid: the full concentration range of 1-ethyl-3-methylimidazolium ethylsulfate and water, J. Phys. Chem. B 115 (9) (2011) 2067−2074.
[74] G. Raabe, J. Khler, Thermodynamical and structural properties of binary mixtures of imidazolium chloride ionic liquids and alcohols from molecular simulation, J. Chem. Phys. 129 (14) (2008) 144503(1)−144503(8).
[75] A. Heintz, R. Ludwig, E. Schmidt, Limiting diffusion coefficients of ionic liquids in water and methanol: a combined experimental and molecular dynamics study, Phys. Chem. Chem. Phys. 13 (8) (2011) 3268−3273.
[76] M.S. Kelkar, E.J. Maginn, Effect of temperature and water content on the shear viscosity of the ionic liquid 1-ethyl-3-methylimidazolium bis (trifluoromethanesulfonyl) imide as studied by atomistic simulations, J. Phys. Chem. B 111 (18) (2007) 4867−4876.
[77] M.S. Kelkar, W. Shi, E.J. Maginn, Determining the accuracy of classical force fields for ionic liquids: atomistic simulation of the thermodynamic and transport properties of 1-ethyl-3-methylimidazolium ethylsulfate ([emim][EtSO4]) and its mixtures with water, J. Ind. Eng. Chem. Res. 47 (23) (2008) 9115−9126.

[78] X. Liu, T.J.H. Vlugt, A. Bardow, Maxwell-Stefan diffusivities in binary mixtures of ionic liquids with DMSO and H_2O, J. Phys. Chem. B 115 (26) (2011) 8506–8517.
[79] A.A.H. Pádua, M.F.C. Gomes, J.N.A. Canongia-Lopes, Molecular solutes in ionic liquids: a structural perspective, Acc. Chem. Res. 40 (11) (2007) 1087–1096.
[80] Y. Wang, W. Jiang, T. Yan, G.A. Voth, Understanding ionic liquids through atomistic and coarse-grained molecular dynamics simulations, Acc. Chem. Res. 40 (11) (2007) 1193–1199.
[81] X. Huang, C.J. Margulis, Y. Li, B.J. Berne, Why is the partial molar volume of CO_2 so small when dissolved in a room temperature ionic liquid? Structure and dynamics of CO_2 dissolved in [bmim$^+$] [PF$^-$], J. Am. Chem. Soc. 127 (50) (2005) 17842–17851.
[82] X. Wu, Z. Liu, S. Huang, W. Wang, Molecular dynamics simulation of room-temperature ionic liquid mixture of [bmim][BF4] and acetonitrile by a refined force field, Phys. Chem. Chem. Phys. 7 (14) (2005) 2771–2779.
[83] O. Borodin, G.D. Smith, W. Henderson, Li^+ cation environment, transport, and mechanical properties of the LiTFSI doped n-methyl-n-alkylpyrrolidinium+TFSI ionic liquids, J. Phys. Chem. B 110 (34) (2006) 16879–16886.
[84] T. Méndez-Morales, J. Carrete, S. Bouzón-Capelo, M. Pérez-Rodríguez, O. Cabeza, L.J. Gallego, L.M. Varela, MD simulations of the formation of stable clusters in mixtures of alkaline salts and imidazolium-based ionic liquids, J. Phys. Chem. B 117 (11) (2013) 3207–3220.
[85] T. Méndez-Morales, J. Carrete, O. Cabeza, O. Russina, A. Triolo, L.J. Gallego, L.M. Varela, Solvation of lithium salts in protic ionic liquids: a molecular dynamics study, J. Phys. Chem. B 118 (3) (2014) 761–770.
[86] A.S. Pensado, A.A.H. Pádua, Solvation and stabilization of metallic nanoparticles in ionic liquids, Angew. Chem. Int. Ed. Engl. 50 (37) (2011) 8683–8687.
[87] C. Cadena, J.L. Anthony, J.K. Shah, T.I. Morrow, J.F. Brennecke, E.J. Maginn, Why is CO_2 so soluble in imidazolium-based ionic liquids? J. Am. Chem. Soc. 126 (16) (2004) 5300–5308.
[88] J.K. Shah, E.J. Maginn, Monte Carlo simulations of gas solubility in the ionic liquid 1-n-butyl-3-methylimidazolium hexafluorophosphate, J. Phys. Chem. B 109 (20) (2005) 10395–10405.
[89] M.J. Monteiro, F.F.C. Bazito, L.J.A. Siqueira, M.C.C. Ribeiro, R.M. Torresi, Transport coefficients, Raman spectroscopy, and computer simulation of lithium salt solutions in an ionic liquid, J. Phys. Chem. B 112 (7) (2008) 2102–2109.
[90] C. Merlet, B. Rotenberg, P.A. Madden, M. Salanne, Computer simulations of ionic liquids at electrochemical interfaces, Phys. Chem. Chem. Phys. 15 (2013) 15781–15792.
[91] M.V. Fedorov, R.M. Lynden-Bell, Probing the neutral graphene-ionic liquid interface: insights from molecular dynamics simulations, Phys. Chem. Chem. Phys. 14 (2012) 2552–2556.
[92] R.M. Lynden-Bell, A.I. Frolov, M.V. Fedorov, Electrode screening by ionic liquids, Phys. Chem. Chem. Phys. 14 (2012) 2693–2701.
[93] C. Pinilla, M.G. Del Pópolo, J. Kohanoff, R.M. Lynden-Bell, Polarization relaxation in an ionic liquid confined between electrified walls, J. Phys. Chem. B 111 (18) (2007) 4877–4884.
[94] A.C.F. Mendona, P. Malfreyt, A.A.H. Pádua, Interactions and ordering of ionic liquids at a metal surface, J. Chem. Theory Comput. 8 (9) (2012) 3348–3355.
[95] C. Pinilla, M.G. Del Pópolo, R.M. Lynden-Bell, J. Kohanoff, Structure and dynamics of a confined ionic liquid. Topics of relevance to dye-sensitized solar cells, J. Phys. Chem. B 109 (38) (2005) 17922–17927.
[96] E.A. Turner, C.C. Pye, R.D. Singer, Use of ab initio calculations toward the rational design of room temperature ionic liquids, J. Phys. Chem. A 107 (13) (2003) 2277–2288.
[97] M. Bühl, A. Chaumont, R. Schurhammer, G. Wipff, Ab initio molecular dynamics of liquid 1,3-dimethylimidazolium chloride, J. Phys. Chem. B 109 (39) (2005) 18591–18599.
[98] Z. Liu, S. Huang, W. Wang, A refined force field for molecular simulation of imidazolium-based ionic liquids, J. Phys. Chem. B 108 (34) (2004) 12978–12989.
[99] Z. Liu, X. Wu, W. Wang, A novel united-atom force eld for imidazolium-based ionic liquids, Phys. Chem. Chem. Phys. 8 (9) (2006) 1096–1104.
[100] J.N. Canongia-Lopes, J. Deschamps, A.H. Pádua, Modeling ionic liquids using a systematic all-atom force field, J. Phys. Chem. B 108 (6) (2004) 2038–2047.

[101] T. Yan, C.J. Burnham, M.G.D. Pópolo, G.A. Voth, Molecular dynamics simulation of ionic liquids: the effect of electronic polarizability, J. Phys. Chem. B 108 (32) (2004) 11877–11881.
[102] C.G. Hanke, S.L. Price, R.M. Lynden-Bell, Intermolecular potentials for simulations of liquid imidazolium salts, Mol. Phys. 99 (10) (2001) 801–809.
[103] J. de Andrade, E.S. Bes, H. Stassen, Computational study of room temperature molten salts composed by 1-alkyl-3-methylimidazolium cations force-field proposal and validation, J. Phys. Chem. B 106 (51) (2002) 13344–13351.
[104] J.K. Shah, J.F. Brennecke, E.J. Maginn, Thermodynamic properties of the ionic liquid 1-n-butyl-3-methylimidazolium hexafluorophosphate from Monte Carlo simulations, Green Chem. 4 (2) (2002) 112–118.
[105] J.N. Canongia-Lopes, A.A.H. Pádua, Molecular force field for ionic liquids composed of triflate or bistriflylimide anions, J. Phys. Chem. B 108 (43) (2004) 16893–16898.
[106] C. Cadena, Q. Zhao, R. Snurr, E.J. Maginn, Molecular modeling and experimental studies of the thermodynamic and transport properties of pyridinium-based ionic liquids, J. Phys. Chem. B 110 (6) (2006) 2821–2832.
[107] S.U. Lee, J. Jung, Y.-K. Han, Molecular dynamics study of the ionic conductivity of 1-n-butyl-3-methylimidazolium salts as ionic liquids, Chem. Phys. Lett. 406 (4–6) (2005) 332–340.
[108] S. Niu, Z. Cao, S. Li, T. Yan, Structure and transport properties of the LiPF6 doped 1-ethyl-2,3-dimethyl-imidazolium hexafluorophosphate ionic liquids: a molecular dynamics study, J. Phys. Chem. B 114 (2) (2010) 877–881.
[109] C. Merlet, B. Rotenberg, P.A. Madden, P.-L. Taberna, P. Simon, Y. Gogotsi, M. Salanne, On the molecular origin of supercapacitance in nanoporous carbon electrodes, Nat. Mater. 11 (2012) 306–310.

SUBCHAPTER 1.6

Green Aspects of Ionic Liquids

Aurora M. Rubio[1], Francisca Tomás-Alonso[1], Jesús Hernández Fernández[1], Antonia Pérez de los Ríos[1], Francisco José Hernández Fernández[2]
[1]Department of Chemical Engineering, Regional Campus of International Excellence "Campus Mare Nostrum," University of Murcia, Campus de Espinardo, Murcia, Spain
[2]Department of Chemical and Environmental Engineering, Regional Campus of International Excellence "Campus Mare Nostrum," Technical University of Cartagena, Campus La Muralla, Cartagena, Murcia, Spain

1. INTRODUCTION

Ionic liquids (ILs) are universally recognized as a clean alternative to conventional volatile organic solvents, mainly because of their negligible vapor pressure. For safe use of these new solvents, it is necessary to know their toxicity and biodegradability. Although these parameters have not yet been determined for many ILs, their negligible volatility reduces potential exposures, being that direct skin contact and ingestion are the only possible ways of contamination. Several experimental tests with biological systems are used to evaluate the ILs' toxicity, such as acetylcholinesterase (AChE) inhibition assay, toxicity tests toward bacterium (e.g., *Vibrio fischeri*), green algae (e.g., *Pseudokirchneriella subcapitata*), cell cultures, plants (e.g., wheat, *Triticum aestivum*), fish (e.g., *Danio rerio*), crustaceans (e.g., *Daphnia magna*), and animals (e.g., frog, *Rana nigromaculata*) [1].

Different degrees of toxicity have been reported from ILs, compared to that of chemicals currently used as solvent in chemical industry, which could be explained by the enormous variety of ILs. Most of the investigated ILs are irritants and have similar toxicity to the classic organic solvents [2]. However, some studies have shown that it is possible to design nontoxic and biodegradable ILs by a suitable selection of the constituent cation and anion. In the next sections, an overview of the most commonly used tests for evaluating the IL's toxicity and biodegradability is given and some guidelines are included for the design of nontoxic and biodegradable ILs.

2. TOXICITY IN VITRO USING THE ENZYME AChE

In vitro assays permit the toxicity determination outside a living organism in controlled environmental conditions. AChE is an essential enzyme for the nervous system of higher organisms, and is usually selected in these assays. The active center of the enzyme is highly conservative among organisms. The actuating mechanism of the many insecticides and pesticides (organophosphates and carbamates) is based on the inhibition of the active center of the enzyme, and this inhibition has important consequences for human health.

AChE catalyzes the hydrolysis of the neurotransmitter acetylcholine, obtaining choline and acetate, degrading the acetylcholine of the medium (see Eqn (1)). When AChE is inhibited and the acetylcholine is not destroyed, this produces nerve hyperactivity that may cause death of the individual.

$$\text{Acetylcholine} + H_2O \xrightarrow{\text{Acetylcholinesterase}} \text{Choline} + \text{Acetate} \tag{1}$$

Toxicity in vitro using the enzyme AChE by the spectrophotometric method of Ellman [3] involves the measure of the enzyme activity by colorimetry at 412 nm. The enzyme hydrolyzes acetylthiocholine to thiocholine that reacts with 5,5′-dithio-bis-2-nitrobenzoic acid obtaining a colored spice compound that adsorbs at 412 nm.

Toxicity is usually expressed as the concentration of reducing agent that produces 50% of the initial luminance (EC_{50}) [4] and is typically expressed in logarithmic terms (log EC_{50}). Thereby, the higher the EC_{50} value, the lower the toxicity of the compound. Related to this, in vitro assays only provide indications of the potential danger, but do not imply an effect overall organism. The chemical needs to reach the target site with the purpose of causing an effect in organisms.

Arning et al. [5] accomplished a study on the AChE inhibition by different ILs. The study took into account the influence of the head group and the side chain, and also regioselective and general structural considerations. Anion species were found to be inactive in the AChE inhibition assays. From the 19 anion species studied, only a significant inhibition was found for fluoride, hexafluoroantimonate, and hexafluorophosphate, with IC_{50} (half maximal inhibitory concentration) logarithmic values of 575 μM (F^-), 219 μM (SbF_6^-), and 145 μM (PF_6^-), respectively, and the 1-dodecylsulfate anion was found to be a weak (log $IC_{50} = 912$ μM) inhibitor of the enzyme's activity. The head group effect could be modulated to reduce the AChE inhibition by choosing polar, nonaromatic head groups or incorporating polar hydroxy, ether, or nitrile functions into the side chains connected to the cationic core structure. The authors identified the dimethylaminopyridinium and the quinolinium head groups as very strong inhibitors of the enzyme, with log IC_{50} values of 0.6, 0.99, 0.5, and <0 μM for 4-(dimethylamino)-1-butylpyridinium chloride, 4-(dimethylamino)-1-ethylpyridinium bromide, 4-(dimethylamino)-1-hexylpyridinium chloride, and 1-octylquinolinium bromide, respectively. In contrast, the morpholinium head group was found to be only weakly inhibiting or even inactive with log IC_{50} values around 3 μM.

3. BIOACCUMULATION

The octanol–water partition coefficient (*P*) of a substance is defined as the ratio of molar concentrations (*C*) of a chemical in 1-octanol and water in dilute solution, as follows [6]:

$$K_{i,ow} \equiv \frac{C_i^o}{C_i^w} \approx \frac{w_i^o}{w_i^w} \frac{1 - w_i^w}{1 - w_i^o} \frac{\rho^o}{\rho^w} \tag{2}$$

where C_i^o is the molar concentration of the IL dissolved in the octanol rich phase, and C_i^w the molar concentration of the IL dissolved in the water rich phase. This coefficient is commonly used as a quantitative measure of the ability of substances to cross biological membranes. Therefore, a greater partition coefficient implies a greater potential for bioaccumulation and better penetration of the compounds through the skin, increasing toxicity. The reduced exposure to ILs is not only because of the negligible vapor pressure of IL but also to the possible lower bioaccumulation of IL. This parameter has been measured from a wide range of ILs and, in general, is significantly lower than that from conventional organic solvents because of the ionic nature of ILs [7].

As can be seen in Table 1, in which results are grouped by log EC_{50} values from highest to lowest toxicity, the octanol–water partition coefficient values (log *P*) for ILs

Table 1 Molecular Weight (*M*), Octanol–Water Partition Coefficients (log *P*) and Toxicity Expressed as log EC_{50} (ppm) for Several Ionic Liquids and Other Common Volatile Organic Compounds

Compound	*M*	log *P*	log EC_{50}
$[C_8C_1Im][Br]$	275.19	~0.8	0.07
$[C_8C_1Pyr][Br]$	286.19	~0.8	0.25
$[C_6C_1Im][Br]$	247.14	~0.15	0.81
o-Xylene	106.18	3.12	0.97
$[C_6C_1Pyr][Br]$	258.14	~0.15	1.48
Phenol	94.12	1.48	1.49
Toluene	92.15	2.73	1.50
Methyl isobutyl ketone	100.16	1.31	1.90
$[C_4C_1Pyr][N(CN_2)_2]$	216.22	~−2.4	1.99
Benzene	78.12	2.13	2.03
$[C_4C_1Pyr][Br]$	230.09	~−2.4	2.12
Ethylene glycol	62.07	1.2	2.79
$[C_4C_1Im][Cl]$	174.63	−2.4	2.95
$[C_4C_1Im][N(CN_2)_2]$	205.22	~−2.4	2.99
Chloroform	119.38	1.97	3.08
$[C_4C_1Im][Br]$	219.08	−2.48	3.35
Dichloromethane	84.93	1.25	3.40
Acetone	58.08	−0.24	4.29
Methanol	32.04	−0.74	5.00

obtained from Lide [8] are lower than the values for common volatile organic compounds (VOCs) obtained from Ropel et al. [7]. Butyl-substituted ILs have much lower octanol–water partition coefficient values ($\log P \approx -2.4$) than octyl and hexyl substituted ILs ($\log P \approx 0.08$) and therefore are more water soluble. Thus, those ILs with longer alkyl chains on the cations were more toxic, even more toxic than commonly used industrial solvents. These values agreed with the EC_{50} results obtained by Docherty and Kulpa [9] for ILs and by Kaiser and Palabrica [10] for VOCs for 30 min acute toxicity tests to Photobacterium phosphoreum. The higher the hydrophobicity of the IL, the greater is its capacity to penetrate biological membranes and the higher is its toxicity.

Another application of this coefficient was made by Ortuño et al. [10] who determined ion partition coefficients of imidazolium cations between water and the membrane of an ion selective electrode based on the IL $[C_4C_1Im][PF_6]$. These authors found a relationship between the partition coefficient's values obtained and the $\log EC_{50}$ toxicity values for imidazolium chlorides to a phytoplankton Selenastrum capricornutum measured by Matzke et al. [11]. As the values of both measures were similar, the partition coefficients could be used for predicting the toxicity level of the ILs tested.

4. TOXICITY IN MAMMALIAN CELLS

In vitro toxicity of a substance can be studied in mammalian cells, such as leukemia rat cells (IPC-81), glioma rat cells (C6), and human colon carcinoma cells (CaCo-2) [12]. This last type of cells in vitro is a good model of intestinal epithelial cells.

4.1 Leukemia Rat Cells (IPC-81)

Ishiyama et al. [13] developed a colorimetric assay with the monosodium salt of 2-(4-iodophenyl)-3-(4-nitrophenyl)-5-(2,4-disulfophenyl)-2H-tetrazolium (WST-1) to determine substance's toxicity. The assays involved 96-well microplates, containing in each plate blanks (no cells), controls (without toxics), and two series of dilution of the tested substance with multiple replicates each. Cells are added to the plates at a given concentration (cells/mL) and cultured for 48 h in the tested substance. After 44 h, a few mL of reagent (WST-1) (diluted 1:4 with phosphate buffer) are added and cells are incubated 4 h more.

The ability to reduce WST-1 reagent, that determines the viable cells, is observed photometrically at 450 nm. Cell survival is calculated as the percentage of viable cells (measured as the reduction of WST-1) comparing with control cells, as follows:

$$\%\ \text{Survival} = \frac{\text{Absorbance of problem cells}}{\text{Absorbance of control cells}} \times 100 \tag{3}$$

Stasiewicz et al. [14] studied the toxicity on IPC-81 promyelocytic leukemia rat cells of ILs composed of 1-alkoxymethyl-3-hydroxypyridinium cations and acesulfamate, saccharinate, and chloride anions. The authors found that the longer the alkoxymethyl chain, the higher inhibition is suffered by the enzyme at effective concentrations. This can be proved by the EC_{50} values obtained, being 3.1×10^3 M, 4.2×10^3 M, and 3.1×10^3 M for a $-C_3H_7$ chain in acesulfamate, saccharinate, and chloride salts, respectively, whereas for a $-C_{11}H_{23}$ chain, these values decreased to 0.03×10^3 M, 0.03×10^3 M, and 0.04×10^3 M, respectively.

4.2 Glioma Rat Cells (C6)

The toxicity of substances in mammalian cells can also be determined by using glioma rat cells (C6). Cell survival was performed in the same way as IPC-81 cells, but, in this case, cells were cultured for 24 h before the substance's addition, with the total incubation time of 72 h (including 4 h with WST-1) [12,14]. Ranke et al. [12] studied the sorption, distribution, and cytotoxicity of a series of 1-alkyl-3-methyl imidazolium tetrafluoroborates on the same cells, and quantified the alkyl chain length influence by linear regression analysis. They found the ion pair formed by the anion and cation reduced the anion effect on the IL toxicity. According to the log EC_{50} values (>3 μM for $[C_4C_1Im][PF_6]$ and $[C_nC_1Im][PF_6]$, 3.08 μM for $[C_7C_1Im][PF_6]$, 2.76 μM for $[C_8C_1Im][PF_6]$, 1.90 μM for $[C_9C_1Im][PF_6]$, and 1.69 μM for $[C_{10}C_1Im][PF_6]$) toxicity was found to decrease with the *n*-alkyl chain length for $[C_nC_1Im][PF_6]$ ILs. Thus, for the same anion, a different cation may vary the IL's toxicity.

4.3 Human Colon Carcinoma Cells (CaCo-2)

In order to determine the toxicity of a substance from the CaCo-2 cells, a number of these cells are added to each well of a plate. After 3 days, when cells are in the exponential growth phase, toxics are added to the medium obtaining a determined concentration range. On the fourth day the medium with toxics is removed and a fresh medium with the colorimetric reagent, [3-(4,5-dimethylthiazol-2-yl)-2,5-diphenyl tetrazolium] bromide is added [15]. This soluble reagent produces a yellow solution. When the reagent contacts living cells for 4 h and the succinate dehydrogenase enzyme is added, a blue-violet insoluble compound (formazan) is obtained. The absorbance of this compound can be measured at 570 nm [16]. Likewise, the cell survival percentage is obtained according to the toxic concentration, and dose–response curves and EC_{50} are obtained.

Frade et al. [17] made a toxicological evaluation of ILs on human colon cancerous cells (CaCo-2). According to their results, the presence of a benzyl group does not seem to contribute to nontoxic $[-C_1Im]$ based ILs but, for example, the introduction of a COOH group led to a great reduction of $[C_{10}C_1Im]$ toxicity. They concluded that $[C_4C_1Im]$, $[C_2OHC_1Im]$, and $[C_4C_1Pyrr]$ are potential candidates for less toxic and human friendly ILs, along with other combinations with different cations that were very promising. On the other hand, the anion produces a significant variation, like the $[PF_6]$, which produces a higher influence than other common anions like $[NTf_2]$ and [DCA].

5. ECOTOXICITY

Ecotoxicity tests are commonly used to detect acute or chronic effects of substances on representative organisms, especially aquatic, like marine luminescent bacteria (*V. fischeri*), water fleas (*D. magna*), algae (*P. subcapitata*, Scenedesmus magna, Chlorella vulgaris, Skeletoma), and fish (*D. rerio*, Oncorhynchus mykiss, Pimephales promelas, Brachydanio rerio, Oryzias latipes, Cyprinodon variegatus, Lepomis macrochirus) [18]. Such tests for bacteria and other microorganisms have the advantage of offering similar biochemical pathways to those of higher organisms, short life cycles, and a quick response to the medium changes.

5.1 Microtox® Assay

For several years now, the Microtox® assay (ISO 11348-3, 1998) has been used to evaluate toxicity in environmental samples. The Microtox® system bioassay provides a rapid means of determining acute toxicity of environmental samples and pure compounds by measuring the light emission from the luminescent bacterium *V. fischeri*.

The blue-green light emitting reaction in these bacteria involves reduced riboflavin phosphate ($FMNH_2$) oxidation with oxygen and an aliphatic aldehyde [19]. This reaction is catalyzed by the luciferase enzyme. The overall reaction is shown below:

$$\begin{aligned} FMNH_2 + O_2 + R\text{-}CHO \rightarrow FMN + H_2O + R - COOH \\ + h\nu(\lambda_{max} = 490{-}505\ nm) \end{aligned} \tag{4}$$

Bacterial bioluminescence reaction is related to cellular respiration, so that a decrease of the bioluminescence indicates a decrease in cellular respiration. On exposure to toxic substances, the *V. fischeri* light emission is reduced proportionally to the toxicity of the sample, expressing toxicity with EC_{50} values [4].

The suspension of bacteria must be incubated for an hour. After being tempered, 100 μL aliquots are prepared and 100 μL of various toxic concentrations are added to each sample. After 15 min, the luminescence values are measured and the EC_{50} values are obtained.

To perform this test the osmotic balance, salinity, temperature, and pH of the medium (ranged 6–8) have to be controlled. Thus, it can be assured that any decrease in bacteria light emission is due exclusively to the effect of contaminants [4].

The Microtox® assay has many advantages, such as sensitivity, discriminant capacity, reproducibility, and ease of application for organic and inorganic contaminants. Moreover, it is faster and cheaper than other biological assays [4].

The Microtox® assay is widely used for determining the toxicity of single compounds, for monitoring industrial effluents in environmental water quality surveys, and in sediment contamination studies [20].

In recent years, acute toxicity tests using bacteria have acquired great importance. Toxicity data obtained with Microtox® assays correlate well with acute toxicities obtained with standard toxicity tests [9,21–23]. Therefore, toxicity in *V. Fischeri* can be applied to predict toxicity in other aquatic organisms [23–27]. In fact, the Microtox® test done with *V. Fischeri* has shown greater sensitivity and better fit to the results than assays done with fish and *D. magna*.

As an example, toxicity data on *V. fischeri* and *D. magna* were obtained by Couling et al. [20] to get correlative and predictive equations to assess the factors that govern the toxicity of a range of different ILs. It has been found that the ILs with shorter alkyl substituents on the cation normally have lower toxicities [2,28–30]. It can be seen in Table 2 that ILs with longer alkyl chains have higher toxicities. As an example, the $\log EC_{50}$ values obtained by Couling et al. [20] for $[C_1C_1C_1C_1N][Br]$ and $[C_2C_2C_2C_2N][Br]$ were >2 mmol/L, which can be compared with the value obtained for $[C_4C_4C_4C_4N][Br]$ which was 0.27 mmol/L. Even changing one of the alkyl substituents, as in $[C_6C_2C_2C_2N][Br]$ case ($\log EC_{50} = -0.54$), the toxicity is increased.

It was also found that the introduction of oxygenated groups on the alkyl chains, such as ether and ester, leads to a decrease of the toxicity of guanidinium and imidazolium compounds. With respect to the effect of the different cations, they found that it is possible to recognize that the phosphonium-based ILs (e.g., $[C_4C_4C_4C_4P][Br]$ $\log EC_{50} = -0.29$ mmol/L) seem to be more toxic when compared to the analog ammonium based ILs, with the same anion and alkyl chains ($[C_4C_4C_4C_4N][Br]$ $\log EC_{50} = 0.27$ mmol/L).

Table 2 Experimental Toxicity Results Obtained by Docherty et al. [20] for *V. fischeri* Expressed as $\log EC_{50}$ (mmol/L) for Several Ionic Liquids and Other Common Organic Solvents

Ionic Liquids	$\log EC_{50}$	Organic Solvents	$\log EC_{50}$
$[C_4C_1C_1NPy][Br]$	−0.68	4-Dimethylaminopyridine	−0.41
$[C_6C_2C_2C_2N][Br]$	−0.54	3,5-Dimethylpyridine	−0.36
$[C_4C_4C_4C_4P][Br]$	−0.29	Pyridine	0.87
$[C_4C_4C_4C_2P][C_2C_2PO_4]$	0.07	1-Bromobutane	0.95
$[C_4C_4C_4C_4N][Br]$	0.27	Sodium dicyanamide	1.72
$[C_4C_1Im][NTf_2]$	0.39	1-Chlorobutane	1.92
$[C_{14}C_6C_6C_6P][Br]$	0.41	Sodium bromide	2.29
$[Chol][NTf_2]$	1.15		
[Chol][Cl]	>2.0		
$[C_2C_2C_2C_2N][Br]$	>2.0		
$[C_1C_1C_1C_1N][Br]$	>2.0		

Values ordered from higher to lower toxicity.

Comparing with the toxicity values for traditional industrial solvents, many of the ILs tested were less toxic to *V. fischeri*. However, some ILs are still more toxic than common organic solvents (see Table 2).

5.2 Toxicity to *Daphnia magna*

Daphnia magna are freshwater crustaceans that live in lakes and ponds, feeding on microscopic algae. Daphnia are often used as model organisms in standard toxicity bioassays used by regulatory agencies (e.g., United States Environmental Protection Agency, European Organization for Economic Cooperation and Development, Association française de Normalisation, Deutsches Institut für Normung) because of their efficiency and sensitivity to a wide variety of pollutants.

These organisms have a great sensitivity to toxic substances, being able to detect, for example, 0.005 mg of mercury in water and even lower concentrations of many pesticides in industrial wastes. Specifically, toxicity tests with *D. magna* can be used to determine the lethality potential of pure chemicals, drinking water, domestic and industrial waste water, surface water or groundwater, among others.

Bernot et al. [31] studied the acute effects of imidazolium based ILs on survival of *D. magna* and their chronic effects on a number of first-brood neonates, total number of neonates, and average brood size. Imidazolium based ILs were found to be more toxic to *D. magna* than Na^+-based salts but less toxic than other common chemicals, such as chlorine and ammonia (see Table 3). Toxicity was apparently related to the imidazolium cation, and not to the IL anion. The toxicity of these ILs was comparable to that of chemicals currently used in manufacturing and disinfection processes (e.g., ammonia and phenol).

Table 3 Lethal Concentrations of Different Ionic Liquids to *Daphnia magna* in 48-h Acute Toxicity Bioassays

Ionic Liquids	LC_{50}
Chlorine	0.12–0.15
Ammonia	2.9–6.93
$[C_4C_1Im][Br]$	8.03
$[C_4C_1Im][Cl]$	14.80
$[C_4C_1Im][PF_6]$	19.91
$[C_4C_1Im][BF_4]$	10.68
Phenol	10–17
Benzene	356–620
Methanol	3289
Acetonitrile	3600
$[Na][BF_4]$	4765.75
$[Na][PF_6]$	9344.81

Toxicity values expressed as LC_{50} (ppm) from higher to lower toxicity.

6. TOXICITY IN ANIMALS

Toxicity in animals is determined by a deadly effect after a period of treatment. In the case of mammals, it is reduced to 24 h, whereas for fish the exposure time can be up to 96 h.

To determine toxicity, a number of animals are divided into groups. Each group is exposed to a specific chemical concentration by intraperitoneal injection (for mammals) or directly in water (for fish). Dosages are based on preliminary tests and toxic effects on the animals are observed.

Pretti et al. [32] analyzed acute toxicity toward zebrafish (*D. rerio*) of several ILs with different anions and cations. These authors found a reduction in general activity, loss of equilibrium, erratic swimming, staying motionless at mid-water level for prolonged periods, and skin alterations. Toxicity was expressed as the mean lethal concentration (LC_{50}), defined as the value capable of killing the 50% of tested fish in the bath after a continuous exposure period of 96 h. Imidazolium, pyridinium, and pyrrolidinium based ILs showed LC_{50} values higher than 100 mg/L so these ILs can be considered as non-highly lethal toward zebrafish. On the other hand, the ammonium salts AMMOENG 100™ and AMMOENG 130™ showed LC_{50} values remarkably lower than that reported for organic solvents and tertiary amines.

For mammals, the results show the acute exposure degree necessary to cause harm to the animal. It can be noted by various adverse clinical signs, such as hyperactivity or abnormal posture. The acute toxicity is expressed as the mean lethal dose (LD_{50}), known as the dosage that kills 50% of the tested organisms [33].

Bailey et al. [34] studied the effects of exposure of mice to a commonly used IL, [C_4C_1Im][Cl], because of the potential for human exposure because of water or soil contamination from industrial effluent or accidental spills. After exposure to the IL, with one of four treatments: vehicle control, 113 mg/kg/day [C_4C_1Im][Cl], 169 mg/kg/day [C_4C_1Im][Cl], or 225 mg/kg/day [C_4C_1Im][Cl], fetal weight decreased and malformations were more numerous at the highest doses. Additionally, maternal weight gain decreased from 10.67 g in the vehicle controls to 9.28 g in the mice treated with 225 mg/kg/day doses. Even simple IL representatives, such as [C_4C_1Im][Cl], a commonly used IL, may possibly have adverse effects.

7. BIODEGRADABILITY

Biodegradability is directly related to the substance's potential for accumulation and persistence in the environment. The most commonly used tests are the modified Sturm and closed bottle tests (OECD 301 B and D, respectively), the DOC Die-Away Test (OECD 301 A), and the CO_2 headspace test (ISO 14593) [35]. Compounds that achieve a biodegradation level higher than 60% are referred to as "easily degradable" [36]. It has been found that certain ILs are biodegradable, especially if they have an ester group present in the alkyl side chain of the cation [28,36,37]. As an example, the biodegradation percentage obtained for the ILs [C_2OCOC_1Im][Br] and [C_2OCOC_1Im][BF_4] were 48% and 59%, respectively, which means they were readily biodegradable [35]. In later studies [36], [C_4C_1Im][C_8SO_4] underwent a small amount of biodegradation (25%) after 28 days, while other ILs ([C_4C_1Im][Br], [C_4C_1Im][Cl], [C_4C_1Im][BF_4], [C_4C_1Im][PF_6], [C_4C_1Im][N(CN)$_2$], and [C_4C_1Im][NTf_2]) remained largely intact (less than 5% biodegradation). Generally, modifications of the anion do not represent changes on biodegradability, but the octyl sulfate anion proved to be an exception.

Biodegradation data for ILs are scarce yet comparing them with toxicity studies. However, it is worthy to note that new biodegradability reports are leading to applications of biodegradable ILs in synthetic chemistry, such as obtaining some IL from biorenewable materials, like fructose, that may be applied as recyclable solvents [38].

8. FUTURE CHALLENGES

As commented above, different degrees of toxicity have been reported from ILs compared to that of chemicals currently used as solvent in chemical industry, which could be explained by the enormous variety of ILs. The studies of IL's toxicity have shown their toxicities vary depending on their cationic and anionic composition. Not only is the toxicity of the ILs because of the IL's compositions but also it is due to the degradation products. In fact' the IL based on PF_6 anions can be degraded in the presence of water and form HF [39]. Evaluations of the risk posed by ILs to the environment, compared to traditional industrial solvents, must consider not only toxicity but also this presumably reduced exposure to ILs. In this sense, it should be noted that, in contrast to conventional solvents, the negligible vapor pressure usually associated with ILs would result in lower emissions and consequently in a reduced exposure. At any rate, deeper investigations are necessary to evaluate further risk assessment. For that, a broader set of test methods should be applied, including studies focusing on exposition pathways as well as on bioaccumulation and degradation processes. From the studies carried out, structural information has been obtained for a rational design of safer ILs. The structure—activity relationship and the quantitative correlation structure-activity relationship, can be used as predictive models to qualitatively or quantitatively relate the chemical structure of a substance (i.e., IL) with its properties (i.e., toxicity, biodegradability) by the application of statistical tools. The development of models applied to toxicity and biodegradability tests is very useful not only for designing new nontoxic and biodegradable ILs, but also in the analysis of the toxicological mechanisms. Moreover, it must be considered that models and analysis results must be continuously improved updating the proposed models with acquisition of new toxicological data of ILs.

REFERENCES

[1] M. Álvarez-Guerra, A. Irabien, Design of ionic liquids: an ecotoxicity (*Vibrio fischeri*) discrimination approach, Green Chem. 13 (2011) 1507–1516.

[2] K.M. Docherty, C.F. Kulpa Jr., Toxicity and antimicrobial activity of imidazolium and pyridinium ionic liquids, Green Chem. 7 (2005) 185–189.

[3] G.L. Ellman, D.K. Courtney, V. Andres, R.M. Featherstone, A new and rapid colorimetric determination of acetylcholinesterase activity, Biochem. Pharmacol. 7 (1961) 88–95.

[4] F. Onorati, M. Mecozzi, Effects of two diluents in the Microtox® toxicity bioassay with marine sediments, Chemosphere 54 (2004) 679–687.

[5] J. Arning, S. Stolte, A. Böschen, F. Stock, W.R. Pitner, U. Welz-Biermann, B. Jastorff, J. Ranke, Qualitative and quantitative structure activity relationships for the inhibitory effects of cationic head groups, functionalised side chains and anions of ionic liquids on acetylcholinesterase, Green Chem. 10 (2008) 47–58.

[6] A.P. De los Ríos, F.J. Hernández-Fernández, F. Tomás-Alonso, M. Rubio, D. Gómez, G. Víllora, On the importance of the nature of the ionic liquids in the selective simultaneous separation of the substrates and products of a transesterification reaction through supported ionic liquid membranes, J. Membr. Sci. 307 (2008) 233–238.

[7] L. Ropel, L.S. Belveze, S.N.V.K. Aki, M.A. Stadtherr, J.F. Brennecke, Octanol–water partition coefficients of imidazolium-based ionic liquids, Green Chem. 7 (2005) 83–90.

[8] D.A. Lide, CRC Handbook of Chemistry and Physics, eighty fifth ed., CRC Press, Cleveland, OH, 2004.

[9] K.L.E. Kaiser, V.S. Palabrica, *Photobacterium Phosphoreum* toxicity data index, Water Qual. Res. J. Can. 26 (1991) 361–431.

[10] J.A. Ortuño, M. Cuartero, M.S. García, M.I. Albero, Response of an ion-selective electrode to butylmethylimidazolium and other ionic liquid cations. Applications in toxicological and bioremediation studies, Electrochim. Acta 55 (2010) 5598–5603.

[11] M. Matzke, S. Stolte, K. Thiele, T. Juffernholtz, J. Arning, J. Ranke, U. Welz-Biermann, B. Jastorff, The influence of anion species on the toxicity of 1-alkyl-3-methylimidazolium ionic liquids observed in an (eco)toxicological test battery, Green Chem. 9 (2007) 1198–1207.

[12] J. Ranke, K. Mölter, F. Stock, U. Bottin-Weber, J. Poczobutt, J. Hoffmann, B. Ondruschka, J. Filser, B. Jastorff, Biological effects of imidazolium ionic liquids with varying chain lengths in acute *Vibrio Fischeri* and WST-1 cell viability assays, Ecotoxicol. Environ. Saf. 58 (2004) 396–404.

[13] M. Ishiyama, H. Tominaga, M. Shiga, K. Sasamoto, Y. Ohkura, K. Ueno, M. Watanabe, Novel cell proliferation and cytotoxicity assays using a tetrazolium salt that produces a water-soluble formazan dye, In Vitro Toxicol. 8 (1995) 187–190.

[14] M. Stasiewicz, E. Mulkiewicz, R. Tomczak-Wandzel, J. Kumirska, E.M. Siedlecka, M. Goebiowski, J. Gajdus, M. Czerwicka, P. Stepnowski, Assessing toxicity and biodegradation of novel, environmentally benign ionic liquids (1-alkoxymethyl-3-hydroxypyridinium chloride, saccharinate and acesulfamates) on cellular and molecular level, Ecotoxicol. Environ. Saf. 71 (2008) 157–165.

[15] T. Mossman, Rapid colorimetric assay for cellular growth and survival: application to proliferation and cytotoxicity assays, J. Immunol. Methods 65 (1983) 55–63.

[16] V. Jaitely, A. Karatas, A.T. Florence, Water-immiscible room temperature ionic liquids (RTILs) as drug reservoirs for controlled release, Int. J. Pharm. 354 (2008) 168–173.

[17] R.F.M. Frade, A.A. Rosatella, C.S. Marques, L.C. Branco, P.S. Kulkarni, N.M.M. Mateus, C.A.M. Afonso, C.M.M. Duarte, Toxicological evaluation on human colon carcinoma cell line (Caco-2) of ionic liquids based on imidazolium, guanidinium, ammonium, phosphonium, pyridinium and pyrrolidinium cations, Green Chem. 11 (2009) 1660–1665.

[18] R. Boluda, J.F. Quintanilla, J.A. Bonilla, E. Sáez, M. Gamón, Application of the Microtox® test and pollution indices to the study of water toxicity in the Albufera Natural Park (Valencia, Spain), Chemosphere 46 (2002) 355–369.

[19] J.W. Lin, Y.F. Chao, S.F. Weng, Nucleotide sequence of the LuxC gene encoding fatty acid reductase of the lux operon from *Photobacterium Leiognathi*, Biochem. Biophys. Res. Commun. 191 (1993) 314–318.

[20] D.J. Couling, R.J. Bernot, K.M. Docherty, J.K. Dixon, E.J. Maginn, Assessing the factors responsible for ionic liquid toxicity to aquatic organisms via quantitative structure–property relationship modeling, Green Chem. 8 (2006) 82–90.
[21] M.W. Toussaint, T.R. Sheed, D. Van, W.H.A. Schalie, Comparison of standard acute toxicity tests with rapid-screening toxicity tests, Environ. Toxicol. Chem. 14 (1995) 907–915.
[22] M. Weideborg, E.A. Vik, G.D. Øfjord, O. Kjønnø, Comparison of three marine screening tests and four Oslo and Paris commission procedures to evaluate toxicity of offshore chemicals, Environ. Toxicol. Chem. 16 (1997) 384–389.
[23] M.I. Arufe, J. Arellano, M.J. Moreno, C. Sarasquete, Toxicity of a commercial herbicide containing terbutryn and triasulfuron to seabream (*Sparus Aurata* L.) larvae: a comparison with the Microtox® test, Ecotoxicol. Environ. Saf. 59 (2004) 209–216.
[24] D.J.W. Blum, R.E. Speece, Determining chemical toxicity to aquatic species, Environ. Sci. Technol. 24 (1990) 284–293.
[25] H.F. Chen, S.S.Q. Hee, Ketone EC50 values in the Microtox test, Ecotoxicol. Environ. Saf. 30 (1995) 120–123.
[26] J.M. Ribo, K.L.E. Kaiser, Effects of selected chemicals to photo-luminescent bacteria and their correlations with acute and sublethal effects on other organisms, Chemosphere 12 (1983) 1421–1442.
[27] Y.H. Zhao, Y.B. He, L.S. Wang, Predicting toxicities of substituted aromatic hydrocarbons to fish by *Daphnia magna* or *Photobacterium phosphoreum*, Toxicol. Environ. Chem. 51 (1995) 191–195.
[28] N. Gathergood, M.T. Garcia, P.J. Scammells, Biodegradable ionic liquids. Part I. Concept, preliminary targets and evaluation, Green Chem. 6 (2004) 166–175.
[29] P. Luis, A. Garea, A. Irabien, Quantitative structure–activity relationships (QSARs) to estimate ionic liquids ecotoxicity EC50 (*Vibrio fischeri*), J. Mol. Liq. 152 (2010) 28–33.
[30] S.P.M. Ventura, C.S. Marques, A.A. Rosatella, C.A.M. Afonso, F. Goncalves, J.A.P. Coutinho, Toxicity assessment of various ionic liquid families towards *Vibrio fischeri* marine bacteria, Ecotoxicol. Environ. Saf. 76 (2012) 162–168.
[31] R.J. Bernot, M.A. Brueseke, M.A. Evans-White, G.A. Lamberti, Acute and chronic toxicity of imidazolium-based ionic liquids on *Daphnia magna*, Environ. Toxicol. Chem. 24 (2005) 87–92.
[32] C. Pretti, C. Chiappe, D. Pieraccini, M. Gregori, F. Abramo, G. Monni, L. Intorre, Acute toxicity of ionic liquids to the zebrafish (Danio Rerio), Green Chem. 8 (2006) 238–240.
[33] M. Yu, S.H. Wang, Y.R. Luo, Y.W. Han, X.Y. Li, B.J. Zhang, J.J. Whang, Effects of the 1-alkyl-3-methylimidazolium bromide ionic liquids on the antioxidant defense system of *Daphnia magna*, Ecotoxicol. Environ. Saf. 72 (2009) 1798–1804.
[34] M.M. Bailey, M.B. Townsend, P.L. Jernigan, J. Sturdivant, W.L. Hough-Troutman, J.F. Rasco, R.P. Swatloski, R.D. Rogers, R.D. Hood, Developmental toxicity assessment of the ionic liquid 1-butyl-3-methylimidazolium chloride in CD-1 mice, Green Chem. 10 (2008) 1213–1217.
[35] D. Coleman, N. Gathergood, Biodegradation studies of ionic liquids, Chem. Soc. Rev. 39 (2010) 600–637.
[36] M.T. Garcia, N. Gathergood, P.J. Scammells, Biodegradable ionic liquids. Part II. Effect of the anion and toxicity, Green Chem. 7 (2005) 9–14.
[37] N. Gathergood, P.J. Scammells, M.T. Garcia, Biodegradable ionic liquids. Part III. The first readily biodegradable ionic liquids, Green Chem. 8 (2006) 156–160.
[38] S.T. Handy, M. Okello, G. Dickenson, Solvents from bio-renewable sources: ionic liquids based on fructose, Org. Lett. 5 (2003) 2513–2515.
[39] A.P. de los Ríos, F.J. Hernández-Fernández, D. Gómez, M. Rubio, G. Víllora, Enhancement of activity and selectivity in lipase-catalyzed transesterification in ionic liquids by the use of additives, J. Chem. Technol. Biotechnol. 82 (2007) 882–887.

CHAPTER 2

Physicochemical Basis of IL Effects on Separation and Transformation Processes: From Equilibrium to Reactivity

M. Canle L., María Isabel Fernández Pérez, Juan Arturo Santaballa
Chemical Reactivity and Photoreactivity Group, Department of Physical Chemistry and Chemical Engineering I, University of A Coruña, A Coruña, Spain

1. INTRODUCTION

In addition to their tunable structure and properties, and their behavior as solvents [1], treated in other chapters of this book, ionic liquids (ILs) show the following physicochemical properties: thermal stability, high electrical conductivity, large electrochemical window, high polarity, and negligible vapor pressure [2]. These properties make them ideal green media to attempt, for example, novel syntheses [3] following the principles of green chemistry [4], or new processes for pollution reduction, both in gas and aqueous medium, including extraction equilibria [1], catalyzed reactions [5], electrochemical [6] and photochemical processes [7], and even radiation-initiated reactions [8,9]. However, the complexity of intermolecular interactions in ILs makes the interpretation of their characteristics at a molecular level difficult [10]. Large efforts have been put into understanding the role of ILs as solvents, and different aspects of chemical reactivity in ILs have been reviewed [11–16], with emphasis mainly on unexpected processes, compared with conventional solvents, or on the efficiency of reactions in ILs.

At the time of writing this, the CAS Registry contains more than 85 million organic and inorganic chemical substances, of which only approximately 310,000 are inventoried and/or regulated [17]. This registry is updated with approximately 15,000 substances daily. Many of these substances end up in the environment, foodstuffs, or different kinds of materials around us, threatening ecosystems and health [18]. This clearly shows the urgency to seek suitable technologies for extraction, separation, transformation, and elimination of chemicals. ILs might help in this task [19], but a higher level of comprehension of their equilibrium processes and a deeper understanding of chemical reactivity (i.e., reaction mechanisms) in these media will be necessary. Different experimental, computational, and methodological tools to study reaction mechanisms have been described [20–23].

We describe the physical basis of different effects that ILs can induce on chemical systems, whether on chemical equilibria or reactivity. We do not attempt a

Ionic Liquids in Separation Technology
http://dx.doi.org/10.1016/B978-0-444-63257-9.00002-X
Copyright © 2014 Elsevier B.V.
All rights reserved.

comprehensive or critical review of IL reactivity, but rather emphasize some aspects of the observed effects and put forward some examples, hoping this will stimulate further studies.

2. SOLVATION BY ILs

Solvent changes can alter chemical processes to different extents, affecting both their equilibria and rate constants [24]. Solvent effects having a static origin, in systems that are close to equilibrium, can be understood within the classical transition state theory (CTST). Alternatively, solvent effects could have a dynamic origin, especially in strongly dipolar media, to do with slow solvent relaxation/reorientation processes, which could even become rate limiting, and lead to CTST failure. For the sake of simplicity, we only deal herein with the first: static effects taking place in situations where the system is close to equilibrium and the CTST holds.

Solvent effects can be particularly important in comparisons between standard solvents and ILs. Usually, strong effects are observed in processes involving ions or dipoles, or in those in which charge development or disappearance takes place. Solvent effects on reaction rates have been classically treated through a variety of approaches: as a function of the permittivity of the medium [25] and considering the ionizing power of the solvent through the Grunwald—Winstein equation [26]. The polarity of ILs has been estimated in different ways (e.g., using solvatochromic dyes [27,28], or through other treatments, such as the use of Kamlet—Taft parameters) [29]. Multiparameter linear energy—density relationships have also been used to model IL solvent effects [30]. However, none of these approaches has been generalized, their applicability depending on each specific case.

Performing reactions (or simple extractions) in mixtures of ILs with water might offer different advantages. It is important to take into account the change in miscibility of ILs with water as their structure changes, even within the same family of ILs. Thus, for example, $[C_nC_1Im]^+[BF_4]^-$ with an alkyl chain of one to five C-atoms is miscible with water at room temperature, but those with six or more C-atoms separate in a different phase in water [31]. This effect may be used for reducing the presence of undesired substances (e.g., pollutants), in different media, or to facilitate the separation of products from reagents or catalysts in chemical reactions.

Entropy plays a fundamental role in equilibrium and reactivity, and it may be strongly affected by solvation, especially in reactions involving ions or highly polar species. One has to consider whether on going from reagents to products (for equilibrium processes) or from reagents to the transition state (TS; for reaction rates), the system gains or loses entropy (ΔS^0, $\Delta S^{\neq}$), and relate this to solvation: (ΔS^0, $\Delta S^{\neq}$) > 0 generally implies desolvation, whereas (ΔS^0, $\Delta S^{\neq}$) < 0 usually implies a solvation increase.

Ionic pairing takes place to different extents in concentrated ionic solutions. Ion pairs can be classified into contact ion pairs (CIPs), solvent-separated ion pairs (SIPs), or solvent-shared ion pairs (SSIPs) [32]. Evidence has been found both in room temperature ionic liquids (RTILs) [33] and weakly polar solvents [34] for ion pairing in ILs. Application of the ion-pair model to ILs would justify that introduction of reagents into the medium may lead to changes in the type of ion pairing, therefore affecting reaction rates due, for example, to polarity effects in CIPs or to missing solvation shells in SIPs or SSIPs.

To account for the change in activity coefficients of the different species involved in equilibria and on rate constants, usually the ionic strength of the medium Eqn (1)

$$I = \sum_i \frac{1}{2} C_i z_i^2 \tag{1}$$

is kept constant, where C_i is the concentration of the different ions present in the medium, and z_i is their charge. This would be of little use in the case of ILs, when used as solvents, because I would be governed by the ionic strength of the solvent itself that, even for (1:1) ILs, would be much higher than the tiny changes due to chemical processes. Therefore, in most cases, I changes would be irrelevant, but it would have to be taken into account when comparing equilibrium or rate constants with those of similar processes taking place in conventional solvents.

3. SOLVENT EFFECTS ON REACTION RATES

Different models have been developed to correlate reaction rates to explain solvent effects, based on solvent properties. In general, these require prior knowledge of certain properties of ILs, such as their relative permittivity or cohesive pressure, as described later. Unfortunately, these are rarely available at the time of writing this, making it necessary to perform measurements that are not always straightforward. Four different situations can be distinguished that may be applied to reactivity in ILs:

- Reactions between neutral apolar molecules depend on cohesive (internal) pressure, known as Hildebrand solubility parameter (δ^2), rather than on solvent polarity. This measures the energy required to solvate a molecule of solute, in other words, the energy needed to break attractive forces within the solvent, separating its molecules (or ions in the case of ILs). It is calculated according to Eqn (2):

  $$\delta^2 = \frac{\Delta U_v}{V_m} = \frac{\Delta H_v - RT}{M/\rho} \tag{2}$$

  where ΔU_v is the vaporization energy per unit molar volume (V_m), ΔH_v is the vaporization enthalpy at a given temperature T, and M and ρ are the molecular mass and density of the solvent, respectively. Because the solubilities of solute and solvent are interdependent on each other, a good solvent should have a δ value that is close to that of the solute [24]. The following simplifications are assumed: (1) van der Waals forces are the only intermolecular interactions in the system, (2) the heat of mixing accounts for all ideality deviations, (3) solute–solvent interactions can be estimated as a geometrical mean of the respective interactions of the pure components of the mixture, and (4) the solution is dilute, and the rate constant of a bimolecular reaction

$$\mathrm{A} + \mathrm{B} \rightarrow \mathrm{Products}$$

in a solvent S can be expressed as a function of the δ^2 Hildebrand solubility parameter and the molar volumes Eqn (3):

$$\ln(k/k_0) = \frac{1}{RT}\left[\delta_S^2\left(V_A + V_B - V^{\neq}\right) + \left(V_A\delta_A^2 + V_B\delta_B^2 - V^{\neq}\delta^{\neq 2}\right) + 2\delta_S\left(V^{\neq}\delta^{\neq} - V_A\delta_A - V_B\delta_B\right)\right] \tag{3}$$

where k_0 would be the rate constant in an ideal solution.
Assuming the molar volume for the TS can be approached by the addition of the molar volumes of the reagents ($V^{\neq} = V_A + V_B + \ldots$), and that $\delta^{\neq}$ can be estimated as in Eqn (4):

$$\delta^{\neq} = \sqrt{\frac{V_A\delta_A^2 + V_B\delta_B^2}{V^{\neq}}} \tag{4}$$

the reaction rate can be expressed as in Eqn (5):

$$\ln(k/k_0) = \frac{1}{RT}\left[2\delta_S\left(\sqrt{V^{\neq}\left(V_A\delta_A^2 + V_B\delta_B^2\right)} - V_A\delta_A - V_B\delta_B\right)\right] \tag{5}$$

Therefore, prediction of the rate would make it necessary to know or obtain the Hildebrand solubility parameters for both reagents and the corresponding IL as a solvent. Values of ΔH_v have been measured, and δ values for different ILs: $[C_nC_1Im]^+[NTf_2]^-$, $[C_nC_1Im]^+$, $[C_nPyrr]^+$, and $[C_nN]^+[BETI]^-$. ILs showed δ values in the range 16.3–26.5 $(J/cm^3)^{1/2}$, similar to those for small alcohols. These values are dependent on the counteranion of the IL [13,35]. δ values have also been estimated for various $[C_nmim]^+$ with counteranions $[Cl]^-$, $[BF_4]^-$, $[PF_6]^-$, $[TfO]^-$, $[Tf_2N]^-$, and $[C_nOSO_3]^-$ [36].

- Reactions between neutral, dipolar molecules: in this case, the main intermolecular solute–solvent interactions are between dipoles. For this case, Kirkwood [37], considering only electrostatic forces between solute and solvent, estimated the standard Gibbs free energy of solvation to transfer a dipolar spherical molecule (μ) with radius r from gas phase to a continuous medium of relative permittivity ε_r as Eqn (6):

$$\Delta G^0_{\mathrm{solv}} = \frac{-N_A\mu^2}{4\pi\varepsilon_0 r^3}\,\frac{\varepsilon_r - 1}{2\varepsilon_r + 1} \tag{6}$$

where ε_0 is vacuum's permittivity and N_A is Avogadro's number. Applying this to the CTST for a bimolecular reaction:

$$\mathrm{A} + \mathrm{B} \rightarrow \text{Products}$$

the dependence between the rate constant and the polarity of the medium and the dipole moments of the reactants can be obtained Eqn (7):

$$\ln k = \ln k_0 - \frac{N_A}{4\pi\varepsilon_0 RT}\frac{\varepsilon_r - 1}{2\varepsilon_r + 1}\left(\frac{\mu_A^2}{r_A^3} + \frac{\mu_B^2}{r_B^3} - \frac{\mu^{\neq 2}}{r^{\neq 3}}\right) \quad (7)$$

where k_0 is the rate constant in a condensed medium with $\varepsilon_r = 1$.
A further development of this model leads to Eqn (8) [38]:

$$\ln k = \ln k_0 - \frac{3N_A}{32\pi\varepsilon_0 RT}\left(\frac{2}{\varepsilon_r} - 1\right)\left(\frac{\mu_A^2}{r_A^3} + \frac{\mu_B^2}{r_B^3} - \frac{\mu^{\neq 2}}{r^{\neq 3}}\right) \quad (8)$$

Both in Eqns (7) and (8), the reaction rate will increase with the permittivity of the medium in processes in which the TS is more dipolar than any of the reagents.

A further development led to Eqn (9) [39]:

$$\ln k = \ln k_\infty - \frac{2N_A}{4\pi\varepsilon_0 RT}\frac{\mu_A\mu_B}{\varepsilon_r r^3} \quad (9)$$

where k_∞ is the rate constant in a medium of infinite permittivity, and $r = r_A + r_B$ is the distance between the centers of both dipolar reagents, taken as spherical. Equation (7) involves plotting ln k versus $(\varepsilon_r - 1)/(2\varepsilon_r + 1)$, whereas Eqns (8) and (9) involve plotting ln k versus $(1/\varepsilon_r)$ to predict the effect of the polarity of the medium. In any case, proper knowledge of the dielectric properties of ILs is required. ε_r Is the most common way of measuring polarity. Conventional solvents with $\varepsilon_r < 9$ are typically considered nonpolar, those for which $9 < \varepsilon_r < 15$ are moderately polar, those for which $15 < \varepsilon_r < 30$ are regarded as polar, and if $\varepsilon_r > 50$ the solvent is highly polar. In the case of ILs, different techniques have been used to measure ε_r [13], finding values $9 < \varepsilon_r < 15$, that would allow to characterize ILs as moderately polar. Some $\varepsilon_r > 30$, and up to approximately 57, have been found for protic ILs [13]. For example, ε_r values have been estimated for various $[C_n\mathrm{mim}]^+$ ILs with counter-anions $[Cl]^-$, $[BF_4]^-$, $[PF_6]^-$, $[TfO]^-$, $[Tf_2N]^-$, and $[C_nOSO_3]^-$. ILs with a common anion showed an ε_r that correlated well with the molecular volumes. Both the nature of the anion and the length of the alkyl chain of the cation significantly contribute to ε_r [36].

- Reactions between neutral, dipolar molecules and ions. For a reaction

$$\mathrm{A}^{z_A} + \mathrm{B}^{z_B} \rightarrow \text{Prods}$$

the rate constant in a medium of I = 0 M ionic strength has been derived as Eqn (10) [40]:

$$\ln k = \ln k_0 + \frac{1}{4\pi\varepsilon_0}\left[\frac{N_A e^2}{2RT}\left(\frac{1}{\varepsilon_r} - 1\right)\left(\frac{z_A^2}{r_A} + \frac{z_B^2}{r_B} - \frac{(z_A + z_B)^2}{r^{\neq}}\right) + \frac{3N_A}{8RT}\left(\frac{2}{\varepsilon_r} - 1\right)\left(\frac{\mu_A^2}{r_A^3} + \frac{\mu_B^2}{r_B^3} - \frac{\mu^{\neq 2}}{r^{\neq 3}}\right)\right] \tag{10}$$

where k is the rate constant in a medium of relative permittivity ε_r and k_0 is the rate constant in a medium of $\varepsilon_r = 1$. If the reaction takes place between two dipoles, only the last term holds, but in a process between an ion and a dipole both terms need to be included. In general, ln k is linearly dependent on $(1/\varepsilon_r)$, with the obtained slope depending on radii and dipole moments. Quantitative deductions from Eqn (10) are restricted, because parameters $r^{\neq}$ and $\mu^{\neq}$ need to be estimated from the radii and characteristics of the reagents.

Alternatively, the rate constant has been derived as a function of ε_r for a reaction between an ion and a dipole ($A^{z_A} + B \rightarrow$ Prods) as in Eqn (11) [41]:

$$\ln k = \ln k_\infty + \frac{N_A z_A e \mu_B}{4\pi\varepsilon_0 RT \varepsilon_r r_{AB}^2} \tag{11}$$

where k_∞ is the rate constant in a medium of infinite permittivity and r_{AB} is the distance between the centers of the ion and the dipole, taken as spherical. From this expression, ln k is linearly dependent on $(1/\varepsilon_r)$, the obtained slope being positive for cations and negative for anions.

- Reactions between ions. These may be diffusion-controlled processes, where $\Delta G^{\neq} \approx 0$ and the rate depends exclusively on the diffusion of ions to get together (strictly, a physical process dependent on the viscosity of the medium), or reactions with covalent bond making (breaking). In this last case, the rate constant can be estimated as in Eqn (12) for an infinitely diluted solution:

$$\ln k = \ln k_0 + \frac{1}{4\pi\varepsilon_0}\frac{N_A e^2}{2RT}\left(\frac{1}{\varepsilon_r} - 1\right)\left(\frac{z_A^2}{r_A} + \frac{z_B^2}{r_B} - \frac{(z_A + z_B)^2}{r^{\neq}}\right) \tag{12}$$

4. EFFECT OF ILs ON THE ENERGETICS OF THE REACTING SYSTEM

ILs have been used mainly in two main different ways: as extraction media in gas–liquid or liquid–liquid equilibria, and as reaction media in different kinds of nonequilibrium chemical reactions. Both phenomena, equilibrium and change (reaction), are the two

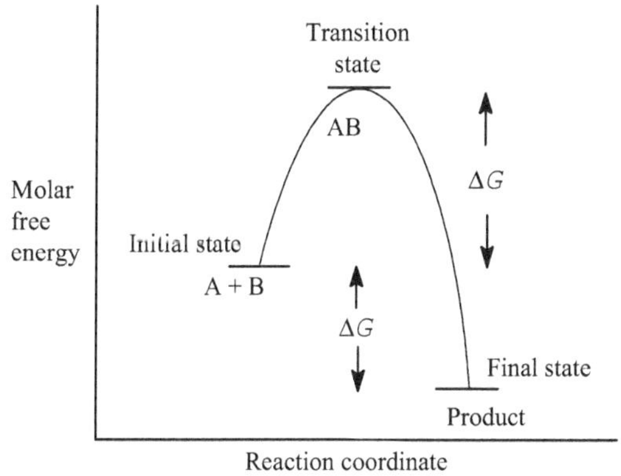

Figure 1 Relationship between equilibrium and change (reaction reactivity) for a bimolecular chemical reaction.

fundamental ways in which chemistry expresses itself, and their intrinsic relationship for a process taking place at constant pressure can be understood through Figure 1.

A system in an initial state (IS) is transformed into a final state (FS) through an intermediate TS, the difference in free energy between FS and IS defines the equilibrium, and the difference in energy between TS and IS accounts for the kinetics of the process (change). Figure 1 refers to a bimolecular reaction, but the corresponding energy profile for unimolecular or even termolecular reactions would be entirely similar. ΔG° might be either exoergonic (<0, spontaneous) or endoergonic (>0, not spontaneous), whereas $\Delta G^{\neq}$ will be endoergonic in all cases. The reaction rate for the process depends on the magnitude of $\Delta G^{\neq}$: large values imply low reaction rates, and small values imply high reaction rates. Also, as a general pattern, when reaction rates approach diffusion control [42], the selectivity of a process decreases as the reactivity increases [43].

Figure 2 shows, in a simplified manner, how the solvating capacity of ILs may influence reactivity (previously described): better solvating media reduce the free energy of both IS and TS, although not in the same proportion. As a consequence, a $\Delta\Delta G^{\neq}$ is generated, with a magnitude that depends on the differential stabilization of IS and TS. Therefore, one of the possible reasons for the observed effects of ILs on reactivity simply rests on alterations of free energy that can be easily determined through experimental measurements [21]. A similar way of reasoning leads to the conclusion that an alteration $\Delta\Delta G^{\circ}$ in the free energy of equilibrium should arise, which may also be experimentally determined.

Therefore, provided a set of enough data is available, the effects of solvation on both equilibrium and reactivity alterations in ILs could be predicted beforehand.

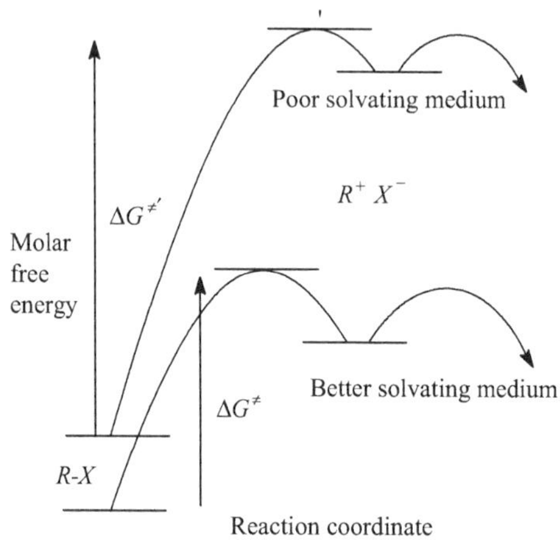

Figure 2 Effect of solvation changes on the energetics of a chemical reaction.

5. ACID/BASE BEHAVIOR

Acid–base processes are among the most important in chemistry, especially when dealing with protic solvents. Rarely, a reaction mechanism does not include some acid–base step. The acidity/basicity of compounds dissolved in ILs tends to follow a behavior much similar to the one observed in the gas phase, which can be easily estimated through computational chemistry [44,45]. Because the anion of the IL is usually chosen as a weak Brønsted and Lewis base, its conjugate acid is much more acidic than H_3O^+, and an acid dissolved in such an IL may have a lower pK_a and be less dissociated than in water [46]. In general, as a solvent, ILs can act both as Lewis and Brønsted acids, and this opens new reactive channels not always present in water.

It would be relevant to accurately know the pK_a of ILs when acting as Brønsted acids, but this has proved not to be an easy task [46].

A general approach to determine pK_as in ILs has been put forward, and the pK_as of 18 carbon-centered indicator acids in four ILs were determined [47]. A major problem is the fact that pK_as are defined for water, according to Eqns (13) and (14):

$$BH^+ + H_2O \rightleftarrows B + H_3O \tag{13}$$

$$K_a = \frac{a_B \cdot a_{H_3O^+}}{a_{BH^+}} \tag{14}$$

where a stands for activity, and the activity of water, a_{H_2O}, is not included as it is unity in its standard state.★ Obviously, this is not the case in ILs, and difficulties arise to establish

★ The standard state $a_{H_2O} = 1$ M for water as a solvent is an infinitely dilute solution of water in itself. It relates to the mole fraction scale, as opposed to the molarity scale, more convenient for solutes in reactivity studies.

comparisons with literature pK_as. However, a range of Brønsted acid ILs is available, covering a large range of the pK_a scale. For example, the zwitterionic ILs $[(C_nSO_3)C_1Im]$ and $[(C_nSO_3)Phe_3P]$ proved to be versatile solvent catalysts in different acid-promoted organic reactions: Fischer esterification, alcohol dehydrodimerization, and pinacol/benzopinacol rearrangement [48]. For example, the imidazolium ILs, $[(C_nCOOH)C_1Im]^+X^-$ ($n = 1$, three; $X = BF_4^-$, $CF_3SO_3^-$) and $[(C_nCOOH)_2Im]^+X^-$ ($n = 1$, three; $X = BF_4^-$, $CF_3SO_3^-$), exhibited extensive hydrogen bonding and an acidity constant $1.33 < pK_a < 4.59$ at 298 K [49]. The pK_as of $[C_nC_1Im]^+$, determined in dimethyl sulfoxide [50] and H_2O [51], were in the following range: $21 < pK_a < 24$. Thus, knowledge of the acidity of protic ILs has allowed control of the movement of protons between a metal center at a catalyst and amines, even at a second coordination sphere, using an IL with pK_a matching that of the catalyst [52].

It is also relevant to be able to estimate the acidity of reagents when dissolved in ILs. For example, in a specifically acid- (or base-) catalyzed process, the slope of a linear correlation of log k versus $pK_a(BH^+)$ Brønsted correlation is a measurement of the degree of proton transfer at the TS, allowing us to determine the kind of mechanism [53]. A general procedure to measure absolute pK_as in ILs has been developed. The pK_as of 18 carbon-centered acid indicators were determined in four different ILs: $[C_nC_1Im]^+[NTf_2]^-$, $[C_nC_1Im]^+[OTf]^-$, $[C_4C_1Pyrr]^+[NTf_2]^-$, and $[C_4C_1C_1Im]^+[NTf_2]^-$. This procedure allows direct comparison of relative acidities in ILs with those in conventional solvents, and even in the gas phase [47].

Quantum mechanic/molecular mechanic studies have allowed estimation of the reaction free energies and associated pK_as for proton transfer from protonated tryptophan ($HTrp^+$) to $[C_4C_1Im]^+[PF_6]^-$ and $[C_4C_1Im]^+[BF_4]^-$. pK_a values of 16.5 and 21.5 were found for proton transfer from $HTrp^+$ to PF_6^- and BF_4^- anions in $[C_4C_1Im]^+[PF_6]^-$ and $[C_4C_1Im]^+[BF_4]^-$, respectively, indicating a highly unfavorable deprotonation compared with aqueous solutions. Dissociation of the protonated anions into HF and BF_3 or PF_5 was also observed [54].

When the pK_a of a compound (Org) is not available in an IL, it can be estimated from the pK_a of the compound itself and the pK_a of the conjugate acid of the anion (An) of the IL, both in H_2O, according to Eqns (15)–(18):

$$\mathrm{OrgH^+ + H_2O \rightleftarrows Org + H_3O^+} \quad pK_a\left(\mathrm{OrgH^+}\right)_{aq} \tag{15}$$

$$\mathrm{AnH^+ + H_2O \rightleftarrows An + H_3O^+} \quad pK_a\left(\mathrm{AnH^+}\right)_{aq} \tag{16}$$

$$\mathrm{OrgH^+ + An \rightleftarrows Org + AnH^+} \quad pK_a\left(\mathrm{OrgH^+}\right)_{IL} \tag{17}$$

$$pK_a\left(\mathrm{OrgH^+}\right)_{IL} = \frac{pK_a\left(\mathrm{OrgH^+}\right)_{aq}}{pK_a\left(\mathrm{AnH^+}\right)_{aq}} \tag{18}$$

The values thus obtained are a rough approach, considering the interactions between IL species and the problem compound are similar than those in water [46].

It is interesting to consider the role of bases in ILs. Because of the intrinsic characteristics of ILs, where proton donors are generally absent, a base may remain in its unprotonated form or undergo protonation by other chemical species that are added to the medium. Base-catalytic effects have been observed, for example, in *O*-acetylation reactions of sugars and alcohols in $[C_nC_1Im]^+$ $[N(CN)_2]^-$ and in $[C_nC_1Im]^+$ $[CH_3COO]^-$, but not in $[C_4C_1Im]^+$ $[(CF_3SO_2)_2N]^-$, which the authors interpret as evidence for a general base catalysis process operating in the first two. This opens the possibility of accessing base-catalysis processes in ILs by tuning the basicity of the anion [46].

6. CONCLUSION

A quick overview of the physicochemical basis of the effects of ILs on chemical equilibria and reactivity shows a huge amount of effort is needed before processes in ILs are fully understood to a level similar to those in conventional solvents. It will be difficult to optimize, from a technological point of view, separation or transformation processes using ILs without a proper knowledge of underlying effects, such as those described herein.

ACKNOWLEDGMENT

We thank the *Xunta de Galicia*, Spain, for financial support through project CN2012/210 (*Rede Galega de Líquidos Iónicos*, ReGaLIs).

REFERENCES

[1] J.D. Holbrey, R.D. Rodgers, Green chemistry and ionic liquids: Synergies and Ironies, in: R.D. Rodgers, K.R. Seddon (Eds.), Ionic Liquids. Industrial Applications to Green Chemistry, American Chemical Society, Washington, 2002, pp. 2–14.
[2] S. Zhang, N. Sun, X. He, X. Lu, X. Zhang, Physical properties of ionic liquids: database and evaluation, J. Phys. Chem. Ref. Data 35 (2006) 1475.
[3] C.C. Weber, A.F. Masters, T. Maschmeyer, Structural features of ionic liquids: consequences for material preparation and organic reactivity, Green Chem. 15 (2013) 2655–2679.
[4] P.T. Anastas, J.C. Warner, Green Chemistry, Theory and Practice Oxford University Press, New York, 1998.
[5] V.I. Parvulescu, C. Hardacre, Catalysis in ionic liquids, Chem. Rev. 107 (2007) 2615–2665.
[6] I.M. AlNashef, M.A. Matthews, J.W. Weidner, Electrochemistry: electrochemically generated superoxide ion in ionic liquids: applications to Green chemistry, in: R.D. Rodgers, K.R. Seddon (Eds.), Ionic Liquids as Green Solvents. Progress and Prospects, American Chemical Society, Washington, 2003, pp. 509–525.
[7] C.M. Gordon, A.J. McLean, M.J. Muldoon, I.R. Dunkin, Photochemistry in ionic liquids, in: R.D. Rodgers, K.R. Seddon (Eds.), Ionic Liquids. Industrial Applications to Green Chemistry, American Chemical Society, Washington, 2002, pp. 428–443.

[8] P. Neta, D. Behar, J. Grodkowski, Pulse radiolysis studies of reaction kinetics in ionic liquids, in: R.D. Rodgers, K.R. Seddon (Eds.), Ionic Liquids as Green Solvents. Progress and Prospects, American Chemical Society, Washington, 2003, pp. 397–409.
[9] J.F. Wishart, Radiation chemistry of ionic liquids: reactivity of primary species, in: R.D. Rodgers, K.R. Seddon (Eds.), Ionic Liquids as Green Solvents. Progress and Prospects, American Chemical Society, Washington, 2003, pp. 381–396.
[10] H. Weingärtner, Understanding ionic liquids at the molecular level: facts, problems, and controversies, Angew. Chem. Int. Ed. 47 (2008) 654–670.
[11] C. Chiappe, M. Malvaldi, C.S. Pomelli, Ionic liquids: solvation ability and polarity, Pure Appl. Chem. 81 (2009) 767–776.
[12] C. Chiappe, D. Pieraccini, Ionic liquids: solvent properties and organic reactivity, J. Phys. Org. Chem. 18 (2005) 275–297.
[13] J.P. Hallett, T. Welton, Room-temperature ionic liquids: solvents for synthesis and catalysis. 2, Chem. Rev. 111 (2011) 3508–3576.
[14] H. Olivier-Bourbigou, L. Magna, D. Morvan, Ionic liquids and catalysis: recent progress from knowledge to applications, Appl. Catal. A 373 (2010) 1–56.
[15] A. Stark, Ionic Liquid Structure-induced Effects on Organic Reactions, Topics in Current Chemistry, Springer, Berlin, 2010, 41–81.
[16] T. Welton, Room-temperature ionic liquids. Solvents for synthesis and catalysis, Chem. Rev. 99 (1999) 2071–2083.
[17] ACS, http://www.cas.org/content/counter, Columbus, Ohio, 2014 (accessed 10.02.14).
[18] R.E. Ney, Where Did that Chemical Go? : A Practical Guide to Chemical Fate and Transport in the Environment, Van Nostrand Reinhold, New York, 1990.
[19] J. Ma, X. Hong, Application of ionic liquids in organic pollutants control, J. Environ. Manage. 99 (2012) 104–109.
[20] T.W. Bentley, Investigation of reaction mechanisms by product studies, in: H. Maskill (Ed.), The Investigation of Organic Reactions and their Mechanisms, Blackwell Publishing Ltd., Oxford, 2006, pp. 18–45.
[21] M.L. Canle, H. Maskill, J.A. Santaballa, Experimental methods for investigating kinetics, in: H. Maskill (Ed.), The Investigation of Organic Reactions and their Mechanisms, Blackwell Publishing Ltd., Oxford, 2006, pp. 46–64.
[22] J.A. Santaballa, H. Maskill, M. Canle L, The relationship betweeen mechanism and rate law, in: H. Maskill (Ed.), The Investigation of Organic Reactions and their Mechanisms, Blackwell Publishing Ltd., Oxford, 2006, pp. 79–103.
[23] P.R. Schreiner, Computational chemistry and the elucidation of mechanism, in: H. Maskill (Ed.), The Investigation of Organic Reactions and their Mechanisms, Blackwell Publishing Ltd., Oxford, 2006, pp. 167–197.
[24] C. Reichardt, Solvents and Solvent Effects in Organic Chemistry, third ed., Wiley-VCH Verlag, Weinheim, 2003.
[25] K.J. Laidler, Chemical Kinetics, third ed., Harper & Row, New York, U.S.A, 1987.
[26] E. Grunwald, S. Winstein, The correlation of solvolysis rates, J. Am. Chem. Soc. 70 (1948) 846–854.
[27] S.N.V.K. Aki, J.F. Brennecke, A. Samanta, How polar are room-temperature ionic liquids? Chem. Commun. (2001) 413–414.
[28] A.J. Carmichael, K.R. Seddon, Polarity study of some 1-alkyl-3-methylimidazolium ambient-temperature ionic liquids with the solvatochromic dye, Nile Red, J. Phys. Org. Chem. 13 (2000) 591–595.
[29] M.A.A. Rani, A. Brant, L. Crowhurst, A. Dolan, M. Lui, N.H. Hassan, J.P. Hallett, P.A. Hunt, H. Niedermeyer, J.M. Perez-Arlandis, M. Schrems, T. Welton, R. Wildinga, Understanding the polarity of ionic liquids, Phys. Chem. Chem. Phys. 13 (2011) 16831–16840.
[30] A. Cerda-Monje, A. Aizman, R.A. Tapia, C. Chiappe, R. Contreras, Solvent effects in ionic liquids: empirical linear energy–density relationships, Phys. Chem. Chem. Phys. 14 (2012) 10041–10049.
[31] M.J. Earle, K.R. Seddon, Ionic liquids. Green solvents for the future, Pure Appl. Chem. 72 (2000) 1391–1398.

[32] Y. Marcus, Ion Solvation, Wiley, Chichester, 1985.
[33] K.J. Fraser, E.I. Izgorodina, M. Forsyth, J.L. Scotta, D.R. MacFarlane, Liquids intermediate between "molecular" and "ionic" liquids: liquid ion pairs? Chem. Commun. (2007) 3817–3819.
[34] T. Köddermann, C. Wertz, A. Heintz, R. Ludwig, Ion-pair formation in the ionic liquid 1-ethyl-3-methylimidazolium Bis-(triflyl)imide as a function of temperature and concentration, ChemPhysChem. 7 (2006) 1944–1949.
[35] H. Luo, G.A. Baker, S. Dai, Isothermogravimetric determination of the enthalpies of vaporization of 1-alkyl-3-methylimidazolium ionic liquids, J. Phys. Chem. B 112 (2008) 10077–10081.
[36] T. Singh, A. Kumar, Static dielectric constant of room temperature ionic liquids: internal pressure and cohesive energy density approach, J. Phys. Chem. B 112 (2008) 12968–12972.
[37] J.G. Kirkwood, Theory of solutions of molecules containing widely separated charges, J. Chem. Phys. 2 (1934) 351.
[38] K.J. Laidler, P.A. Landskroener, The influence of the solvent on reaction rates, Trans. Faraday Soc. 52 (1956) 200–210.
[39] E.S. Amis, Coulomb's law and the quantitative interpretation of reaction rates, J. Chem. Educ. 29 (1952) 337.
[40] K.J. Laidler, Chemical Kinetics, third ed., Harper and Row, Hilversum, 1987.
[41] E.S. Amis, J.F. Hinton, Solvent Effects on Chemical Phenomena, vol. 1, Academic Press, London, 1973.
[42] H. Mayr, A.R. Ofial, The reactivity–selectivity principle: an imperishable myth in organic chemistry, Angew. Chem. Int. Ed. 45 (2006) 1844–1854.
[43] P. Müller, Glossary of terms used in physical organic chemistry (IUPAC recommendations 1994), Pure Appl. Chem. 66 (1994) 1077–1184.
[44] M. Canle, I. Demirtas, A. Freire, H. Maskill, M. Mishima, Base strengths of substituted tritylamines, N-alkylanilines, and tribenzylamine in aqueous solution and the gas phase: steric effects upon solvation and resonance interactions, Eur. J. Org. Chem. (2004) 5031–5039.
[45] J.Z. Davalos, J. Gonzalez, R. Ramos, D. Hnyk, J. Holub, J.A. Santaballa, M. Canle L., J.M. Oliva, Acidities of closo-1-COOH-1,7-C2B10H11 and Amino Acids Based on Icosahedral Carbaboranes, J. Phys. Chem. A 118 (2014) 2788–2793. http://dx.doi.org/10.1021/jp412400q.
[46] D.R. MacFarlane, S.A. Forsyth, Acids and bases in ionic liquids, in: R.D. Rodgers, K.R. Seddon (Eds.), Ionic Liquids as Green Solvents. Progress and Prospects, American Chemical Society, Washington, 2003, pp. 264–276.
[47] H. Deng, X. Li, Y. Chu, J. He, J.-P. Cheng, Standard pKa scales of carbon-centered indicator acids in ionic liquids: effect of media and structural implication, J. Org. Chem. 77 (2012) 7291–7298.
[48] A.C. Cole, J.L. Jensen, I. Ntai, K.L.T. Tran, K.J. Weaver, D.C. Forbes, J. James, H. Davis, Novel brønsted acidic ionic liquids and their use as dual solvent-catalysts, J. Am. Chem. Soc. 124 (2002) 5962–5963.
[49] Z. Fei, D. Zhao, T.J. Geldbach, R. Scopelliti, P.J. Dyson, Brønsted acidic ionic liquids and their zwitterions: synthesis, characterization and pKa determination, Chemistry 10 (2004) 4886–4893.
[50] R.W. Alder, P.R. Allen, S.J. Williams, Stable carbenes as strong bases, J. Chem. Soc. Chem. Commun. (1995) 1267–1268.
[51] T.L. Amyes, S.T. Diver, J.P. Richard, F.M. Rivas, K. Toth, Formation and stability of N-heterocyclic carbenes in water: the carbon acid pK_a of imidazolium cations in aqueous solution, J. Am. Chem. Soc. 126 (2004) 4366–4374.
[52] D.H. Pool, M.P. Stewart, M. O'Hagan, W.J. Shaw, J.A.S. Roberts, R.M. Bullock, D.L. DuBois, Acidic ionic liquid/water solution as both medium and proton source for electrocatalytic H_2 evolution by $Ni(P_2N_2)_2^{2+}$ complexes, Proc. Natl. Acad. Sci. 109 (2012) 15634–15639.
[53] H. Maskill, The Physical Basis of Organic Chemistry, Oxford University Press, Oxford, U.K., 1993.
[54] M. Klähn, A. Seduraman, P. Wu, Proton transfer between tryptophan and ionic liquid solvents studied with molecular dynamics simulations, J. Phys. Chem. B 115 (2011) 8231–8241.

CHAPTER 3

Liquid-Phase Extraction and Microextraction

Elena Fernández, Lorena Vidal
Departamento de Química Analítica, Nutrición y Bromatología e Instituto Universitario de Materiales, Universidad de Alicante, Alicante, Spain

1. INTRODUCTION

The most common sentence that scientists read in bibliography about ionic liquids (ILs) in analytical chemistry is that "*Ionic liquids are a group of organic salts with melting points below 100 °C. They possess unique properties as their immeasurable low vapour pressure, good chemical and thermal stability, nonflammability, high ionic conductivity, and a wide electrochemical potential window. These properties make them widely used in every chemistry field. The negligible volatility and the adjustable hydrophobicity, polarity and selectivity propose them as excellent* "green" *solvents to replace harmful organic solvents, mainly in extraction and separation techniques.*"

The green aspect is highlighted in any publication related to ILs and analytical chemistry; however, some studies prove the toxicity of some ILs for the aqueous media [1,2]. This aspect is also described by Patel et al. [3], confirming that toxicity has been proved, but in addition, to achieve a sustainable procedure, the impacts on human health, environment, economy, and society should be also investigated. Studies about the perspectives of ILs as environmentally benign [4] or assessing the greenness of the synthesis of ILs compared to that of organic molecular solvents [5] can be found in the literature. As a result, the synthesis of ILs presents higher environmental impact than the synthesis of organic solvents, an important aspect to be considered. Other important aspect is the hydrofluoric acid generated in the decomposition of ILs containing halogenated compounds [6,7] making syringes unusable, or the shortened life of the liquid chromatography (LC) columns due to high viscosity of ILs. Therefore, at this point, we should wonder why ILs are used as solvents for extraction. First, the final application of the method determines the global environmental impact of the procedure; for example, if gas chromatography–mass spectrometry (GC-MS) is selected as the analysis technique, its use likely presents higher environmental impact than the few microliters of IL used in the liquid-phase microextraction (LPME) technique. Another property of ILs is their negligible vapor pressure, thereby minimizing environmental contamination through evaporation, and thus being safer for the operator, and facilitating recycling compared with volatile organic solvents. Furthermore, from the extraction point of view, the use of ILs allows bigger droplet volumes and higher stirring speeds, resulting in higher

Ionic Liquids in Separation Technology
http://dx.doi.org/10.1016/B978-0-444-63257-9.00003-1
Copyright © 2014 Elsevier B.V.
All rights reserved.

extraction and therefore better analytical parameters. In addition, ILs are called "designer" solvents because of the 10^{18} possible combinations [8] that provides ILs with different properties designed for many applications. These different properties enable the extraction of organic compounds, polar and nonpolar, and metals [3,7,9–28]. From an analytical point of view, the use of ILs in extraction techniques provides more benefits than harms, and additional, larger studies are needed before firm conclusions can be drawn. However, more criticism should be shown when the green aspect of a newly developed method or technique using ILs is highlighted.

As described above, the term "ionic liquid" is used to describe a broad class of salts with a melting point below 100 °C, and room temperature ionic liquids (RTILs) describe a subset of ILs that are liquid at room temperature (25 °C) [16]. The first IL discovered was ethanolammonium nitrate (melting point of 52–55 °C) reported by Gabriel and Weiner in 1888 [29]. The first RTIL discovered was ethylammonium nitrate (melting point approximately 13–14 °C) reported by Paul Walden in 1914 [30]. However, the first air- and moisture-stable imidazolium salts were developed by Wilkes and Zaworotko in 1992, who synthesized stable RTILs containing weakly complexing anions, such as tetrafluoroborate ([BF_4]) [31]. Since then, the number of publications on ILs has increased exponentially as is shown in Figure 1(A). The marriage of ILs and extraction techniques was perfect because of the immiscibility of some ILs with water, allowing the formation of biphasic systems, and the high solubility for organic and inorganic compounds. In the late 1990s, Huddleston et al. [32] successfully used for first time ILs as solvents for liquid–liquid extraction (LLE) of substituted benzene derivatives. At the same time that ILs started spreading into different fields of science, LPME techniques appeared in 1996 [33,34], and because of their well-known, powerful properties, the number of publications has increased since then (Figure 1(B)). The synergic effect of combining ILs and LPME techniques was confirmed by the increasing number of publications in the last decade (Figure 1(C)), being almost half of the LPME publications. Another indicator of the interest awakened by the scientific community is the numerous excellent reviews about ILs involving extraction techniques that have been published since 2005 [3,7,9–28]. Specific reviews of ILs have focused on analytical chemistry [9,13,15,16,22,25]; general applications [3,10,12]; sample preparation [7,14]; extraction and microextraction techniques [11,17,21,23,26]; organic compounds [18]; total-metal and metal-speciation analysis [19]; bioactive compounds [24], and environmental [27], biological [28], and food samples [20,27].

Returning to the aspect of "designer solvents" and the 10^{18} possible combinations [8], it is important to point out that although this property is highlighted in almost every publication related to ILs and extraction techniques, the most common ILs used for LLEs and LPMEs are those few shown in Figure 2. The common cations are imidazolium, pyridinium, pyrrolidinium, tetraalkylammonium, and tetraalkylphosphonium, and the most common anions (inorganic and organic) are halides, [BF_4], hexafluorophosphate ([PF_6]), trifluoroacetate, trifluoromethanesulfonate, tris(pentafluoroethyl)trifluorophosphate

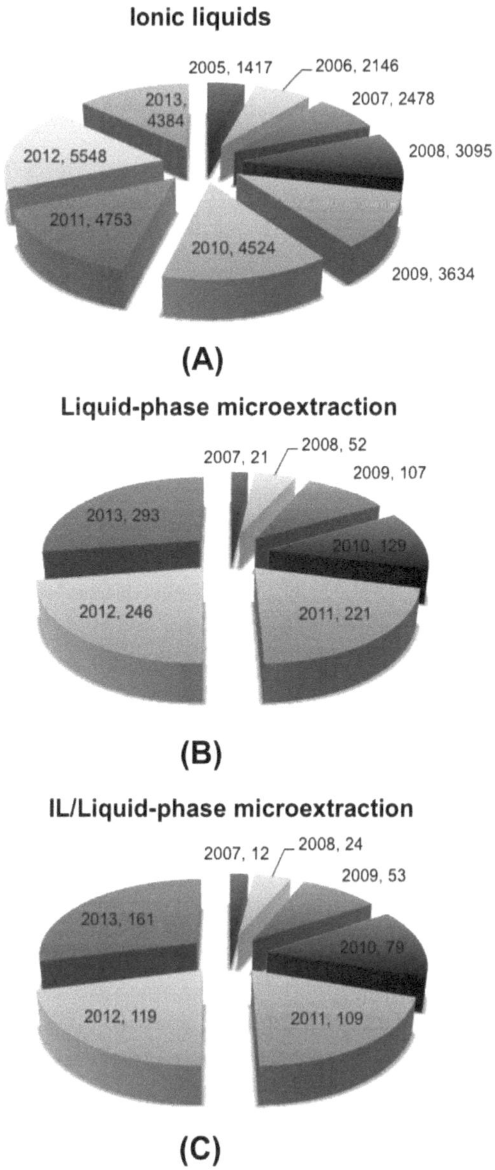

Figure 1 ***Number of studies published last years related with ILs and LPME. (A) Keyword: Ionic liquids; (B) keyword: Liquid-phase microextraction; (C) Keywords: Ionic liquids and liquid-phase microextraction.*** *(SciFinder® 2013.)*

([FAP]), and bis[(trifluoromethyl)sulfonyl]imide ([NTf_2]). ILs with halide anions present higher water solubility than the other anions and are used mainly for LLE or with the in situ IL dispersive liquid–liquid microextraction technique (in situ IL-DLLME) [35,36]. The ILs primarily and most used for microextraction techniques are those of the imidazolium family with [BF_4] and [PF_6] anions [32] due to their low solubility in water and hydrophobicity values, making these ILs excellent for the extraction of

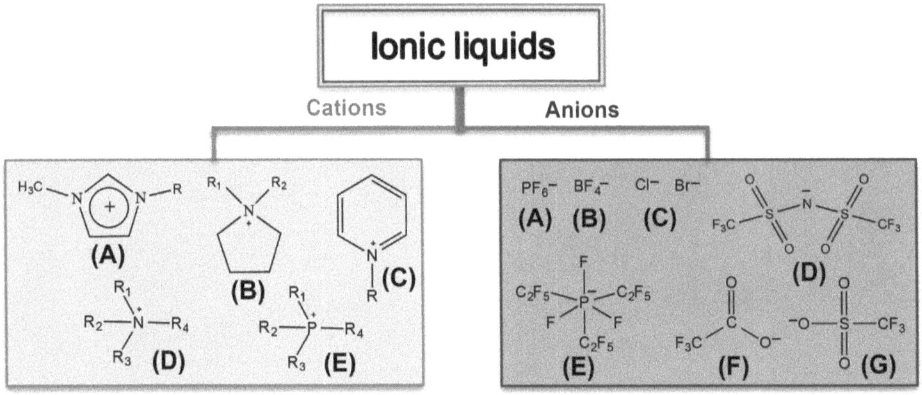

Figure 2 ***Ionic liquids commonly used for LLE and LPME. Cations: (A) imidazolium; (B) pyrrolidinium; (C) pyridinium; (D) tetraalkylammonium; (E) tetraalkylphosphonium. Anions: (A) hexafluorophosphate; (B) tetrafluoroborate: (C) halides; (D) NTf$_2$; (E) FAP; (F) trifluoroacetate; (G) OTf.***

organic compounds [32] and complexed metals [37]. However, when longer alkyl chains of the cation than 1-butyl-3-methylimidazolium ([C_4mim]) are selected, such as 1-hexyl-3-methylimidazolium ([C_6mim]) and 1-methyl-3-octylimidazolium ([C_8mim]), the viscosity value increases; this is optimum for the single-drop microextraction (SDME) technique but makes filling of the syringe difficult and broadens peaks in LC or capillary electrophoresis (CE). ILs with the [NTf_2] anion were introduced in late 1900s [37]. Imidazolium ILs with [NTf_2] present higher hydrophobicity and lower viscosity than the [BF_4] and [PF_6] analogs, thereby avoiding dissolution problems in large volumes or long-term extractions and providing higher enrichment factors. In addition, the lower viscosity provides better conductivity for an electrochemical detection [38]. Yao et al. [39] proved that the extraction ability of [FAP]-based ILs that are significantly more hydrophobic and hydrolytically stable permitted them to be used in the sampling of large volumes of aqueous solutions without dissolution or loss of the IL. In that study, different cations were tested with the [FAP] anion, and the highest enrichment factors were obtained with the trihexyl(tetradecyl)phosphonium [FAP] ([$C_{14}C_6C_6C_6P$][FAP]) IL for compounds with high molecular weight and fused rings, whereas high enrichment factors were obtained for smaller, more polar molecules using the [C_6mim][FAP] IL [39].

The negligible volatility can be considered one of the most important advantages to implementing ILs as solvents in extraction techniques. However, this advantage can be a drawback from the analysis point of view. The compatibility of ILs with LC, CE, and spectroscopic techniques (ultraviolet/visible [UV/Vis] spectrophotometry, spectrofluorimetry, electrothermal or flame atomic absorption spectrometry, and inductively coupled plasma-optical emission spectroscopy [ICP-OES] or ICP-mass spectrometry

[ICP-MS]) made these systems optimum for the analysis. Nevertheless, the low vapor pressure limited their direct analysis by GC, and different systems were developed for that purpose [40–43]. The coupling of IL with GC not only improves analytical parameters (i.e., limits of detection) but also avoids the interference of organic solvent wide peaks, allowing the determination of a wide range of analytes with different polarities and boiling points [17]. The first system designed by Aguilera-Herrador et al. [40] in 2008 proposed a removable interface that enabled the introduction of the extracted analytes, contained in IL, into the GC, while preventing the IL from entering the column (Figure 3). The system developed by Zhao et al. [41] proposed the exposition of the IL drop in the injection port, and then it was retracted into the syringe. For this system an improvement in the injection system was required and the upper diameter of the split inlet liner of the GC instrument had

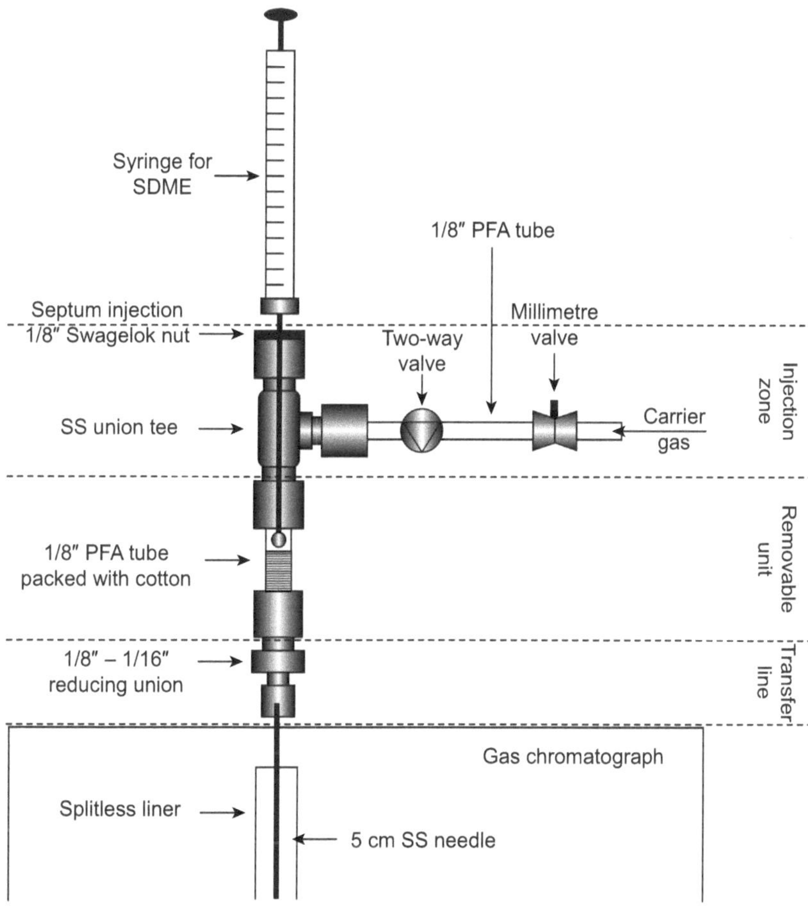

Figure 3 ***Schematic diagram of the interface developed for the direct introduction of ionic liquid containing the extracted analytes. SDME, single-drop microextraction; PFA, perfluoroalkoxi; SS, stainless steel.*** *(Reprinted with permission from Ref. [40]. Copyright (2008) American Chemical Society.)*

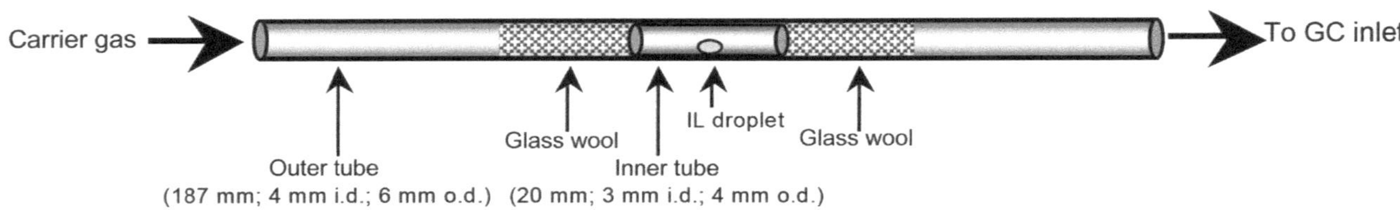

Figure 4 ***Scheme of the thermal desorption device designed for the direct use of IL-SDME-GC. o.d., outer diameter; i.d., inner diameter.*** *(Reprinted from Ref. [43], Copyright (2009), with permission from Elsevier.)*

to be enlarged [41]. Other modification was carried out by the same research group by placing a small glassy tube in the sample-injection part of the GC to prevent the IL from entering the column [42]. An external option for direct coupling was proposed by Chisvert et al. [43]. The approach was based on thermal desorption of the analytes from the IL droplet to the GC system by using a robust and commercially available thermodesorption system. For this purpose, a two-glass tube concentrically disposed system was designed (Figure 4).

The majority of the conventional detection systems (i.e., LC-UV/MS, GC-MS, or ICP-OES/MS) are slow, expensive, and bulky, so analytical instrumentation used for detection has not achieved miniaturization to the same extent as sample preparation methods. Furthermore, ILs have some disadvantages when chromatographic techniques are used. For example, special devices are needed when ILs are injected in GC, as just reviewed, and shorter column life and resolution problems are challenges for LC with ILs. For this reason, Fernández et al. [38] have proposed the use of electrochemical sensors as an attractive detection option for IL-LPME techniques. Screen-printed electrodes (SPELs) [44] are inexpensive, mass-produced, disposable devices that are ideal for low-volume sample analysis. The combination of IL-LPME with SPLE (Figure 5) is

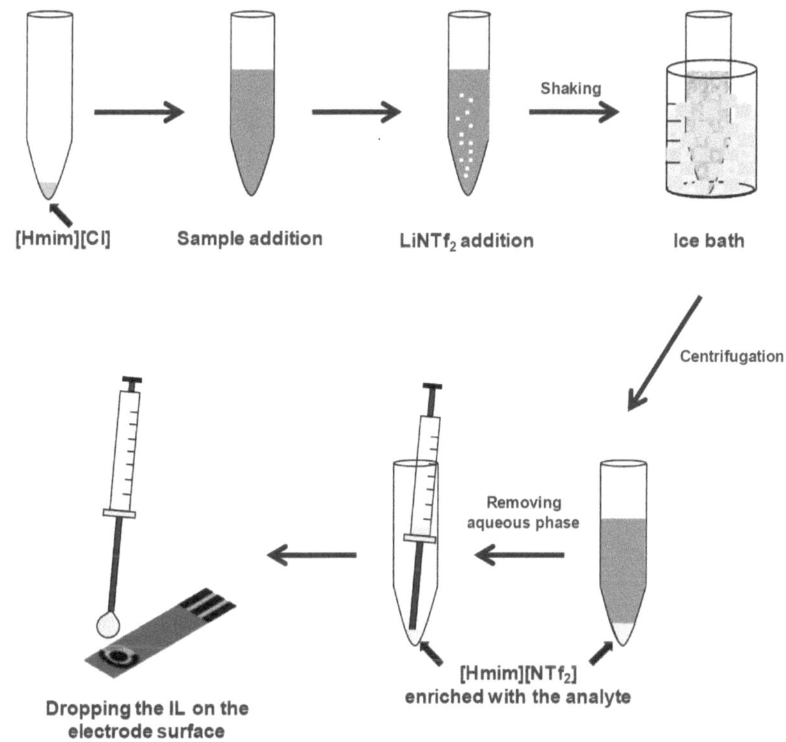

Figure 5 ***In situ IL-DLLME coupled with screen-printed graphite electrodes.*** *(Reprinted from Ref. [38] with kind permission from Springer Science and Business Media. Copyright (2013) Springer.)*

presented as a step forward in the development of portable and economical systems available to any laboratory.

As an overview, the latest trends on ILs and extraction techniques have been the use of different ILs to improve LPME techniques [39], the development of new techniques such as in situ IL-DLLME [35,36], and the coupling of LPME with new detection systems [38]. However, almost all the ILs used in extraction techniques until now contain halide, mainly fluoride atoms, in the anion and are thus toxic for the aquatic media [1,2]. Efforts for the synthesis of more environmental benign ILs have been carried out [45–47], but just a few have been used as solvents for extraction techniques. Therefore, new research studies should address developing methods where environmental friendly ILs are used and where the slow, bulky, and expensive instrumentation is replaced by portable systems that allow measurement on site.

This chapter focuses on the recent efforts of ILs in LLE and LPME techniques since the extensive volume of scientific literature involving ILs. For this reason, referring all the publications has been a difficult task, and only the more representative and recent reports were selected.

2. LIQUID–LIQUID EXTRACTION

Nowadays, there are many sample preparation methods available for cleanup and concentration of samples; however, the LLE technique is still one of the most widely used and just replaced in the last years by its miniaturized techniques. One of the main drawbacks of this technique is the large amount of toxic organic solvents needed, and for this reason, ILs were first proposed to replace such organic solvents in LLE. In 1998 Huddleston et al. [32] used $[C_4C_1Im][PF_6]$ as a solvent for LLE to study the partitioning of simple, substitutive benzene derivatives in ILs, comparing the results with the octanol–water partitioning coefficient. This study demonstrated that those substances with charged groups or strong hydrogen-bonding moieties were extracted to a lesser extent than similar neutral or nonpolar species. However, the extraction of charged groups was modulated with pH adjustment [32,48]. The first LLE of metals was carried out by Dai et al. [37] in 1999, where strontium nitrate was successfully removed from aqueous phase with a crown ether dissolved in different ILs. The distribution coefficients for the IL extraction systems were several orders of magnitude better than those observed for extraction systems based on organic solvents [37]. In contrast, the partition coefficients of 40 compounds with various functionalities (including organic acids, organic bases, amino acids, antioxidants, and neutral compounds) between $[C_4C1Im][PF_6]$ and water, and between $[C_4C1Im][PF_6]$ and heptane, were reported by Carda-Broch et al. [49].

Many organic and inorganic compounds, small and big molecules, polar and nonpolar compounds, and charged and noncharged compounds have been successfully extracted by IL-LLE since then. As a representative example of the wide range of extraction of ILs, small molecules such as ethanol [50] and lactic acid [51]; nanoparticles and nanorods [52]; quantum dots [53]; flame retardants [54]; or simple and complex biomolecules such as amino acids [55–57], penicillin G [58], cytochrome *c* [59,60], double-stranded DNA [61], and hemoglobin [62] have been successfully extracted. Regarding to sample matrices, numerous simple and complex matrices have been studied during these 15 years, including environmental [27], food [20,27], and biological [28] samples as the most tested. However, other matrices such as fuel and gasoline [63,64] have been also investigated.

Among the different ILs, two groups of ILs optimum for extraction can be distinguished, namely, the RTILs with melting points around 25 °C and a special class named task-specific ionic liquids (TSILs). From RTILs, many combinations are possible, but the imidazolium family cations with [BF_4], [PF_6], [NTf_2], and [FAP] anions are the most widely used. Nevertheless, other polyaromatic quinolinium-based ILs [64] and magnetic ILs [65] used as solvents for extraction can be found in the literature. TSILs comprise ILs that can be designed and synthesized based on specific requirements, a key point to developing ILs for task-specific extraction [66,67]. These ILs can act as both solvents and extractants containing amino, hydroxyl, thioether, thiourea, carboxylic, and olefin groups. TSILs with thiourea, thioether, and urea groups were first developed by Visser et al. for the extraction of Hg^{2+} and Cd^{2+} from aqueous solutions [66,67]. Significant increases in the extraction efficiency were observed for these two metal ions. Hg^{2+} was also extracted by another kind of TSILs that contained a polyether-linked bis-imidazolium cation [68]. Americium ions were extracted by an IL bearing 2-hydroxbenzylamine units [69]. Alternatively, organic compounds have also been extracted by TSILs such as the lysine-rich protein cytochrome *c* that was quantitatively extracted into hydroxyl group–containing ILs with dicyclohexano-18-crown-6 (DCH18C6) through the formation of supramolecular complexation [59]. The same group developed a new task-specific IL ([$18C_6C_1Im$][PF_6]) by attaching the crown-ether moiety to the imidazolium cation and a hydroxyethyl group–containing IL ([C_2OHC_1Im][NTf_2]). The mixture prepared by dissolving [$18C_6C_1Im$][PF_6] in [C_2OHC_1Im][NTf_2] provided similar extraction behavior to that using DCH18C6 as the extractant [60]. Functional amino acid ILs were used as solvent and selector in chiral LLE of the L-enantiomer of the amino acid from a mixture of L-enantiomer and D-enantiomer in ethyl acetate [70], and for selective extraction of α-tocopherol [71].

As described in the Introduction, the immiscibility of some ILs with water enables the formation of biphasic systems; therefore, ILs have been used as extractant solvents in LLE and LPME techniques. For this reason, the ILs mainly used in those techniques are highly hydrophobic. However, hydrophilic ILs have been used in the named aqueous two-phase systems (ATPSs). ATPSs are defined as systems usually formed by two incompatible polymers or one polymer and one salt above a certain concentration [72,73]. However, combinations of ILs with inorganic salts are able to form ATPSs [74].

The IL-based ATPS systems have the advantage of lower price, more ILs available, and lower viscosity over the biphasic systems using hydrophobic ILs [72]. Imidazolium, pyridinium, tetraalkylammonium, and tetraalkylphosphonium are used as cations, and halides (Cl, Br), [BF_4], and acetate, among others, are used as anions. Gutowski et al. [74] reported for the first time ATPSs formed by some hydrophilic ILs in the presence of inorganic salts that were used for the extraction of small organic molecules (e.g., methanol, propanol, butanol, pentanol). Generally, IL-based ATPS have been particularly suitable for the extraction of biomolecules such as proteins [75–77], penicillin G [78–80], testosterone [81], and codeine and papaverine [82].

The use of ILs as solvents for the LLE technique not only provide lower environmental contamination through evaporation and wider extraction range than conventional organic solvents, but also the reduction of extractant volume solvent (ILs) up to 1 mL or even lower (e.g., 500 μL). Hence, the generation of significant amounts of pollutants is being reduced, but some of its drawbacks still remain, such as manual operation or slow extraction times, what turns LPME into the foremost techniques of sample preparation. This fact can be seen comparing Table 1 that shows some analytical applications of IL-LLE with the number of studies about LPME published in the past three years (Tables 2–7).

3. LIQUID-PHASE MICROEXTRACTION

Liquid-phase microextraction can be defined as a miniaturization of LLE technique where the volume of the extractant phase is equal or below 100 μL [26]. The term solvent microextraction (SME) also can be found in the literature for the same purpose [26]. The main advantages of the miniaturized extraction techniques are the ease of operation, low cost, small amount of sample required, extremely low or even no solvent consumption, high enrichment factors, and availability to any laboratory [26,193–196]. In addition, the use of ILs as the extractant phase in LPME has helped to overcome problems associated with LPME techniques using organic solvents (i.e., stability, nonvolatility, and reproducibility) [17]; allowed extraction of a wider range of analytes; and enabled the development of new methods, including temperature-controlled IL dispersive liquid–liquid microextraction (TC-IL-DLLME) [197] and in situ IL-DLLME [35,36]. Numerous developments and applications about ILs and LPME techniques have appeared in the past 10 years. This fact is reflected in the numerous and excellent reviews recently published [16,17,19,22,23,25,27,28]. However and according to Kokosa [26], there is a negative fact because new names with new acronyms appear when each scientist tends to vary a general method. Therefore, LPME literature contains a confusing array of more than 100 acronyms [26]. Hence, classifying all the developed methods into a few groups is a difficult task. Therefore, in this chapter LPME techniques have been classified in three main groups: (1) Single-drop microextraction (SDME), (2)

Table 1 Representative Analytical Applications of *IL-LLE*

Analyte(s)	Sample(s)	IL (Amount)	Extraction Time	Comments	Analytical Technique	Coefficient of Variation (Spiking Level)	Limit of Detection	Ref.
Amino acids	Pharmaceutical samples and fermentation broth	$[C_4C_1Im][PF_6]$ (1 mL)	15 min	Crown ether to bind ammonium group and enhance the extraction. Back-extraction into an aqueous phase. Fluorescence determination after derivatization with *o*-phthalaldehyde. Study the effect of foreign substances.	Fluorimetry	1–7 (4.99–112 mg L^{-1})	$8.51 \cdot 10^{-7}$–$1.46 \cdot 10^{-6}$ mol L^{-1}	[55]
Phenols	Water	$[C_4C_1Im][PF_6]$ (1 mL)	30 min	Vortex agitation. Application of IL as both extractant and electrolyte.	Linear scan voltammetry	—	—	[83]
Al	Dialysis concentrates	$[C_4tmsIm][PF_6]$ (7.5 mL)	—	DTBSF as chelating agent. Study the effect of coexisting ions.	UV spectrophotometry	1.2 (4.0 μg L^{-1})	0.06 μg L^{-1}	[84]
Hg	Water	$[C_4C_1Im][PF_6]$ (0.7 g)	2 min	Flow injection-cold vapor atomic absorption spectrometry. Hg^{2+} was extracted as Hg-2-(5-bromo-2-pyridylazo)-5-diethylaminophenol. Back-extraction into an aqueous phase.	CV-AAS	2.8 (1 μg L^{-1})	2.3 ng L^{-1}	[85]
Cd	Plastic food packaging materials	$[C_4C_1Im][PF_6]$ (0.7 g)	5 s	On-line system. 2-(5-bromo-2-pyridylazo)-5-diethylaminophenol was used as complexant. The RTIL rich-phase was retained in a microcolumn filled with silica gel and then eluted with ethanol (0.5 mol L^{-1} nitric acid).	FAAS	3.9 (10 μg L^{-1})	6 ng g^{-1}	[86]
Ni	Water, rice flour, and black tea	$[C_4C_1Im][PF_6]$ (0.5 mL)	15 min	Conventional digestion of the samples. 1-(2-Pyridylazo)-2-naphthol as complexant agent was dissolved in IL. Back-extraction into an aqueous phase. Two reference materials were analyzed.	FAAS	3.1 (50 μg L^{-1})	12.5 μg L^{-1}	[87]

(*Continued*)

Table 1 Representative Analytical Applications of *IL-LLE*—cont'd

Analyte(s)	Sample(s)	IL (Amount)	Extraction Time	Comments	Analytical Technique	Coefficient of Variation (Spiking Level)	Limit of Detection	Ref.
Zn, Cd, Cu, and Fe	Hydrochloride solutions	[MTOA][Cl] [C_8C_1Im] [BF_4] (1 mL)	~35 min	Absence of chelating agents. Six different ILs [MTOA][Cl], [C_8C_1Im][BF4], [C_8C_1Im][PF6], [C_8C_1Im][NTf_2], [C_4C_1Im][PF_6], and [C_4C_1Im][NTf_2] were studied.	AAS	≤3 (–)	–	[88,89]
Rutin and quercetin	Herbal medicines	[C_4C_1Im] [Cl]	5 min	Herbal medicines were previously dried in a vacuum oven. Pressurized liquid extraction.	LC-CD	2.6 (–)	$1.1 \cdot 10^{-2}$ and $3.8 \cdot 10^{-3}$ mg L^{-1}	[90]
Tanshinone I, cryptotanshinone, and tanshinone IIA	Roots	[C_6C_1Im][Cl] (10 mL, 2 mol L^{-1})	3 min	Roots were previously dried. On-line microwave-assisted extraction. Single-factor and multivariate optimization. Comparison with reference extraction methods.	LC-DAD	0.63–2.16 (–)	0.009–0.014 mg g^{-1}	[91]
Anthraquinones	Aloe leaves	[C_4C_1Im] [BF_4] (1 mL)	10 min	IL/salt aqueous two-phase extraction. Use of Na_2SO_4 salt. Recycled IL via reverse extraction. Study of interferences.	UV/Vis spectrophotometry	2.4 (–)	0.0038 mg mL^{-1}	[92]
Sulfonamides	Water, milk, honey, and meat	[C_4C_1Im] [BF_4] (1 mL)	~12 h	IL/salt aqueous two phase extraction system. Use of $(NH_4)_2SO_4$ salt. One variable at time and multivariate optimizations. Food samples previously deproteinized.	LC-UV/Vis	2.52–3.18 (–)	0.15–0.3 ng mL^{-1}	[93]
Au and Pt	Water	[C_8dC_1Im] [NTf_2] [C_8C_1Im] [NTf_2] (0.75–0.8 g)	–	Au(III) was extracted as an $AuCl_4^-$. Pt(IV) was extracted as $PtCl_6^{2-}$ or $Pt(SCN)_6^{2-}$.	UV/Vis spectrophotometry	–	$5.0 \cdot 10^{-6}$ mol L^{-1}	[94]

Table 2 Overview of Recent *IL-SDME* Methods

Analyte(s)	Sample(s)	IL (Amount)	Mode	Extraction Time (min)	Comments	Analytical Technique	EF/Coefficient of Variation (Spiking Level)	Limit of Detection	Ref.
Trimethylamine	Fish	$[C_6C_1Im][PF_6]$ (20 μL)	HS-SDME	2	QD dispersed in the IL for fluorimetric detection. Sample heated at 50–60 °C.	Fluorometry	−/3.5 (0.5 mg L^{-1})	0.014 mg L^{-1}	[95]
Colchicine	Lanzhou lily	$[C_4C_1Im][PF_6]$ (2.4 nL)	DI-SDME	10	On-line coupled SDME with CE. Analysis of the aqueous extract of Lanzhou lily.	CE-UV/Vis	41/6.4 (5 μg mL^{-1})	0.25 μg mL^{-1}	[96]
Cu	Water, tea and defatted milk	$[C_4C_1Im][PF_6]$ (20 μL)	DI-SDME	10	DDTC as chelating agent. Study of the effect of coexisting ions. Previous digestion of tea and milk samples.	UV/Vis	33/3.4 (20 μg L^{-1})	0.15 μg L^{-1}	[97]
Escherichia coli, *Serratia marcescens*	Water	$[C_4C_1Im][PF_6]$ (1 μL)	DI-SDME	10	Platinum nanoparticles within IL enhanced extraction efficiency. IL as MALDI MS matrix. Detection of biomarker protein peaks in 10^6 cfu mL^{-1} bacterial concentration.	MALDI-TOF/MS	–	–	[98]
2,4,6-Tricholoroanisole	Water, wine	$[C_6C_1Im][NTf_2]$ (2 μL)	HS-SDME	30	Sample heated at 30 °C.	IMS	−/1.4 (10 ng L^{-1})	0.2 ng L^{-1}	[99]
2,4,6-Tricholoroanisole	Wine	$[C_6C_1Im][NTf_2]$ (2 μL)	HS-SDME	30	Sample heated at 30 °C.	MCC-IMS	−/<6 (1 ng L^{-1})	0.01 ng L^{-1}	[100]
Methylcyclopentadienyl-manganese tricarbonyl	Water, gasoline	$[C_4C_1Im][PF_6]$ (2.5 μL)	HS-SDME	30	Sample heated at 85 °C.	ETAAS	124/4.8 (5 μg L^{-1})	10 ng L^{-1}	[101]
Camphor, *trans*-anethole	Liquorice tablets	$[C_4C_1Im][PF_6]$ (12 μL)	HS-SDME	25	Sample heated at 80 °C. Comparison with USA extraction using n-hexane as extractant.	LC-UV/Vis	−/2.7, 3.4 (−)	9.77, 0.0195 μg L^{-1}	[102]

(Continued)

Table 2 Overview of Recent *IL-SDME* Methods—cont'd

Analyte(s)	Sample(s)	IL (Amount)	Mode	Extraction Time (min)	Comments	Analytical Technique	EF/Coefficient of Variation (Spiking Level)	Limit of Detection	Ref.
Sulfonamides	Water	$[C_8C_1Im][PF_6]$ (10 μL)	DI-SDME	20	–	LC-VWD	–/2.8–9.9 (5, 50 ng mL^{-1})	0.5–2 ng mL^{-1}	[103]
Volatile and semivolatile compounds	Seeds	$[C_8C_1Im][PF_6]$ (1.5 μL)	HS-SDME	3	Sample pretreatment extracting essential oil from seeds. Reflux for 30 min before drop exposition.	GC-MS	–/1.1–3.5 (–)	–	[104]
Musk fragrances	Water	$[C_8C_1Im][PF_6]$ (1 μL)	HS-SDME	45	Sample heated at 60 °C.	GC-MS/MS	/2-6 (1 ng mL^{-1})	0.010–0.030 ng mL^{-1}	[105]
Musk fragrances	Sewage Sludge	$[C_8C_1Im][PF_6]$ (1 μL)	HS-SDME	45	Lyophilization and pressurized liquid extraction of sludges. Sample heated at 60 °C.	GC-MS/MS	–/3–7 (10 ng g^{-1})	1–3 ng g^{-1}	[106]
Flavor compounds	Fruit juice	$[C_6C_1Im][BF_4]$ (12.5 μL)	HS-SDME	5	Previous centrifugation of fruit juice and dilution with water. Heating at 80 °C and ultrasonic nebulization for 20 min.	GC-MS	–/– (–)	0.0044–0.099 μg mL^{-1}	[107]
Cd	Water, rice	$[C_4C_1Im][PF_6]$ (20 μL)	DI-SDME	10	APDC as chelating agent.	W-coil ETAAS	33/5.2 (0.5 μg L^{-1})	0.015 μg L^{-1}	[108]

Table 3 Overview of Recent *IL-HF-LPME* Methods

Analytes	Sample	Mode	SLM	Acceptor Phase	Extraction Time	Comments	Analytical Technique	EF/Coefficient of Variation (Spiking Level)	Limit of Detection	Ref.
Chlorophenols	Water	Three phase	$[C_8C_1Im][PF_6]$	NaOH solution	60 min	–	LC-VWD	–/4.3–5.9 (5 μg L^{-1})	0.5–1 μg L^{-1}	[109]
Aliphatic and aromatic hydrocarbons	Water	Three phase	$[C_4C_1Im][PF_6]$	Toluene	40 min	Comparison with two-phase HF-LPME and SPME.	GC-MS	53–210/4–11 (5 μg L^{-1})	1–7 ng L^{-1}	[110]
Pb and Ni	Water and oyster tissue	Two phase	$[C_6C_1Im][PF_6]$	$[C_6C_1Im][PF_6]$	15, 12 min	APDC as chelating agent. Study of the effect of coexisting ions.	ETAAS	76, 60/5, 4.2 (0.5 μg L^{-1})	0.02, 0.03 μg L^{-1}	[111]
Sulfonamides	Water	Three phase	$[C_8C_1Im][PF_6]$	NaOH solution	8 h	SBME configuration.	LC-UV	58–135/0.5–4.8 (5 μg L^{-1})	0.1–0.4 μg L^{-1}	[112]
Phenols	Water	Three phase	$[C_4C_1Im][PF_6]$	NaOH solution	20 min	SBME. Comparison with 1-octanol as intermediate solvent and IL-HF-LPME using a microsyringe to hold the fiber. Extraction at 60 °C.	LC-UV/Vis	81–158/5.2–8.3 (–)	0.01–0.1 μg L^{-1}	[113]
BTEXs	Water	Two phase	$[C_4C_1Im][PF_6]$	$[C_4C_1Im][PF_6]$	30 min	SBME configuration.	GC-FID	–/1.3–5.4 (0.2 mg L^{-1})	2.7–4.0 μg L^{-1}	[114]
Cd	Water, hair and tea	Two phase	$[C_4C_1Im][PF_6]$	$[C_4C_1Im][PF_6]$	15 min	APDC as chelating agent. SBME configuration. Study of the effect of coexisting ions.	TS-FF-AAS	90/4.7 (0.5 ng mL^{-1})	9 ng L^{-1}	[115]
Ultraviolet filters	Water	Two phase	$[C_6C_1Im][FAP]$	$[C_6C_1Im][FAP]$	50 min	–	LC-UV	25–221/1.1–8.2 (50 ng mL^{-1})	0.3–0.5 ng mL^{-1}	[116]

Table 4 Overview of Recent Classical *IL-DLLME* Methods

Analyte(s)	Sample(s)	IL (Amount)	Dispersant (Volume)	Comments	Analytical Technique	EF/Coefficient of Variation (Spiking Level)	Limit of Detection	Ref.
Permethrin and biphenthrin	Water	$[C_8C_1Im][PF_6]$ (50 μL)	$[C_4C_1Im][BF_4]$ (300 μL)	IL/IL-DLLME.	LC-UV	−/6.58, 7.78 (10 μg L^{-1})	0.28, 0.83 μg L^{-1}	[117]
Hexabromocyclododecane diastereomers	Water	$[C_8C_1Im][PF_6]$ (40 μL)	$[C_2C_1Im][BF_4]$ (400 μL)	IL/IL-DLLME.	LC-ESI-MS/MS	−/4.1−6.7 (0.73−7.69 μg L^{-1})	0.12−0.22 μg L^{-1}	[118]
Triclosan and triclocarban	Water	$[C_8C_1Im][PF_6]$ (50 μL)	$[C_4C_1Im][BF_4]$ (300 μL)	IL/IL-DLLME.	LC-ESI-MS/MS	−/6.41, 5.40 (10, 2 μg L^{-1})	0.35, 0.23 μg L^{-1}	[119]
Cu	Water	$[C_6C_1Im][NTf_2]$ (65 mg)	Acetone (550 μL)	TMK as chelating agent. Multivariate optimization. Study of the effect of coexisting ions.	FAAS	136.6/3.3 (10 μg L^{-1})	0.45 μg L^{-1}	[120]
Hg species	Cosmetics	$[C_6C_1Im][PF_6]$ (52 mg)	Methanol (500 μL)	APDC as chelating agent. Study of the effect of coexisting ions.	LC-ICP-MS	115−760/2.3−7.4 (1 ng mL^{-1})	1.3−7.2 ng L^{-1}	[121]
DDT and metabolites	Water	$[C_6C_1Im][PF_6]$ (70 μL)	Acetone (600 μL)	−	LC-UV	−/5.27−6.73 (5 μg L^{-1})	0.33−0.63 μg L^{-1}	[122]
Tetrabromobisphenol A	Water	$[C_6C_1Im][PF_6]$ (70 μL)	Acetonitrile (500 μL)	−	LC−ESI-MS/MS	−/6.95 (5 μg L^{-1})	0.06 μg L^{-1}	[123]
Emerging organic contaminants	Water	$[HNH_2C_1Pyrr][FAP]$ (30 μL)	Methanol (500 μL)	Comparison with in situ IL-DLLME.	LC-UV	0.81−379.3/ 1.5−5.8 (−)	0.1−55.1 μg L^{-1}	[124]
Se	Water and garlic	$[C_{14}C_6C_6C_6P][Cl]$ (CYPHOS® IL 101) (50 mg)	Methanol (100 μL)	USA extraction pretreatment of inorganic Se species from freeze-dried garlic samples. Reduction of Se(VI) to Se(IV) before determinations. APDC as chelating agent. On-line IL-DLLME. Study of the effect of coexisting ions.	ETAAS	20/5.1 (0.5 μg L^{-1})	15 ng L^{-1}	[125]
Phenolic compounds	Cosmetics	$[C_8C_1Im][PF_6]$ (80 μL)	Acetone (900 μL)	USA extraction pretreatment. Back-extraction into alkaline solution.	CE-UV	18.0−60.1/ 1.1−12.8 (0.05−5 μg mL^{-1})	5−100 ng mL^{-1}	[126]

Emodin and metabolites	Urine	$[C_6C_1Im][PF_6]$ (50 μL)	Acetonitrile (300 μL)	Comparison with organic solvent as extractant phase.	LC-UV	63–192/<5 7 (40–100 $\mu g\ L^{-1}$)	0.1–1 $ng\ mL^{-1}$	[127]
Formaldehyde	Water and shampoo	$[C_6C_1Im][PF_6]$ (75 μL)	Ethanol (1 mL)	Derivatization with methyl acetoacetate. Study of the effect of coexisting ions.	Spectrophotometry	158.5/2.5 (0.8 $ng\ mL^{-1}$)	0.02 $ng\ mL^{-1}$	[128]
Coumarins	Roots	$[C_6C_1Im][PF_6]$ (50 μL)	Ethanol (200 μL)	Previous pulverization, dissolution in methanol and filtration of root samples. Comparison with classical DLLME using organic solvent.	LC-UV	130–230/ <7.8 (–)	0.013–0.66 $ng\ mL^{-1}$	[129]
Rhodamine B	Water, liquid soap, matches tip, pencil colored and dyes	$[C_6C_1Im][PF_6]$ (75 mg)	Ethanol (500 μL)	Study of the effect of interferences.	FO-LADS	65.5/1.3 (50 $\mu g\ L^{-1}$)	1.05 $\mu g\ L^{-1}$	[130]
Tl species	Water	$[C_6C_1Im][PF_6]$ (60 mg)	Ethanol (500 μL)	Iodide as chelating agent. $[C_{14}C_6C_6C_6P]$ [Cl] (CYPHOS® IL 101) as ion-pairing reagent. Previous separation of Tl^{+} and Tl^{3+} with anion exchange resin for speciation. Study of the effect of coexisting ions.	ETAAS	100/5,3 (0.4 $\mu g\ L^{-1}$)	3.3 $ng\ L^{-1}$	[131]
As species	Wine	$[C_8C_1Im][PF_6]$ (40 mg)	Methanol (100 μL)	DDTC as chelating agent. On-line IL-DLLME. Study of the effect of coexisting ions.	ETAAS	46/4.7-5.7 (0.2 $\mu g\ L^{-1}$)	5 $ng\ L^{-1}$	[132]
Benzophenone-type UV filters	Water	$[C_8C_1Im][PF_6]$ (40 μL)	Methanol (200 μL)	Up-and-down-shaker-assisted IL-DLLME.	UPLC-DAD	146–260/1–1.8 (50 $ng\ mL^{-1}$)	0.2–1.3 $ng\ mL^{-1}$	[133]
Cr species	Water	$[C_8C_1Im][PF_6]$ (40 μL)	Methanol (1 mL)	DDTC as chelating agent. Analysis of interferences.	FAAS	–/3.4, 4 (80 $\mu g\ L^{-1}$)	1.00, 0.41 $\mu g\ L^{-1}$	[134]
Irbesartan and valsartan	Human urine	$[C_8C_1Im][PF_6]$ (52 mg)	Acetone (300 μL)	–	LC-DAD	–/2.9, 2.6 (–)	1.5, 3.3 $\mu g\ L^{-1}$	[135]
Hg species	Water	$[C_8C_1Im][PF_6]$ (50 mg)	Acetone (500 μL)	Dithionate as chelating agent.	LC-HG-AFS	25/1.4–5.2 (–)	0.016–0.092 $\mu g\ L^{-1}$	[136]

Table 5 Overview of Recent *USA* and *VA-IL-DLLME* Methods

Analyte(s)	Sample(s)	IL (Amount)	Dispersant (Volume)	US/Vortex/MW (Conditions)	Comments	Analytical Technique	EF/Coefficient of Variation (Spiking Level)	Limit of Detection	Ref.
Sulfonamides	Milk powder	$[C_6C_1Im][PF_6]$ (70 μL)	$[C_4min][BF_4]$ (100 μL)	US (100 W, 10 min, 30 °C)	US-IL/IL-DLLME. Comparison with in situ IL-DLLME, US-IL-DLLME and TC-IL-DLLME.	LC-DAD	−/1.0−3.5 (100 μg Kg^{-1})	2.94−16.7 μg Kg^{-1}	[137]
Benzylurea insecticides	Water	$[C_6C_1Im][PF_6]$ (70 μL)	Acetonitrile (300 μL)	US (1 min)	Magnetic retrieval of IL.	LC-UV/Vis	261−302/2.7−4.9 (−)	0.05−0.15 μg L^{-1}	[138]
Hg species	Water and fish tissues	TOMAS (30 μL)	−	US (45 W, 2 min)	Microwave-assisted digestion of reference organic samples. Ice-water bath after sonication. Study of interferences.	CV-AAS	310, 200/4, 6 (1 ng mL^{-1}, 5 ng mL^{-1})	0.03, 0.4 ng mL^{-1}	[139]
Pesticides and metabolites	Soil	$[C_6C_1Im][PF_6]$ (117.5 mg)	Methanol (418 μL)	Vortex (1 min, centrifugation after 8 min)	USA extraction pretreatment of soil samples. Multivariate optimization.	LC-FD	−/−	0.02−27.07 ng g^{-1}	[140]
Benzophenone-type UV filters	Water	$[C_4C_1Im][PF_6]$ (30 μL)	Methanol (15 μL)	Vortex (4 min)	Multivariate optimization.	LC-UV	18.9−26.8/<8.0 (−)	1.9−6.4 ng mL^{-1}	[141]
Rh	Water and roses leaves	$[C_8C_1Im][NTf_2]$ (30 mg)	−	US (2 min)	5-Br-PADAP as chelating agent. Previous digestion of rose samples. Heating to form the complex. Study of the effect of coexisting ions.	FAAS	29.1/1.63 (200 ng mL^{-1})	0.37 ng mL^{-1}	[142]
Aromatic amines	Water	$[C_4C_1Im][PF_6]$ (100 μL)	Methanol (750 μL)	US (200 W, 25 min)	Comparison with USA-DLLME using organic solvents.	LC-UV	>50/3.27−4.73 (5 μg L^{-1})	0.015−0.026 ng mL^{-1}	[143]
Linear alkylbenzene sulfonates	Water	$[C_6C_1Im][PF_6]$ (72 μL)	−	US (30 s, 25 °C)	Multivariate optimization. Study of the effect of coexisting ions.	Spectrophotometry	79.1/2 (40 ng mL^{-1})	0.37 ng mL^{-1}	[144]
Pyrethroid pesticides	Honey	$[C_8C_1Im][NTf_2]$ (60 μL)	Methanol (200 μL)	US (2 min)	Diluted honey samples with water. Comparison with classical IL-DLLME and TC-IL-DLLME.	LC-DAD	506−515/4.3−5.7 (−)	0.21−0.38 μg L^{-1}	[145]

Celastrol	Urine	$[C_6C_1Im][PF_6]$ (45 μL)	Methanol (100 μL)	US (200 W, 1 min)	Diluted urine in HCl solution. Ice-water bath after sonication.	LC-DAD	110/0.43 (−)	1.6 μg L^{-1}	[146]
Fungicides	Red wine	$[C_6C_1Im][PF_6]$ (50 μL)	−	US (75 W, 5 min)	Ice-water bath after sonication.	LC-DAD	100−200/5.1−9.9 (50 μg L^{-1})	2.8−16.8 μg L^{-1}	[147]
Biogenic amines	Beer	$[C_4C_1Im][PF_6]$ (30 μL)	−	US (100 W, 1 min, 30 °C)	Previous derivatization with a fluorescence probe. Comparison with classical IL-DLLME.	LC-FD	−/2.1−3.4 (−)	0.25−50 ng mL^{-1}	[148]
Ultraviolet filters	Water	$[C_6C_1Im]$[FAP] (60 μL)	−	US (270 W, 12 min, RT)	Multivariate optimization.	LC-UV	−/2.6−6.6 (10 ng mL^{-1})	0.5−1 ng mL^{-1}	[149]
Ultraviolet filters	Water	$[C_6C_1Im]$[FAP] (20 μL)	Methanol (100 μL)	US (320 W, 3 min)	−	LC-UV	354−464/4.0−6.3 (50 ng mL^{-1})	0.2−5.0 ng mL^{-1}	[150]
Fluoroquinolones	Water	$[C_8C_1Im][PF_6]$ (65 mg)	Methanol (400 μL)	US (5 min)	Vortex mixing before sonication. Ice-water bath after sonication.	LC-FD	122−205/4.8−9.4 (10−250 ng L^{-1})	0.8−13 ng L^{-1}	[151]
Phytocides	Soil	$[C_4C_1Im][NTf_2]$ (70 μL)	Methanol (300 μL)	US (2 min)	USA extraction pretreatment of soil samples.	LC-UV	−/0.93-2.43 (6.6,16.0 μg mL^{-1})	0.05−0.31 μg mL^{-1}	[152]
Co	Nutritional supplement	$[C_6C_1Im][PF_6]$ (80 mg)	−	US (600 W, 3 min)	Extraction pretreatment of solid samples. 1-nitroso-2-Naphtol as chelating agent. Study of the effect of coexisting ions.	ETAAS	60/4.7 (0.5 μg L^{-1})	5.4 ng L^{-1}	[153]
Fungicides	Water	$[C_8C_1Im][PF_6]$ (40 μL)	−	US (60 KHz, 15 min, 25 °C)	−	LC-VWD	88-137/3.1-5.4 (50 ng mL^{-1})	0.73−2.2 ng mL^{-1}	[154]
Pharmaceuticals	Water	$[C_8C_1Im][PF_6]$ (85 mg)	Acetonitrile (500 μL)	US (4 min)	Vortex mixing before sonication. Ice-water bath after sonication.	LC-MS/MS	255−340/1.1−9.4 (1−710 ng L^{-1})	0.2−60 ng L^{-1}	[155]
Antidepressant drugs	Water	$[C_6C_1Im]$[FAP] (20 μL)	Methanol (200 μL)	US (1 min)	USA-IL-DLLME combined with vortex-assisted micro-SPE.	LC-UV/Vis	17−43/1.5−7.8 (4 ng mL^{-1})	0.3−1 μg L^{-1}	[156]

(Continued)

Table 5 Overview of Recent *USA* and *VA-IL-DLLME* Methods—cont'd

Analyte(s)	Sample(s)	IL (Amount)	Dispersant (Volume)	US/Vortex/MW (Conditions)	Comments	Analytical Technique	EF/Coefficient of Variation (Spiking Level)	Limit of Detection	Ref.
As species	Urine and blood	$[C_4C_1Im][PF_6]$ (200 mg)	–	Vortex (2 min)	APDC as chelating agent. Flow-injection system. Back-extraction into HCl solution. Reducing As(V) to As(III) with KI and ascorbic acid in HCl solution for total As determination. Study of the effect of coexisting ions.	HG-AAS	–/3.2 (–)	5 ng L^{-1}	[157]
Clozapine and metabolites	Urine and serum	$[C_2C_1Im][NTf_2]$ (25 μL, off-line) (8 μL in-vial)	–	Vortex (30 s)	Off-line and in-vial extractions.	CE-DAD	80–200/–	3–11 μg L^{-1} (urine) 6–55 μg L^{-1}(serum)	[158]
Organophosphorus pesticides	Apple and pear	$[C_8C_1Im][PF_6]$ (62.3 mg)	Methanol (1 mL)	Vortex (1 min, 2000 rpm)	Fruity juices were prepared and diluted with water before extraction.	LC-UV	–/2.3–5.7 (–)	0.061–0.73 μg Kg^{-1}	[159]
Nifurtimox and benznidazole	Human plasma	$[C_8C_1Im][PF_6]$ (13.5 μL)	Methanol (32.4 μL)	Vortex (6 min)	Removing proteins before extraction.	LC-UV	31.24–39.2/0.76–2.52 (5–10 μg mL^{-1})	15.7, 26.5 ng mL^{-1}	[160]
Glucocorticoids	Water	$[C_4C_1Im][PF_6]$ (200 μL)	Triton X-100 (500 μL, 0.05% v/v)	Vortex (3 min)	Ice-water bath after extraction.	LC-UV	99.85/1.57–1.81 (1 ng mL^{-1})	4.11–9.19 ng mL^{-1}	[161]
Antihypertensive	Rat serum	$[C_4C_1Im][PF_6]$ (60 μL)	Acetone (120 μL)	Vortex (1 min)	–	LC-DAD	4.64–4.92/0.9–7.1 (0.1–20 μg L^{-1})	0.012–0.020 μg mL^{-1}	[162]

Table 6 Overview of Recent *TC-IL-DLLME* and *CIAME* Methods

Analyte(s)	Sample(s)	IL (Amount)	Dispersant (Volume)	Heating (°C)	Cooling (min)	Comments	Analytical Technique	EF/Coefficient of Variation (Spiking Level)	Limit of Detection	Ref.
Ag	Hair and water	$[C_6C_1Im][PF_6]$ (50 μL)	–	77	20	Disulfiram as chelating agent. Previous acid digestion of hair samples. Study of the effect of coexisting ions.	GF-AAS	120/4.5 (20 ng L^{-1})	5.2 ng L^{-1}	[163]
Carbamate pesticides	Water	$[C_6C_1Im][PF_6]$ (40 μL)	–	70	20	–	LC-DAD	–/1–1.8 (–)	0.45–1.40 μg L^{-1}	[164]
Bisphenol A, 4-n-Nonylphenol and 4-tert-octylphenol	Water	$[C_8C_1Im\ PF_6]$ (65 μL)	–	80	20	Comparison with SPE.	LC-FD	–/4.6–5.5 (–)	0.23–0.48 μg L^{-1}	[165]
Pyrethroid insecticides	Water	$[C_8C_1Im][PF_6]$ (40 μL)	–	80	30	–	LC-UV	50/2.0–3.4 (20 μg L^{-1})	0.34–0.48 μg L^{-1}	[166]
Phenols	Water	$[C_8C_1Im\ PF_6]$ (50 μL)	Ethanol (700 μL)	60	20	–	LC-UV	334–371/2.1–3.7 (–)	0.27–0.68 μg L^{-1}	[167]
Phthalate esters and pyrethroid insecticides	Water	$[C_8C_1Im\ PF_6]$ (40 μL)	–	90	–	–	LC-UV	–/2.2–5.9 (10 μg L^{-1})	0.23–0.47 μg L^{-1}	[168]
Pb	Blood	$[C_4C_1Im][PF_6]$ (45 μL)	–	75	10	MW-assisted digestion of blood samples. APDC as chelating agent. Multivariate optimization. Study of the effect of coexisting ions.	FAAS	93/4.3 (10 μg L^{-1})	0.13 μg L^{-1}	[169]
Cr species	Water and urine	$[C_6C_1Im][PF_6]$ (25 μL)	–	45	10	APDC as chelating agent. Study of the effect of coexisting ions.	ETAAS	42, 43/3.05, 4.24 (100 ng L^{-1})	5.40, 2.45 ng L^{-1}	[170]
Fungicides	Water	$[C_8C_1Im][PF_6]$ (55 μL)	–	90	30	–	LC-UV	–/3.7–5.9 (–)	0.32–0.79 μg L^{-1}	[171]
Ultraviolet filters	Water	$[C_6C_1Im][FAP]$ (20 μL)	–	50	20	–	LC-UV	240–350/4.9–7.9 (100 ng mL^{-1})	0.3–5.0 ng mL^{-1}	[172]
Bisphenol A and 2-naphthol	Water	$[C_8C_1Im][PF_6]$ (35 μL)	Acetone (1 mL)	60	3	CIAME.	LC-DAD	112, 186/2.3, 4.1 (20 ng mL^{-1})	0.58, 0.86 ng mL^{-1}	[173]
Phthalate esters	Water	$[C_8C_1Im][PF_6]$ (32 μL)	Acetonitrile (750 μL)	50	3	CIAME.	LC-VWD	174–212[(1)]/2.2–3.7 (20 ng mL^{-1})	0.68–1.36 ng mL^{-1}	[174]
Zn	Water, flour, potato and apple	$[C_6C_1Im][PF_6]$ (95 mg)	Ethanol (650 μL)	50	4	CIAME. 8-hydroxyquinoline as chelating agent. Acid digestion of food samples. Study of the effect of coexisting ions.	FAAS	60/3 (10 μg L^{-1})	0.18 μg L^{-1}	[175]
Phenol	Water	$[C_6C_1Im][PF_6]$ (74 mg)	–	23	2	USA-CIAME. Study of the effect of coexisting ions.	Spectrophotometry	75/2.65 (50 μg L^{-1})	0.86 μg L^{-1}	[176]

Table 7 Overview of Recent *In Situ IL-DLLME* Methods

Analyte(s)	Sample(s)	Hydrophilic IL (Amount)	Ion-Pair Reagent (Amount)	Comments	Analytical Technique	EF/Coefficient of Variation (Spiking Level)	Limit of Detection	Ref.
2,4,6-Trinitrotoluene	Water	$[C_6C_1Im][Cl]$ (32 mg)	$LiNTf_2$ (45.3 mg)	Multivariate optimization. Electrochemical detection using screen-printed graphite electrodes.	DPV	300/5−7 (20−50 μg L^{-1})	7 μg L^{-1}	[38]
DNA	DNA aqueous solutions	$[C_{16}POHIm][Br]$ (0.5 mg)	$LiNTf_2$ (Ratio 1:1, 1 g mL^{-1})	Extraction from matrixes containing albumin and metal ions. Indirect analysis of the aqueous phase.	LC-UV/Vis	−/− (−)	—	[177]
Insecticides	Water	$[C_6C_1Im][Cl]$ (27 mg)	$LiNTf_2$ (1280 μL, 0.03 g mL^{-1})	Ice-water bath before centrifugation.	LC-VWD	264−326/2.9−4.6 (20 μg L^{-1})	0.98−2.54 μg L^{-1}	[178]
Endocrine disrupter phenols	Water	$[C_4C_1Im][Cl]$ (38 μL)	$LiNTf_2$ (340 μL, 0.2 g mL^{-1})	Vortex mixing after $LiNTf_2$ addition.	LC-DAD	140−989/2.8−11 (10 ng mL^{-1})	0.8−4.8 ng mL^{-1}	[179]
Ag	Water	$[C_6C_1Im][BF_4]$ (60 μL, 60 mg μL^{-1})	$NaPF_6$ (800 μL, 120 mg mL^{-1})	TMK as chelating agent. Heating before $NaPF_6$ addition. Ice bath before centrifugation. Study of the effect of coexisting ions.	Spectrophotometry	95.5/1.8 (50 ng mL^{-1})	0.4 ng mL^{-1}	[180]
Organic contaminants	Marine sediments	$[C_{16}C_1Im][Br]$ (4 mL, 40 mM)	$LiNTf_2$ (92 μL, 0.5 g mL^{-1})	MW-assisted extraction pretreatment with $[C_{16}C_1Im][Br]$. Heating and vortex mixing after $LiNTf_2$ addition.	LC-DAD	5.4−24/3.8−19 (4−110 μg L^{-1})	Limit of quantification 0.04−1.00 mg Kg^{-1}	[181]
Chlorophenols	Water	$[C_8C_1Im][Cl]$ (50 mg)	KPF_6 (50 mg)	Vortex mixing after KPF_6 addition.	GC-MS	−/4.5−9.9 (0.5−2 ng mL^{-1})	60−440 ng L^{-1}	[182]
PAHs	Toasted cereals	$[C_{16}C_4Im][Br]$ (1.5 mL, 40 mM)	$LiNTf_2$ (34 μL, 0.5 g mL^{-1})	MW-assisted extraction pretreatment with $[C_{16}C_1Im][Br]$. Vortex mixing after $LiNTf_2$ addition. Freezing (1 h) after centrifugation.	LC-UV-FD	5.6−14/5.5−22 (9−4500 μg Kg^{-1})	0.03−83 μg Kg^{-1}	[183]
Cr species	Water and lixiviates of plastic toys	$[C_8C_1Im][Cl]$ (100 μL, 1 M)	$LiNTf_2$ (100 μL, 1 M)	APDC as chelating agent. Triton X-114 as antisticking agent. Addition of EDTA for Cr(VI) determination.	ETAAS	152, 301/8.1, 7.2 (0.05 μg L^{-1})	2 ng L^{-1}	[184]

Active constituents	Fruits	$[C_4C_1Im][BF_4]$ (150 μL)	NH_4PF_6 (800 mg)	MW irradiation before NH_4PF_6 addition. Comparison with US and soxhlet extractions.	LC-DAD	−/2.19−5.94 (0.12−6.47 μg mg^{-1})	28.561−33.878 μg g^{-1}	[185]
Phenylurea pesticides	Water	$[C_6C_1Im][Cl]$ (34 mg)	$LiNTf_2$ (360 μL, 0.162 g mL^{-1})	US irradiation after $LiNTf_2$ addition. Multivariate optimization.	LC-VWD	244−268/2.4−3.5 (5 μg L^{-1})	0.06−0.08 μg L^{-1}	[186]
Triazine herbicides	Water	$[C_4C_1Im][BF_4]$ (40 μL)	$LiNTf_2$ (500 μL, 0.2 g mL^{-1})	MW irradiation after salt addition.	LC-UV	103−132/1.6−6.2 (40 μg L^{-1})	0.52−1.3 μg L^{-1}	[187]
Cd and Pb	Water and lixiviates of plastic toys	$[C_8C_1Im][Cl]$ (100 μL, 1 M)	$LiNTf_2$ (100 μL, 1 M)	APDC as chelating agent.	ETAAS	280/6.5, 7.3 (10, 0.2 μg L^{-1})	0.2, 3 ng L^{-1}	[188]
Acaricides	Water	$[C_8C_1Im][Cl]$ (40 mg)	KPF_6 (43−45 mg)	US after salt addition. Comparison with US-IL-DLLME.	LC-VWD	100.9−278.3/1.7−3.8 (50 μg L^{-1})	0.54−1.71 μg L^{-1}	[189]
Pyrethroids	Water	$[C_8C_1Im][Cl]$ (20 mg)	$LiNTf_2$ (830 μL, 0.03 g mL^{-1})	Ice-water bath before centrifugation. Multivariate optimization.	LC-VWD	134−153/0.58−1.89 (−)	0.02−0.18 μg L^{-1}	[190]
Cu	Water	$[C_6C_1Im][Cl]$ (45 μL, 0.45 g mL^{-1})	$LiNTf_2$ (100 μL, 0.2 g mL^{-1})	DDTC as chelating agent. Study of the effect of coexisting ions.	ETAAS	200/4 (0.7 μg L^{-1})	4 ng L^{-1}	[191]
Nonpolar heterocyclic aromatic amines	Beef burgers	$[C_6C_1Im][BF_4]$ (200 μL, 18 mg mL^{-1})	$NaPF_6$ (240 μL, 120 mg mL^{-1})	Beef burgers were cooked, the outer crust was lyophilized and mixed with n-heptane for the elimination of fat material. Defatted samples were mixed with NaOH, centrifuged and the remaining aqueous phase was ready for microextraction.	LC-DAD-FD	19−30/5.4−10.9 (−)	0.04−0.3 ng g^{-1}	[192]

hollow-fiber liquid-phase microextraction (HF-LPME), and (3) dispersive liquid–liquid microextraction (DLLME).

3.1 Single-Drop Microextraction

Single-drop microextraction was the first LPME technique developed by Liu and Dasgupta [33] and by Jeannot and Cantwell [34]. The former introduced the drop-in-drop system wherein one organic drop was suspended inside a flowing aqueous drop from which the analyte was extracted [33]. The latter suspended an organic drop at the end of a Teflon rod that was immersed in a stirred aqueous sample solution. After the extraction, the organic phase was sampled with a microsyringe and injected into the GC for quantification [34]. The inconvenience of using different devices for extraction and posterior injection in the analytical system was solved with the use of a conventional GC syringe, working as a support of the droplet and for the injection in the analytical system afterward [198]. SDME uses a few microliters of solvent held as a single drop on the tip of a syringe. The droplet can be either disposed to the headspace or directly immersed in the sample. The analytes within the sample partition to the droplet and after the extraction, the droplet is retracted back and transferred to the analytical instrument for analysis. Depending on the position of the drop, the SDME technique was defined as follows: (1) headspace SDME (HS-SDME) [199] or (2) direct immersion SDME (DI-SDME) [198]. HS-SDME is suitable for the extraction of volatile and semivolatile compounds; in contrast, DI-SDME is appropriate for the extraction of nonvolatile compounds. The volume of the drop typically used with organic solvents is around 1–3 μL; however, the higher viscosity of the ILs enables bigger drops (e.g., 3–10 μL). The syringe usually used with ILs is an LC syringe with a flat tip, in contrast to the bevel tip of the GC syringes used with organic solvents. The flat tip prevents the formation of the drop; therefore, different systems using a short 3-mm-long tube [200] or a bell-mouthed [201] have been developed. These systems also help to maximize the contact area between the needle and the drop, allowing bigger volumes (e.g., 20 μL).

Liu et al. [200] described for the first time the use of ILs as solvents in LPME in 2003. $[C_4C_1Im][PF_6]$, $[C_6C_1Im][PF_6]$, and $[C_8C_1Im][PF_6]$ ILs were used as extractant phases for the extraction of polycyclic aromatic hydrocarbons (PAHs). HS-SDME and DI-SDME were studied, providing better extraction depending on the volatility of PAHs. $[C_8C_1Im][PF_6]$ provided better enrichment factors because of its higher hydrophobicity and viscosity, allowing larger drop volume and a more stable drop compared with 1-octanol. Numerous developments and application have been published since then, and Table 2 shows an overview of the past three years.

Even when ILs are used, the volume is restricted to the stability of the drop on the syringe tip; therefore, the SDME technique has experimented important modifications. Cruz-Vera et al. eliminated the syringe and proposed the use of ILs in dynamic liquid-phase microextraction (dLPME) for the efficient extraction of nonsteroidal anti-inflammatory drugs [202] and phenothiazine derivatives [203] in urine. The volumes and flow rates of different solutions involved in the extraction process were controlled by the designed automatic flow configurations. The volume of IL was increased up to 50 μL, and high viscosity of the IL was avoided by diluting it with acetonitrile before its introduction into the flow system. Another dynamic system named IL-based cycle-flow SDME was developed by Xia et al. for preconcentration of Co, Hg, and Pb from biological and environmental samples [204]. This system extracted analytes by exposing an IL droplet (2.5 μL) to a flowing stream of sample. $[C_4C_1Im][PF_6]$ IL was the extraction solvent, and 1-(2-pyridylazo)-2-naphthol reagent was used as both extractant and chemical modifier. The extraction efficiency and preconcentration factor were higher than in static conditions because continuous contact between the IL phase and the fresh flowing sample solution was achieved. Sarafraz-Yazdi et al. proposed the IL-based submerged SDME technique for the determination of aromatic amines in environmental water samples [205]. A 15 μL drop of $[C_6C_1Im][PF_6]$ IL was injected slowly at the conical bottom of the glass tube, under the sample solution, forming a spherical droplet. Then, the glass tube was sonicated, and after this process, the spherical drop of the IL was withdrawn by a 3-mm-long polytetrafluoroethylene tube. The results show that the proposed method presents a lower limit of detection and better repeatability values and also requires a much lower extraction time than SDME or HF-LPME.

ILs combined with other reagents or materials, such as surfactants or quantum dots (QD), have been proposed as extractant phases in SDME. Carrillo-Carrión et al. proposed the use of $[C_6C_1Im][PF_6]$ IL modified with CdSe/ZnS QD for the combination of IL-HS-SDME and QD-based fluorimetric detection [95]. Trimethylamine from fish samples was extracted and preconcentrated directly onto a (QD)IL microdrop. Then, the microdrop was transferred to a microcuvette with 300 mL of acetonitrile, and the fluorescence was recorded. Yao et al. [206] studied micellar ILs as extraction solvents in HS-SDME to perform the extraction of 17 aromatic compounds. Two different micellar solutions were formed by dissolving 1-decyl-3-methylimidazolium bromide $[C_{10}C_1Im][Br]$ IL and traditional surfactant sodium dodecyl sulfate (SDS) in $[C_4C_1Im][Cl]$ IL. Extraction efficiency increased with the micellar systems compared with the net $[C_4C_1Im][Cl]$. The highest sensitivity was obtained with the $[C_4C_1Im][Cl]$-$[C_{10}C_1Im][Br]$ micellar solution for PAHs with high molecular weight and fused rings, whereas the $[C_4C_1Im][Cl]$-SDS micellar solution was proven to be more sensitive for smaller, more polar molecules.

Microwave radiation has been used for enhancement of the extraction. Vidal et al. proposed a one-step and in situ sample preparation method coupling microwave and SDME (MW-HS-SDME) [207]. A home-made experimental setup with a domestic microwave oven was used for the extraction of chlorobenzenes from water samples. The chlorobenzenes were extracted directly onto a $[C_6C_1Im][PF_6]$ IL single-drop in headspace under the aid of microwave radiation. Limits of detection one order of magnitude below than those obtained without microwave heating [208] demonstrated the favorable effect of the microwave radiation upon SDME.

The SDME technique suffered from many drawbacks, such as the small surface area of the droplet that confines the amount of the analyte and the extraction efficiency, instability and ease of dislodgement from the tip of the syringe, long times to reach equilibrium, and poor reproducibility. On the contrary, DLLME overcomes all these disadvantages and therefore is the LPME technique widely used nowadays.

3.2 Hollow-Fiber Liquid-Phase Microextraction

Single-drop microextraction has gained a widespread interest since its appearance and has obtained an undoubted relevance as start point of miniaturized liquid-phase extraction techniques. However, some problems related to the instability of the hanging drop, which easily dislodged from the syringe during extraction, needed to be solved. In 1999, Pedersen–Bjergaard and Rasmussen [209] introduced the HF-LPME technique that easily addressed this shortcoming through the stabilization of the extractant phase by using a porous membrane. In HF-LPME, a hydrophobic solvent is immobilized in the pores of a polymeric hollow fiber, usually made of polypropylene, thereby forming a supported liquid membrane (SLM). The final acceptor phase is held within the lumen of the fiber and is thereby protected and physically separated from the sample by the SLM. Two extraction modes, two phase and three phase, are described for HF-LPME [210]. In two-phase HF-LPME, the extractant solvent is immobilized in the pores of the fiber and introduced into its lumen, whereas in the three-phase mode, the lumen of the fiber is filled with an aqueous solution [209,210]. Three-phase HF-LPME is more suitable for polar and ionizable compounds. The manipulation of the pH in the donor and acceptor phases, the use of carriers, and the application of an electric field [211] (i.e., electromembrane extraction) have been used in most cases to improve transport mechanisms and enhance extraction efficiencies. In general, the fiber can be in contact with the sample by two different manners, namely, held by a microsyringe or directly introduced into the sample with both sealed ends, leading to a configuration called solvent bar microextraction (SBME) [212].

Compared with SDME, HF-LPME presents some remarkable advantages [210,213,214]: (1) Higher stirring rates can be used because the problem of the dislodgment of the drop is avoided, although undesirable air bubbles can be generated on the surface of the fiber at very high stirring rates. (2) The contact area between the extractant phase and the sample is higher, favoring the mass transfer. (3) The hollow fiber acts as a protection barrier for the extractant phase; therefore, dirty samples can be analyzed and still obtain very clean extracts. (4) The fiber can be disposed of after a single use because its low cost, avoiding carryover effects between extractions.

The selection of the appropriate solvent within the pores of the fiber is a critical task of the technique. Some important aspects to consider are as follows [214]: (1) immiscibility with water to prevent losses by dissolution, (2) capability to be strongly immobilized in the pores of the fiber, (3) low volatility to avoid losses by evaporation, (4) good extractability and high partition coefficients of the target analytes, (5) low viscosity to ensure high diffusion coefficients across the liquid membrane, and (6) compatibility with the selected analytical technique in two-phase HF-LPME. Although organic solvents such as toluene, 1-octanol, undecane, and dihexylether fulfill all these requirements and have successfully been used in HF-LPME [210], their losses by evaporation are a weak point of the technique and an important source of irreproducibility between extractions.

ILs are an interesting and useful alternative to those traditional solvents for the impregnation of the fiber in HF-LPME. Previous studies [215,216] demonstrated the feasibility of using ILs based on 1-n-alkyl-3-methylimidazolium cations to obtain stable SLMs, even under vigorous stirring conditions, because of properties such as relative high viscosities, interfacial tensions, and low solubility in water, depending on the anion. These studies also investigated the mechanisms that regulate solute transport across this SLM. The applicability of an SLM based on ILs in HF-LPME was demonstrated by Peng et al. [109] who used $[C_8C_1Im][PF_6]$ as intermediary solvent in three-phase HF-LPME for the extraction of chlorophenols from different water samples. Parameters affecting the extraction such as the effect of the alkalinity of aqueous acceptor solution and the acidity of the sample, salt addition, stirring rate, extraction time, and the presence of humic acids were evaluated. They also showed the effective immobilization of the IL in the pores of the hollow fiber by scanning electron microscopic images. Afterward, Basheer et al. [110] presented a three-phase HF-LPME in which the $[C_4C_1Im][PF_6]$ IL was used as intermediary solvent and an organic solvent was the final acceptor phase. Thereby, aromatic and aliphatic hydrocarbons could be determined in storm water using a GC-MS system. Although different IL—organic solvent combinations were explored, $[C_4C_1Im][PF_6]$ and toluene provided the best combination. The use of ILs in two-phase HF-LPME was successfully presented by Abulhassani et al. [111] who used $[C_6C_1Im]$ $[PF_6]$ for the extraction of ammonium pyrrolidine dithiocarbamate (APDC) complexes of lead and nickel from water and biological samples and ETAAS as the detection system.

Table 3 shows an overview of the most relevant studies using ILs in HF-LPME. As can be observed, organic and inorganic analytes are mainly extracted from aqueous samples, although more complex matrixes such as biological tissues, hair, and tea have also been studied. In some studies, the hollow fiber is maintained in the needle tip of a syringe, whereas the SBME configuration has been used in other studies [112—115,217].

Otherwise, it notable that different applications of ILs in HF-LPME have been reported in the literature. As an example, Liu et al. [218] proposed the use of a water-miscible IL as the acceptor phase additive in a methodology named IL-three phase HF-PLME. They proved that the presence of $[C_4C_1Im][Cl]$ in the aqueous acceptor phase enhanced the extraction of some acidic, alkaline, and neutral compounds. Zeng et al. [219] used the [C_4C_1Im][BF_4] IL as an enhancement reagent for chromium speciation in two-phase HF-LPME, with octanol as the extractant phase. Although the enhancement mechanism was unknown, it was proven that the presence of the IL improved the extraction efficiency of Cr(VI).

Finally, the trioctylmethylammonium chloride ([MTOA][Cl]) IL has been used as carrier mediated in three-phase HF-LPME to facilitate the mass transfer into the organic phase of phenylarsenic compounds, forming ion pairs [220]. Several cationic carriers were proven, but IL showed the best performance.

3.3 Dispersive Liquid–Liquid Microextraction

3.3.1 Classical Dispersive Liquid–Liquid Microextraction

Dispersive liquid–liquid microextraction was proposed for the first time in 2006 [221]. Assadi and coworkers [221] presented a novel microextraction technique based on a ternary component solvent system in which a water-immiscible organic solvent was dispersed in fine drops into the aqueous sample with the aid of an organic disperser agent. A cloudy solution was formed because of the cosolvency of the dispersant with the other two phases, leading to a great contact surface area. After extraction, phases were separated by centrifugation, and the enriched organic phase that sedimented in the bottom of the centrifuge tube was collected and analyzed. This simple methodology presents the characteristics of LPME techniques, such as being inexpensive, with low solvents and sample consumption, and has a negligible environmental and human health impact because of the reduction of wastes generated and exposition to high amounts of harmful organic compounds. In addition, DLLME offers the following remarkable advantages [222,223]: (1) fast achievement of the equilibrium state because the infinitely large contact surface between the fine drops of the extractant solvent and the aqueous phase; (2) short extraction times; (3) high enrichment factors; and (4) problems associated with other LPME techniques, such as the dislodgement of the drop in SDME and the manipulation of the fiber in HF-LPME are avoided.

DLLME has gained rapid and widespread recognition, attracting the interest of the scientific community and even coming to dominate LPME research publications in recent years [26]. However, the original configuration of DLLME [221], termed as classical DLLME, suffers from some drawbacks or limitations that are in continuous revision: (1) harmful organic solvents are used as extractants, and they must be denser than water; (2) emulsification requires a dispersant solvent that competes with the extractant solvent for the analyte, thereby reducing extraction efficiency; and (3) centrifugation is necessary to separate phases after microextraction. Numerous modifications of classical DLLME have been proposed to overcome the above-mentioned weak points of the technique and develop efficient and easier approaches. One of the most representative modifications is the use of alternative extractant solvents, such as those lighter than water [224] or ILs [225].

The use of ILs in classical DLLME (classical IL-DLLME) was described for the first time by Liu et al. [226] in 2009. ILs present low miscibility with water, good extractability of organic and inorganic compounds, and higher density than water; therefore, they seemed to be perfect candidates to act as solvents in DLLME. Liu et al. used $[C_6C_1Im][PF_6]$ as the extractant phase and methanol as the disperser solvent for the determination of heterocyclic insecticides in water samples. Since then, a huge number of studies have used classical IL-DLLME for the determination of both organic and inorganic analytes. Table 4 summarizes information about the most representative studies of the past three years using classical IL-DLLME. As can be seen, in the majority of the cases, analytes are extracted from aqueous samples, although more complex matrixes such as urine, saliva, and cosmetics have also been explored. LC followed by UV/Vis, diode-array detector (DAD), fluorescence detector (FD), or MS is by far the most commonly selected analytical technique. ILs are compatible with reverse phase mobile phases; therefore, they can directly be injected in the LC system without any further modification. However, the direct injection of ILs sometimes involves problems related to their high viscosity, such as broader chromatographic peaks. Some studies use a final dilution step to reduce the viscosity of the acceptor phase before its analysis and to ensure the suitability with the instrument.

The centrifugation step, needed for phases separation, is considered the most time-consuming part of the analysis and is thus one of the weakness of DLLME. Some studies have attempted to avoid this step by developing alternative strategies to break the emulsification. Cruz-Vera et al. [227] proposed an approach named in-syringe IL-DLLME for the determination of nonsteroidal anti-inflammatory drugs in urine, in which a plastic syringe is used as the microextraction unit. After extraction, the IL phase was self-sedimented in the wider part of the syringe. Then, the plunger of the plastic syringe was moved, the sample was removed, and the IL phase containing the analytes was easily recovered from the syringe tip [227]. Despite the proposed methodology presented benefits related to the elimination of the centrifugation, some aspects, such as the extraction recoveries and the precision of the procedure, must be improved [227].

The use of alternative disperser agents vs traditional agents (i.e., methanol, acetone, acetonitrile, or ethanol) has also spurred the interest of some research groups that have proposed hydrophilic ILs to assist classical IL-DLLME. Zhao et al. [117–119] used $[C_2C_1Im][BF_4]$ and $[C_4C_1Im][BF_4]$ for the emulsification of the hydrophobic IL in a methodology called ionic liquid/ionic liquid dispersive liquid–liquid microextraction (IL/IL-DLLME). Via this method not only the organic extractant solvent but also the dispersant are avoided, so the exposure to harmful volatile compounds was notably reduced.

3.3.2 Ultrasound and Vortex Assisted IL-DLLME

Ultrasound-assisted ionic liquid dispersive liquid–liquid microextraction (USA-IL-DLLME) is based on the use of ultrasounds energy (US) to facilitate the dispersion of the IL into the sample and enhance the formation of the cloudy state [228]. The use of US to assist IL-DLLME allows the elimination of the disperser solvent of the procedure; however, in most cases, it is also used to ensure the effective formation of the micro-droplets. Zhou et al. [229] and Li et al. [230] proposed, almost at the same time, the use of US in IL-DLLME. The former extracted aromatic amines from water samples using acetonitrile as the disperser agent, whereas the latter used USA-IL-DLLME for the determination of cadmium, but obviating the organic disperser solvent. In both cases, the $[C_6C_1Im][PF_6]$ IL was selected as the extractant phase.

Table 5 shows a general overview of the most significative methods of the past three years using USA-IL-DLLME. Besides organic solvents, the use of a hydrophilic IL as disperser agent has also been described in USA-IL-DLLME [137]. Zhang et al. [138] presented a novel approach based on the magnetic retrieval of the IL using unmodified magnetic nanoparticles (MNPs) of Fe_3O_4. After USA-IL-DLLME, the MNPs were added to the mixture and vigorously shaken by vortex. Thereafter, MNPs were sedimented in the bottom of the test tube using two magnets, and the aqueous phase was retrieved. Acetonitrile was then injected to desorb the IL and the analytes from the nanoparticles using US energy. They emphasized the advantages of their method in regard to avoiding the centrifugation step and the ease of manipulation of MNPs. Stanisz et al. [139] have recently proposed the application of USA-IL-DLLME using a TSIL for the determination of mercury species in water samples. The novelty of this work is that the use of the TSIL (i.e., methyltrioctylammonium thiosalycilate, TOMAS) as the extractant phase enabled the direct extraction of mercury species without the need of a chelating agent.

The US energy offers advantages when improving the speed and efficiency of the extractions in IL-DLLME. However, we can mention some drawbacks, such as the formation of stable emulsions, the lack of uniformity of US between individual samples, and the possibility of analyte degradation [228]. As an alternative mixing mode, vortex methodology has been applied to assist IL-DLLME with the aim of enhancing the dispersion of the extractant phase but overcoming the problems of USA-IL-DLLME. Vortex-assisted ionic liquid dispersive liquid–liquid microextraction (VA-IL-DLLME) avoids the generation of free radicals that normally appear when US energy is used [228]. In addition, the dispersion under vortex mixing is unstable; therefore, phase separation is easily accomplished. Finally, another important point to consider is that vortex methodology is more cost-effective than US bath or US probe [228].

The appearance of VA-IL-DLLME in the literature is quite recent. In 2011, Asensio Ramos et al. [140] used vortex methodology to assist the extraction of pesticides in soil, using the $[C_6C_1Im][PF_6]$ IL as the extractant phase and methanol as the disperser solvent.

Almost simultaneously, Ye et al. [141] used VA-IL-DLLME for the determination of UV filters in water samples. Despite its latest appearance, VA-IL-DLLME has been used in several studies for the determination of organic and inorganic analytes in different samples, such as water, fruit, soil, urine, and blood (Table 5).

3.3.3 Temperature-Assisted Extraction

Temperature-controlled ionic liquid dispersive liquid—liquid microextraction was presented by Q. Zhou et al. in 2008 [197], representing the first time that ILs were used in DLLME. This novel microextraction methodology is based on the use of a temperature increase for the complete dispersion or solubilization of the extractant phase into the sample. Thereafter, when the mixture is cooled in an ice-water or ice bath, the solution becomes turbid, and the phases are finally separated by centrifugation. Temperatures between 50 and 90 °C are normally required for IL dissolution, where the upper limit is obviously established by the temperature of water evaporation. Extraction time is generally considered as the cooling time and is the most time-consuming part of the methodology. Cooling times between 10 and 30 min are usually used, although shorter extraction times have been reported when disperser agents are included.

M. Baghdadi and F. Shemirani [231] proposed, almost at the same time as Q. Zhou et al. [197], an IL-based microextraction named cold-induced aggregation microextraction (CIAME). The general procedure and basic principles of CIAME are the same as those of TC-IL-DLLME, although these authors named them differently. In both cases, temperature changes are used to modify IL solubility, and a cooling step is needed for the formation of the droplets of the extractant phase from a homogeneous solution. Table 6 shows general characteristics of the most relevant studies of the past three years using TC-IL-DLLME and CIAME. In general, water samples have been analyzed to determine both organic and inorganic analytes and using hexafluorophophate-based ILs as extractant phases.

3.3.4 *In Situ* Ionic Liquid Dispersive Liquid—Liquid Microextraction

Simultaneously, M. Baghdadi and F. Shemirani [35] and C. Jao and J. Anderson [36] proposed a novel IL-DLLME methodology based on the in situ formation of the extractant phase. Although the former authors named the technique in situ solvent formation microextraction (ISFM) [35], the terms IL-based in situ DLLME or in situ IL-DLLME have more frequently been adopted. In situ IL-DLLME is based on the formation of the extractant phase into sample solution via a metathesis reaction between a water-miscible IL and an ion exchange reagent, to form a water-immiscible IL (Figure 5). Briefly, in the general procedure, the water-miscible IL is dissolved into the sample containing the analytes. Then, the ion exchange salt is added; a cloudy solution is immediately formed, and the extraction of analytes is simultaneously produced to the ion exchange reaction. Finally, phases are separated by centrifugation, and the enriched phase can be analyzed with the selected analytical technique. Among the advantages that in situ IL-DLLME methodology offers, we highlight that the dispersion of the extractant IL takes place via metathesis reaction; therefore, the disperser agent is not needed. Thus, the competition between the IL and the disperser solvent is avoided. Moreover, additional devices such as vortex or US are not necessary to assist the emulsification. As can be seen in Table 7, in situ IL-DLLME has mainly been used for the determination of organic and inorganic analytes in water samples. More complex matrixes such as cereals, marine sediment fruits, tea, and blood have also been explored, with a combination of different pretreatments. Recently, extraction of DNA from water matrixes containing albumin protein or metal ions has been tested using in situ DLLME [177], proving the advantages of this methodology over other DNA extraction protocols in terms of speed, low consumption of solvents, and high extraction efficiency. Recently, a novel methodology combining in situ IL-DLLME with the electrochemical detection using SPELs has been presented [38]. Considering the electrochemical properties of ILs and the low volume of the IL phase formed during in situ IL-DLLME, SPELs were perfectly compatible candidates for analyzing the IL drop after microextraction without any further modification. Thus, these studies used miniaturized systems both in sample preparation and in the detection stage, and they avoided the disadvantages of chromatographic techniques (e.g., special devices or interfaces in GC or shorter column life and resolution problems in LC). The feasibility of the proposed methodology was demonstrated through the determination of 2,4,6-trinitrotoluene in water samples (Figure 5).

LIST OF ABBREVIATIONS

AAS Atomic absorption spectroscopy
APDC Ammonium pyrrolidine dithiocarbamate
ATPS Aqueous two-phase system
5-Br-PADAP 2-(5-bromo-2-pyridylazo)-5-Diethylaminophenol
BTEXs Benzene, toluene, ethyl-benzene, and *o*-, *m*-, and *p*-xylenes
CD Chemiluminescence detector
CE Capillary electrophoresis
CIAME Cold-induced aggregation microextraction
CV-AAS Cold vapor atomic absorption spectrometry
DAD Diode-array detector
DDT Dichlorodiphenyltrichloroethane
DDTC Sodium diethyldithiocarbamate
DI-SDME Direct immersion single-drop microextraction
DLLME Dispersive liquid–liquid microextraction
DPV Differential pulse voltammetry
DTBSF 3,5-Ditertbutylsalicylfluorone
EDTA Ethylenediaminetetraacetic acid
ESI-MS/MS Electrospray ionization-mass spectrometry/mass spectrometry
ETAAS Electrothermal atomic absorption spectrometry

FAAS Flame atomic absorption spectroscopy
FD Fluorescence detector
FID Flame ionization detector
FO-LADS Fiber optic-linear array detection spectrophotometry
GC Gas chromatography
GF-AAS Graphite furnace-atomic absorption spectrometry
HF-LPME Hollow-fiber liquid-phase microextraction
HG-AAS Hydride generation-atomic absorption spectrometry
HG-AFS Hydride generation-atomic fluorescence spectrometry
HS-SDME Headspace single-drop microextraction
ICP-MS Inductively coupled plasma-mass spectrometry
ICP-OES Inductively coupled plasma-optical emission spectroscopy
IL Ionic liquid
IL/IL-DLLME Ionic liquid/Ionic liquid dispersive liquid–liquid microextraction
IMS Ion mobility spectrometry
In situ IL-DLLME In situ ionic liquid dispersive liquid–liquid microextraction
KPF_6 Potassium hexafluorophosphate
LC Liquid chromatography
$LiNTf_2$ Lithium bis[(trifluoromethyl)sulfonyl]imide
LLE Liquid–liquid extraction
LPME Liquid-phase microextraction
MALDI-TOF/MS Matrix-assisted laser desorption/ionization-time of flight/mas spectrometry
MCC Multicapillary column
MNP Magnetic nanoparticle
MS Mass spectrometry
MW Microwaves
$NaPF_6$ Sodium hexafluorophosphate
NH_4PF_6 Ammonium hexafluorophosphate
OPA *o*-Phthalaldehyde
PAHs Polycyclic aromatic hydrocarbons
QD Quantum dots
RTIL Room temperature ionic liquid
SBME Solvent bar microextraction
SDME Single-drop microextraction
SDS Sodium dodecyl sulfate
SLME Supported liquid membrane
SPE Solid-phase extraction
SPEL Screen-printed electrode
SPME Solid-phase microextraction
TC-IL-DLLME Temperature-controlled ionic liquid dispersive liquid–liquid microextraction
TMK 4,4′-bis(dimethylamino)thiobenzophenone
TS-FF-AAS Thermospray-flame furnace-absorption atomic spectrometry
TSIL Task-specific ionic liquid
UPLC Ultraperformance liquid chromatography
US Ultrasound energy
USA Ultrasound assisted
USA-IL-DLLM Ultrasound-assisted ionic liquid dispersive liquid–liquid microextraction
UV Ultraviolet

VA-IL-DLLME Vortex-assisted ionic liquid dispersive liquid—liquid microextraction
Vis Visible
VWD Variable wavelength detector

IONIC LIQUIDS

[C_2C_1Im][BF_4] 1-Ethyl-3-methylimidazolium tetrafluoroborate
[C_2C_1Im][NTf_2] 1-Ethyl-3-methylimidazolium bis[(trifluoromethyl)sulfonyl]imide
[C_4C_1Im][NTf_2] 1-Butyl-3-methylimidazolium bis[(trifluoromethyl)sulfonyl]imide
[C_4C_1Im][Cl] 1-Butyl-3-methylimidazolium chloride
[C_4C_1Im][PF_6] 1-Butyl-3-methylimidazolium hexafluorophosphate
[C_4C_1Im][BF_4] 1-Butyl-3-methylimidazolium tetrafluoroborate
[C_4tmsIm][PF_6] 1-Butyl-3-trimethylsilylimidazolium hexafluorophosphate
[C_6C_1Im][NTf_2] 1-Hexyl-3-methylimidazolium bis[(trifluoromethyl)sulfonyl]imide
[C_6C_1Im][Cl] 1-Hexyl-3-methylimidazolium chloride
[C_6C_1Im][PF_6] 1-Hexyl-3-methylimidazolium hexafluorophosphate
[C_6C_1Im][BF_4] 1-Hexyl-3-methylimidazolium tetrafluoroborate
[C_6C_1Im][FAP] 1-Hexyl-3-methylimidazolium tris(pentafluoroethyl)trifluorophosphate
[C_8C_1Im][NTf_2] 1-Methyl-3-octylimidazolium bis[(trifluoromethyl)sulfonyl]imide
[C_8C_1Im][Cl] 1-Methyl-3-octylimidazolium chloride
[C_8C_1Im][PF_6] 1-Methyl-3-octylimidazolium hexafluorophosphate
[C_8C_1Im][BF_4] 1-Methyl-3-octylimidazolium tetrafluorophosphate
[C_8C_1Im][NTf_2] 1,2-Dimethyl-3-octylimidazolium bis[(trifluoromethyl)sulfonyl]imide
[$C_{10}C_1$Im][Br] 1-Decyl-3-methylimidazolium bromide
[C_{16}POHIm][Br] 1-(1,2-Dihydroxypropyl)-3-hexadecylimidazolium bromide
[$C_{16}C_4$Im][Br] 1-Hexadecyl-3-butylimidazolium bromide
[$C_{16}C_1$Im][Br] 1-Hexadecyl-3-methylimidazolium bromide
[$C_8C_8C_8C_1$N] [Cl] Methyltrioctylammonium chloride
[HNH_2 C_1Pyrr][FAP] 1-(6-Amino-hexyl)-1-methylpyrrolidinium
tris(pentafluoroethyl) Trifluorophosphate
[$C_{14}C_6C_6C_6$P] [Cl] (CYPHOS® IL 101)Tetradecyl(trihexyl)phosphonium chloride
[$C_8C_8C_8C_1$N] [SSal] Methyltrioctylammonium thiosalycilate
[$C_{14}C_6C_6C_6$P] Trihexyl(tetradecyl)phosphonium
[OTf] Trifluoromethanesulfonate

ACKNOWLEDGMENT

The authors would like to thank the Spanish Ministry of Science and Innovation (project n. CTQ2011-23968), Generalitat Valenciana (Spain) (projects n. ACOMP/2013/072 and PROMETEO/2013/038) and University of Alicante (Spain) (project n. GRE12-45) for the financial support. E. Fernández also thanks Generalitat Valenciana for her fellowship.

REFERENCES

[1] D. Zhao, Y. Liao, Z. Zhang, Toxicity of ionic liquids, clean — soil, air, Water 35 (2007) 42—48.
[2] T.P.T. Pham, C.-W. Cho, Y.-S. Yun, Environmental fate and toxicity of ionic liquids: a review, Water Res. 44 (2010) 352—372.

[3] D.D. Patel, J.-M. Lee, Applications of ionic liquids, Chem. Rec. 12 (2012) 329–355.
[4] D. Ott, D. Kralisch, A. Stark, Perspectives of ionic liquids as environmentally benign substitutes for molecular solvents, in: Handbook of Green Chemistry, Wiley-VCH, Weinheim, 2010, pp. 315–340.
[5] M. Deetlefs, K.R. Seddon, Assessing the greenness of some typical laboratory ionic liquid preparations, Green Chem. 12 (2010) 17–30.
[6] J.G. Huddleston, A.E. Visser, W.M. Reichert, H.D. Willauer, G.A. Broker, R.D. Rogers, Characterization and comparison of hydrophilic and hydrophobic room temperature ionic liquids incorporating the imidazolium cation, Green Chem. 3 (2001) 156–164.
[7] R. Liu, J. Liu, Y. Yin, X. Hu, G. Jiang, Ionic liquids in sample preparation, Anal. Bioanal. Chem. 393 (2009) 871–883.
[8] A.J. Carmichael, K.R. Seddon, Polarity study of some 1-alkyl-3-methylimidazolium ambient-temperature ionic liquids with the solvatochromic dye, Nile Red, J. Phys. Org. Chem. 13 (2000) 591–595.
[9] M. Koel, Ionic liquids in chemical analysis, Crit. Rev. Anal. Chem. 35 (2005) 177–192.
[10] J. Liu, J.Å. Jönsson, G. Jiang, Application of ionic liquids in analytical chemistry, TrAC Trends Anal. Chem. 24 (2005) 20–27.
[11] H. Zhao, S. Xia, P. Ma, Use of ionic liquids as "green" solvents for extractions, J. Chem. Technol. Biotechnol. 80 (2005) 1089–1096.
[12] S. Pandey, Analytical applications of room-temperature ionic liquids: a review of recent efforts, Anal. Chim. Acta 556 (2006) 38–45.
[13] X. Han, D.W. Armstrong, Ionic liquids in separations, Acc. Chem. Res. 40 (2007) 1079–1086.
[14] M. Tobiszewski, A. Mechlińska, B. Zygmunt, J. Namieśnik, Green analytical chemistry in sample preparation for determination of trace organic pollutants, TrAC Trends Anal. Chem. 28 (2009) 943–951.
[15] R.J. Soukup-Hein, M.M. Warnke, D.W. Armstrong, Ionic liquids in analytical chemistry, Annu. Rev. Anal. Chem. 2 (2009) 145–168.
[16] P. Sun, D.W. Armstrong, Ionic liquids in analytical chemistry, Anal. Chim. Acta 661 (2010) 1–16.
[17] E. Aguilera-Herrador, R. Lucena, S. Cárdenas, M. Valcárcel, The roles of ionic liquids in sorptive microextraction techniques, TrAC Trends Anal. Chem. 29 (2010) 602–616.
[18] C.F. Poole, S.K. Poole, Extraction of organic compounds with room temperature ionic liquids, J. Chromatogr. A 1217 (2010) 2268–2286.
[19] E.M. Martinis, P. Berton, R.P. Monasterio, R.G. Wuilloud, Emerging ionic liquid-based techniques for total-metal and metal-speciation analysis, TrAC Trends Anal. Chem. 29 (2010) 1184–1201.
[20] A. Martín-Calero, V. Pino, A.M. Afonso, Ionic liquids as a tool for determination of metals and organic compounds in food analysis, TrAC Trends Anal. Chem. 30 (2011) 1598–1619.
[21] Q. Zhao, J.L. Anderson, Task-specific microextractions using ionic liquids, Anal. Bioanal. Chem. 400 (2011) 1613–1618.
[22] Z. Tan, J. Liu, L. Pang, Advances in analytical chemistry using the unique properties of ionic liquids, TrAC Trends Anal. Chem. 39 (2012) 218–227.
[23] V. Vičkačkaitė, A. Padarauskas, Ionic liquids in microextraction techniques, Cent. Eur. J. Chem. 10 (2012) 652–674.
[24] B. Tang, W. Bi, M. Tian, K.H. Row, Application of ionic liquid for extraction and separation of bioactive compounds from plants, J. Chromatogr. B 904 (2012) 1–21.
[25] M.D. Joshi, J.L. Anderson, Recent advances of ionic liquids in separation science and mass spectrometry, RSC Adv. 2 (2012) 5470–5484.
[26] J.M. Kokosa, Advances in solvent-microextraction techniques, TrAC Trends Anal. Chem. 43 (2013) 2–13.
[27] L. Ruiz-Aceituno, M.L. Sanz, L. Ramos, Use of ionic liquids in analytical sample preparation of organic compounds from food and environmental samples, TrAC Trends Anal. Chem. 43 (2013) 121–145.
[28] L.B. Escudero, A. Castro, E.M. Martinis, R.G. Wuilloud, Bioanalytical separation and preconcentration using ionic liquids, Anal. Bioanal. Chem. 405 (2013) 7597–7613.
[29] S. Gabriel, J. Weiner, Ueber einige abkömmlinge des propylamins, Ber 21 (1888) 2669–2679.

[30] P. Walden, Molecular weights and electrical conductivity of several fused salts, Bull. Acad. Imp. Sci. St. Petersburg (1914) 405–422.
[31] J.S. Wilkes, M.J. Zaworotko, Air and water stable 1-ethyl-3-methylimidazolium based ionic liquids, J. Chem. Soc. Chem. Commun. (1992) 965–967.
[32] J.G. Huddleston, H.D. Willauer, R.P. Swatloski, A.E. Visser, R.D. Rogers, Room temperature ionic liquids as novel media for " clean " liquid-liquid extraction, Chem. Commun. (1998) 1765–1766.
[33] H. Liu, P.K. Dasgupta, Analytical chemistry in a drop. Solvent extraction in a microdrop, Anal. Chem. 68 (1996) 1817–1821.
[34] M.A. Jeannot, F.F. Cantwell, Solvent microextraction into a single drop, Anal. Chem. 68 (1996) 2236–2240.
[35] M. Baghdadi, F. Shemirani, In situ solvent formation microextraction based on ionic liquids: a novel sample preparation technique for determination of inorganic species in saline solutions, Anal. Chim. Acta 634 (2009) 186–191.
[36] C. Yao, J.L. Anderson, Dispersive liquid-liquid microextraction using an in situ metathesis reaction to form an ionic liquid extraction phase for the preconcentration of aromatic compounds from water, Anal. Bioanal. Chem. 395 (2009) 1491–1502.
[37] S. Dai, Y.H. Ju, C.E. Barnes, Solvent extraction of strontium nitrate by a crown ether using room-temperature ionic liquids, Dalt. Commun. 2 (1999) 1201–1202.
[38] E. Fernández, L. Vidal, J. Iniesta, J.P. Metters, C.E. Banks, A. Canals, Screen-printed electrode-based electrochemical detector coupled with in-situ ionic-liquid-assisted dispersive liquid-liquid micro-extraction for determination of 2,4,6-trinitrotoluene, Anal. Bioanal. Chem. 406 (2014) 2197–2204.
[39] C. Yao, W.R. Pitner, J.L. Anderson, Ionic liquids containing the tris(pentafluoroethyl)trifluorophosphate anion: a new class of highly selective and ultra hydrophobic solvents for the extraction of polycyclic aromatic hydrocarbons using single drop microextraction, Anal. Chem. 81 (2009) 5054–5063.
[40] E. Aguilera-Herrador, R. Lucena, S. Cardenas, M. Valcárcel, Direct coupling of ionic liquid based single-drop microextraction and GC/MS, Anal. Chem. 80 (2008) 793–800.
[41] F.-Q. Zhao, J. Li, B.-Z. Zeng, Coupling of ionic liquid-based headspace single-drop microextraction with GC for sensitive detection of phenols, J. Sep. Sci. 31 (2008) 3045–3049.
[42] F. Zhao, S. Lu, W. Du, B. Zeng, Ionic liquid-based headspace single-drop microextraction coupled to gas chromatography for the determination of chlorobenzene derivatives, Microchim. Acta 165 (2009) 29–33.
[43] A. Chisvert, I.P. Román, L. Vidal, A. Canals, Simple and commercial readily-available approach for the direct use of ionic liquid-based single-drop microextraction prior to gas chromatography determination of chlorobenzenes in real water samples as model analytical application, J. Chromatogr. A 1216 (2009) 1290–1295.
[44] J.P. Metters, R.O. Kadara, C.E. Banks, New directions in screen printed electroanalytical sensors: an overview of recent developments, Analyst 136 (2011) 1067–1076.
[45] N. Gathergood, M.T. Garcia, P.J. Scammells, Biodegradable ionic liquids: Part I. Concept, preliminary targets and evaluation, Green Chem. 6 (2004) 166–175.
[46] M.T. Garcia, N. Gathergood, P.J. Scammells, Biodegradable ionic liquids: Part II. Effect of the anion and toxicology, Green Chem. 7 (2005) 9–14.
[47] N. Gathergood, P.J. Scammells, M.T. Garcia, Biodegradable ionic liquids: Part III. The first readily biodegradable ionic liquids, Green Chem. 8 (2006) 156–160.
[48] A.E. Visser, R.P. Swatloski, R.D. Rogers, pH-Dependent partitioning in room temperature ionic liquids provides a link to traditional solvent extraction behavior, Green Chem. 2 (2000) 1–4.
[49] S. Carda-Broch, A. Berthod, D.W. Armstrong, Solvent properties of the 1-butyl-3-methylimidazolium hexafluorophosphate ionic liquid, Anal. Bioanal. Chem. 375 (2003) 191–199.
[50] D.L. Zhang, Y.F. Deng, C.B. Li, J. Chen, Separation of ethyl acetate – ethanol azeotropic mixture using hydrophilic ionic liquids, Ind. Eng. Chem. Res. 47 (2008) 1995–2001.

[51] J. Martak, S. Schlosser, Extraction of lactic acid by phosphonium ionic liquids, Sep. Purif. Technol. 57 (2007) 483–494.
[52] G.-T. Wei, Z. Yang, C.-Y. Lee, H.-Y. Yang, C.R.C. Wang, Aqueous-organic phase transfer of gold nanoparticles and gold nanorods using an ionic liquid, J. Am. Chem. Soc. 126 (2004) 5036–5037.
[53] T. Nakashima, T. Kawai, Quantum dots-ionic liquid hybrids: efficient extraction of cationic CdTe nanocrystals into an ionic liquid, Chem. Commun. (2005) 1643–1645.
[54] H. Lateef, S.M. Grimes, R. Morton, L. Mehta, Extraction of components of composite materials: ionic liquids in the extraction of flame retardants from plastics, J. Chem. Technol. Biotechnol. 545 (2008) 541–545.
[55] S.V. Smirnova, I.I. Torocheshnikova, A.A. Formanovsky, I.V. Pletnev, Solvent extraction of amino acids into a room temperature ionic liquid with dicyclohexano-18-crown-6, Anal. Bioanal. Chem. 378 (2004) 1369–1375.
[56] J.-J. Wang, Y.-C. Pei, Y. Zhao, S.-J. Zhang, Selective extraction of bioproducts by ionic liquids, Chin. J. Chem. 23 (2005) 662–664.
[57] J. Wang, Y. Pei, Y. Zhao, Z. Hu, Recovery of amino acids by imidazolium based ionic liquids from aqueous media, Green Chem. 7 (2005) 196–202.
[58] M. Matsumoto, T. Ohtani, K. Kondo, Comparison of solvent extraction and supported liquid membrane permeation using an ionic liquid for concentrating penicillin G, J. Membr. Sci. 289 (2007) 92–96.
[59] K. Shimojo, N. Kamiya, F. Tani, H. Naganawa, Y. Naruta, M. Goto, Functional conversion of cytochrome c in ionic liquids via crown ether complexation, Anal. Chem. 78 (2006) 7735–7742.
[60] F. Kubota, Y. Koyanagi, K. Nakashima, K. Shimojo, N. Kamiya, M. Goto, Extraction of cytochrome c by a functionalized ionic liquid containing a crown ether, Solvent Extr. Res. Dev. 14 (2007) 115–120.
[61] J.-H. Wang, D.-H. Cheng, X.-W. Chen, Z. Du, Z.-L. Fang, Direct extraction of double-stranded DNA into ionic liquid 1-butyl-3-methylimidazolium hexafluorophosphate and its quantification, Anal. Chem. 79 (2007) 620–625.
[62] D.-H. Cheng, X.-W. Chen, Y. Shu, J.-H. Wang, Selective extraction/isolation of hemoglobin with ionic liquid 1-butyl-3-trimethylsilylimidazolium hexafluorophosphate ($BtmsimPF_6$), Talanta 75 (2008) 1270–1278.
[63] Y. Nie, C. Li, A. Sun, H. Meng, Z. Wang, Extractive desulfurization of gasoline using imidazolium-based phosphoric ionic liquids, Energy & Fuels 20 (2006) 2083–2087.
[64] J.D. Holbrey, I. López-Martin, G. Rothenberg, K.R. Seddon, G. Silvero, X. Zheng, Desulfurisation of oils using ionic liquids: selection of cationic and anionic components to enhance extraction efficiency, Green Chem. 10 (2008) 87–92.
[65] N. Deng, M. Li, L. Zhao, C. Lu, S.L. de Rooy, I.M. Warner, Highly efficient extraction of phenolic compounds by use of magnetic room temperature ionic liquids for environmental remediation, J. Hazard. Mater. 192 (2011) 1350–1357.
[66] A.E. Visser, R.P. Swatloski, W.M. Reichert, R. Mayton, S. Sheff, A. Wierzbicki, J.H. Davis Jr., R.D. Rogers, Task-specific ionic liquids for the extraction of metal ions from aqueous solutions, Chem. Commun. (2001) 135–136.
[67] A.E. Visser, R.P. Swatloski, W.M. Reichert, R. Mayton, S. Sheff, A. Wierzbicki, J.H. Davis Jr., R.D. Rogers, Task-specific ionic liquids incorporating novel cations for the coordination and extraction of Hg^{2+} and Cd^{2+}: synthesis, characterization, and extraction studies, Environ. Sci. Technol. 36 (2002) 2523–2529.
[68] J.D. Holbrey, A.E. Visser, S.K. Spear, W.M. Reichert, R.P. Swatloski, G.A. Broker, R.D. Roger, Mercury(II) partitioning from aqueous solutions with a new, hydrophobic ethylene-glycol functionalized bis-imidazolium ionic liquid, Green Chem. 5 (2003) 129–135.
[69] A. Ouadi, B. Gadenne, P. Hesemann, J.J.E. Moreau, I. Billard, C. Gaillard, S. Mekki, G. Moutiers, Task-specific ionic liquids bearing 2-hydroxybenzylamine units: synthesis and americium-extraction studies, Chem. Eur. J. 12 (2006) 3074–3081.

[70] F. Tang, Q. Zhang, D. Ren, Z. Nie, Q. Liu, S. Yao, Functional amino acid ionic liquids as solvent and selector in chiral extraction, J. Chromatogr. A 1217 (2010) 4669–4674.
[71] X. Ni, H. Xing, Q. Yang, J. Wang, B. Su, Z. Bao, Y. Yang, Q. Ren, Selective liquid–liquid extraction of natural phenolic compounds using amino acid ionic liquids: a case of α-tocopherol and methyl linoleate separation, Ind. Eng. Chem. Res. 51 (2012) 6480–6488.
[72] Z. Li, Y. Pei, H. Wang, J. Fan, J. Wang, Ionic liquid-based aqueous two-phase systems and their applications in green separation processes, TrAC Trends Anal. Chem. 29 (2010) 1336–1346.
[73] S. Oppermann, F. Stein, U. Kragl, Ionic liquids for two-phase systems and their application for purification, extraction and biocatalysis, Appl. Microbiol. Biotechnol. 89 (2011) 493–499.
[74] K.E. Gutowski, G.A. Broker, H.D. Willauer, J.G. Huddleston, R.P. Swatloski, J.D. Holbrey, R.D. Rogers, Controlling the aqueous miscibility of ionic liquids: aqueous biphasic systems of water-miscible ionic liquids and water-structuring salts for recycle, metathesis, and separations, J. Am. Chem. Soc. 125 (2003) 6632–6633.
[75] S. Dreyer, U. Kragl, Ionic liquids for aqueous two-phase extraction and stabilization of enzymes, Biotechnol. Bioeng. 99 (2008) 1416–1424.
[76] Y. Pei, J. Wang, K. Wu, X. Xuan, X. Lu, Ionic liquid-based aqueous two-phase extraction of selected proteins, Sep. Purif. Technol. 64 (2009) 288–295.
[77] S. Dreyer, P. Salim, U. Kragl, Driving forces of protein partitioning in an ionic liquid-based aqueous two-phase system, Biochem. Eng. J. 46 (2009) 176–185.
[78] Y. Jiang, H. Xia, C. Guo, I. Mahmood, H. Liu, Phenomena and mechanism for separation and recovery of penicillin in ionic liquids aqueous solution, Ind. Eng. Chem. Res. 46 (2007) 6303–6312.
[79] Y. Jiang, H. Xia, C. Guo, I. Mahmood, H. Liu, Enzymatic hydrolysis of penicillin in mixed ionic liquids/water two-phase system, Biotechnol. Prog. 23 (2007) 829–835.
[80] Y. Jiang, H. Xia, J. Yu, C. Guo, H. Liu, Hydrophobic ionic liquids-assisted polymer recovery during penicillin extraction in aqueous two-phase system, Chem. Eng. J. 147 (2009) 22–26.
[81] C. He, S. Li, H. Liu, K. Li, F. Liu, Extraction of testosterone and epitestosterone in human urine using aqueous two-phase systems of ionic liquid and salt, J. Chromatogr. A 1082 (2005) 143–149.
[82] S. Li, C. He, H. Liu, K. Li, F. Liu, Ionic liquid-based aqueous two-phase system, a sample pretreatment procedure prior to high-performance liquid chromatography of opium alkaloids, J. Chromatogr. B 826 (2005) 58–62.
[83] K.S. Khachatryan, S.V. Smirnova, I.I. Torocheshnikova, N.V. Shvedene, A.A. Formanovsky, I.V. Pletnev, Solvent extraction and extraction-voltammetric determination of phenols using room temperature ionic liquid, Anal. Bioanal. Chem. 381 (2005) 464–470.
[84] Z. Li, N. Lu, X. Zhou, Q. Song, Extraction spectrophotometric determination of aluminum in dialysis concentrates with 3,5-ditertbutylsalicylfluorone and ionic liquid 1-butyl-3-trimethylsilylimidazolium hexafluorophosphate, J. Pharm. Biomed. Anal. 43 (2007) 1609–1614.
[85] E.M. Martinis, P. Bertón, R.A. Olsina, J.C. Altamirano, R.G. Wuilloud, Trace mercury determination in drinking and natural water samples by room temperature ionic liquid based-preconcentration and flow injection-cold vapor atomic absorption spectrometry, J. Hazard. Mater. 167 (2009) 475–481.
[86] E.M. Martinis, R.A. Olsina, J.C. Altamirano, R.G. Wuilloud, On-line ionic liquid-based preconcentration system coupled to flame atomic absorption spectrometry for trace cadmium determination in plastic food packaging materials, Talanta 78 (2009) 857–862.
[87] S. Dadfarnia, A.M.H. Shabani, M.S. Bidabadi, A.A. Jafari, A novel ionic liquid/micro-volume back extraction procedure combined with flame atomic absorption spectrometry for determination of trace nickel in samples of nutritional interest, J. Hazard. Mater. 173 (2010) 534–538.
[88] F.J. Hernández-Fernández, A.P. de los Ríos, A. Ginestá, S. Sánchez-Segado, L.J. Lozano, J.I. Moreno, C. Godínez, Use of ionic liquids as green solvents for extraction of Zn^{2+}, Cd^{2+}, Fe^{3+} and Cu^{2+} from aqueous solutions, Chem. Eng. Trans. 21 (2010) 631–636.
[89] A.P. de los Ríos, F.J. Hernández-Fernández, F.J. Alguacil, L.J. Lozano, A. Ginestá, I. García-Díaz, S. Sánchez-Segado, F.A. López, C. Godínez, On the use of imidazolium and ammonium-based ionic liquids as green solvents for the selective recovery of Zn(II), Cd(II), Cu(II) and Fe(III) from hydrochloride aqueous solutions, Sep. Purif. Technol. 97 (2012) 150–157.

[90] H. Wu, M. Chen, Y. Fan, F. Elsebaei, Y. Zhu, Determination of rutin and quercetin in Chinese herbal medicine by ionic liquid-based pressurized liquid extraction-liquid chromatography-chemiluminescence detection, Talanta 88 (2012) 222–229.
[91] S. Gao, W. Yu, X. Yang, Z. Liu, Y. Jia, H. Zhang, On-line ionic liquid-based dynamic microwave-assisted extraction-high performance liquid chromatography for the determination of lipophilic constituents in root of Salvia miltiorrhiza Bunge, J. Sep. Sci. 35 (2012) 2813–2821.
[92] Z. Tan, F. Li, X. Xu, Isolation and purification of aloe anthraquinones based on an ionic liquid/salt aqueous two-phase system, Sep. Purif. Technol. 98 (2012) 150–157.
[93] J. Han, Y. Wang, Y. Liu, Y. Li, Y. Lu, Y. Yan, L. Ni, Ionic liquid-salt aqueous two-phase extraction based on salting-out coupled with high-performance liquid chromatography for the determination of sulfonamides in water and food, Anal. Bioanal. Chem. 405 (2013) 1245–1255.
[94] N. Papaiconomou, S. Génand-Pinaz, J.-M. Leveque, S. Guittonneau, Selective extraction of gold and platinum in water using ionic liquids. A simple two-step extraction process, Dalt. Trans. 42 (2013) 1979–1982.
[95] C. Carrillo-Carrión, B.M. Simonet, M. Valcárcel, (CdSe/ZnS QDs)-ionic liquid-based headspace single drop microextraction for the fluorimetric determination of trimethylamine in fish, Analyst 137 (2012) 1152–1159.
[96] Q. Wang, H. Qiu, J. Li, H. Han, X. Liu, S. Jiang, Novel approach to improve the detection of colchicine via online coupling of ionic liquid-based single-drop microextraction with capillary electrophoresis, J. Sep. Sci. 34 (2011) 594–600.
[97] X. Wen, Q. Deng, J. Guo, Ionic liquid-based single drop microextraction of ultra-trace copper in food and water samples before spectrophotometric determination, Spectrochim, Acta Part A 79 (2011) 1941–1945.
[98] F. Ahmad, H.-F. Wu, Characterization of pathogenic bacteria using ionic liquid via single drop microextraction combined with MALDI-TOF MS, Analyst 136 (2011) 4020–4027.
[99] I. Márquez-Sillero, E. Aguilera-Herrador, S. Cárdenas, M. Valcárcel, Determination of 2,4,6-tricholoroanisole in water and wine samples by ionic liquid-based single-drop microextraction and ion mobility spectrometry, Anal. Chim. Acta 702 (2011) 199–204.
[100] I. Márquez-Sillero, S. Cárdenas, M. Valcárcel, Direct determination of 2,4,6-tricholoroanisole in wines by single-drop ionic liquid microextraction coupled with multicapillary column separation and ion mobility spectrometry detection, J. Chromatogr. A 1218 (2011) 7574–7580.
[101] M. Rahmani, M. Kaykhaii, Determination of methylcyclopentadienyl-manganese tricarbonyl in gasoline and water via ionic-liquid headspace single drop microextraction and electrothermal atomic absorption spectrometry, Microchim. Acta 174 (2011) 413–419.
[102] X. He, F. Zhang, Y. Jiang, An improved ionic liquid-based headspace single-drop microextraction-liquid chromatography method for the analysis of camphor and trans-anethole in compound liquorice tablets, J. Chromatogr. Sci. 50 (2012) 457–463.
[103] X. Guo, D. Yin, J. Peng, X. Hu, Ionic liquid-based single-drop liquid-phase microextraction combined with high-performance liquid chromatography for the determination of sulfonamides in environmental water, J. Sep. Sci. 35 (2012) 452–458.
[104] J. Yang, H. Wei, C. Yu, Y. Shi, H. Zhang, Extraction of the volatile and semivolatile compounds in seeds of cuminum cyminum l. using hydrodistillation followed by headspace-ionic liquid-based single-drop microextraction, Chromatographia 75 (2012) 1435–1443.
[105] L. Vallecillos, E. Pocurull, F. Borrull, Fully automated ionic liquid-based headspace single drop microextraction coupled to GC-MS/MS to determine musk fragrances in environmental water samples, Talanta 99 (2012) 824–832.
[106] L. Vallecillos, F. Borrull, E. Pocurull, Determination of musk fragrances in sewage sludge by pressurized liquid extraction coupled to automated ionic liquid-based headspace single-drop microextraction followed by GC-MS/MS, J. Sep. Sci. 35 (2012) 2735–2742.
[107] C. Jiang, S. Wei, X. Li, Y. Zhao, M. Shao, H. Zhang, A. Yu, Ultrasonic nebulization headspace ionic liquid-based single drop microextraction of flavour compounds in fruit juices, Talanta 106 (2013) 237–242.

[108] X. Wen, Q. Deng, J. Wang, S. Yang, X. Zhao, A new coupling of ionic liquid based-single drop microextraction with tungsten coil electrothermal atomic absorption spectrometry, Spectrochim, Acta Part A 105 (2013) 320–325.
[109] J.-F. Peng, J.-F. Liu, X.-L. Hu, G.-B. Jiang, Direct determination of chlorophenols in environmental water samples by hollow fiber supported ionic liquid membrane extraction coupled with high-performance liquid chromatography, J. Chromatogr. A 1139 (2007) 165–170.
[110] C. Basheer, A.A. Alnedhary, B.S. Madhava Rao, R. Balasubramanian, H.K. Lee, Ionic liquid supported three-phase liquid-liquid-liquid microextraction as a sample preparation technique for aliphatic and aromatic hydrocarbons prior to gas chromatography-mass spectrometry, J. Chromatogr. A 1210 (2008) 19–24.
[111] J. Abulhassani, J.L. Manzoori, M. Amjadi, Hollow fiber based-liquid phase microextraction using ionic liquid solvent for preconcentration of lead and nickel from environmental and biological samples prior to determination by electrothermal atomic absorption spectrometry, J. Hazard. Mater. 176 (2010) 481–486.
[112] Y. Tao, J.-F. Liu, X.-L. Hu, H.-C. Li, T. Wang, G.-B. Jiang, Hollow fiber supported ionic liquid membrane microextraction for determination of sulfonamides in environmental water samples by high-performance liquid chromatography, J. Chromatogr. A 1216 (2009) 6259–6266.
[113] L. Guo, H.K. Lee, Ionic liquid based three-phase liquid-liquid-liquid solvent bar microextraction for the determination of phenols in seawater samples, J. Chromatogr. A 1218 (2011) 4299–4306.
[114] X. Ma, M. Huang, Z. Li, J. Wu, Hollow fiber supported liquid-phase microextraction using ionic liquid as extractant for preconcentration of benzene, toluene, ethylbenzene and xylenes from water sample with gas chromatography-hydrogen flame ionization detection, J. Hazard. Mater. 194 (2011) 24–29.
[115] C. Zeng, Y. Hu, J. Luo, Ionic liquid-based hollow fiber supported liquid membrane extraction combined with thermospray flame furnace AAS for the determination of cadmium, Microchim. Acta 177 (2011) 53–58.
[116] D. Ge, H.K. Lee, Ionic liquid based hollow fiber supported liquid phase microextraction of ultraviolet filters, J. Chromatogr. A 1229 (2012) 1–5.
[117] R.-S. Zhao, X. Wang, F.-W. Li, S.-S. Wang, L.-L. Zhang, C.-G. Cheng, Ionic liquid/ionic liquid dispersive liquid-liquid microextraction, J. Sep. Sci. 34 (2011) 830–836.
[118] R.-S. Zhao, X. Wang, L.-L. Zhang, S.-S. Wang, J.-P. Yuan, Ionic liquid/ionic liquid dispersive liquid–liquid microextraction, a new sample enrichment procedure for the determination of hexabromocyclododecane diastereomers in environmental water samples, Anal. Methods 3 (2011) 831–836.
[119] R.-S. Zhao, X. Wang, J. Sun, C. Hu, X.-K. Wang, Determination of triclosan and triclocarban in environmental water samples with ionic liquid/ionic liquid dispersive liquid-liquid microextraction prior to HPLC-ESI-MS/MS, Microchim. Acta 174 (2011) 145–151.
[120] R. Khani, F. Shemirani, B. Majidi, Combination of dispersive liquid–liquid microextraction and flame atomic absorption spectrometry for preconcentration and determination of copper in water samples, Desalination 266 (2011) 238–243.
[121] X. Jia, Y. Han, C. Wei, T. Duan, H. Chen, Speciation of mercury in liquid cosmetic samples by ionic liquid based dispersive liquid–liquid microextraction combined with high-performance liquid chromatography-inductively coupled plasma mass spectrometry, J. Anal. At. Spectrom. 26 (2011) 1380–1386.
[122] R.-S. Zhao, L.-L. Zhang, X. Wang, Dispersive liquid-phase microextraction using ionic liquid as extractant for the enrichment and determination of DDT and its metabolites in environmental water samples, Anal. Bioanal. Chem. 399 (2011) 1287–1293.
[123] R.-S. Zhao, S.-S. Wang, C.-G. Cheng, L.-L. Zhang, X. Wang, Rapid enrichment and sensitive determination of tetrabromobisphenol A in environmental water samples with ionic liquid dispersive liquid-phase microextraction prior to HPLC–ESI-MS–MS, Chromatographia 73 (2011) 793–797.

[124] C. Yao, T. Li, P. Twu, W.R. Pitner, J.L. Anderson, Selective extraction of emerging contaminants from water samples by dispersive liquid-liquid microextraction using functionalized ionic liquids, J. Chromatogr. A 1218 (2011) 1556–1566.
[125] E.M. Martinis, L.B. Escudero, P. Berton, R.P. Monasterio, M.F. Filippini, R.G. Wuilloud, Determination of inorganic selenium species in water and garlic samples with on-line ionic liquid dispersive microextraction and electrothermal atomic absorption spectrometry, Talanta 85 (2011) 2182–2188.
[126] C. Zhou, S. Tong, Y. Chang, Q. Jia, W. Zhou, Ionic liquid-based dispersive liquid-liquid microextraction with back-extraction coupled with capillary electrophoresis to determine phenolic compounds, Electrophoresis 33 (2012) 1331–1338.
[127] J. Tian, X. Chen, X. Bai, Comparison of dispersive liquid-liquid microextraction based on organic solvent and ionic liquid combined with high-performance liquid chromatography for the analysis of emodin and its metabolites in urine samples, J. Sep. Sci. 35 (2012) 145–152.
[128] M. Arvand, E. Bozorgzadeh, S. Shariati, M.A. Zanjanchi, Ionic liquid-based dispersive liquid-liquid microextraction for the determination of formaldehyde in wastewaters and detergents, Environ. Monit. Assess. 184 (2012) 7597–7605.
[129] L.H. Li, H.F. Zhang, S. Hu, X.H. Bai, S. Li, Dispersive liquid–liquid microextraction coupled with high-performance liquid chromatography for determination of coumarin compounds in Radix Angelicae Dahuricae, Chromatographia 75 (2012) 131–137.
[130] M. Taziki, F. Shemirani, B. Majidi, Robust ionic liquid against high concentration of salt for preconcentration and determination of rhodamine B, Sep. Purif. Technol. 97 (2012) 216–220.
[131] L.B. Escudero, P. Berton, E.M. Martinis, R.A. Olsina, R.G. Wuilloud, Dispersive liquid-liquid microextraction and preconcentration of thallium species in water samples by two ionic liquids applied as ion-pairing reagent and extractant phase, Talanta 88 (2012) 277–283.
[132] L.B. Escudero, E.M. Martinis, R.A. Olsina, R.G. Wuilloud, Arsenic speciation analysis in monovarietal wines by on-line ionic liquid-based dispersive liquid-liquid microextraction, Food Chem. 138 (2013) 484–490.
[133] Y.-C. Ku, M.-I. Leong, W.-T. Wang, S.-D. Huang, Up-and-down shaker-assisted ionic liquid-based dispersive liquid-liquid microextraction of benzophenone-type ultraviolet filters, J. Sep. Sci. 36 (2013) 1470–1477.
[134] S. Wen, J. Wu, X. Zhu, Room temperature ionic liquid-based dispersive liquid–liquid microextraction combined with flame atomic absorption spectrometry for the speciation of chromium(III) and chromium(VI), J. Mol. Liq. 180 (2013) 59–64.
[135] Z. Li, F. Chen, X. Wang, C. Wang, Ionic liquids dispersive liquid-liquid microextraction and high-performance liquid chromatographic determination of irbesartan and valsartan in human urine, Biomed. Chromatogr. 27 (2013) 254–258.
[136] X. Song, M. Ye, X. Tang, C. Wang, Ionic liquids dispersive liquid-liquid microextraction and HPLC-atomic fluorescence spectrometric determination of mercury species in environmental waters, J. Sep. Sci. 36 (2013) 414–420.
[137] S. Gao, X. Yang, W. Yu, Z. Liu, H. Zhang, Ultrasound-assisted ionic liquid/ionic liquid-dispersive liquid-liquid microextraction for the determination of sulfonamides in infant formula milk powder using high-performance liquid chromatography, Talanta 99 (2012) 875–882.
[138] J. Zhang, M. Li, M. Yang, B. Peng, Y. Li, W. Zhou, H. Gao, R. Lu, Magnetic retrieval of ionic liquids: fast dispersive liquid-liquid microextraction for the determination of benzoylurea insecticides in environmental water samples, J. Chromatogr. A 1254 (2012) 23–29.
[139] E. Stanisz, J. Werner, H. Matusiewicz, Mercury species determination by task specific ionic liquid-based ultrasound-assisted dispersive liquid-liquid microextraction combined with cold vapour generation atomic absorption spectrometry, Microchem. J. 110 (2013) 28–35.
[140] M. Asensio-Ramos, J. Hernández-Borges, T.M. Borges-Miquel, M.Á. Rodríguez-Delgado, Ionic liquid-dispersive liquid-liquid microextraction for the simultaneous determination of pesticides and metabolites in soils using high-performance liquid chromatography and fluorescence detection, J. Chromatogr. A 1218 (2011) 4808–4816.

[141] L. Ye, J. Liu, X. Yang, Y. Peng, L. Xu, Orthogonal array design for the optimization of ionic liquid-based dispersive liquid-liquid microextraction of benzophenone-type UV filters, J. Sep. Sci. 34 (2011) 700–706.
[142] E. Molaakbari, A. Mostafavi, D. Afzali, Ionic liquid ultrasound assisted dispersive liquid-liquid microextraction method for preconcentration of trace amounts of rhodium prior to flame atomic absorption spectrometry determination, J. Hazard. Mater. 185 (2011) 647–652.
[143] D. Han, H. Yan, K.H. Row, Ionic liquid-based dispersive liquid-liquid microextraction for sensitive determination of aromatic amines in environmental water, J. Sep. Sci. 34 (2011) 1184–1189.
[144] M. Arvand, E. Bozorgzadeh, S. Shariatib, M.A. Zanjanchia, Trace determination of linear alkylbenzene sulfonates using ionic liquid based ultrasound-assisted dispersive liquid–liquid microextraction and response surface methodology, Anal. Methods 4 (2012) 2272–2277.
[145] J. Zhang, H. Gao, B. Peng, S. Li, Z. Zhou, Comparison of the performance of conventional, temperature-controlled, and ultrasound-assisted ionic liquid dispersive liquid-liquid microextraction combined with high-performance liquid chromatography in analyzing pyrethroid pesticides in honey samples, J. Chromatogr. A 1218 (2011) 6621–6629.
[146] J.-N. Sun, Y.-P. Shi, J. Chen, Ultrasound-assisted ionic liquid dispersive liquid-liquid microextraction coupled with high performance liquid chromatography for sensitive determination of trace celastrol in urine, J. Chromatogr. B 879 (2011) 3429–3433.
[147] S. Wang, L. Ren, Y. Xu, F. Liu, Application of ultrasound-assisted ionic liquid dispersive liquid-phase microextraction followed high-performance liquid chromatography for the determination of fungicides in red wine, Microchim. Acta 173 (2011) 453–457.
[148] K.-J. Huang, C.-X. Jin, S.-L. Song, C.-Y. Wei, Y.-M. Liu, J. Li, Development of an ionic liquid-based ultrasonic-assisted liquid-liquid microextraction method for sensitive determination of biogenic amines: application to the analysis of octopamine, tyramine and phenethylamine in beer samples, J. Chromatogr. B 879 (2011) 579–584.
[149] D. Ge, H.K. Lee, A new 1-hexyl-3-methylimidazolium tris(pentafluoroethyl)trifluorophosphate ionic liquid based ultrasound-assisted emulsification microextraction for the determination of organic ultraviolet filters in environmental water samples, J. Chromatogr. A 1251 (2012) 27–32.
[150] Y. Zhang, H.K. Lee, Ionic liquid-based ultrasound-assisted dispersive liquid-liquid microextraction followed high-performance liquid chromatography for the determination of ultraviolet filters in environmental water samples, Anal. Chim. Acta 750 (2012) 120–126.
[151] M.M. Parrilla, P. Parrilla, M. Martínez, M.D. Gil, Determination of eight fluoroquinolones in groundwater samples with ultrasound-assisted ionic liquid dispersive liquid-liquid microextraction prior to high-performance liquid chromatography and fluorescence detection, Anal. Chim. Acta 748 (2012) 20–27.
[152] S. Dong, Q. Hu, Z. Yang, R. Liu, G. Huang, T. Huang, An ionic liquid-based ultrasound assisted dispersive liquid–liquid microextraction procedure followed by HPLC for the determination of low concentration of phytocides in soil, Microchem. J. 110 (2013) 221–226.
[153] P. Berton, E.M. Martinis, L.D. Martinez, R.G. Wuilloud, Selective determination of inorganic cobalt in nutritional supplements by ultrasound-assisted temperature-controlled ionic liquid dispersive liquid phase microextraction and electrothermal atomic absorption spectrometry, Anal. Chim. Acta 713 (2012) 56–62.
[154] P. Liang, F. Wang, Q. Wan, Ionic liquid-based ultrasound-assisted emulsification microextraction coupled with high performance liquid chromatography for the determination of four fungicides in environmental water samples, Talanta 105 (2013) 57–62.
[155] M.M. Parrilla, P. Parrilla, M. Martínez, M.D. Gil, A. Uclés, Ultrasound-assisted ionic liquid dispersive liquid-liquid microextraction coupled with liquid chromatography-quadrupole-linear ion trap-mass spectrometry for simultaneous analysis of pharmaceuticals in wastewaters, J. Chromatogr. A 1291 (2013) 19–26.
[156] D. Ge, H.K. Lee, Ionic liquid based dispersive liquid-liquid microextraction coupled with micro-solid phase extraction of antidepressant drugs from environmental water samples, J. Chromatogr. A 1317 (2013) 217–222.

[157] H. Shirkhanloo, A. Rouhollahi, H.Z. Mousavi, Ultra-trace arsenic determination in urine and whole blood samples by flow injection-hydride generation atomic absorption spectrometry after preconcentration and speciation based on dispersive liquid-liquid microextraction, Bull. Korean Chem. Soc. 32 (2011) 3923–3927.
[158] M.C. Breadmore, Ionic liquid-based liquid phase microextraction with direct injection for capillary electrophoresis, J. Chromatogr. A 1218 (2011) 1347–1352.
[159] L. Zhang, F. Chen, S. Liu, B. Chen, C. Pan, Ionic liquid-based vortex-assisted dispersive liquid-liquid microextraction of organophosphorus pesticides in apple and pear, J. Sep. Sci. 35 (2012) 2514–2519.
[160] J.M. Padró, M.E. Marsón, G.E. Mastrantonio, J. Altcheh, F. García-Bournissen, M. Reta, Development of an ionic liquid-based dispersive liquid-liquid microextraction method for the determination of nifurtimox and benznidazole in human plasma, Talanta 107 (2013) 95–102.
[161] H. Qin, B. Li, M.S. Liu, Y.L. Yang, Separation and pre-concentration of glucocorticoids in water samples by ionic liquid supported vortex-assisted synergic microextraction and HPLC determination, J. Sep. Sci. 36 (2013) 1463–1469.
[162] R.N. Rao, S.S. Raju, R.M. Vali, Ionic-liquid based dispersive liquid-liquid microextraction followed by high performance liquid chromatographic determination of anti-hypertensives in rat serum, J. Chromatogr. B 931 (2013) 174–180.
[163] G. Absalan, M. Akhond, L. Sheikhian, D.M. Goltz, Temperature-controlled ionic liquid-based dispersive liquid-phase microextraction, preconcentration and quantification of nano-amounts of silver ion by using disulfiram as complexing agent, Anal. Methods 3 (2011) 2354–2359.
[164] Q. Zhou, L. Pang, J. Xiao, Ultratrace determination of carbamate pesticides in water samples by temperature controlled ionic liquid dispersive liquid phase microextraction combined with high performance liquid phase chromatography, Microchim. Acta 173 (2011) 477–483.
[165] Q. Zhou, Y. Gao, G. Xie, Determination of bisphenol A, 4-n-nonylphenol, and 4-tert-octylphenol by temperature-controlled ionic liquid dispersive liquid-phase microextraction combined with high performance liquid chromatography-fluorescence detector, Talanta 85 (2011) 1598–1602.
[166] Q. Zhou, X. Zhang, G. Xie, Preconcentration and determination of pyrethroid insecticides in water with ionic liquid dispersive liquid-phase microextraction in combination with high performance liquid chromatography, Anal. Methods 3 (2011) 356–361.
[167] Q. Zhou, Y. Gao, J. Xiao, G. Xie, Sensitive determination of phenols from water samples by temperature-controlled ionic liquid dispersive liquid-phase microextraction, Anal. Methods 3 (2011) 653–658.
[168] Q. Zhou, X. Zhang, G. Xie, Simultaneous analysis of phthalate esters and pyrethroid insecticides in water samples by temperature-controlled ionic liquid dispersive liquid-phase microextraction combined with high-performance liquid chromatography, Anal. Methods 3 (2011) 1815–1820.
[169] F. Shah, T.G. Kazi, H.I. Afridi, M. Soylak, Temperature controlled ionic liquid-dispersive liquid phase microextraction for determination of trace lead level in blood samples prior to analysis by flame atomic absorption spectrometry with multivariate optimization, Anal. Chim. Acta 101 (2012) 5–10.
[170] S. Sadeghi, A. Zeraatkar Moghaddam, Preconcentration and speciation of trace amounts of chromium in saline samples using temperature-controlled microextraction based on ionic liquid as extraction solvent and determination by electrothermal atomic absorption spectrometry, Talanta 99 (2012) 758–766.
[171] Y. Gao, Q. Zhou, G. Xie, Z. Yao, Temperature-controlled ionic liquid dispersive liquid-phase microextraction combined with HPLC with ultraviolet detector for the determination of fungicides, J. Sep. Sci. 35 (2012) 3569–3574.
[172] Y. Zhang, H.K. Lee, Determination of ultraviolet filters in environmental water samples by temperature-controlled ionic liquid dispersive liquid-phase microextraction, J. Chromatogr. A 1271 (2013) 56–61.
[173] X. Jiang, H. Zhang, X. Chen, Determination of phenolic compounds in water samples by HPLC following ionic liquid dispersive liquid-liquid microextraction and cold-induced aggregation, Microchim. Acta 175 (2011) 341–346.

[174] H. Zhang, X. Chen, X. Jiang, Determination of phthalate esters in water samples by ionic liquid cold-induced aggregation dispersive liquid-liquid microextraction coupled with high-performance liquid chromatography, Anal. Chim. Acta 689 (2011) 137–142.

[175] M. Zeeb, M. Sadeghi, Modified ionic liquid cold-induced aggregation dispersive liquid-liquid microextraction followed by atomic absorption spectrometry for trace determination of zinc in water and food samples, Microchim. Acta 175 (2011) 159–165.

[176] M. Eisapour, F. Shemirani, B. Majidi, M. Baghdadi, Ultrasound assisted cold-induced aggregation: an improved method for trace determination of volatile phenol, Microchim. Acta 177 (2012) 349–355.

[177] T. Li, M.D. Joshi, D.R. Ronning, J.L. Anderson, Ionic liquids as solvents for in situ dispersive liquid-liquid microextraction of DNA, J. Chromatogr. a 1272 (2013) 8–14.

[178] S. Li, H. Gao, J. Zhang, Y. Li, B. Peng, Z. Zhou, Determination of insecticides in water using in situ halide exchange reaction-assisted ionic liquid dispersive liquid-liquid microextraction followed by high-performance liquid chromatography, J. Sep. Sci. 34 (2011) 3178–3185.

[179] J. López-Darias, V. Pino, J.H. Ayala, A.M. Afonso, In-situ ionic liquid-dispersive liquid-liquid microextraction method to determine endocrine disrupting phenols in seawaters and industrial effluents, Microchim. Acta 174 (2011) 213–222.

[180] M. Vaezzadeh, F. Shemirani, B. Majidi, Determination of silver in real samples using homogeneous liquid-liquid microextraction based on ionic liquid, J. Anal. Chem. 67 (2011) 28–34.

[181] B. Delgado, V. Pino, J.L. Anderson, J.H. Ayala, A.M. Afonso, V. González, An in-situ extraction-preconcentration method using ionic liquid-based surfactants for the determination of organic contaminants contained in marine sediments, Talanta 99 (2012) 972–983.

[182] F. Galán-Cano, R. Lucena, S. Cárdenas, M. Valcárcel, Ionic liquid based in situ solvent formation microextraction coupled to thermal desorption for chlorophenols determination in waters by gas chromatography/mass spectrometry, J. Chromatogr. A 1229 (2012) 48–54.

[183] M. Germán-Hernández, V. Pino, J.L. Anderson, A.M. Afonso, A novel in situ preconcentration method with ionic liquid-based surfactants resulting in enhanced sensitivity for the extraction of polycyclic aromatic hydrocarbons from toasted cereals, J. Chromatogr. A 1227 (2012) 29–37.

[184] I. López-García, Y. Vicente-Martínez, M. Hernández-Córdoba, Determination of very low amounts of chromium(iii) and (vi) using dispersive liquid–liquid microextraction by in situ formation of an ionic liquid followed by electrothermal atomic absorption spectrometry, J. Anal. At. Spectrom. 27 (2012) 874–880.

[185] Y. Xiao, H. Zhang, Homogeneous ionic liquid microextraction of the active constituents from fruits of *Schisandra chinensis* and *Schisandra sphenanthera*, Anal. Chim. Acta 712 (2012) 78–84.

[186] J. Zhang, Z. Liang, S. Li, Y. Li, B. Peng, W. Zhou, H. Gao, In-situ metathesis reaction combined with ultrasound-assisted ionic liquid dispersive liquid-liquid microextraction method for the determination of phenylurea pesticides in water samples, Talanta 98 (2012) 145–151.

[187] Q. Zhong, P. Su, Y. Zhang, R. Wang, Y. Yang, In-situ ionic liquid-based microwave-assisted dispersive liquid–liquid microextraction of triazine herbicides, Microchim. Acta 178 (2012) 341–347.

[188] I. López-García, Y. Vicente-Martínez, M. Hernández-Córdoba, Determination of lead and cadmium using an ionic liquid and dispersive liquid-liquid microextraction followed by electrothermal atomic absorption spectrometry, Talanta 110 (2013) 46–52.

[189] B. Peng, X. Yang, J. Zhang, F. Du, W. Zhou, H. Gao, R. Lu, Comparison of two ultrasound-enhanced microextractions combined with HPLC for determining acaricides in water, J. Sep. Sci. 36 (2013) 2196–2202.

[190] C. Yu, S. Zhang, J. Zhang, S. Li, W. Zhou, H. Gao, R. Lu, An in situ ionic liquid dispersive liquid-liquid microextraction method for the detection of pyrethroids by LC-UV in environmental water samples, J. Braz. Chem. Soc. 24 (2013) 1034–1040.

[191] E. Stanisz, A. Zgoła-Grześkowiak, In situ metathesis ionic liquid formation dispersive liquid–liquid microextraction for copper determination in water samples by electrothermal atomic absorption spectrometry, Talanta 115 (2013) 178–183.

[192] L.B. Agudelo Mesa, J.M. Padró, M. Reta, Analysis of non-polar heterocyclic aromatic amines in beefburguers by using microwave-assisted extraction and dispersive liquid-ionic liquid microextraction, Food Chem. 141 (2013) 1694–1701.
[193] A. Sarafraz-Yazdi, A. Amiri, Liquid-phase microextraction, TrAC Trends Anal. Chem. 29 (2010) 1–14.
[194] M.A. Jeannot, A. Przyjazny, J.M. Kokosa, Single drop microextraction—development, applications and future trends, J. Chromatogr. A 1217 (2010) 2326–2336.
[195] A. Jain, K.K. Verma, Recent advances in applications of single-drop microextraction: a review, Anal. Chim. Acta 706 (2011) 37–65.
[196] D. Han, K.H. Row, Trends in liquid-phase microextraction, and its application to environmental and biological samples, Microchim. Acta 176 (2012) 1–22.
[197] Q. Zhou, H. Bai, G. Xie, J. Xiao, Temperature-controlled ionic liquid dispersive liquid phase microextraction, J. Chromatogr. A 1177 (2008) 43–49.
[198] M.A. Jeannot, F.F. Cantwell, Mass transfer characteristics of solvent extraction into a single drop at the tip of a syringe needle, Anal. Chem. 69 (1997) 235–239.
[199] A.L. Theis, A.J. Waldack, S.M. Hansen, M.A. Jeannot, Headspace solvent microextraction, Anal. Chem. 73 (2001) 5651–5654.
[200] J. Liu, G. Jiang, Y. Chi, Y. Cai, Q. Zhou, J.-T. Hu, Use of ionic liquids for liquid-phase microextraction of polycyclic aromatic hydrocarbons, Anal. Chem. 75 (2003) 5870–5876.
[201] C. Ye, Q. Zhou, X. Wang, Improved single-drop microextraction for high sensitive analysis, J. Chromatogr. A 1139 (2007) 7–13.
[202] M. Cruz-Vera, R. Lucena, S. Cárdenas, M. Valcárcel, Ionic liquid-based dynamic liquid-phase microextraction: application to the determination of anti-inflammatory drugs in urine samples, J. Chromatogr. A 1202 (2008) 1–7.
[203] M. Cruz-Vera, R. Lucena, S. Cárdenas, M. Valcárcel, Determination of phenothiazine derivatives in human urine by using ionic liquid-based dynamic liquid-phase microextraction coupled with liquid chromatography, J. Chromatogr. B 877 (2009) 37–42.
[204] L. Xia, X. Li, Y. Wu, B. Hu, R. Chen, Ionic liquids based single drop microextraction combined with electrothermal vaporization inductively coupled plasma mass spectrometry for determination of Co, Hg and Pb in biological and environmental samples, Spectrochim. Acta Part B 63 (2008) 1290–1296.
[205] A. Sarafraz-Yazdi, F. Mofazzeli, Ionic liquid-based submerged single drop microextraction: a new method for the determination of aromatic amines in environmental water samples, Chromatographia 72 (2010) 867–873.
[206] C. Yao, P. Twu, J.L. Anderson, Headspace single drop microextraction using micellar ionic liquid extraction solvents, Chromatographia 72 (2010) 393–402.
[207] L. Vidal, C.E. Domini, N. Grané, E. Psillakis, A. Canals, Microwave-assisted headspace single-drop microextration of chlorobenzenes from water samples, Anal. Chim. Acta 592 (2007) 9–15.
[208] L. Vidal, E. Psillakis, C.E. Domini, N. Grané, F. Marken, A. Canals, An ionic liquid as a solvent for headspace single drop microextraction of chlorobenzenes from water samples, Anal. Chim. Acta 584 (2007) 189–195.
[209] S. Pedersen-Bjergaard, K.E. Rasmussen, Liquid-liquid-liquid microextraction for sample preparation of biological fluids prior to capillary electrophoresis, Anal. Chem. 71 (1999) 2650–2656.
[210] S. Pedersen-Bjergaard, K.E. Rasmussen, Liquid-phase microextraction with porous hollow fibers, a miniaturized and highly flexible format for liquid-liquid extraction, J. Chromatogr. A 1184 (2008) 132–142.
[211] S. Pedersen-Bjergaard, K.E. Rasmussen, Electrokinetic migration across artificial liquid membranes. New concept for rapid sample preparation of biological fluids, J. Chromatogr. A 1109 (2006) 183–190.
[212] X. Jiang, H.K. Lee, Solvent bar microextraction, Anal. Chem. 76 (2004) 5591–5596.
[213] L. Chimuka, E. Cukrowska, M. Michel, B. Buszewski, Advances in sample preparation using membrane-based liquid-phase microextraction techniques, TrAC Trends Anal. Chem. 30 (2011) 1781–1792.

[214] M. Ghambarian, Y. Yamini, A. Esrafili, Developments in hollow fiber based liquid-phase microextraction: principles and applications, Microchim. Acta 177 (2012) 271–294.
[215] R. Fortunato, C.A.M. Afonso, M.A.M. Reis, J.G. Crespo, Supported liquid membranes using ionic liquids: study of stability and transport mechanisms, J. Memb. Sci. 242 (2004) 197–209.
[216] R. Fortunato, C.A.M. Afonso, J. Benavente, E. Rodriguez, J.G. Crespo, Stability of supported ionic liquid membranes as studied by X-ray photoelectron spectroscopy, J. Memb. Sci. 256 (2005) 216–223.
[217] Y. Zhang, R. Wang, P. Su, Y. Yang, Ionic liquid-based solvent bar microextraction for determination of organophosphorus pesticides in water samples, Anal. Methods 5 (2013) 5074–5078.
[218] W. Liu, Z. Wei, Q. Zhang, F. Wu, Z. Lin, Q. Lu, F. Lin, G. Chen, L. Zhang, Novel multifunctional acceptor phase additive of water-miscible ionic liquid in hollow-fiber protected liquid phase microextraction, Talanta 88 (2012) 43–49.
[219] C. Zeng, Y. Lin, N. Zhou, J. Zheng, W. Zhang, Room temperature ionic liquids enhanced the speciation of Cr(VI) and Cr(III) by hollow fiber liquid phase microextraction combined with flame atomic absorption spectrometry, J. Hazard. Mater. 237-238 (2012) 365–370.
[220] X. Guo, B. Chen, M. He, B. Hu, X. Zhou, Ionic liquid based carrier mediated hollow fiber liquid liquid liquid microextraction combined with HPLC-ICP-MS for the speciation of phenylarsenic compounds in chicken and feed samples, J. Anal. At. Spectrom. 28 (2013) 1638–1647.
[221] M. Rezaee, Y. Assadi, M.-R. Milani Hosseini, E. Aghaee, F. Ahmadi, S. Berijani, Determination of organic compounds in water using dispersive liquid-liquid microextraction, J. Chromatogr. A 1116 (2006) 1–9.
[222] M. Rezaee, Y. Yamini, M. Faraji, Evolution of dispersive liquid-liquid microextraction method, J. Chromatogr. A 1217 (2010) 2342–2357.
[223] H. Yan, H. Wang, Recent development and applications of dispersive liquid-liquid microextraction, J. Chromatogr. A 1295 (2013) 1–15.
[224] L. Kocúrová, I.S. Balogh, J. Šandrejová, V. Andruch, Recent advances in dispersive liquid–liquid microextraction using organic solvents lighter than water. A review, Microchem. J. 102 (2012) 11–17.
[225] M.J. Trujillo-Rodríguez, P. Rocío-Bautista, V. Pino, A.M. Afonso, Ionic liquids in dispersive liquid-liquid microextraction, TrAC Trends Anal. Chem. 51 (2013) 87–106.
[226] Y. Liu, E. Zhao, W. Zhu, H. Gao, Z. Zhou, Determination of four heterocyclic insecticides by ionic liquid dispersive liquid-liquid microextraction in water samples, J. Chromatogr. A 1216 (2009) 885–891.
[227] M. Cruz-Vera, R. Lucena, S. Cárdenas, M. Valcárcel, One-step in-syringe ionic liquid-based dispersive liquid-liquid microextraction, J. Chromatogr. A 1216 (2009) 6459–6465.
[228] V. Andruch, M. Burdel, L. Kocúrová, J. Šandrejová, I.S. Balogh, Application of ultrasonic irradiation and vortex agitation in solvent microextraction, TrAC Trends Anal. Chem. 49 (2013) 1–9.
[229] Q. Zhou, X. Zhang, J. Xiao, Ultrasound-assisted ionic liquid dispersive liquid-phase microextraction: a novel approach for the sensitive determination of aromatic amines in water samples, J. Chromatogr. A 1216 (2009) 4361–4365.
[230] S. Li, S. Cai, W. Hu, H. Chen, H. Liu, Ionic liquid-based ultrasound-assisted dispersive liquid–liquid microextraction combined with electrothermal atomic absorption spectrometry for a sensitive determination of cadmium in water samples, Spectrochim. Acta Part B 64 (2009) 666–671.
[231] M. Baghdadi, F. Shemirani, Cold-induced aggregation microextraction: a novel sample preparation technique based on ionic liquids, Anal. Chim. Acta 613 (2008) 56–63.

CHAPTER 4

Ionic Liquids for Separation of Metal Ions and Organic Compounds from Aqueous Solutions

Magdalena Regel-Rosocka, Katarzyna Materna
Poznan University of Technology, Institute of Chemical Technology and Engineering, Poznan, Poland

1. INTRODUCTION

For more than a decade, ionic liquids have been often applied for phase-boundary catalysis, as solvents for organic synthesis and biochemical reactions, as electrolytes for electrochemical devices, and in separation processes. The reason for such an interest comes from almost unlimited possibilities of tunability/design of new ionic liquids for specific applications and needs, insignificant volatility, high thermal stability and ionic conductivity, and broad electrochemical window [1,2].

As for separation processes, application of ionic liquids (ILs) can be divided into some specific systems in which ILs replace typical solvents, ILs are used as new extractants, ILs form a pure extracting phase, or they can be additives to traditional systems [3]. Some examples of the vast variety of cations and anions applied for separation are presented in Table 1 [4−17]. Extraction of aromatic hydrocarbons from an aromatic/aliphatic mixture [18] can be given as an example of well-elaborated separation with ILs, including pilot plant scale. Although cyano-containing ILs can successfully replace conventional solvents in aromatic/aliphatic extraction, giving much better results of separation, they are still too expensive to use them on a large scale.

Regarding separations from aqueous solutions, extraction, adsorption, membrane techniques, or biphasic systems containing ILs are investigated for both low-molecular-weight organic compounds and metal ions [19−21].

Separation with ILs of almost all metal ions has been investigated in various systems. In many cases, imidazolium ILs are alternative solvents that replace typical volatile solvents in an organic phase used for extraction. However, imidazolium, ammonium, and phosphonium ILs are often able to form ion pairs or complexes with solutes and actively separate them from aqueous solutions; they can be considered as prospective solvents and carriers in separation techniques [4,5]. Moreover, task-specific ILs are designed and synthesized for particular purposes (e.g., Hg^{2+} separation from aqueous solutions) [6].

However, even if some flow sheets are proposed, these are mainly separations only on a laboratory scale. Hence, in the light of a variety of the separation systems and the

Ionic Liquids in Separation Technology
http://dx.doi.org/10.1016/B978-0-444-63257-9.00004-3
Copyright © 2014 Elsevier B.V.
All rights reserved.

Table 1 Examples of ILs Used in Separation Processes [4–17]

Cations

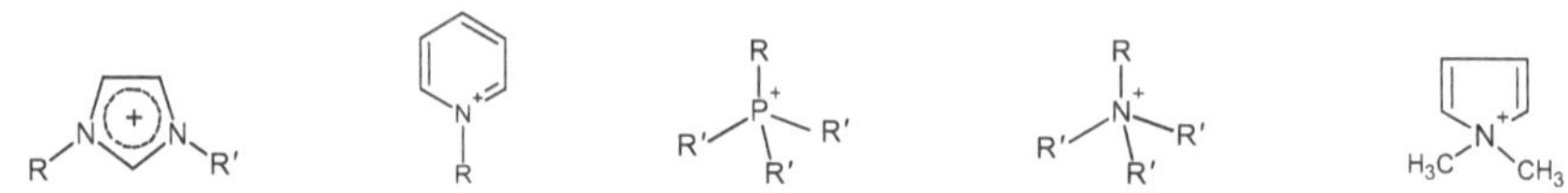

Imidazolium $[C_nC_mIm]$ Alkylpyridinium Phosphonium $[C_nC_nC_nC_mP]$ Ammonium $[C_nC_nC_nC_mN]$ 1,1′-alkylpyrolinium

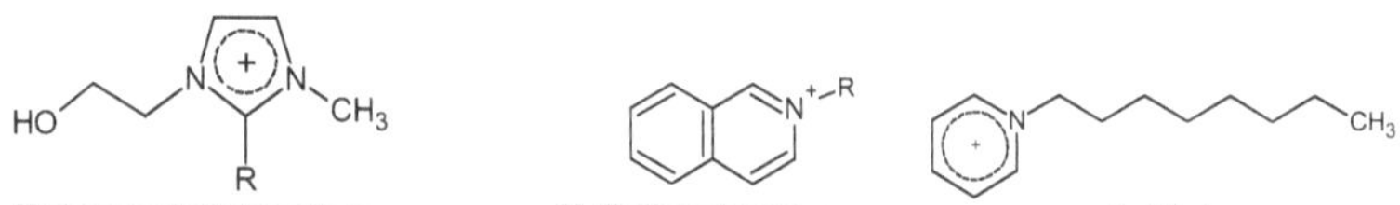

Hydroxymethylimidazolium *N*-alkylisoquinolinium Pyridinium

Task specific cations

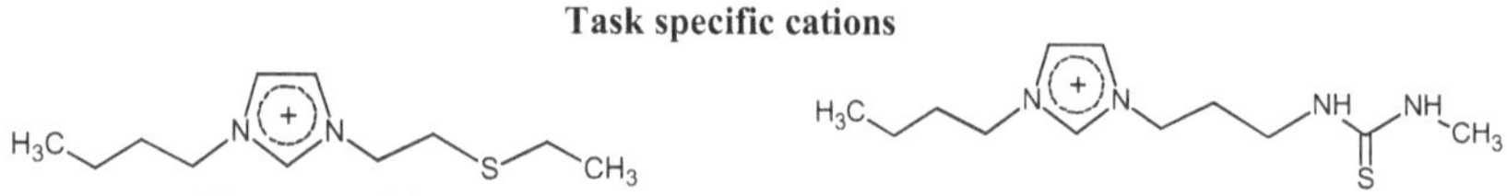

Thioeter-appended IL Thiourea derivative

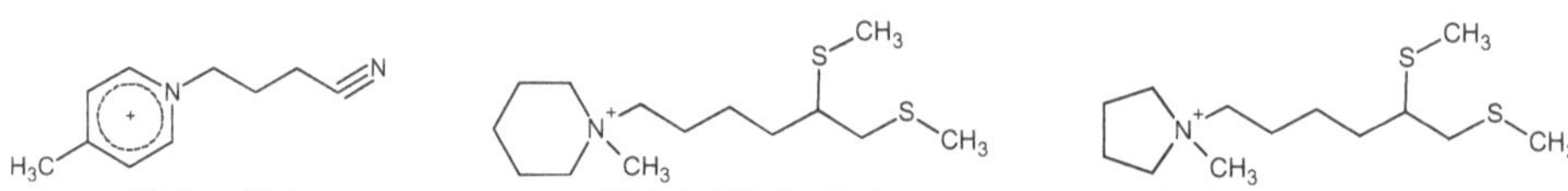

Nitrilepyridinium Methylsulfidepiperidinium Methylsulfidepyrrolidinium

Anions

Cl^- Chloride; Br^- Bromide; SCN^- Thiocyanate; BF_4^- Tetrafluoroborate; PF_6^- Hexafluorophosphate; $CF_3SO_3^-$ Trifluoromethylsulfonate [TfO]; $B(CN)_4^-$ Tetracyanoborate

$CF_3(CF_2)_3SO_3^-$ Nonafluoromethylsulfonate [NfO]; $N(SO_2CF_3)_2^-$ Bis(trifluoromethylsulfonyl)imide [NTf_2]; $N(SO_2CF_2CF_3)_2^-$ Bis(perfluoroethylsulfonyl)imide [BETI]

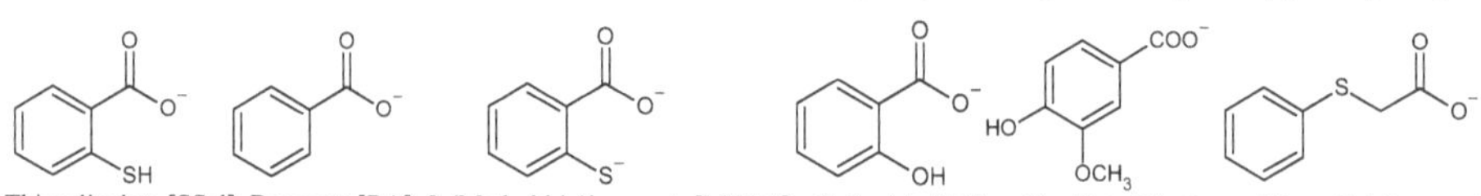

Thiosalicylate [SSal] Benzoate [BA] 2-(Methylthio)benzoate [MTBA] Salicylate [Sal] Vanillate [Van] (Phenylthio)acetate [PTA]

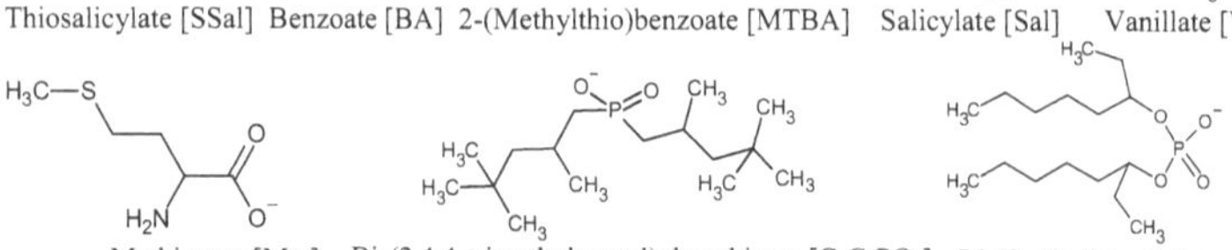

Methionate [Met] Bis(2,4,4-trimethylpentyl)phosphinate $[C_8C_8PO_2]$ Bis(2-ethylhexyl)phosphonate [DEHP]

R, R′, alkyl groups, the same or different; n, m, number of carbons in hydrocarbon chain.

abundance of ILs used, there is a question of the necessity and real opportunities for application of the IL-based separation processes.

In this chapter, the current applications of ILs as solvents and carriers in liquid–liquid extraction, adsorption, membrane techniques, and biphasic systems for metal ions or organic compound separations are presented and discussed in view of their further use and prospective development.

2. LIQUID–LIQUID EXTRACTION

Liquid–liquid extraction is a well-established and described technique, used in industry for more than 60 years [22]. However, as a consequence of industrial and economic development, there are new problems to be solved and challenges to face. Recovery of metal ions from secondary sources and treatment of wastewaters to fulfill the standards of "zero discharge technologies" are the reasons for continuous search for new solutions in separation techniques, where use of ILs seems to give new opportunities.

2.1 Separation of Metal Ions

2.1.1 ILs Replacing Typical Solvents

Among the IL extraction systems, the ones involving imidazolium ionic liquids are the most frequently investigated and reported in the literature for separation of metal ions. Imidazolium-based ILs usually replace volatile organic solvents in conventional systems.

1-Alkyl-3-methylimidazolium hexafluorophosphate $[C_nC_1Im][PF_6]$, tetrafluoroborate $[C_nC_1Im\ BF_4]$, and bis(trifluoromethylsulfonyl)imide $[C_nC_1Im][NTf_2]$ are the most frequently applied for liquid–liquid extraction. At the same time, their physicochemical properties, toxicological aspects, and reactivity are well defined, and much data on them are available in the literature. For example, 1-hexyl-3-methylimidazolium bis(trifluoromethylsulfonyl)imide $[C_6C_1Im][NTf_2]$ is regarded as an International Union of Pure and Applied Chemistry standard IL [23], for which many investigations and measurements have been done.

In many cases, extraction of metal ions into the hydrophobic ionic liquid phase is insignificant because of hydration of metal cations in the aqueous phase and weak affinity of the IL phase to the aqueous one. Thus, extraction of metal ions uses extractants—organic compounds able to react (form complexes or ion pairs) with metal ions and make them more hydrophobic to facilitate their transfer from the aqueous feed to the organic phase. A variety of extractants/ligands diluted in ILs can be categorized as follows: macrocyclic compounds (e.g., pyridinecalix-4-arene, 18-crown-6 ether (18C6), or dicyclohexano-18-crown-6 (DCH18C6)), solvating extractants (tributyl phosphate (TBP), octyl(phenyl)-*N,N*-diisobutylcarbamoylmethylphosphine oxide (CMPO), 1-(2-pyridylazo)-2-naphthol), and TAN (1-(2-thiazolyl)-2-naphthol (PAN)), neutral (*N,N,N′,N′*-tetra(octyl)diglycolamide (TODGA), bis(2-ethylhexyl)2-ethylhexyl phosphonate (DEHEHP, P507)), or acidic (bis(2-ethylhexyl)phosphoric acid (DEHPA, P204)). Some examples of metal ion separations from aqueous solutions with ILs replacing typical solvents are given in Table 2 [7,17,24–48]. ILs replace such molecular solvents as chloroform, dodecane, octan-1-ol, or isooctane. However, it is not always the presence of ionic solvents, instead of the conventional ones, that positively affects extraction behavior of the organic phases. For instance, uranium is extracted more efficiently when

Table 2 Examples of Separation of Metal Ions from Aqueous Solutions with ILs as Replacing Typical Solvents

IL	n	Ligand or Solvent	Extracted Species	Ref.
$[C_nC_1Im]$ $[PF_6]$	4, 6, 8	18C6, DCH18C6	Na^+, Cs^+, Sr^{2+}	[24]
	4–9	Dtb18C6	K^+, Na^+, Rb^+, Cs^+, Li^+	[25]
		DC18C6	Mg^{2+}, Ca^{2+}, Sr^{2+}, Ba^{2+}	[26]
			Pb^{2+}	[26]
	4, 6, 8	Dithizone	Hg^{2+}, Na^+, Ni^{2+}, Co^{2+}, Cd^{2+}, Fe^{3+}	[27]
	4	PAN and TAN	UO_2^{2+}	[28]
	4, 6, 8	TBP	UO_2^{2+}	[29]
	4	TBDA, MBDA	Pu^{4+}, Th^{4+}, Am^{3+}, UO_2^{2+}	[17]
		CMPO/TBP		[30]
	4	CMPO	Ce^{3+}, Eu^{3+}, Y^{3+}	[31,32]
	4, 6	PAN and TAN	Hg^{2+}, Na^+, Ni^{2+}, Co^{2+}, Cd^{2+}, Fe^{3+}	[27]
	4, 6, 8	^tBu [1]CH_2Py, ^tBu [4]CH_2Py	Ag^+ in the presence of Cu^{2+}, Zn^{2+}, Co^{2+}, Ni^{2+}	[33]
	8	DEHEHP	Ce^{4+}, Th^{4+}, RE	[34]
$[C_nC_1Im][NTf_2]$	5–10	DCH18C6	Na^+	[35,36]
	2, 4, 6, 8	*N*-Alkylaza-18-crown-6	Na^+, K^+, Cs^+, Sr^{2+}	[37]
		ether, BOBCalixC6	Cs^+ K^+, Na^+, Sr^{2+}	[38]
		18C6, DC18C6	Sr^{2+}	[39,40]
	2, 5, 6, 8	DC18C6/TBP	Sr^{2+}	[41]
	10	DC18C6/DAAP		
	5, 8, 10	TBP	UO_2^{2+}	[42]
	8	CMPO	UO_2^{2+}	[30]
	10	DEHPA, Cyanex 272	UO_2^{2+}, Am^{3+}, Nd^{3+}, Eu^{3+}	[7]
	4	Htta	Ln^{3+}	[43]
		CMPO	Ce^{3+}, Eu^{3+}, Y^{3+}	[32]
	2, 4, 6	TODGA	La^{3+}, Eu^{3+}, Lu^{3+}	[44]
$[C_nC_1Im][BETI]$	2, 4	DCH18C6	Cs^+, Sr^{2+} in Cl^-	[45]
$[C_{14}C_6C_6C_6P]$ $[NTf_2]$	–	DCH18C6	TcO_4^-, ReO_4^-	[46]
$[(CH_3)_3NCH_2CH_2OMe]$ $[NTf_2]$	–	TBP	UO_2^{2+}	[47]
		CMPO, TODGA	Rh^{3+}	[48]

Table 3 Distribution Ratios of Metal Ions in IL (D_{IL}) and Conventional (D_{conv}) Extraction Systems

Metal Ions	D_{IL}	D_{conv}	Ref.
Cs	23.9	380	[49]
Sr	213	935	[49]
Na	0.11	0.06	[35]
UO_2	1000	100	[17]
UO_2	0.16	3	[42]
La	100	0.01	[44]

TBP is diluted with dodecane than with $[C_nC_1Im][NTf_2]$ [42] (Table 3 [17,35,42,44,49]). In some cases, because of the high viscosity of ILs, cosolvents, such as carbon tetrachloride, 1,2-dichloroethane, *m*-xylene, diisopropylbenzene, phenylcyclohexane, saturated hydrocarbons (e.g., dodecane), and TBP, are added to the organic phase.

An influence of IL structure on partition behavior of metal ions is related to their hydrophobicity, resulting from both alkyl chain length on IL cation and type of anion. It strongly affects not only solubility of metal ions in the organic phase but also extractant solubility in IL, water transfer to the organic phase, and IL solubility in the aqueous phase. Extraction of Cs^+, Rb^+, K^+, Na^+, and Li^+ with crown ethers or calixarene is reported to increase with shortening of the alkyl chain on IL cation [25,38]. However, selection of IL for extraction system should be a compromise between the extraction efficiency and the extractant/ligand solubility in the organic phase.

Sodium extraction with DCH18C6 is an example of positive effect of $[C_nC_1Im][NTf_2]$ presence in the organic phase compared with octan-1-ol, a conventional diluent (Table 3). However, it is noticed that the partitioning of Na^+ is strongly affected by stereochemistry of the crown ether applied, which is not reported in conventional solvents [36]. The mechanism of extraction can be tuned by changing the isomeric form of the extractant; thus, the presence of trans-isomers of DCH18C6 make the neutral complex extraction dominate over the ion-exchange mechanism. The latter mechanism is more environmentally friendly because of no release of IL to the aqueous phase.

Generally, IL systems are reported to demonstrate a tendency to increase distribution ratios and selectivity (Figure 1), resulting from change in extraction mechanism. Conventional extraction of La^{3+} with TODGA in isooctane runs by ion-pair extraction accompanied by anions, whereas replacement of isooctane by $[C_2C_1Im][NTf_2]$ results in a cation-exchange mechanism [44]:

$$Ln^{3+}_{(w)} + 3TODGA_{(o)} + 3[C_nC_1Im]^+_{(o)} \Leftrightarrow LnTODGA^{3+}_{3(o)} + 3[C_nC_1Im]^+_{(w)} \quad (1)$$

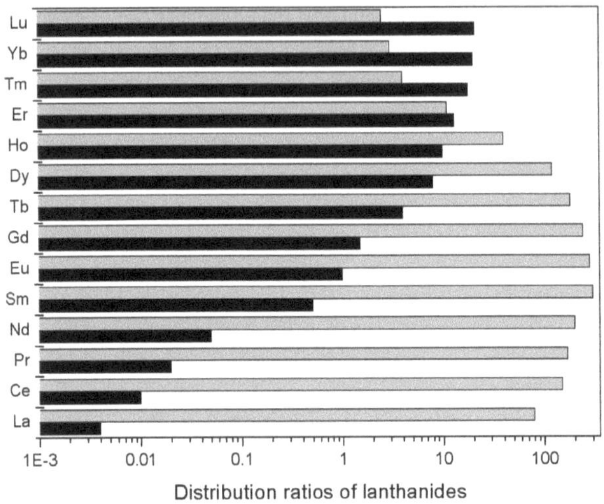

Figure 1 Distribution ratios of lanthanides between TODGA in (■) $[C_2C_1Im][NTf_2]$ and (■) isooctane and the aqueous phase. *(According to Shimojo et al. [44].*

An opposite mechanism is reported for efficient extraction of Ce^{4+} from HNO_3 with pure $[C_8C_1Im][PF_6]$ [50]. The anion-exchange mechanism results in $[PF_6]^-$ release to the aqueous phase:

$$Ce(NO_3)^{2-}_{6(w)} + 2[C_8C_1Im][PF_6]_{(o)} \Leftrightarrow [C_8C_1Im]^+_{(2)} \cdot Ce(NO_3)_{6(o)} + 2[PF_6]^-_{(w)} \quad (2)$$

A simultaneous release of four protons and $[NTf_2]^-$ anion to the aqueous phase (resulting in IL losses) is also reported when lanthanide is extracted by formation of complexes with Htta in $[C_4C_1Im][NTf_2]$ [43].

When DEHEHP is added as an active extractant, the efficiency of the metal ion transport is comparable with the conventional DEHEHP-heptane system [34], and follows the order: $Ce^{4+} > Th^{4+} > RE^{3+}$. DEHEHP both in $[C_8C_1Im][PF_6]$ and heptane takes part in Ce^{4+} extraction, forming a $Ce(HF)(NO_3)_4 \cdot$DEHEHP complex in the organic phase.

As a positive influence on the extraction efficiency of the IL presence in the organic phase, can be given removal of Rh^{3+} from nitric acid solutions with CMPO or TODGA as extractants dissolved with novel methoxyethyl trimethylammonium bis(trifluoromethylsulfonyl)imide IL ($[(CH_3)_3NCH_2CH_2OMe][NTf_2]$) [48] or TcO_4^- extraction from NaOH and/or NH_4OH with DCH18C6 in $[C_{14}C_6C_6C_6P][NTf_2]$ [46]. The use of $[C_{14}C_6C_6C_6P][NTf_2]$ ensures the highest distribution of TcO_4^- both in the presence and absence of the crown ether (DCH18C6) and prefers extraction of TcO_4^- over ReO_4^-. The ion pair transfer according to the reaction:

$$Na^+_{(w)} + TcO^-_{4(w)} + Ce_{(o)} \Leftrightarrow Na \cdot Ce^+TcO^-_{4(o)} \quad (3)$$

is the dominant extraction mechanism. Thus, metal ions are extracted without IL loss (negligible amounts of $[NTf_2]^-$ determined in the aqueous phase) to the aqueous phase. Stepinski et al. [46] indicate that "these solvents may provide the basis for improved approaches to the extraction and recovery of a variety of anions." The enhancement in metal ion extraction (the synergic effect) in several IL systems (Table 3) can be attributed to improvement in the solvation properties of nonpolar organic solvents in the presence of IL, ion-recognition capabilities of complexing ligands, unique ionic solvation environment of the extracted species, and ion-exchange capabilities of ILs [49] (in other words, to a change in the number of ligands coordinating with metal ions).

It is now generally recognized that acid concentration in the aqueous phase strongly influences change in the extraction mechanism in IL systems. Shen et al. [29] reported that ion exchange is favored for UO_2^{2+} extraction at low acidity with diglycolamides (TBDA (*N,N,N′,N′*-tetrabutyl-3-oxapentanediamide) and MBDA (*N,N,N′,N′*-dimethyldibutyl-3-oxapetanediamide)) in $[C_nC_1Im][PF_6]$, whereas neutral uranyl-diamide nitrato complexes are extracted at a high HNO_3 content. Bell and Ikeda [48] arrived at similar conclusions after investigating the organic phase containing TBP in novel ammonium-based ionic liquids for extraction of uranium(VI) from nitric acid.

2.1.2 ILs as Pure Extractants

In several IL extraction systems, no additional ligand is necessary to extract metal ions. Pure imidazolium- or ammonium-based ILs extract efficiently metal ions directly from the aqueous solutions.

Papaiconomou et al. [11] investigated extraction of Cu^{2+}, Hg^{2+}, Ag^+, and Pd^{2+} from aqueous chloride solutions at pH 7 with 10 imidazolium, pyridinium, piperidinium, and pyrrolidinium ionic liquids comprising typical anions (i.e., $[BF_4]^-$, $[NTf_2]^-$, $[TfO]^-$, or $[NfO]^-$) (Table 1). Only mercury has been efficiently extracted ($D > 24$) with imidazolium and pyridinium ionic liquids, whereas the other metal cations are not transferred to the IL phase. Moreover, good extraction abilities of imidazolium ILs with $[NfO^-]$ have been confirmed in the studies on Li^+, Na^+, Cs^+, Ca^{2+}, Sr^{2+}, and La^{3+} extraction [51]. According to the authors, the metal ions with greater charge are more easily transferred to the $[C_nC_1Im][NfO]$ phase and the affinity corresponds with the facility for adsorption onto cation-exchange resins. Metal cations (e.g., La^{3+}) are extracted with pure imidazolium ILs (without additional ligands), according to cation exchange between two phases [51].

Zn^{2+} and Fe^{3+} are transported directly to the imidazolium IL phase, with the extraction efficiency depending on the IL anion. The distribution of metal ions between the IL phase and aqueous chloride solution raises the following order: $[NTf_2]^- < [PF_6]^- < [BF_4]^-$, which corresponds to decreasing hydrophobicity of the anions studied [52,53]. Moreover, the same authors noticed that extraction efficiency increases with lengthening of the alkyl chain of imidazolium cation, unlike in the case of Cs^+ extraction studied by Luo et al. [38]. Depending on the IL type used for extraction, it is likely to recover selectively various metal ions from their mixtures in aqueous solutions. For example, $[C_8C_1Im][BF_4]$ allows Zn^{2+} and Cd^{2+} to be selectively separated from Fe^{3+} and Cu^{2+}, whereas methyltrioctylammonium chloride (known as a commercial extractant, Aliquat 336) removes almost completely Zn^{2+}, Cd^{2+}, and Fe^{3+} from the aqueous feed [53]. The metal ions extracted from HCl solutions are able to form anionic chlorocomplexes that are involved in an extraction mechanism by ion-pair formation with ILs.

Task-specific ionic liquids (TSILs) are a particular type of ILs, functionalized for exact application and high selectivity toward selected metal ions [1,19]. Modification of IL cation with other functionalized groups eliminates the need for use of additional extractant or ligand dissolved in the IL phase. However, the recent results of appending of functional groups to IL anions let us specify their separation properties.

A well-known quaternary ammonium chloride—Aliquat 336, used for extraction for more than 50 years, is a precursor for other ammonium ILs $[C_8C_8C_8C_1N]^+$ already described in the scientific literature [53,54]. Extraction with Aliquat 336 or quaternary phosphonium $[C_{14}C_6C_6C_6P]^+$ cation-based ILs with thiol and thio-ether functionalized anions or aromatic anions (e.g., thiosalicylate, thiocyanate [SCN], 2-[methylthio]benzoate [MTBA], methionate, benzoate [BA], salicylate [SSal], and hexanoate [Hex]) (Table 1) was applied for preconcentration of UO_2^{2+} [13], for Pt^{4+} extraction from chloride solution [14], for removal of Cd^{2+} from natural river matrix [54], or for Fe^{3+}, Cu^{2+}, Ni^{2+}, or Mn^{2+} removal from sulfate, nitrate, or chloride solutions [55] (Table 4 [11–14,27,49–58]). The extraction efficiency of Cd^{2+} from river water decreases in the following order of the ILs used: $[C_8C_8C_8C_1N]$[SSal] (>99.9%) > $[C_8C_8C_8C_1N]$[Hex] (14%) > $[C_8C_8C_8C_1N]$[BA] (11%), whereas Pt^{4+} extraction efficiency tends to diminish with decreasing viscosity of IL in the following order: $[C_{14}C_6C_6C_6P]$[MTBA] (100%) > $[C_8C_8C_8C_1N]$[SSal] (85%) > $[C_{14}C_6C_6C_6P]$[SSal] (76%) > $[C_8C_8C_8C_1N]$[MTBA] (40%). A thiol group appended to the anion of $[C_8C_8C_8C_1N]$[SSal] seems to play a crucial role in Cd^{2+} extraction, and could be considered in this case as TSIL. In addition, this IL is extremely selective toward Cd^{2+} removal from natural river matrix containing Na^+, K^+, Ca^{2+}, and Mg^{2+}, whereas there are no remarkable distribution differences between ILs with other aliphatic and aromatic anions (i.e., [Hex] and [BA]). The extraction properties of the ammonium and phosphonium ILs can be attributed not only to task-specific anions but also physicochemical properties of the IL phase, such as hydrophobicity and viscosity.

$[C_8C_8C_8C_1N]$[Sal] efficiently extracts Fe^{3+} and Cu^{2+} (99% and 89%), which is directly related to the stability constants of corresponding salicylate complexes. The

Table 4 Examples of Separation of Metal Ions from Aqueous Solutions with ILs as Pure Extractants

IL	n	Extracted Species	Ref.
$[C_nC_1Im]$ $[PF_6]$	4, 8	Zn^{2+}, Fe^{3+}, Cd^{2+}, Cu^{2+}	[52,53]
	8	Ce^{4+}, Th^{4+}, RE	[50]
$[C_nC_1Im][NTf_2]$	4, 8	Zn^{2+}, Fe^{3+}, Cd^{2+}, Cu^{2+}	[52,53]
$[C\ C_nC_1Im][NfO]$	4–6	Li^+, Na^+, Cs^+, Ca^{2+}, Sr^{2+}, La^{3+}	[51]
$[diC_2hist18C6][Br]$ $[diC_2hist18C6][NTf_2]$ $[C_nC_1Im\ 18C6][Br]$ $C_nC_1Im\ 18C6][NTf_2]$	4	Cs^+, Sr^{2+}	[49]
TSILs on the base of $[C_nC_1Im]$ $[PF_6]$ thioether, thiourea and urea derivatives	4	Hg^{2+}, Cd^{2+},	[27,56]
$[C_nC_1Im]$ $[PF_6]$, $[C_2OC_2\ C_1Im][PF_6]$, $[(C_2OC_2)_2C_1Im]$ $[PF_6]$ or $[C_nC_1Im][BF_4]$	4, 6, 8	Cu^{2+}, Zn^{2+}, Cr	[57,58]
TSILs on the base of pyridinium or piperidinium cations		Cu^{2+}, Hg^{2+}, Ag^+, Pd^{2+}	[11]
$[C_8C_8C_8C_1N][Cl]$	–	Zn^{2+}, Fe^{3+}, Cd^{2+}	[53]
$[C_8C_8C_8C_1N][SSal]$	–	Cd^{2+}	[54]
		UO_2^{2+}	[13]
		Pd^{2+}	[14]
$[C_8C_8C_8C_1N][Sal]$	–	Fe^{3+}, Ni^{2+}, Cu^{2+}, Mn^{2+},	[55]
$[C_{14}C_6C_6C_6P]$ [MTBA] or $[C_{14}C_6C_6C_6P][SSal]$	–	Pd^{2+}	[14]

mechanism proposed for the extraction of Fe^{3+} may be represented by the following equation [55]:

$$Fe^{3+}_{(w)} + 2[C_8C_8C_8C_1N][HSal]_{(o)} + HSO^-_{4(w)} \Leftrightarrow [C_8C_8C_8C_1N][FeSal_2]_{(o)} + [C_8C_8C_8C_1N][HSO_4]_{(o)} + 2H^+_{(w)} \quad (4)$$

For divalent metals, the mechanism is similar; however, in this case, a formation of metal ion salicylate in the organic phase is expected. [Sal] contacted with acid solution is protonated, but after extraction, the protons are released to the aqueous phase as a result of metal ion complexation by [Sal] anion. Separation factors of Fe^{3+} over other metals, such as Co^{2+}, Ni^{2+}, Mn^{2+}, Cr^{3+}, and Zn^{2+}, are high and in all cases exceed 10. This means that Fe^{3+} can be selectively separated from these solutions.

N-(3-butylimidazolium propyl)aza-18-crown-6 ether bis[(trifluoromethyl)sulphonyl]imide $[C_4C_1Im18C6][NTf_2]$ illustrates TSIL with the IL-cation structure modified with aza-crown ether via covalent bonds [49]. It appears that aza-crown—modified TSILs are less efficient for Cs^{+} and Sr^{2+} extraction than the DCH18C6 or N-octylaza18C6 mixed with $[C_4C_1Im][NTf_2]$ (Table 3). The selectivity of Sr^{2+}/Cs^{+} increases with increasing hydrophobicity of the IL anion [45]. Another type of TSILs, based on thioether, thiourea, and urea derivatives, is involved as carrier in Hg^{2+} and Cd^{2+} liquid—liquid extraction [27,56]. Correspondingly, ILs composed of a functional disulfide group bonded to piperidinium or pyrrolidinium cation show high selectivity toward Hg^{2+} or Cu^{2+} extraction. Moreover, ionic liquids containing a functional nitrile group bonded to pyridinium or piperidinium cation extract completely Ag^{+} and Pd^{2+} from the aqueous phase [11]. The type of functional group, anion and cation ring, strongly influences the extraction abilities and selectivity.

Nevertheless, TSIL and IL use as pure extracting phases is limited because of rather small quantities of these compounds that can be synthesized. In addition, when metal cations are transported according to a cation-exchange mechanism, IL cations are lost to the aqueous phase. For this reason, they could be applied for removal of metal ions in low concentrations or immobilized at any support.

2.1.3 ILs as Extractants in Molecular Solvents

Some ionic liquid extraction systems operate efficiently when IL is diluted with a molecular solvent (Table 5 [8,12,52,53,59—70]). Such an approach helps to overcome the problem of rather small amounts of the ILs synthesized and their high prices.

Extraction of Pd^{2+} with $[C_8C_8C_8C_1N][Cl]$ and $[C_8C_8C_8C_1N][NO_3]$ diluted with chloroform is efficient; however, $[C_8C_8C_8C_1N][Cl]$ provides stronger binding abilities [59]. Transport of the palladium complex to the organic phase is stabilized by the water co-extraction.

The ILs combining quaternary cation and anion of an acidic extractant are called "bifunctional ILs." A synergic effect coming from cation and anion of a bifunctional IL

Table 5 Examples of Separation of Metal Ions from Aqueous Solutions with ILs in Molecular Solvents

IL	Molecular Solvent	Extracted Species	Ref.
$[C_8C_8C_8C_1N][Cl]$ or $[C_8C_8C_8C_1N]$ $[NO_3]$	Chloroform	Pd^{2+}	[59]
$[C_8C_8C_8C_1N]$ [CA-12]	Toluene	Co^{2+}	[60]
$[C_8C_8C_8C_1N]$ $[C_8C_8PO_2]$	Toluene	La^{3+}, Yb^{3+}	[61]
$[C_8C_8C_8C_1N]$[DEHP]or	Dodecane	Eu^{3+}	[12]
$[C_8C_8C_8C_1N]$ [DGA]	Heptane	RE^{3+}	[62]
$[C_{14}C_6C_6C_6P][Cl]$ or $[C_{14}C_6C_6C_6P][C_8C_8PO_2]$	Toluene, kerosene or tributyl phosphate	Fe^{2+}, Zn^{2+}, Co^{2+}, Pd^{2+}, Pt^{4+}	[8,52,53,63—70]

called "inner synergic effect" is reported by some authors [60,62]. An example of such a system is $[C_8C_8C_8C_1N][CA\text{-}12]$ (sec-octylphenoxy acetate) (0.02 M solution in toluene) for Co^{2+} and Ni^{2+} separation from sulfate solutions [60]. Distribution ratios of extracted metal ions increase when $[C_8C_8C_8C_1N][CA\text{-}12]$ IL is used compared with mixture of $[C_8C_8C_8C_1N]_2[SO_4]$ and CA-12 or single solutions of $[C_8C_8C_8C_1N]_2[SO_4]$ or CA-12. The synergy coefficient of Co^{2+} is reported to be as high as almost 110, and the extraction of Co^{2+} from sulfate solution runs through an ion association mechanism:

$$\begin{aligned} &Co^{2+}_{(w)} + [C_8C_8C_8C_1N][CA - 12]_{(o)} \\ &\quad + SO^{2-}_{4(w)} \Leftrightarrow CoSO_4 \cdot [C_8C_8C_8C_1N][CA - 12]_{(o)} \end{aligned} \tag{5}$$

Easy stripping with diluted H_2SO_4 confirms rather weak binding of Co^{2+} to this IL phase.

Extraction of La^{3+} and Yb^{3+} from nitric acid solutions with $[C_8C_8C_8C_1N]$-based ILs does not follow a simple mechanism [61]. After analysis of extraction constants for various extraction reactions of $[C_8C_8C_8C_1N][C_8C_8PO_2]$ and lanthanides (Ln^{3+}), several species may be formed in the organic phase, according to the following reaction:

$$\begin{aligned} &2Ln^{3+}_{(w)} + 2[C_8C_8C_8C_1N][C_8C_8PO_2]_{(o)} \\ &\quad + 6NO^{-}_{3(w)} \Leftrightarrow [C_8C_8C_8C_1N]\left[Ln(NO_3)_3(C_8C_8PO_2)\right]_{(o)} \\ &\quad + [C_8C_8C_8C_1N][NO_3]_{(o)} + \left[Ln(NO_3)_2(C_8C_8PO_2)\right]_{(o)} \end{aligned} \tag{6}$$

The ratio between the species extracted/IL phase depends on their stability and is difficult to evaluate. Ionic liquids are rather soluble in polar diluents (e.g., aromatics), and not miscible with paraffinic diluents. However, some Aliquat 336–based ILs demonstrate good solubility in aliphatic diluents ranging from hexane to hexadecane [12,62]. The ammonium cation combined with anion of well-known acidic extractant, bis(2-ethylhexyl)phosphoric acid $[DEHP]^-$ (shown in Table 1) or bis(2-ethylhexyl) diglycolamic acid $[DGA]^-$, were successfully used for Eu^{3+} or RE^{3+} extraction. Good extraction performance is attributed to the inner synergic effect of IL, and participation of both cation and anion of the IL in the reaction. However, decrease in Eu^{3+} extraction yield is noticed when the organic phase is recycled after pre-equilibration with the aqueous phase [12], whereas for RE extraction, isooctanol as a modifier of the organic phase must be used because of emulsification tendency in the system [62]. The IL decomposes and forms the precursors of the IL—$[C_8C_8C_8C_1N][NO_3]$ and DEHPA. Thus, a procedure to convert the resulting organic phase back into $[C_8C_8C_8C_1N][DEHP]$ is necessary.

The following trihexyl(tetradecyl)phosphonium ILs have been studied as extractants: $[C_{14}C_6C_6C_6P][Cl]$ (Cyphos IL 101), $[C_{14}C_6C_6C_6P][C_8C_8PO_2]$ (Cyphos IL 104),

$[C_{14}C_6C_6C_6P][NTf_2]$ (Cyphos IL 109), and $[C_{14}C_6C_6C_6P][BF_4]$ (Cyphos IL 111), diluted with chloroform, toluene, and alkylene carbonates. In some cases, the organic phase properties were modified with TBP. In fact, many phosphonium ILs used for separation are based on widely known and applied Cytec Industries extractants, such as trialkylphosphine oxides or Cyanex 272; however, there is still little information about their use in extraction processes. They are proposed for extraction of Zn^{2+}, Fe^{3+}, or Fe^{2+} from chloride media [63–66], Co^{2+} separation from Ni^{2+} in chloride solutions [67], and Pd^{2+} separation from mixture of metal chlorides [68–70].

Fe^{3+} can be efficiently separated from Ni^{2+} in 6 M HCl with $[C_{14}C_6C_6C_6P][Cl]$ dissolved in chloroform [63]. The spectroscopic data support the assumption that tetrachloroferrate is formed in the organic phase $[C_{14}C_6C_6C_6P][FeCl_4]$ and simultaneously chloride anion is released to the aqueous phase.

The same mechanism of extraction is confirmed for Zn^{2+} and Co^{2+} removal from HCl solutions with $[C_{14}C_6C_6C_6P][Cl]$. Although extraction with $[C_{14}C_6C_6C_6P]$ $[C_8C_8PO_2]$ runs probably with the formation of bis(2,4,4-trimethylpentyl)phosphinic acid $[HC_8C_8PO_2]$ (known as Cyanex 272) in the organic phase [69]:

$$H^+_{(w)} + PdCl^{2-}_{4(w)} + [C_{14}C_6C_6C_6P][C_8C_8PO_2]_{(o)} \Leftrightarrow [C_{14}C_6C_6C_6P][PdCl_3]_{(o)} + HC_8C_8PO_{2(o)} + Cl^-_{(w)} \quad (7)$$

Extraction efficiency of the phosphonium ILs decreases with increasing hydrophobicity of the anion (similarly to pure imidazolium ILs [52,53]) in the following sequence [65]: $[C_{14}C_6C_6C_6P][Cl] > [C_{14}C_6C_6C_6P][Br] > [C_{14}C_6C_6C_6P][C_8C_8PO_2] > [C_{14}C_6C_6C_6P][BF_4] > [C_{14}C_6C_6C_6P][PF_6] > [C_{14}C_6C_6C_6P][NTf_2]$. The dependence is also observed for the extraction with pure ILs, as it is mentioned for Zn^{2+} and Fe^{3+} extraction with imidazolium salts [52]. Because of low extraction of Zn^{2+}, $[C_{14}C_6C_6C_6P]$ $[NTf_2]$ cannot be considered as an effective extractant in the systems studied. Both $[C_{14}C_6C_6C_6P][Cl]$ and $[C_{14}C_6C_6C_6P][C_8C_8PO_2]$ are considered to be prospective extractants because they transfer metal ions quickly and almost completely to the organic phase.

2.2 Separation of Organic Compounds and Biomolecules

Phenol and phenolic compounds (Table 6 [57,58,71–74]) were extracted quantitatively with $[C_4C_1Im][PF_6]$ [71,74]. 2,4,6-Trinitrophenol is extracted efficiently in its ionic form, whereas resorcinol and pyrocatechol are partitioned much worse [74]. In addition, tyrosol and *p*-hydroxybenzoic acid were removed more efficiently with BF_4 than PF_6 imidazolium ILs [57,58]. Some changes in the structure of the methylimidazolium cation (alkyl group replacement with ether) resulted in slightly better extraction of the phenolic compounds. However, the positive effect on solute solubility in ILs is attributed to the existence of polarizable π electrons present in aromatic compounds

Table 6 Phenols Extracted with Different Ionic Liquids

Name of Phenol	pK_a	IL Used for Extraction	Ref.
Phenol	9.9	$[C_4C_1Im][PF_6]$, $[C_6C_1Im][PF_6]$, $[C_8C_1Im][PF_6]$ $[C_8C_1Im][BF_4]$, $[C_{10}C_1Im][BF_4]$, $[C_6C_6C_6C_6N][DHSS]$, $[C_8C_8C_8C_1N]$ [Sal]	[57,58,71–73]
Tyrosol	10.2	$[C_6C_1Im][PF_6]$, $[C_8C_1Im][PF_6]$ $[C_8C_1Im][BF_4]$, $[C_{10}C_1Im][BF_4]$	[57,58]
p-Hydroxybezoic acid	4.54		
Resorcinol	9.15	$[C_{10}C_1Im][PF_6]$	[71,72,74]
Pyrocatechol	9.85		
2-chlorophenol	8.52		
4-chlorophenol	9.18		
2,4-Dichlorophenol	7.89		
2,4,6-Trichlorophenol	6.23		
2,3,4,5-Tetrachlorophenol	6.35		
Pentachlorophenol	4.70		
2-Nitrophenol	7.23	$[C_6C_1Im][BF_4]$, $[C_8C_1Im][BF_4]$	[72]
4-Nitrophenol	7.14	$[C_6C_1Im][BF_4]$, $[C_8C_1Im][BF_4]$, $[C_6C_6C_6C_6N][DHSS]$, $[C_8C_8C_8C_1N]$ [Sal]	[72,73]
2,4-Dinitrophenol	4.08	$[C_6C_6C_6C_6N][DHSS]$, $[C_8C_8C_8C_1N]$ [Sal]	[73]
2,6-Dinitrophenol	4.15		
2,4,6-Trinitrophenol	0.69	$[C_4C_1Im][PF_6]$, $[C_6C_6C_6C_6N][DHSS]$, $[C_8C_8C_8C_1N]$ [Sal]	[73,74]
Aniline	4.87	$[C_6C_1Im][BF_4]$, $[C_8C_1Im][BF_4]$	[72]
4-Toluidine	5.08		
p-Chloroaniline	3.89		
1-Naphthol	9.85	$[C_6C_6C_6C_6N][DHSS]$, $[C_8C_8C_8C_1N]$ [Sal]	[73,74]
2-Naphthol	9.63		
p-Toluidine	5.07		
3-Nitroaniline	2.47		
Tryptamine hydrochloride	10.2		

[75]. Generally, phenols are better extracted from acidic or neutral solutions, whereas phenyl amines are better extracted from alkaline solutions [71,72]. This behavior results from dissociation of hydroxyl groups of phenols at high pH or protonation of amino groups of phenyl amines at low pH. Obviously, hydrophobicity of the phenolic compounds (the more hydrophobic compound, the better extraction) plays an important role for compound solubility in the hydrophobic IL phase.

Novel ammonium ILs (tetrahexylammonium dihexylsulfosuccinate [$C_6C_6C_6C_6N$][DHSS] and trioctylmethyl ammonium salicylate [$C_8C_8C_8C_1N$][Sal]) are compared with imidazolium ILs [C_6C_1Im][NTf_2] and [$C_{10}C_1Im$][NTf_2] as extraction media for various phenols and aromatic amines [73], and are considered to be more efficient phenol extractants than imidazolium ILs. In contrary to the imidazolium ILs, not only molecular forms of phenols are extracted efficiently because, at pH higher than pK_a, ion-exchange mechanism controls the extraction.

An interesting example of organic separation with ILs is given by Bauman et al. [76] who succeeded in phenol reduction resulting from its partition into IL and simultaneous degradation of phenol by xenobiotic-degrading bacteria. [$C_{14}C_6C_6C_6P$][$C_8C_8PO_2$] was selected as the most biocompatible IL of the tested ones, and correspondingly showed efficient partition of phenol to deliver it to the cell-containing phase.

Both typical extraction and liquid membrane pertraction with phosphonium ionic liquids ([$C_{14}C_6C_6C_6P$][Cl], [$C_{14}C_6C_6C_6P$][Br], [$C_{14}C_6C_6C_6P$][$C_8C_8PO_2$], [$C_{14}C_6C_6C_6P$][NTf_2], and [$C_{14}C_6C_6C_6P$][BF_4]) were successfully applied to recover butyric (BA) and lactic acid (LA) from aqueous solutions [77–82]. Kyuchoukov et al. [83] extracted LA with Aliquat 336 diluted with dodecane or decanol, and indicated that physical extraction probably plays a crucial role in the acid extraction.

Extraction of organic acids with imidazolium ILs is controlled by the hydrophobicity of the acids (values of distribution constant log K_d are roughly correlated with values of octanol/water partition coefficient log P) [84,85]. However, extraction only with imidazolium ILs is not efficient and follows the order of extractability [C_8C_1Im] < [C_4C_1Im] < [C_6C_1Im], which is not clearly explained. The extraction enhancement is noticed when the ILs act as diluents of extractant (e.g., TBP). However, it raises doubts about sense of IL use in such a system because of their still high prices and not significant impact compared with conventional solvents. Although Huddleston et al. [9] suggest that even low partition between aqueous and IL could be economical because of negligible volatility of the organic phase.

Dissociated acid is reported to be extracted with [$C_{14}C_6C_6C_6P$][Cl], according to the ion-exchange mechanism. On the contrary, [$C_{14}C_6C_6C_6P$][$C_8C_8PO_2$] extracts only undissociated acids, which indicates a coordination mechanism via H-bonding. High solubility of water in hydrophobic [$C_{14}C_6C_6C_6P$][$C_8C_8PO_2$] (up to 14.4 wt%) proves reverse micelles formation, which is responsible for water extraction in dual mechanism. In the first step, water is incorporated into reverse micelles, after being included into a hydrated complex of the organic acid with IL. Formation of carboxylic acid/IL complex causes splitting of the reverse micelles with releasing water from them. Also, Wang et al. [85] emphasize the role of water molecule presence in disrupting hydrogen bond association between cation and anion of the imidazolium IL. Consequently, the mobility of ions in IL is enhanced by high water solubility and increases, resulting in easier interactions between cationic form of an acid and anion of IL.

The strong influence of the aqueous phase pH on organic acid extraction efficiency depends on the distribution of cationic (A^+), zwitterionic ($A^{\pm}$), or anionic (A^-) forms of the acid. Ferulic (FA) and caffeic acids (CA) are extracted from aqueous phases with $[C_4C_1Im][PF_6]$ and $[C_6C_1Im][PF_6]$, [86] the most efficiently at a pH lower than 4 (with $[C_4C_1Im][PF_6]$). The authors indicate steric hindrance and hydrogen bond interaction between FA or CA and IL as an effect of chemical structure of the species on extraction efficiency. Nuclear magnetic resonance studies prove formation of strong ionic pairs between cation and anion of the ILs by formation of hydrogen bond between H-2 of the imidazolium ring and the fluorine of $[BF_4]^-$ or $[PF_6]^-$ [87]. The steric hindrance affecting interactions between cationic form of amino acid and IL anion can be the reason for decreasing amino acid (L-tryphophan, L-phenylalanine, L-tyrosine, D-valine, or L-leucine) extraction with increasing alkyl chain length of the imidazolium cation [85]. The more hydrophobic acid extracted the bigger affinity to the organic phase and higher partition coefficient. Moreover, the effective negative charge of IL anion plays an important role in the extraction of organic acids by influencing electrostatic interactions between cationic form of acid and IL anion. The negative charge of $[BF_4]^-$ anions is stronger than of $[PF_6]^-$; therefore, ILs of the former are better extractants of acids in low pH.

What is interesting, but still unemployed on a bigger scale, and maybe a little controversial, is a proposal for in situ extractive fermentation toward LA in the presence of imidazolium ILs. Preliminary results on toxicity of the ILs to *Lactobacillus rhamnosus* show that the bacterium grows and produces lactate in the presence of C_1Im ILs [84].

Vijayaraghavan et al. [88] proved 1-butyl-1-methylpyrrolidinium bis(trifluoromethylsulfonyl)imide $[C_4C_1Pyrr][NTf_2]$ efficiently extracts in four extraction steps (>98%), using acid blue and acid red (coming from the leather industry). The IL could be regenerated by scrubbing dyes with a mixture of isopropanol and water (1:1). However, some unfavorable loss of IL to the aqueous phase is noticed. Thus, the authors propose recovery of IL ions by means of ion-exchange resins without presentation of the exact results for the complete regeneration procedure.

Two different dyes (cationic methylene blue and anionic methyl orange) are extracted with $[C_8C_8C_8C_1P][SCN]$, dicyanoamide [BA], and hexanate [89]. The authors notice that an IL anion and a type of dye affect extraction efficiency. For instance, strong proton-accepting abilities of thiocyanate cause excellent efficiency of methyl orange extraction, whereas distribution coefficients for cationic dye are significantly lower. Moreover, recovery of methylene blue from the organic phase was difficult and inefficient, whereas more than 90% of methyl orange is stripped with 0.1 M HCl. Pei et al. [72] concluded that hydrophobic interactions between a dye and IL significantly affected extraction of such anionic dyes as methyl orange, eosin yellow, or orange G with typical imidazolium ILs (e.g., $[C_4C_1Im][PF_6]$, $[C_6C_1Im][PF_6]$, and $[C_8C_1Im][PF_6]$).

The most popular imidazolium IL—$[C_4C_1Im][PF_6]$—is used for extraction of Acid Yellow RN, Weak Acid Brilliant Blue RAW, Reactive Black KN-G2RC, and Reactive Yellow M-5R [90]. Acid dyes are extracted quantitatively only with imidazolium IL, but recovery and regeneration of the IL is difficult and that can eliminate the IL from the separation process. However, because extraction of reactive dyes is insufficient, dicyclohexyl-18-crown-6 (DCH-18C6) in $[C_4C_1Im][PF_6]$ is used to enhance the extraction efficiency.

Finally, interactions between imidazolium ILs and solvatochromic dyes are investigated to find out polarity and nucleophilicity of 1,3-dialkyl and 1,2,3-trisubstituted imidazolium ILs. These studies revealed that the ILs studied (e.g., popular $[C_4C_1Im][PF_6]$) are unique in their high polarity and, simultaneously, low nucleophilicity [91].

Among biomolecules extracted with ILs are penicillin G [92], amoxicillin and ampicillin [93], and erythromycin-A [94]. Penicillin G liquid–liquid extraction with $[C_4\text{-}C_8C_1Im][PF_6]$ is inefficient. However, use of trioctylmethylammonium chloride ($[C_8C_8C_8C_1N][Cl]$) gives good results, but stripping from the loaded organic phase is difficult.

Antibiotic extraction with $[C_8C_1Im][BF_4]$ or $[C_4C_1Im][PF_6]$ depends on pH of the aqueous phase and the structure of the solute; however, more studies should be done to say more about usefulness of ILs for biomolecule separation [92,94]. Better extraction at higher values of pH ($>pK_a$) suggests ion-pair formation between charged antibiotics and charged imidazolium ILs. Moreover, extraction of erythromycin-A is reduced at high pH (>10) due to worse solubility of the antibiotic in the IL. In that case, it can be advantageous for stripping of erythromycin; however, the authors proposed no exact stripping procedure.

Because bioprocesses have been becoming more popular for the past decade for production of chemicals and fuels, there has been a necessity for efficient separation of bioproducts (e.g., alcohols) from the fermentation broth. Distillation is a conventional method for their separation from water; however, this method is energy consuming. Recently, imidazolium ILs have been proposed for liquid–liquid extraction or pervaporation aimed at butan-1-ol (BuOH) separation [23,95]. $[C_8C_1Im][PF_6]$ is more efficient than $[C_4C_1Im][PF_6]$ for BuOH extraction from BuOH-water solution because of higher hydrophobicity of the former IL [95]. Higher BuOH distribution coefficients in $[C_8C_1Im][PF_6]$ are a consequence of reduced mutual solubilities of water and IL, resulting from bigger molecular dimensions of the alkylimidazolim cation. As a consequence, $[C_8C_1Im][PF_6]$ performs better BuOH extractive selectivity over water.

Problems of high IL viscosity are overcome by elevation of separation temperature. Although BuOH solubilities in $[C_8C_1Im][PF_6]$ are similar to solubilities in octanol, ILs are proposed to replace volatile organic solvents because their nonvolatility is considered to be economical for extraction systems even if the partition coefficients are not high [9].

On the basis of ternary phase diagrams, $[C_6C_1Im][NTf_2]$ (with more hydrophobic anion than $[PF_6]^-$), presented by Chapeaux et al. [23], is predicted to separate easily BuOH from water, whereas ethanol extraction would need unreasonably high solvent/feed ratios. Thus, BuOH could be separated from water using $[C_6C_1Im][NTf_2]$ in multistage extraction.

The separation efficiency of alcohols from water by imidazolium-based ILs depends on the alcohol alkyl chain length that influences both their solubility in IL phase and their miscibility with water. With increasing hydrophobicity of an alcohol, van der Waals interactions between an alcohol and IL become more important than hydrogen bonding. As a consequence, the solubility of an alcohol in IL decreases and its miscibility with water is reduced. It results in smaller separation coefficients in ethanol/water than BuOH/water systems, where for low (2 wt%) BuOH concentrations in the feed, separation factors reach 100 [23].

Another solution for BuOH separation from water is proposed by Yu et al. [96], who applied imidazolium tetrafluoroborates with hydroxyl groups (1-(2-hydroxyethyl)-3-methylimidazolium $[C_2OH\ C_1Im][BF_4]$ or 1-(2-hydroxyethyl)-2,3-dimethylimidazolium $[C_2OHC_1C_1Im][BF_4]$ tetrafluoroborates—shown in Table 1). These ILs are proposed to extract efficiently water from the BuOH/water mixture. Better selectivity of water extraction by $[C_2OHC_1C_1Im][BF_4]$ than $[C_2OHC_1Im][BF_4]$ is explained by the substitution of C_2 in imidazolium ring with CH_3 instead of H. Therefore, formation of a hydrogen bond between the cation and alcohol is reduced, which results in a decrease in mutual solubility of alcohol and IL. On the other hand, water solubility in the IL increases and leads to better extraction capacity than in $[C_2OHC_1Im][BF_4]$. The two ILs are proposed as potential candidates to separate BuOH and water by liquid–liquid extraction; however, it could not be realized on an industrial scale without elaboration of reasonable regeneration of the ILs. Probably, removal of water from the ILs would be the most difficult and energy-consuming step of the process.

3. SUPPORTED LIQUID MEMBRANES

Among the membrane techniques using ILs for metal ion or organic compound separation—supported liquid membranes, polymer inclusion membranes and pervaporation are reported.

Looking for environmentally friendly and efficient separation systems for metal ions, many authors propose membrane techniques, particularly liquid membranes, that are divided according to their structure into the following: bulk liquid membranes (BLMs), supported liquid membranes (SLMs), and emulsion liquid membranes (ELMs). BLMs are the simplest system of liquid membranes; however, because of their significant thickness, the amount of the solute transported is too small. In ELM systems, problems with emulsion stability are reported.

SLMs are favored because, due to a small volume of organic phase applied in the membrane support, they make possible and advantageous the use of expensive extractants. SLMs are formed by impregnation of porous supports with organic solvents. On the other hand, the possibility of organic phase leakage makes them less selective and attractive for industry. Polymer inclusion membranes (PIMs)—a particular type of SLMs—are reported to be more stable, mechanically stronger, and chemically resistant. SLMs can be used for pervaporation, a technique for mixture separation by partial vaporization of the mixture components through porous or nonporous membrane.

3.1 Separation of Metal Ions

SLMs impregnated with ILs are called supported ionic liquid membranes (SILMs) and are used for separation in different configurations (e.g., polymer inclusion membranes and pseudo-emulsion membrane strip dispersion).

The letter involves one module (for simultaneous extraction and stripping) in which the aqueous feed is contacted with pseudo-emulsion phase. The pseudo-emulsion phase is formed as a consequence of dispersion of a stripping phase in the organic one (e.g., IL containing). The liquid phases are separated from each other with a porous membrane. After extraction stripping, the pseudo-emulsion phase should separate quickly when mixing is stopped. This technique was successfully used for Cr^{6+} separation with $[C_{14}C_6C_6C_6P][Cl]$ as a carrier [97]. Such a separation performed in a hollow fiber (HF) module is called pseudo-emulsion hollow fiber strip dispersion and was applied for Cr^{3+} recovery from alkaline solutions with $[C_8C_8C_8C_1N][Cl]$ [98] and iron(III) from acid solutions with ammonium IL based on commercial primary amine Primene JMT $[PJMTH]_2[SO_4]$ [99].

SILMs—hydrophilic polyamide membranes—were impregnated with $[C_8C_8C_8C_1N]$ $[Cl]$ to be used as a membrane for Zn^{2+}, Cd^{2+}, Cu^{2+}, and Fe^{3+} selective separation from acidic solution to various receiving phases. Different pH of the receiving phases used allowed the metal ions to be selectively pertracted through the SILM [100].

PIMs containing Aliquat 336 as a carrier included in PVC matrix separate efficiently Co^{2+} from Ni^{2+} acidic solution (7 M HCl) [101] or Cd^{2+} from HCl solution [102,103]. A decrease in transport of metal ions at less than 40% of $[C_8C_8C_8C_1N][Cl]$ content in the membrane is attributed to the absence of micropores in the membrane and too tight structure of the matrix. On the other hand, the reduction in Cd^{2+} flux results probably from destructive effect of the metal ions on the polyvinyl chloride (PVC) membrane [102].

Transport through PIMs is strongly affected by membrane composition, type of carrier, composition of feed, or type of stripping phase. Because some phosphonium

ILs are used as efficient extractants, they are also investigated as metal ion carriers in PIMs ($[C_{14}C_6C_6C_6P][Cl]$, $[C_{14}C_6C_6C_6P][C_8C_8PO_2]$) [104,105]. The results obtained for extraction are also confirmed for PIMs with cellulose triacetate (CTA) matrix; $[C_{14}C_6C_6C_6P][Cl]$ more efficiently transports Fe^{3+} than Zn^{2+} from HCl solutions. Effectiveness of $[C_{14}C_6C_6C_6P][Cl]$ in Zn^{2+} permeation through PVC membrane depends on the carrier content. When the membrane contains less than 30% of carrier, the transport of Zn^{2+} is reduced.

New polyvinylidene difluoride (PVDF) polymer inclusion membranes contain both bifunctional ILs, such as phosphonium $[C_{14}C_6C_6C_6P][C_8C_8PO_2]$ or ammonium $[C_8C_8C_8C_1N][C_8C_8PO_2]$, $[C_8C_8C_8C_1N][DEHP]$ as Cr^{6+} carriers and imidazolium ILs ($[C_4C_1Im][BF_4]$, $[C_4C_1Im][PF_6]$, $[C_8C_1Im][BF_4]$, and $[C_8C_1Im][PF_6]$) as plasticizers [106,107]. The most selective transport of Cr^{6+} from a mixture of Fe^{3+}, Zn^{2+}, Co^{2+}, and Cu^{2+} in hydrochloric acid is provided by a membrane containing 0.4 mmol/g $[C_{14}C_6C_6C_6P][C_8C_8PO_2]$ and $[C_8C_1Im][BF_4]$ in PVDF matrix. Transport parameters of Cr^{6+} are always higher for $[C_8C_8C_8C_1N]^+$ carriers than for $[C_{14}C_6C_6C_6P]^+$ $[C_8C_8PO_2]^-$.

3.2 Separation of Organic Compounds

As penicillin G liquid–liquid extraction is inefficient [92], especially for pharmaceutical, analytical, or laboratory applications, it can be replaced with supported liquid membranes that join extraction and stripping in one device. Penicillin G transport across SILM with $[C_8C_8C_8C_1N]][Cl]$ as a carrier is efficient and proceeds according to the anion exchange mechanism.

In BLM and SLM with Aliquat 336, $[C_8C_1Im][PF_6]$, and $[C_4C_1Im][PF_6]$ as active carriers, there is reported a phenomenon of reverse micelles formation in the IL phase [108–110]. Loss in extraction selectivity in SLM separations indicates that the transport mechanism is different than in liquid–liquid extraction. Formation of water microenvironments inside the organic phase reduces SLM selectivity because of the existence of nonselective environments for solute transport. The presence of water in the organic phase enhances extraction of water-soluble solutes with low affinity to hydrophobic ILs, and leads to a smaller contribution of the ion-exchange mechanism to overall transport process.

Martak and Schlosser [79] proposed a new mechanism of LA pertraction through SILM-containing phosphonium ILs, assuming reverse micelle existence in the IL phase. The mechanism involves splitting of the reverse micelles at the feed/SILM interface, formation of IL-LA-H_2O–hydrated complexes, transport of the complexes through SILM, splitting of the hydrated complexes at the SILM/stripping phase interface, formation of reverse micelles by free IL molecules at the stripping interface, and transport of reverse micelles through SILM.

Stability of SILMs is a crucial issue for their further applications. Not many investigations on membrane stability are reported; however, some authors indicate that

SLMs containing phosphonium or ammonium ILs for carboxylic acid extraction have good membrane stability. For example, after eight cycles of PIM use, 90% of permeation is still maintained [82]. The SLM containing $[C_{14}C_6C_6C_6P][C_8C_8PO_2]$ to transport lactic acid in a spiral flat sheet module is also stable [79].

Because pervaporation is considered the most prospective technique for the recovery of organic compounds from aqueous solutions, some authors propose to apply ILs for separation of alcohols, carboxylic acids, or phenols to improve pervaporation performance [10,95,96]. ILs, attractive media for pervaporation because of their nonvolatility, would not permeate through a membrane, thus increasing solute separation selectivity and mass fluxes. A nanofiltration ceramic membrane (pore size, 0.9 nm) impregnated with a novel hydrophobic IL (tetrapropylammonium tetracyanoborate) (Figure 2(A)) was used as an SILM for propane-1,3-diol pervaporation [10]. The preliminary results show that propane-1,3-diol transport is preferred when the membrane is impregnated with IL because water permeation through hydrophobic IL is more difficult than through a nonimpregnated ceramic membrane. Meanwhile, the solute flux decreases in relation to the nonimpregnated membrane. However, the selectivity of propane-1,3-diol (a valuable intermediate of fermentation processes used for production of polymers) increases more than two orders of magnitude compared with one order of magnitude slower permeation.

Another attitude to pervaporation with ILs is reported for BuOH separation. The polydimethylsiloxane membrane (MEM-100, PDMS) was not impregnated with the IL (Figure 2(B)), but the separation was performed from a solution of $[C_8C_1Im][PF_6]$-BuOH-water [95]. The selectivity of BuOH separation is higher than in the system without IL. However, the flux is reduced, probably because of the concentration polarization enhanced by high viscosity of the feed solution containing IL.

$[C_4C_1Im][PF_6]$ can be applied as an active layer on a surface of a hydrophobic membrane (Figure 2(C)) to extract and concentrate acetic acid that is further permeated through the membrane (e.g., PDMS) [96]. The IL improves both the acetic acid flux and selectivity by rejection of water and acid concentration before pervaporation both from the model and real antibiotic industrial effluent.

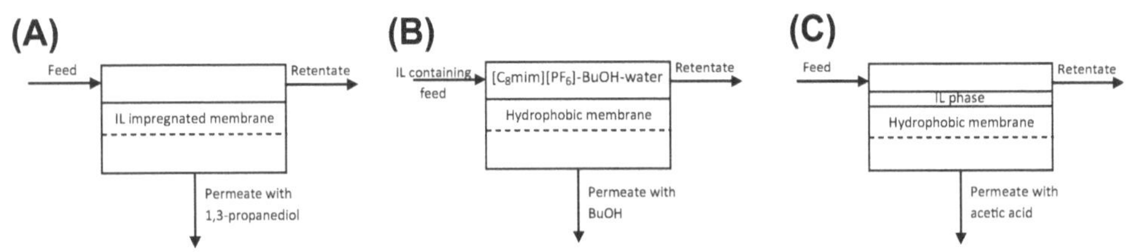

Figure 2 Pervaporation systems containing ILs: (A) membrane impregnated with IL, (B) IL mixed with feed, and (C) active layer of IL on a membrane surface.

4. AQUEOUS BIPHASIC SYSTEMS

Aqueous biphasic systems (ABS) are formed when two polymers (polymer—polymer) or a polymer and an inorganic salt (polymer—salt) are dissolved in water above a certain critical concentration. These systems result in two phases with a water content of 80% [111,112]. Phase separation occurs on mixing at an appropriate temperature in which phases are formed with different compositions. Solutes allocate themselves between the two aqueous phases, depending on their relative affinity for each phase. More recently, aqueous mixtures of organic salts, including hydrophilic ionic liquids (ILs), and inorganic salts have been shown to form biphasic systems above the critical concentration (salt—salt ABS) [113,114]. In the past decade, ionic liquids have shown to be a possible alternative to be used in the preparation of ABS and partitioning studies [115].

One type of aqueous biphasic system is formed when a polymer, such as polyethylene glycol (PEG) and a salt, are dissolved together in water. A wide variety of polymer/salt combinations exist, usually with Na^+, K^+, or NH_4^+ salts of anions like OH^-, CO_3^{2-} SO_4^{2-}, and PO_4^{3-}. The phase behavior of ABS is influenced by the salt type and concentration. A sufficiently high concentration of salt in a single polymer-water system can induce phase separation to yield a salt-rich, polymer-poor bottom phase that coexists with a salt-poor, polymer-rich top phase [111]. The relative effectiveness of various salts in promoting phase separation is observed to follow the Hofmeister series, which is a classification of ions, based on their salting-out ability [116]. The contribution of the anion is more important than that of the cation in determining the effectiveness of a particular salt. The multivalent anions like HPO_4^{2-} and SO_4^{2-} are most effective in inducing phase separation with PEG. Rogers and coworkers [117] found that the salting-out ability of a given inorganic salt was directly correlated to its Gibbs-free energy of hydration. The salting-out ability of the anions follows the ordering $PO_4^{3-} > CO_3^{2-} > SO_4^{2-} > OH^-$. The calculated ΔG_{hyd} values for these anions are PO_4^{3-} (−2835 kJ/mol) > CO_3^{2-} (−1300 kJ/mol) > SO_4^{2-} (−1145 kJ/mol) > OH^- (−345 kJ/mol) [118]. The salting-out effects are additive, and the cations have a contribution as well with the same argument as offered for the anions. When comparing the salts having the same anion, the one whose cation has a more negative ΔG_{hyd} value is better at salting out. The slight difference caused by varying the cation or anion in the salt used to form an ABS, the polymer molecular weight and hydrophobicity, the pH, and the temperature can be used to fine-tune the phase incompatibility in a reproducible and predicable model [119].

In ABS, the major component in each of the two phases is water, which means the possibility of completely eliminating the use of volatile organic compounds during separation. In addition, PEG-rich phases in PEG-ABSs appear to be tunable. Their phase characteristics can be changed to match the hydrophobicity and water content of several organic solvents [120,121]. The number of variables necessary to fully define an optimized process is significantly increased over traditional solvent extraction systems. The choice of polymer and its molecular weight, the choice of phase-forming salt, the relative concentration of each component, the system pH, and temperature all affect the exact phase composition and the partitioning experiment.

Initial uses for PEG-based ABS focused on the separation of biomolecules and proteins such that the biomaterials could be salted-out into the nondenaturing PEG phase. Then, the focus on separations expanded the use of ABS to include the partitioning of organic solutes and metal ions in various PEG/salt systems. Small organic molecules typically have a high affinity for the PEG-rich phase corresponding to the solutes octan-1-ol/water log P values, a common descriptor for solute hydrophobicity [122]. More complex organic materials can also be partitioned to the polymer-rich phase, as indicated in a report in which a PEG-based ABS was used to remove lignin for paper-pulping applications [123].

The first separation study using ILs was done by Huddleston and coworkers, [9] who investigated the partitioning of substituted aryl molecules between a water phase and a $[C_4C_1Im][PF_6]$ phase, and observed a close similarity of distribution ratios in the IL/aqueous system and their distribution ratios in an octan-1-ol-water system [9]. The report by Huddleston and coworkers was followed by studies investigating organic compound separations using ILs via different techniques, including liquid–liquid extraction [24], supported liquid membrane separation [124], and salt–salt aqueous biphasic extraction [113]. Most water-soluble room temperature ionic liquids tend to be water-destructuring salts (chaotropic salts) and are capable of being salted out by water-structuring salts (kosmotropic salts) forming aqueous biphasic systems.

A new class of IL-based ABS can be formed when a hydrophilic IL and a certain salt or a certain water-soluble polymer are combined in an aqueous solution. Therefore, in the case of polymer-salt ABS, both the polymer and salt can be replaced by a particular IL to form a new IL–polymer or IL–salt ABS (Figure 3). In recent years, definitely more

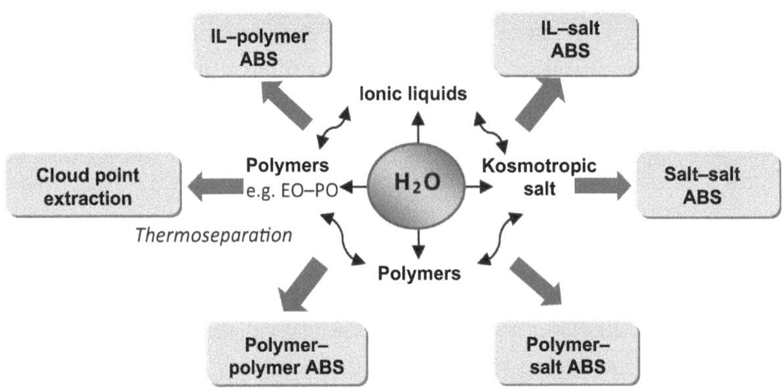

Figure 3 Aqueous-based systems using ionic liquids and polymers.

reports in the literature could be found on the IL–salt ABS [115] than IL–polymer ABS [125–130].

In aqueous solutions containing ionic liquid and inorganic salt, the driving force for phase separation is the competition between the ionic liquid and inorganic salt for water molecules. The higher affinity of the inorganic salt for water induces a migration of water away from the ions of the ionic liquid and toward those of the inorganic salt. This decreases the hydration of the ionic liquid and reduces its solubility in water. As a result, a phase rich in the salted-out ionic liquid separates from the rest of the solution. The inorganic salts typically used are ammonium, potassium, or sodium salts of multiply charged anions, such as phosphate, sulfate, carbonate, or citrate [131].

4.1 Separation of Organic Compounds

Reports in the literature concern the separation of alkaloids, pharmaceuticals, amino acids and proteins, steroid hormones, short-chain alcohol homologues, pesticides, and aromatic and phenolic compounds in ABS formation (Table 7 [132–152]). The applications using room temperature ionic liquids include extraction studies with a significant number of low-molecular-mass compounds (e.g., tryptophan [139], opium alkaloids [132], and testosterone and epitestosterone [136,137]). Typical partition coefficients fall into the range of 10–125 with recoveries higher than 85% in a single extraction. Jiang et al. [142] extracted penicillin with efficiency ~96% from an imidazolium-terminated poly(ethylene glycol) system. In this case, the penicillin was recovered from the polymer phase in a second aqueous biphasic system formed by adjusting the pH to slightly basic and adding 1-butyl-3-methylimidazolium hexafluorophosphate. This allowed recycling of the imidazolium-terminated polymer [142]. Because of the low solubility of proteins in water-insoluble ionic liquids and loss of activity, ABS-containing water-soluble ionic liquids are suitable for the isolation of proteins [134,135,143]. Selected proteins have been recovered from ionic liquid–inorganic salt biphasic systems with an efficiency of 75–100% and with retention of biological activity. Moscoso and coworkers [152] demonstrated that the combination of K_2HPO_4 with methylsulfate-based ILs allows the separation efficiency of pentachlorophenol higher than 99%.

4.2 Separation of Metal Ions

ILs have also been of significant interest as substitutes for traditional solvents in separation of metal ions from aqueous media. Hydrophobic ILs form a two phase with water, where the IL is the lower phase. However, in the absence of a ligand capable of forming an organophilic complex, most metal ions preferentially partition to the upper aqueous phase of a biphasic IL/aqueous system. Even when a metal complex is sufficiently organophilic to partition into the IL phase, the partitioning equilibria do not necessarily simulate those found in conventional liquid–liquid extraction systems. In this case, the

Table 7 Ionic Liquid-Based ABS Reported in the Literature for the Extraction of Organic Compounds Using the Selected ILs

Ionic Liquid	Salting-out Species	Solute	Ref.
$[C_4C_1Im][Cl]$	K_2HPO_4	Codeine, papaverine	[132]
		Chloroamphenicol	[133]
		BSA	[134]
		Cytochrome C, myoglobin, ovalbumin, hemoglobin, methanol, ethanol, propanol, butanol, pentanol, hexanol	[135]
		Testosterone, epitestosterone	[136]
		Phenol, nitrobenzene, 4-nitrophenol	[137]
	K_3PO_4	Nicotine, caffeine	[138]
		L-Tryptophan	[139]
		Testosterone, epitestosterone	[136]
		Vanillin	[140]
		Phenol, nitrobenzene, 4-nitrophenol	[137]
	NaH_2PO_4	Penicillin	[141]
	K_2CO_3	Testosterone, epitestosterone	[136]
		Phenol, nitrobenzene, 4-nitrophenol	[137]
$[C_4C_1Im][Br]$	Na_2HPO_4	Penicillin	[142]
	K_2HPO_4	BSA, trypsin, cytochrome C, γ-globulin	[143]
	$K_3C_6H_5O_7$	L-Tryptophan, L-phenylalanine, L-tyrosine, L-leucine, L-valine	[144]
$[C_4C_1Im][BF_4]$	NaH_2PO_4	Penicillin	[145]
		Acetylspiramycin	[146]
		Azithromycin, mydecamycin	[147]
		Tetracycline	[148]
	Na_2CO_3	Roxithromycin	[149]
		Azithromycin, mydecamycin	[147]
	$(NH_4)_2SO_4$	Sulfadimidine	[150]
		Azithromycin	[147]
	Na_2SO_4	Azithromycin, mydecamycin	[147]
$[C_4C_1Im][CF_3SO_3]$	K_3PO_4	Caffeine	[138]
		L-Tryptophan	[151]
$[C_4C_1Im][C_1SO_3]$	K_2HPO_4	Pentachlorophenol	[152]

equilibria have been shown to change depending on the complexant or the IL cation and anion [7,153].

Cation exchange processes with exceptionally high distribution ratios [17,39] have been the most commonly reported metal ion extraction mechanism for biphasic IL systems, where transfer of a cationic metal-ligand complex into the IL phase is accompanied by a corresponding loss of IL cations to the aqueous phase. The extraction of anionic complexes with the release of IL anions to the aqueous phase was also studied [43].

The main problem with the cation and anion exchange equilibria in IL-based systems is that IL components are transferred to the aqueous phase, which pollutes the aqueous phase and can make it difficult to subsequently recover the extracted metal ion. Dietz et al. [154] have shown that the mode of metal ion extraction by crown ethers can be shifted from cation exchange to extraction of a neutral ion-pair complex by increasing the lipophilicity of the IL cation. Formation of such neutral metal-crown ether—anion complexes is the conventional transport mechanism with crown ether extractants in organic solvents. However, in this case, the extracted complex displayed a different inner sphere coordination environment than is observed in conventional organic solvents [154].

Akama et al. [155] and Akama and Sali [156] used ABS consisting of tetrabutylammonium bromide and ammonium sulfate to extract Cr^{6+} from Cr^{3+}, and Cd^{2+} from Co^{2+}, Cu^{2+}, Fe^{3+}, and Zn^{2+}. In both cases, the extraction mechanism relied on the formation of ion pairs between the metal cation and the ionic liquid anion [155,156]. The authors proved that it is possible to use this ABS combined with atomic absorption analytical methods to determine traces of Cr^{6+} in wastewater samples. The determination limit found was equal to 60 mg/L and recovery efficiency was 90% [156]. In relation to the traditional extraction, the described method is easy and does not require the use of organic solvents.

5. REGENERATION AND REUSE OF ILS

Extracted metal cations become part of ionic liquid phase as a result of binding the ligands contained in the organic phase, and IL cation exchange. The strong bonding between metal cation and IL anion causes difficulties in IL regeneration. In addition, the loss of IL cation to the aqueous phase is not advantageous for high cost and environmental impact [157]. However, the change in the Sr^{2+} and Cs^{+} extraction mechanism from cation exchange to extraction of neutral complexes, observed with increasing alkyl chain on imidazolium cation [154], reduces IL loss to the aqueous phase.

In most extraction systems, the regeneration is run by stripping and washing of the organic phase with various aqueous solutions. However, some stripping phases cause IL cation loss [50]:

$$[C_8C_1Im]_2 \cdot Ce(NO_3)_{6(o)} \Leftrightarrow 2[C_8C_1Im]^+_{(w)} + Ce^{4+}_{(w)} + 6NO^-_{3(w)} \qquad (8)$$

Alternatively, addition of organophilic species ($NaBPh_4$, KPF_6) to control a transfer of imidazolium cations to the aqueous phase can overcome the loss of IL cation [38,50]. The presence of these salts in the stripping phase decreases the loss of IL cation by 24%. As a result, stripping in these systems is easier than for the cation-exchange mechanism, because it is enough to reverse the equilibrium of extraction reactions by use of $H(NTf_2)$ or KPF_6 aqueous solutions:

$$[C_8C_1Im]_2 \cdot Ce(NO_3)_{6(o)} + 2[PF_6]^-_{(w)} \Leftrightarrow 2[C_8C_1Im][PF_6]_{(o)} + Ce^{4+}_{(w)} + 6NO^-_{3(w)} \quad (9)$$

At the same time, pure cerium as CeF_3 nanoparticles or $Ce_2(SO_4)_3$ solutions are obtained as products after regeneration of the $[C_8C_1Im][PF_6]$ phase. Recycling of the crown ether-IL phase loaded with metal ions can proceed by simple pH change. For example, at low pH, protonated aza crown releases 98% Cs^+ and Sr^{2+}, and can be reused for extraction after deprotonation with base [37].

The studies on stripping and regeneration of phosphonium IL phases have revealed that Zn^{2+} or Pd^{2+} can be successfully stripped from the loaded organic phase with sulfuric acid or ammonia solutions, respectively. The ability to reuse the $[C_{14}C_6C_6C_6P]]$ $[Cl]$/toluene mixture in several cycles of Zn^{2+} or Pd^{2+} extraction stripping has been proved; however, Zn^{2+} recovery from the organic phase needs three steps [66,68]. After five cycles of extraction (one step) and stripping (two steps with 0.5 M NH_4OH), the percentage extraction of Pd^{2+} does not change and remains 98% and 55% for the feed containing 0.1 and 3 M HCl, respectively [68,69]. Easy regeneration of the IL phase with nitric acid is reported for bifunctional ILs, such as $[C_8C_8C_8C_1N][DEHP]$. Repeated loading/stripping experiments prove only slight extraction changes and negligible loss of IL to the aqueous phase.

Recovery of ionizable compounds (e.g., ferulic and caffeic acids or L-tryptophan amino acid or phenols) from imidazolium ILs is realized by pH changes using NaOH solution as a stripping agent [57,86,102]. The procedure covered washing with water until neutral pH and IL drying. Finally, the ILs can be reused for extraction. Also, $[C_8C_1Im]$ $[BF_4]$ and $[C_6C_1Im][PF_6]$ can efficiently remove phenolic compounds after three extraction-stripping cycles [57]. On the basis of the results obtained by Yan-Ying et al. [86] a simple scheme of extraction procedure, presented in Figure 4, can be proposed.

However, it is generally recognized that the presence of certain solutes (e.g., crown ethers) in hydrophobic ILs, even in the absence of metal ion in the extraction system, enhances the solubility of the ionic liquid in acidic aqueous media [158]. In addition, the solubility of imidazolium ILs in the aqueous solutions is reported to be at 25 °C and 2, 6.7, and 1.8 wt% of $[C_4C_1Im][PF_6]$, $[C_6C_1Im][BF_4]$, and $[C_8C_1Im][BF_4]$, respectively [72]. It is also known that $[BF_4]^-$ or $[PF_6]^-$-based ILs are liable to degradation in contact with an aqueous phase, especially with high acid concentrations [57,159,160]. Thus, Swatloski et al. [160] propose to consider the list of nontoxic pharmaceutically

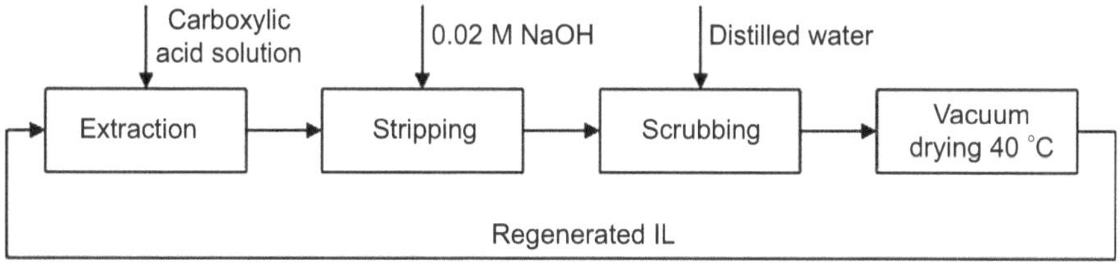

Figure 4 Scheme of ferulic and caffeic acid extraction procedure, according to Yan-Ying et al. [86].

acceptable anions when designing ILs as solvents for extraction. Use of $[BF_4]$ or $[PF_6^-]$ ILs must be considered cautiously, and the question about their "greenness" and possibility of successful replacement of the traditional organic solvents by ILs should be clearly answered.

Not only stripping can be an efficient method for regeneration of the IL phases. Also, direct electrodeposition is proposed as an interesting and effective way of metal recovery from the IL phase due to the wide electrochemical window of ILs and their ion exchange abilities [59]. Palladium can be deposited directly from ammonium ILs ($[C_8C_8C_8C_1][Cl]$ and $[C_8C_8C_8C_1][NO_3]$) in one step or after chloroform distillation from the IL phase. However, electrodeposition after molecular solvent distillation must be performed at temperatures higher than 373 K because of the high viscosity of the $[C_8C_8C_8C_1][Cl]$ containing palladium. The metallic palladium, decontaminated from fission products as a result of extraction with ILs and electrodeposition, is a valuable final product of the separation process.

$[C_4C_1Pyrr][NTf_2]$ is proposed to be regenerated by scrubbing acidic dyes with mixture of isopropanol and water (1:1) [88]. Some loss of IL to the aqueous phase can be avoided by use of ion exchange resins. A similar approach is presented by Pei et al. [72] who precipitate or back extract from imidazolium ILs acidic dyes with solutions of HCl, propan-1-ol, or mixture of NaOH in isopropanol. However, none of the authors presents results for reuse of these ILs.

Furthermore, Stasiewicz [161] proposes ozonation for regeneration of didecyldimethylammonium tetrafluoroborate loaded with food dyes, such as Sunset Yellow, Cochineal Red, Allura Red, and Brilliant Blue. According to the studies, it is enough to ozonate the IL phase loaded with the dyes for 30 min to obtain more than 90% of dye decomposition. Despite the ammonium, IL is resistant to ozonation and could be successfully reused for extraction; the author presents no results of extraction with regenerated IL.

Molecular distillation has been proposed as a promising new method for regeneration of ionic liquid containing solvents. This method is elaborated for the separation and purification of thermally unstable or highly viscous and low vapor pressure liquids, [162] without threat of thermal decomposition of organic compounds. The liquid is distributed as a thin film on the evaporating surface, and, because of vacuum production, it is evaporated at temperatures significantly lower than 250 °C. Because the distance for the vapors between the evaporator and condenser is short, the rate of mass transfer is relatively high. Almost 90% of BA can be evaporated this way from the loaded IL-dodecane phase. A great advantage of the method lies in reduction of chemical consumption in the stripping step.

6. SUMMARY

Ionic liquids represent a fascinating new class of solvents with unusual physical and chemical properties, which can be used for separation of metal ions and organic compounds from aqueous solutions. Although recently their "greenness" is questioned because of possible hydrolysis with formation of toxic HF or partial loss to the aqueous phase, they are constantly interesting and important compounds for liquid–liquid extraction, supported liquid membranes, and formation of aqueous biphasic systems. Furthermore, mutual solubility of imidazolium ILs and aqueous phase (i.e., problems with extraction but also [when the solubility is too high] with loss of ILs) must be pointed out as their limitations. Thus, it is important to achieve a compromise between IL hydrophobicity and its extraction power. Their potential seems to lie in the inner synergic effect enhancing the extraction efficiency, enormous possibilities to design TSILs for specific uses, and separation process intensification.

Imidazolium ILs are still the best described and applied for a variety of separation systems replacing conventional volatile solvents, being pure extractants (L–L extraction, SILM, ABS) or extractants in molecular solvents. However, recently, some interesting applications for ammonium and phosphonium ILs have been also developed.

Apart from small amounts of synthesized ILs and their solubility in the aqueous phase, there is one more issue that must be indicated as limiting the ILs use for separation processes, which is hydrolysis of some ILs in contact with aqueous acid solutions.

Finally, regeneration of ILs is an important issue affecting their greenness and possibilities of use on a large scale. It seems that electrowinning, molecular distillation, or ozonation regenerates the IL phase more efficiently than typical stripping.

Most studies in IL separation systems are still developed on a laboratory scale, and ILs are likely to be rather applied for special separation processes for small amounts of aqueous solutions. Their specific ionic solvation environment, ion-exchange capabilities, and design possibilities make them interesting for rare, noble, or unique compound removal.

ACKNOWLEDGMENT

The author (KM) would like to acknowledge the National Science Centre (Poland) for its financial support through project DEC-2011/01/B/ST5/06349. The work was supported by the grant No. 32/369/14/DS-PB.

REFERENCES

[1] C.O. Ania, J. Pernak, F. Stefaniak, E. Raymundo-Piñero, F. Béguin, Polarization-induced distortion of ions in the pores of carbon electrodes for electrochemical capacitors, Carbon 47 (2009) 3158–3166.

[2] F. Walkiewicz, K. Materna, A. Kropacz, A. Michalczyk, R. Gwiazdowski, T. Praczyk, J. Pernak, Multifunctional long-alkyl-chain quaternary ammonium azolate based ionic liquids, New J. Chem. 34 (2010) 2281–2289.

[3] I. Billard, Ionic liquids: new hopes for efficient lanthanide/actinide extraction and separation? in: J.-C.G. Bünzli, V.K. Pecharsky (Eds.), Handbook on the Physics and Chemistry of Rare Earths, Elsevier, 2013, pp. 213–273.

[4] C.J. Bradaric, A. Downard, C. Kennedy, A.J. Robertson, Y. Zhou, Industrial preparation of phosphonium ionic liquids, Green Chem. 5 (2003) 143–152.

[5] R.E. Del Sesto, C. Corley, A. Robertson, J.S. Wilkes, Tetraalkylphosphonium-based ionic liquids, J. Organomet. Chem. 690 (2005) 2536–2542.

[6] A.E. Visser, J.D. Holbrey, R.D. Rogers, Hydrophobic ionic liquids incorporating n-alkylisoquinilinium cations and their utilization in liquid–liquid separations, Chem. Commun. (2001) 2484–2485.

[7] V.A. Cocalia, M.P. Jensen, J.D. Holbrey, S.K. Spear, D. Stepinski, R.D. Rogers, Identical extraction behaviour and coordination of trivalent or hexavalent f-element cations using ionic liquid and molecular solvents, Dalton Trans. 11 (2005) 1966–1971.

[8] L. Hernandez, L.E. Hernandez, F. Legorreta, M. Avila-Rodriguez, Study of extraction of Zn(II) from ammoniacal media with ionic liquids, IOP Conf. Ser. Mater. Sci. Eng. 45 (2012) 1–4.

[9] J.G. Huddleston, H.D. Willauer, R.P. Swatloski, A.E. Visser, R.D. Rogers, Room temperature ionic liquids as novel media for "clean" liquid–liquid extraction, Chem. Commun. (1998) 1765–1766.

[10] P. Izak, M. Kocherling, U. Kragl, Solute transport from aqueous mixture through supported ionic liquid membrane by pervaporation, Desalination 199 (2006) 96–98.

[11] N. Papaiconomou, J.M. Lee, J. Salminen, M. von Stosch, J.M. Prausnitz, Selective extraction of copper, mercury, silver, and palladium ions from water using hydrophobic ionic liquids, Ind. Eng. Chem. Res. 47 (2008) 5080–5086.

[12] A. Rout, K.A. Venkatesan, T.G. Srinivasan, P.R. Vasudeva Rao, Ionic liquid extractants in molecular diluents: extraction behavior of europium(III) in quarternary ammonium-based ionic liquids, Sep. Purif. Technol. 95 (2012) 26–31.

[13] M. Srncik, D. Kogelnig, A. Stojanovic, W. Körner, R. Krachler, G. Wallner, Uranium extraction from aqueous solutions by ionic liquids, Appl. Radiat. Isot. 67 (12) (2009) 2146–2149.

[14] A. Stojanovic, D. Kogelnig, L. Fischer, S. Hann, M. Galanski, M. Groessl, R. Krachler, B.K. Keppler, Phosphonium and ammonium ionic liquids with aromatic anions: synthesis, properties, and platinum extraction, Aust. J. Chem. 63 (2010) 511–524.

[15] A. Stojanovic, B.K. Keppler, Ionic liquids as extracting agents for heavy metals, Sep. Sci. Technol. 47 (2012) 189–203.

[16] A.E. Visser, J.D. Holbrey, R.D. Rogers, Room temperature ionic liquids as alternatives to traditional organic solvents in solvent extraction, Proc. Int. Solvent Extr. Conf., Cape Town (2002). ISBN: 1-919783-24-5.

[17] A.E. Visser, R.D. Rogers, Room-temperature ionic liquids: new solvents for f-element separations and associated solution chemistry, J. Solid State Chem. 171 (2003) 109–113.

[18] G.W. Meindersma, A.B. De Haan, Cyano-containing ionic liquids for the extraction of aromatic hydrocarbons from an aromatic/aliphatic mixture, Sci. China Chem. 55 (8) (2012) 1488–1499.

[19] A.P. Abbott, G. Frisch, J. Hartley, K.S. Ryder, Processing of metals and metal oxides using ionic liquids, Green Chem. 13 (2011) 471–481.

[20] M. Branicka, Removal of food dyes with 1-methyl-3-methyloxymethylimidazolium tetrafluoroborate, Przem. Chem. 85 (8–9) (2006) 574–576.

[21] H. Zhao, S. Xia, P. Ma, Use of ionic liquids as "green" solvents for extractions, J. Chem. Technol. Biotechnol. 80 (2005) 1089–1096.

[22] J. Rydberg, M. Cox, C. Musicas, G.R. Coppin, Solvent Extraction and Practice, Taylor & Francis, 2004. E-book at MyiLibrary (18.10.2013).
[23] A. Chapeaux, L.D. Simoni, T.S. Ronan, M.A. Stadtherr, J.F. Brennecke, Extraction of alcohols from water with 1-hexyl-3-methylimidazolium bis(trifluoromethylsulfonyl)imide, Green Chem. 10 (2008) 1301–1306.
[24] A.E. Visser, R.P. Swatloski, R.D. Rogers, pH-Dependent partitioning in room temperature ionic liquids provides a link to traditional solvent extraction behavior, Green Chem. 2 (2000) 1–4.
[25] S. Chun, S.V. Dzyuba, R.A. Bartsch, Influence of structural variation in room-temperature ionic liquids on the selectivity and efficiency of competitive alkali metal salt extraction by a crown ether, Anal. Chem. 73 (2001) 3737–3741.
[26] R.A. Bartsch, S. Chun, S.V. Dzyuba, Ionic liquids as novel diluents for solvent extraction of metal salts by crown ethers, in: R.D. Rogers, K.R. Seddon (Eds.), Ionic Liquids. Industrial Applications for Green Chemistry, vol. 818, ACS, Washington, 2002, pp. 58–68. ISBN13: 978-08-41237-89-6.
[27] A.E. Visser, R.P. Swatloski, W.M. Reichert, R. Mayton, S. Sheff, A. Wierzbicki, J.H. Davis, R.D. Rogers, Task-specific ionic liquids for the extraction of metal ions from aqueous solutions, Chem. Commun. (2001) 135–136.
[28] P. Giridhar, K.A. Venkatesan, T.G. Srinivasan, P.R. Vasudewa Rao, Extraction of uranium(VI) from nitric acid medium by 1.1 M tri-n-butylphosphonate in ionic liquid diluents, J. Radioanal. Nucl. Chem. 265 (2005) 31–38.
[29] Y. Shen, X. Tan, L. Wang, W. Wu, Extraction of the uranyl ion from the aqueous phase into anionic liquid by diglycolamide, Sep. Purif. Technol. 78 (2011) 298–302.
[30] A.E. Visser, M.P. Jensen, I. Laszak, K.L. Nash, G.R. Choppin, R.D. Rogers, Uranyl coordination environment in hydrophobic ionic liquids: an in situ investigation, Inorg. Chem. 42 (7) (2003) 2197–2199.
[31] K. Nakashima, F. Kubota, T. Maruyama, M. Goto, Ionic liquids as a novel solvent for lanthanide extraction, Anal. Sci. 19 (2003) 1097–1098.
[32] K. Nakashima, F. Kubota, T. Maruyama, M. Goto, Feasibility of ionic liquids as alternative separation media for industrial solvent extraction processes, Ind. Eng. Chem. Res. 44 (2005) 4368–4372.
[33] K. Shimojo, M. Goto, Solvent extraction and stripping of silver ions in room-temperature ionic liquids containing calixarenes, Anal. Chem. 76 (2004) 5039–5044.
[34] Y. Zuo, Y. Liu, J. Chen, D.Q. Li, Extraction and recovery of cerium(IV) along with fluorine(I) from bastnasite leaching liquor by DEHEHP in $[C_8mim]PF_6$, J. Chem. Technol. Biotechnol. 84 (2009) 949–956.
[35] M.L. Dietz, D.C. Stepinski, A ternary mechanism for the facilitated transfer of metal ions into room-temperature ionic liquids (RTILs): implications for the "greenness" of RTILs as extraction solvents, Green Chem. 7 (2005) 747–750.
[36] M.L. Dietz, S. Jakab, K. Yamato, R.A. Bartsch, Stereochemical effects on the mode of facilitated ion transfer into room-temperature ionic liquids, Green Chem. 10 (2008) 174–176.
[37] H. Luo, S. Dai, P.V. Bonnesen, Solvent extraction of Sr^{2+} and Cs^{+} based on room-temperature ionic liquids containing monoaza-substituted crown ethers, Anal. Chem. 76 (2004) 2773–2779.
[38] H. Luo, S. Dai, P.V. Bonnesen, A.C. Buchanan III, J.D. Holbrey, N.J. Bridges, R.D. Rogers, Extraction of cesium ions from aqueous solutions using calix[4]arene-bis(tert-octylbenzo-crown-6) in ionic liquids, Anal. Chem. 76 (2004) 3078–3083.
[39] M.L. Dietz, J.A. Dzielawa, Ion-exchange as a mode of cation transfer into room-temperature ionic liquids containing crown ethers: implications for the "greenness" of ionic liquids as diluents in liquid–liquid extraction, Chem. Commun. (2001) 2124–2125.
[40] M.P. Jensen, J.A. Dzielawa, P. Rickert, M.L. Dietz, EXAFS investigations of the mechanism of facilitated ion transfer into a room-temperature ionic liquid, J. Am. Chem. Soc. 124 (2002) 10664–10665.
[41] D.C. Stepinski, M.P. Jensen, J.A. Dzielawa, M.L. Dietz, Synergistic effects in the facilitated transfer of metal ions into room-temperature ionic liquids, Green Chem. 7 (2005) 151–158.
[42] M.L. Dietz, D.C. Stepinski, Anion concentration-dependent partitioning mechanism in the extraction of uranium into room-temperature ionic liquids, Talanta 75 (2008) 598–603.

[43] M.P. Jensen, J. Neuefeind, J.V. Beitz, S. Skanthakumar, L. Soderholm, Mechanisms of metal ion transfer into room-temperature ionic liquids: the role of anion exchange, J. Am. Chem. Soc. 125 (2003) 15466–15473.
[44] K. Shimojo, K. Kurahashi, H. Naganawa, Extraction behavior of lanthanides using a diglycolamide derivative TODGA in ionic liquids, Dalton Trans. 37 (2008) 5083–5088.
[45] H. Luo, S. Dai, P.V. Bonnesen, T.J. Haverlock, B.A. Moyer, A.C. Buchanan III, A striking effect of ionic-liquid anions in the extraction of Sr^{2+} and Cs^{+} by dicyclohexano-18-crown-6, Solvent Extr. Ion Exch. 24 (2006) 19–31.
[46] D.C. Stepinski, G.F. Vandegrift, I.A. Shkrob, J.F. Wishart, K. Kerr, M.L. Dietz, D.T.D. Qadah, S.L. Garvey, Extraction of tetra-oxo anions into a hydrophobic, ionic liquid-based solvent without concomitant ion exchange, Ind. Eng. Chem. Res. 49 (2010) 5863–5868.
[47] T.J. Bell, Y. Ikeda, The application of novel hydrophobic ionic liquids to the extraction of uranium(VI) from nitric acid medium and a determination of the uranyl complexes formed, Dalton Trans. 40 (2011) 10125–10130.
[48] T.J. Bell, Y. Ikeda, Efficient extraction of Rh(III) from nitric acid medium using a hydrophobic ionic liquid, Dalton Trans. 41 (2012) 4303–4305.
[49] H. Luo, S. Dai, P.V. Bonnesen, A.C. Buchanan III, Separation of fission products based on ionic liquids: task-specific ionic liquids containing an aza-crown ether fragment, J. Alloys Compd. 418 (2006) 195–199.
[50] Y. Zuo, Y. Liu, J. Chen, D.Q. Li, The separation of cerium(IV) from nitric acid solutions containing thorium(IV) and lanthanides(III) using pure $[C_8mim]PF_6$ as extracting phase, Ind. Eng. Chem. Res. 47 (2008) 2349–2355.
[51] N. Kozonoi, Y. Ikeda, Extraction mechanism of metal ion from aqueous solution to the hydrophobic ionic liquid, 1-butyl-3-methylimidazolium nonafluorobutanesulfonate, Monatsh. Chem. 138 (2007) 1145–1151.
[52] A.P. de los Rios, F.J. Hernandez-Fernandez, S. Sanchez-Segado, L.J. Lozano, J.I. Moreno, C. Godinez, Selective separation of Zn(II) and Fe(III) from acidic media using ionic liquids as sole extraction agents, Chem. Eng. Trans. 21 (2010) 625–630.
[53] A.P. de los Ríos, F.J. Hernández-Fernández, F.J. Alguacil, L.J. Lozano, A. Ginestá, I. García-Díaz, S. Sánchez-Segado, F.A. López, C. Godínez, On the use of imidazolium and ammonium-based ionic liquids as green solvents for the selective recovery of Zn(II), Cd(II), Cu(II) and Fe(III) from hydrochloride aqueous solutions, Sep. Purif. Technol. 97 (2012) 150–157.
[54] D. Kogelnig, A. Stojanovic, M. Galanski, M. Groessl, F. Jirsa, R. Krachler, B.K. Keppler, Greener synthesis of new ammonium ionic liquids and their potential as extracting agents, Tetrahedron Lett. 49 (17) (2008) 2782–2785.
[55] V.M. Egorov, D.I. Djigailo, D.S. Momotenko, D.V. Chernyshov, I.I. Torocheshnikova, S.V. Smirnova, I.V. Pletnev, Task-specific ionic liquid trioctylmethylammonium salicylate as extraction solvent for transition metal ions, Talanta 80 (3) (2010) 1177–1182.
[56] A.E. Visser, R.P. Swatloski, W.M. Reichert, R. Mayton, S. Sheff, A. Wierzbicki, J.H. Davis Jr, R.D. Rogers, Task-specific ionic liquids incorporating novel cations for the coordination and extraction of Hg^{2+} and Cd^{2+}: synthesis, characterization, and extraction studies, Environ. Sci. Technol. 36 (2002) 2523–2529.
[57] S.T.M. Vidal, M.J.N. Correia, M.M. Marques, M.R. Ismael, M.T.A. Reis, Studies on the use of ionic liquids as potential extractants of phenolic compounds and metal ions, Sep. Sci. Technol. 39 (2004) 2155–2169.
[58] M.J.N. Correia, M.M. Marques, M.R. Ismael, M.T.A. Reis, Ionic liquids as extractants of phenolic compounds and metal ions: influence of the type of ionic liquid, Proc. Int. Solvent Extr. Conf., Beijing (2005). ISBN: 7-900692-02-9.
[59] P. Giridhar, K.A. Venkatesan, T.G. Srinivasan, P.R. Vasudeva Rao, Extraction of fission palladium by Aliquat 336 and electrochemical studies on direct recovery from ionic liquid phase, Hydrometallurgy 81 (2006) 30–39.
[60] X. Sun, Y. Ji, L. Zhang, J. Chen, D. Li, Separation of cobalt and nickel using inner synergistic extraction from bifunctional ionic liquid extractant (Bif-ILE), J. Hazard. Mater. 182 (2010) 447–452.

[61] V.V. Belova, A.A. Voshkin, N.S. Egorova, A.I. Kholkin, Extraction of rare earth metals from nitrate solutions with a binary extractant based on Cyanex 272, Russ. J. Inorg. Chem. 55 (2010) 629–633.
[62] H. Yang, W. Wang, H. Cui, D. Zhang, Y. Liu, J. Chen, Recovery of rare earth elements from simulated fluorescent powder using bifunctional ionic liquid extractants (Bif-ILEs), J. Chem. Technol. Biotechnol. 87 (2011) 198–205.
[63] D. Kogelnig, A. Stojanovic, F. Jirsa, W. Körner, R. Krachler, B.K. Keppler, Transport and separation of iron(III) from nickel(II) with the ionic liquid trihexyl(tetradecyl)phosphonium chloride, Sep. Purif. Technol. 72 (1) (2010) 56–60.
[64] B. Marszalkowska, M. Regel-Rosocka, Ł. Nowak, M. Wisniewski, Quaternary phosphonium salts as effective extractants of zinc(II) and iron(III) ions from acidic pickling solutions, Pol. J. Chem. Technol. 12 (4) (2010) 1–5.
[65] L. Nowak, M. Regel-Rosocka, B. Marszałkowska, M. Wisniewski, Removal of Zn(II) from chloride acidic solutions with hydrophobic quaternary salts, Pol. J. Chem. Technol. 12 (3) (2010) 24–28.
[66] M. Regel-Rosocka, Extractive removal of zinc(II) from chloride liquors with phosphonium ionic liquids/toluene mixtures as novel extractants, Sep. Purif. Technol. 66 (2009) 19–24.
[67] P. Rybka, M. Regel-Rosocka, Nickel(II) and cobalt(II) extraction from chloride solutions with quaternary phosphonium salts, Sep. Sci. Technol. 47 (2012) 1296–1302.
[68] A. Cieszynska, M. Wisniewski, Extraction of palladium(II) from chloride solutions with Cyphos IL 101/toluene mixtures as novel extractant, Sep. Purif. Technol. 73 (2010) 202–207.
[69] A. Cieszynska, M. Wisniewski, Selective extraction of palladium(II) from hydrochloric acid solutions with phosphonium extractants, Sep. Purif. Technol. 80 (2) (2011) 385–389.
[70] A. Cieszynska, M. Wisniewski, Extractive recovery of palladium(II) from hydrochloric acid solutions with Cyphos IL 104, Hydrometallurgy 113–114 (2012) 79–85.
[71] E. Bekou, D.D. Dionysiou, R.-Y. Qian, G.D. Botsaris, Extraction of chlorophenols from water using room temperature ionic liquids, in: R.D. Rogers, K.R. Seddon (Eds.), Ionic Liquids as Green Solvents, vol. 856ACS, Washington, 2003, pp. 544–560. ISBN13: 978-08-41238-56-5.
[72] Y.C. Pei, J.J. Wang, Z.P. Xuan, J. Fan, M. Fan, Factors affecting ionic liquids based removal of anionic dyes from water, Environ. Sci. Technol. 41 (14) (2007) 5090–5095.
[73] V.M. Egorov, S.V. Smirnova, I.V. Pletnev, Highly efficient extraction of phenols and aromatic amines into novel ionic liquids incorporating quaternary ammonium cation, Sep. Purif. Technol. 63 (2008) 710–715.
[74] K.S. Khachatryan, S.V. Smirnova, I.I. Torocheshnikova, N.V. Shvedene, A.A. Formanovsky, I.V. Pletnev, Solvent extraction and extraction-voltammetric determination of phenols using room temperature ionic liquid, Anal. Bioanal. Chem. 381 (2005) 464–470.
[75] J. McFarlane, W.B. Ridenour, H. Luo, R.D. Hunt, D.W. DePaoli, R.X. Ren, Room temperature ionic liquids for separating organics from produced water, Sep. Sci. Technol. 40 (2005) 1245–1265.
[76] M.D. Bauman, A.J. Daugulis, P.G. Jessop, Phosphonium ionic liquids for degradation of phenol in a two-phase partitioning bioreactor, Appl. Microbiol. Biotechnol. 67 (2005) 131–137.
[77] J. Martak, S. Schlosser, Pertraction of organic acids through liquid membranes containing ionic liquids, Desalination 199 (2006) 518–520.
[78] J. Martak, S. Schlosser, Extraction of lactic acid by phosphonium ionic liquids, Sep. Purif. Technol. 57 (2007) 483–494.
[79] J. Martak, S. Schlosser, S. Vlckova, Pertraction of lactic acid through supported liquid membranes containing phosphonium ionic liquid, J. Membr. Sci. 318 (2008) 298–310.
[80] J. Martak, S. Schlosser, M. Blahušiak, Mass-transfer in pertraction of butyric acid by phosphonium ionic liquids and dodecane, Chem. Pap. 65 (5) (2011) 608–619.
[81] M. Matsumoto, A. Panigrahi, Y. Murakami, K. Kondo, Effect of ammonium- and phosphonium-based ionic liquids on the separation of lactic acid by supported ionic liquid membranes (SILMs), Membranes 1 (2011) 98–108.
[82] M. Matsumoto, Y. Murakami, Y. Minamidate, K. Kondo, Separation of lactic acid through polymer inclusion membranes containing ionic liquids, Sep. Sci. Technol. 47 (2012) 354–359.

[83] G. Kyuchoukov, D. Yankov, J. Albet, J. Molinier, Mechanism of lactic acid extraction with quaternary ammonium chloride (Aliquat 336), Ind. Eng. Chem. Res. 44 (2005) 5733–5739.
[84] M. Matsumoto, K. Mochiduki, K. Fukunishi, K. Kondo, Extraction of organic acids using imidazolium-based ionic liquids and their toxicity to *Lactobacillus rhamnosus*, Sep. Purif. Technol. 40 (2004) 97–101.
[85] J.J. Wang, Y.C. Pei, Y. Zhao, S.J. Zhang, Recovery of amino acids by imidazolium based ionic liquids from aqueous media, Green Chem. 7 (2005) 196–202.
[86] Y. Yan-Ying, Z. Wei, C. Shu-Wen, Extraction of ferulic acid and caffeic acid with ionic liquids, Chin. J. Anal. Chem. 35 (12) (2007) 1726–1730.
[87] A.D. Headley, N.M. Jackson, The effect of the anion on the chemical shifts of the aromatic hydrogen atoms of liquid 1-butyl-3-methylimidazolium salts, J. Phys. Org. Chem. 15 (1) (2002) 52–55.
[88] R. Vijayaraghavan, N. Vedaraman, M. Surianarayanan, D.R. MacFarlane, Extraction and recovery of azo dyes into an ionic liquid, Talanta 69 (5) (2006) 1059–1062.
[89] X. Chen, F. Li, C. Asumana, G. Yu, Extraction of soluble dyes from aqueous solutions with quaternary ammonium-based ionic liquids, Sep. Purif. Technol. 106 (2013) 105–109.
[90] C. Li, B. Xin, X. Wenguo, Q. Zhang, Study on the extraction of dyes into a room-temperature ionic liquid and their mechanisms, J. Chem. Technol. Biotechnol. 82 (2007) 196–204.
[91] M.J. Muldoon, M.G. Charles, I.R. Dunkin, Investigations of solvent–solute interactions in room temperature ionic liquids using solvatochromic dyes, J. Chem. Soc. Perkin Trans. 2 (2001) 433–435.
[92] M. Matsumoto, T. Ohtani, K. Kondo, Comparison of solvent extraction and supported liquid membrane permeation using an ionic liquid for concentrating penicillin G, J. Membr. Sci. 289 (2007) 92–96.
[93] A. Soto, A. Arce, M.K. Khoshkbarchi, Partitioning of antibiotics in a two-liquid phase system formed by water and a room temperature ionic liquid, Sep. Purif. Technol. 44 (2005) 242–246.
[94] S.G. Cull, J.D. Holbrey, V. Vargas-Mora, K.R. Seddon, G.J. Lye, Room-temperature ionic liquids as replacements for organic solvents in multiphase bioprocess operations, Biotechnol. Bioeng. 69 (2) (2000) 227–233.
[95] A.G. Fadeev, M.M. Meagher, Opportunities for ionic liquids in recovery of biofuels, Chem. Commun. (2001) 295–296.
[96] J. Yu, H. Li, H. Liu, Recovery of acetic acid over water by pervaporation with a combination of hydrophobic ionic liquids, Chem. Eng. Commun. 193 (2006) 1422–1430.
[97] F.J. Alguacil, M. Alonso, F. Lopez, A. Lopez Delgado, Pseudo-emulsion membrane strip dispersion (PEMSD) pertraction of chromium(VI) using Cyphos IL101ionic liquid as carrier, Environ. Sci. Technol. 44 (2010) 7504–7508.
[98] F.J. Alguacil, M. Alonso, F.A. Lopez, A. Lopez-Delgado, Application of pseudo-emulsion based hollow fiber strip dispersion (PEHFSD) for recovery of Cr(III) from alkaline solutions, Sep. Purif. Technol. 66 (2009) 586–590.
[99] F.J. Alguacil, M. Alonso, F.A. Lopez, A. Lopez-Delgado, I. Padilla, H. Tayibi, Pseudo-emulsion based hollow fiber with strip dispersion pertraction of iron(III) using $(PJMTH^{+})_2(SO_4^{2-})$ ionic liquid as carrier, Chem. Eng. J. 157 (2010) 272–366.
[100] A.P. de los Ríos, F.J. Hernández-Fernández, L.J. Lozano, S. Sánchez-Segado, A. Ginestá-Anzola, C. Godínez, F. Tomas-Alonso, J. Quesada-Medina, On the selective separation of metal ions from hydrochloride aqueous solution by pertraction through supported ionic liquid membranes, J. Membr. Sci. 444 (2013) 469–481.
[101] A.H. Blitz-Raith, R. Paimin, R.W. Cattrall, S.D. Kolev, Separation of cobalt(II) from nickel(II) by solid-phase extraction into Aliquat 336 chloride immobilized in poly(vinyl chloride), Talanta 71 (2007) 419–423.
[102] L. Wang, W. Shen, Chemical and morphological stability of Aliquat 336/PVC membranes in membrane extraction: a preliminary study, Sep. Purif. Technol. 46 (2005) 51–62.
[103] J. Xu, L. Wang, W. Shen, R. Paimin, X. Wang, The influence of the interior structure of Aliquat 336/PVC membranes to their extraction behavior, Sep. Sci. Technol. 39 (15) (2004) 3527–3539.

[104] D. Kogelnig, A. Regelsberger, A. Stojanovic, F. Jirsa, R. Krachler, B.K. Keppler, A polymer inclusion membrane based on the ionic liquid trihexyl(tetradecyl)phosphonium chloride and PVC for solid–liquid extraction of Zn(II) from hydrochloric acid solution, Monatsh. Chem. 142 (2011) 769–772.
[105] M. Regel-Rosocka, Ł. Nowak, M. Wiśniewski, Removal of zinc(II) and iron ions from chloride solutions with phosphonium ionic liquids, Sep. Sci. Technol. 97 (2012) 158–163.
[106] L. Guo, Y. Liu, C. Zhang, J. Chen, Preparation of PVDF-based polymer inclusion membrane using ionic liquid plasticizer and Cyphos IL 104 carrier for Cr(VI) transport, J. Membr. Sci. 372 (2011) 314–321.
[107] L. Guo, J. Zhang, D. Zhang, Y. Liu, Y. Deng, J. Chen, Preparation of poly(vinylidene fluoride-co-tetrafluoroethylene)-based polymer inclusion membrane using bifunctional ionic liquid extractant for Cr(VI) transport, Ind. Eng. Chem. Res. 51 (2012) 2714–2722.
[108] I.M. Coelhoso, T.F. Moura, J.P.S.G. Crespo, M.J.T. Carrondo, Transport mechanisms in liquid membranes with ion exchange carriers, J. Membr. Sci. 108 (1995) 231–244.
[109] R. Fortunato, C.A.M. Afonso, M.A.M. Reis, J.G. Crespo, Supported liquid membranes using ionic liquids: study of stability and transport mechanisms, J. Membr. Sci. 242 (2004) 197–209.
[110] R. Fortunato, M.J. Gonzalez-Munoz, M. Kubasiewicz, S. Luquec, J.R. Alvarez, C.A.M. Afonso, I.M. Coelhoso, J.G. Crespo, Liquid membranes using ionic liquids: the influence of water on solute transport, J. Membr. Sci. 249 (2005) 153–162.
[111] P.-A. Albertsson, Partition of Cell Particles and Macromolecules, third ed., Wiley, New York, 1986.
[112] B.Y. Zaslavsky, Aqueous Two-phase Systems: Physical Chemistry and Bioanalytical Applications, Marcel Dekker, New York, 1995.
[113] K.E. Gutowski, G.A. Broker, H.D. Willauer, J.G. Huddleston, R.P. Swatloski, J.D. Holbrey, R.D. Rogers, Controlling the aqueous miscibility of ionic liquids: aqueous biphasic systems of water-miscible ionic liquids and water-structuring salts for recycle, metathesis, and separations, J. Am. Chem. Soc. 125 (2003) 6632–6633.
[114] N.J. Bridges, K.E. Gutowski, R.D. Rogers, Investigation of aqueous biphasic systems formed from solutions of chaotropic salts with kosmotropic salts (salt–salt ABS), Green Chem. 9 (2007) 177–183.
[115] M.G. Freire, A.F.M. Cláudio, J.M.M. Araújo, J.A.P. Coutinho, I.M. Marrucho, J.N. Canongia Lopes, L.P.N. Rebelo, Aqueous biphasic systems: a boost brought about by using ionic liquids, Chem. Soc. Rev. 41 (2012) 4966–4995.
[116] K.P. Ananthapadmanabhan, E.D. Goddard, Aqueous biphase formation in polyethylene oxide-inorganic salt systems, Langmuir 3 (1) (1987) 25–31.
[117] R.D. Rogers, A.H. Bond, C.B. Bauer, J. Zhang, S.T. Griffin, Metal ion separations in polyethylene glycol-based aqueous biphasic systems: correlation of partitioning behavior with available thermodynamic hydration data, J. Chromatogr. B Biomed. Appl. 680 (1–2) (1996) 221–229.
[118] Y. Marcus, Thermodynamics of solvation of ions. Part 5-Gibbs free energy of hydration at 298.15 K, J. Chem. Soc. Faraday Trans. 87 (1991) 2995–2999.
[119] R.D. Rogers, A.R. Bond, J. Zhang, C.B. Bauer, Polyethylene glycol based-aqueous biphasic systems as technetium-99m generators, Appl. Radiat. Isot. 47 (5–6) (1996) 497–499.
[120] R.D. Rogers, A.H. Bond, C.B. Bauer, Polyethylene Glycol-based Aqueous Biphasic System for Liquid/Liquid Extraction of Environmentally Toxic Heavy Metals, in: D.H. Logsdail, M.J. Slater (Eds.), Solvent Extraction in the Process Industries, Proceedings of ISEC 93, vol. 3Elsevier Applied Science, London, 1993, p. 1641.
[121] R.D. Rogers, M.A. Eiteman, Aqueous Biphasic Separation: Biomolecules to Metal Ions, Plenum Press, New York, 1995.
[122] H.D. Willauer, J.G. Huddleston, S.T. Griffin, R.D. Rogers, Partitioning of aromatic molecules in aqueous biphasic systems, Sep. Sci. Technol. 34 (6–7) (1999) 1069–1090.
[123] H.D. Willauer, J.G. Huddleston, R.D. Rogers, Investigation of aqueous biphasic systems for the separation of lignins from cellulose in the paper pulping process, J. Chromatogr. B 743 (1–2) (2000) 127–135.
[124] L.C. Branco, J.G. Crespo, C.A.M. Afonso, Highly selective transport of organic compounds by using supported liquid membranes based on ionic liquids, Angew. Chem. Int. Ed. 41 (15) (2002) 2771–2773.

[125] Z.P. Visak, J.N. Canongia Lopes, L.P.N. Rebelo, Ionic liquids in polyethylene glycol aqueous solutions: salting-in and salting-out effects, Monatsh. Chem. 138 (2007) 1153–1157.
[126] C. Wu, J. Wang, Y. Pei, H. Wang, Z. Li, Salting-out effect of ionic liquids on poly(propylene glycol) (PPG): formation of PPG + ionic liquid aqueous two-phase systems, J. Chem. Eng. Data 55 (2010) 5004–5008.
[127] M.G. Freire, J.F.B. Pereira, M. Francisco, H. Rodriguez, L.P.N. Rebelo, R.D. Rogers, J.A.P. Coutinho, Insight into the interactions that control the phase behaviour of new aqueous biphasic systems composed of polyethylene glycol polymers and ionic liquids, Chem. Eur. J. 18 (2012) 1831–1839.
[128] M.T. Zafarani-Moattar, S. Hamzehzadeh, S. Nasiri, A new aqueous biphasic system containing polypropylene glycol and a water-miscible ionic liquid, Biotechnol. Prog. 28 (2012) 146–156.
[129] Z. Li, X. Liu, Y. Pei, J. Wang, M. He, Design of environmentally friendly ionic liquid aqueous two-phase systems for the efficient and high activity extraction of proteins, Green Chem. 14 (2012) 2941–2950.
[130] X. Liu, Z. Li, Y. Pei, H. Wang, J. Wang, Liquid–liquid equilibria for (cholinium-based ionic liquids + polymers) aqueous two-phase systems, J. Chem. Thermodyn. 60 (2013) 1–8.
[131] C.F. Poole, S.K. Poole, Extraction of organic compounds with room temperature ionic liquids, J. Chromatogr. A 1217 (2010) 2268–2286.
[132] S.H. Li, C.Y. He, H.W. Liu, K. Li, F. Liu, Ionic liquid-based aqueous two-phase system, a sample pretreatment procedure prior to high-performance liquid chromatography of opium alkaloids, J. Chromatogr. B 826 (1) (2005) 58–62.
[133] J.A. Han, Y. Wang, C.L. Yu, C.X. Li, Y.S. Yan, Y. Liu, L.A. Wang, Separation, concentration and determination of chloramphenicol in environment and food using an ionic liquid/salt aqueous two-phase flotation system coupled with high-performance liquid chromatography, Anal. Chim. Acta 685 (2011) 138–145.
[134] Z. Du, Y.L. Yu, J.H. Wang, Extraction of proteins from biological fluids by use of an ionic liquid/aqueous two-phase system, Chem. Eur. J. 13 (7) (2007) 2130–2137.
[135] M.J. Ruiz-Angel, V. Pino, S. Carda-Broch, A. Berthod, Solvent systems for countercurrent chromatography: an aqueous two phase liquid system based on a room temperature ionic liquid, J. Chromatogr. A 1151 (1–2) (2007) 65–73.
[136] C.Y. He, S.H. Li, H.W. Liu, K. Li, F. Liu, Extraction of testosterone and epitestosterone in human urine using aqueous two-phase systems of ionic liquid and salt, J. Chromatogr. A 1082 (2005) 143–149.
[137] D.L. Zhang, Y.F. Deng, J. Chen, Enrichment of aromatic compounds using ionic liquid and ionic liquid-based aqueous biphasic systems, Sep. Sci. Technol. 45 (5) (2010) 663–669.
[138] M.G. Freire, C.M.S.S. Neves, I.M. Marrucho, J.N. Canongia Lopes, L.P.N. Rebelo, J.A.P. Coutinho, High-performance extraction of alkaloids using aqueous two-phase systems with ionic liquids, Green Chem. 12 (2010) 1715–1718.
[139] C.M.S.S. Neves, S.P.M. Ventura, M.G. Freire, I.M. Marrucho, J.A.P. Coutinho, Evaluation of cation influence on the formation and extraction capability of ionic-liquid-based aqueous biphasic systems, J. Phys. Chem. B 113 (2009) 5194–5199.
[140] A.F.M. Claudio, M.G. Freire, C.S.R. Freire, A.J.D. Silvestre, J.A.P. Coutinho, Extraction of vanillin using ionic-liquid-based aqueous two-phase systems, Sep. Purif. Technol. 75 (1) (2010) 39–47.
[141] Q.F. Liu, J. Yu, W.L. Li, X.S. Hu, H.S. Xia, H.Z. Liu, P. Yang, Partitioning behavior of penicillin G in aqueous two phase system formed by ionic liquids and phosphate, Sep. Sci. Technol. 41 (12) (2006) 2849–2858.
[142] Y. Jiang, H. Xia, J. Yu, C. Guo, H. Liu, Hydrophobic ionic liquids-assisted polymer recovery during penicillin extraction in aqueous two-phase system, Chem. Eng. J. 147 (1) (2009) 22–26.
[143] Y.C. Pei, J.J. Wang, K. Wu, X.P. Xuan, X.J. Lu, Ionic liquid-based aqueous two-phase extraction of selected proteins, Sep. Purif. Technol. 64 (3) (2009) 288–295.
[144] M.T. Zafarani-Moattar, S. Hamzehzadeh, Partitioning of amino acids in the aqueous biphasic system containing the water-miscible ionic liquid 1-butyl-3-methylimidazolium bromide and the water-structuring salt potassium citrate, Biotechnol. Prog. 27 (4) (2011) 986–997.

[145] Y.Y. Jiang, H.S. Xia, C. Guo, I. Mahmood, H.Z. Liu, Phenomena and mechanism for separation and recovery of penicillin in ionic liquids aqueous solution, Ind. Eng. Chem. Res. 46 (19) (2007) 6303–6312.
[146] Y. Wang, J.A. Han, X.Q. Xie, C.X. Li, Extraction of trace acetylspiramycin in real aqueous environments using aqueous two-phase system of ionic liquid 1-butyl-3-methylimidazolium tetrafluoroborate and phosphate, Cent. Eur. J. Chem. 8 (6) (2010) 1185–1191.
[147] J. Han, Y. Wang, W.B. Kang, C.X. Li, Y.S. Yan, J.M. Pan, X.Q. Xie, Phase equilibrium and macrolide antibiotics partitioning in real water samples using a two-phase system composed of the ionic liquid 1-butyl-3-methylimidazolium tetrafluoroborate and an aqueous solution of an inorganic salt, Microchim. Acta 169 (2010) 15–22.
[148] C.H. Ma, L. Wang, Y.S. Yan, G.B. Che, Y.S. Yin, R.Z. Wang, D.Y. Li, Extraction of tetracycline via ionic liquid two-phase system, Chem. Res. Chin. Univ. 25 (2009) 832–835.
[149] C.X. Li, J. Han, Y. Wang, Y.S. Yan, X.H. Xu, J.M. Pan, Extraction and mechanism investigation of trace roxithromycin in real water samples by use of ionic liquid–salt aqueous two-phase system, Anal. Chim. Acta 653 (2) (2009) 178–183.
[150] C. Yu, J. Han, Y. Wang, Y. Yan, S. Hu, Y. Li, C. Ma, Ionic liquid/ammonium sulfate aqueous two-phase system coupled with HPLC extraction of sulfadimidine in real environmental water samples, Chromatographia 74 (5–6) (2011) 407–413.
[151] S.P.M. Ventura, C.M.S.S. Neves, M.G. Freire, I.M. Marrucho, J. Oliveira, J.A.P. Coutinho, Evaluation of anion influence on the formation and extraction capacity of ionic-liquid-based aqueous biphasic systems, J. Phys. Chem. B 113 (27) (2009) 9304–9310.
[152] F. Moscoso, F.J. Deive, J.M.S.S. Esperanca, A. Rodriquez, Pesticide removal from aqueous solutions by adding salting out agents, Int. J. Mol. Sci. 14 (2013) 20954–20965.
[153] V.A. Cocalia, K.E. Gutowski, R.D. Rogers, The coordination chemistry of actinides in ionic liquids: a review of experiment and simulation, Coord. Chem. Rev. 250 (7–8) (2006) 755–764.
[154] M.L. Dietz, J.A. Dzielawa, I. Laszak, B.A. Young, M.P. Jensen, Influence of solvent structural variations on the mechanism of facilitated ion transfer into room-temperature ionic liquids, Green Chem. 5 (6) (2003) 682–685.
[155] Y. Akama, M. Ito, S. Tanaka, Selective separation of cadmium from cobalt, copper, iron(III) and zinc by water-based two-phase system of tetrabutylammonium bromide, Talanta 53 (2000) 645–650.
[156] Y. Akama, A. Sali, Extraction mechanism of Cr(VI) on the aqueous two-phase system of tetrabutylammonium bromide and $(NH_4)_2SO_4$ mixture, Talanta 57 (2002) 681–686.
[157] M.L. Dietz, Ionic liquids as extraction solvents: where do we stand? Sep. Purif. Technol. 41 (2006) 2047–2063.
[158] P.G. Rickert, D.C. Stepinski, D.J. Rausch, R.M. Bergeron, S. Jakab, M.L. Dietz, Solute-induced dissolution of hydrophobic ionic liquids in water, Talanta 72 (2007) 315–320.
[159] I. Billard, A. Ouadi, C. Gaillard, Liquid–liquid extraction of actinides, lanthanides, and fission products by use of ionic liquids: from discovery to understanding, Anal. Bioanal. Chem. 400 (2011) 1555–1566.
[160] R.P. Swatloski, J.D. Holbrey, R.D. Rogers, Ionic liquids are not always green: hydrolysis of 1-butyl-3-methylimidazolium hexafluorophosphate, Green Chem. 5 (2003) 361–363.
[161] M. Stasiewicz, The recovery of food dyes from waste waters by extraction with hydrophobic ionic liquids, Przem. Chem. 89 (11) (2010) 1516–1519.
[162] M. Blahušiak, S. Schlosser, J. Cvengroš, J. Marták, New approach to regeneration of an ionic liquid containing solvent by molecular distillation, Chem. Pap. 65 (5) (2011) 603–607.

CHAPTER 5

Separation of Organic Acids Through Liquid Membranes Containing Ionic Liquids

Aulia I. Pratiwi, Michiaki Matsumoto
Department of Chemical Engineering and Materials Science, Doshisha University, Japan

1. INTRODUCTION

When liquids that are immiscible with the feed and receiving phases are used as membrane materials, they are referred to as liquid membranes (LMs). LMs work as a semi-permeable barrier between the feed and the receiving phases. Since the 1960s, LM technology has been applied to gas separation, recovery of metallic ions, removal of organic compounds, and recovery of fermentation products. Furthermore, LMs also have been combined with biotechnological processes.

In general, membrane separation techniques have many advantages: the concepts and operation of these techniques are simple; they are modular and easy to scale up; and they consume low amounts of energy [1]. Furthermore, it is possible to tune the selectivity of LMs using chemical modification and to avoid the large consumption of harmful organic solvents [2].

As illustrated in Figure 1, LMs can be divided into three forms, depending on their configuration. Configurations include bulk liquid membranes (BLMs), emulsion liquid

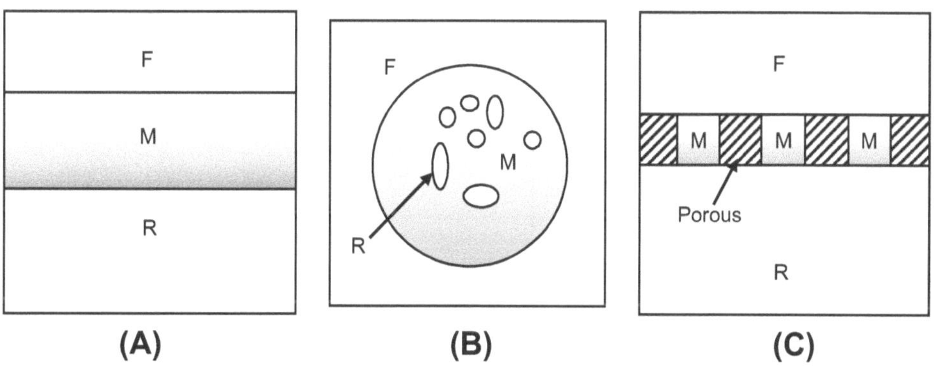

Figure 1 Liquid membranes: (A) bulk, (B) emulsion and (C) supported. F, feed; M, membrane; R, receiving.

Ionic Liquids in Separation Technology
http://dx.doi.org/10.1016/B978-0-444-63257-9.00005-5
Copyright © 2014 Elsevier B.V.
All rights reserved.

membranes (ELMs), and supported liquid membranes (SLMs). In this review, the aqueous feed and receiving phases and an organic, water-immiscible LM phase are explained.

A BLM consists of bulk aqueous feed and receiving phases separated by a bulk organic liquid phase. This organic liquid phase can be arranged in several ways [3]; examples of these arrangements are displayed in Figure 2. In the initial stage of the LM, only the

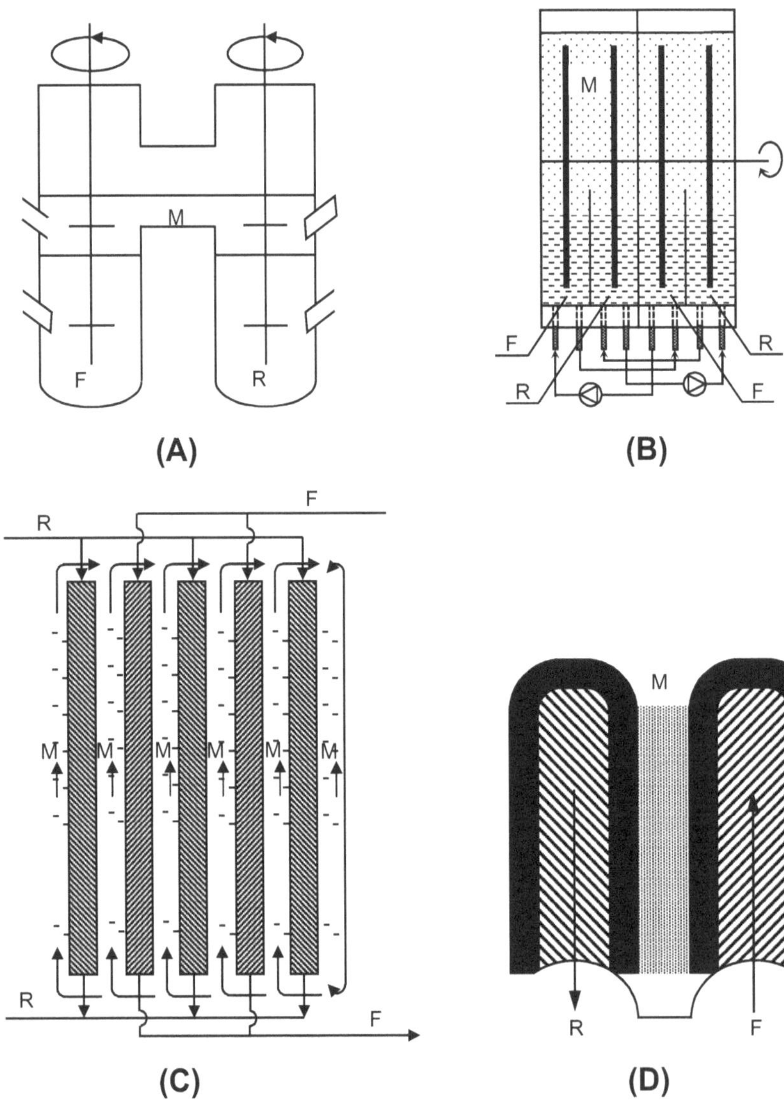

Figure 2 Bulk liquid membranes: (A) layered, (B) rotating disc, (C) creeping film and (D) flowing. F, feed; M, membrane; R, receiving.

layered BLM (Figure 2(A)) has been considered. Some research noted that the BLM's transport and selectivity is not suitable for practical applications. In the past decade, many technologies have been developed for LMs in an attempt to make them practically viable. These developments, such as rotating disc contactors (Figure 2(B)), creeping film contactors (Figure 2(C)) and flowing LM contactors (Figure 2(D)) must be included in the BLM group. Further details regarding these contactors can be found in Ref. [3].

The ELM was invented in 1968 by Li [4], who separated hydrocarbons with oil-in-water-in-oil emulsions. For water-in-oil-in-water emulsions, the aqueous receiving phase is emulsified in an organic LM containing a lipophilic surfactant. This water-in-oil emulsion is then dispersed in an aqueous feed solution to form multiple (water-in-oil)-in-water emulsions. Solutes then typically are transferred from the external feed to the internal receiving phases through an organic LM. The main advantage of the ELM lies in its high permeation flux, which is a result of a very high interfacial area per unit volume. However, the ELM has two primary disadvantages. The first is the complex behavior of its multiple emulsions: emulsion breakage, entrainment, and osmotic swelling. The second is that the stable water-in-oil emulsion must be broken down to recover the internal phase.

The SLM consists of bulk aqueous feed and receiving phases. These phases are separated by thin, porous, solid supports whose pores are impregnated with a membrane organic liquid. A flat-sheet SLM is useful on a laboratory scale, but its specific area is too small for industrial applications. To allow for this, spiral-wound and hollow-fiber SLMs were developed for practical application.

Although SLMs have such advantages as simple scale-up, low energy requirements, low capital and operating costs, and higher separation factors, concerns about their stability and long-term performance exist. These concerns are due to impregnation of the membrane pore with conventional organic solvents, which eventually deteriorate because of liquid vaporization, dissolution into contact phases, and displacement from the porous structure under a low-pressure gradient [5].

New LM technologies have been developed to improve the membrane stability of SLM systems. These alterations include a gel LM, a polymer inclusion membrane (PIM), and pseudo-emulsion-based hollow fiber strip dispersion. These technologies are considered modifications to SLMs [6] and are shown as such in Figure 3.

To improve the stability of SLMs, ionic liquids (ILs) have been used comprehensively as a stable impregnated liquid in the SLM process. Their utilization allows the physical properties of ILs, such as negligible volatility, relatively high viscosity, and interfacial tension, to be used to advantage.

Many different research groups have reviewed research regarding supported IL membrane (SILM) technology [7–11]. This chapter reviews the recent progress regarding liquid separations using IL membranes, and our research on lactic and succinic acids (Figure 4) permeation through SILM and PIM is introduced. Although gas

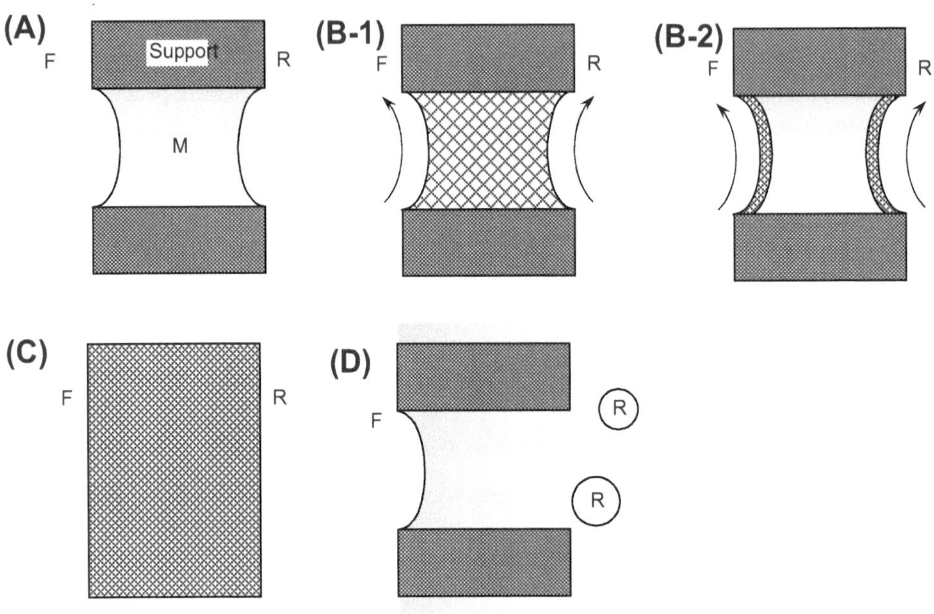

Figure 3 Supported liquid membranes: (A) normal type, (B-1) whole gel, (B-2) gel surface, (C) polymer inclusion and (D) pseudo-emulsion-based strip dispersion. F, feed; M, membrane; R, receiving.

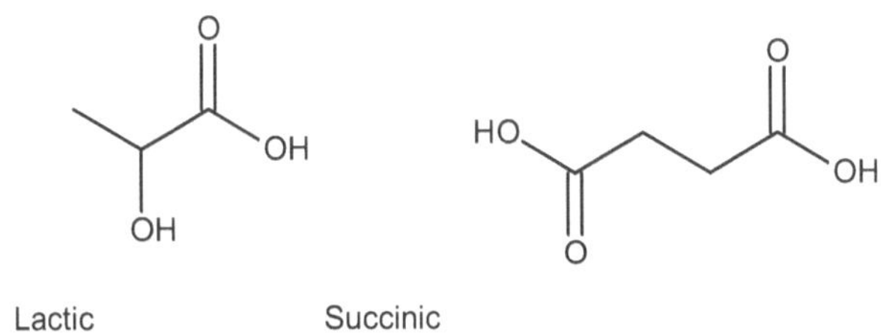

Figure 4 Structures of lactic acid (left) and succinic acid (right).

separations using IL membranes also have been studied in great depth, this research is not of concern here. Table 1 summarizes the studies of the application of ILs to LM separation since 2011.

2. BULK IONIC LIQUID MEMBRANES

Ng et al. [12] and Lakshmi et al. [13] performed phenol transport experiments in bulk liquid membrane containing 1-methyl-3-butylimizazolium [C_4C_1Im] chloride [Cl], hexafluorophosphate [PF_6], tetrafluoroborate [BF_4], bis(trifluoromethylsulfonyl)imide [NTf_2], tris(pentafluoroethyl)trifluorophosphate [FAP] as the membrane IL. In these

Table 1 Liquid Separation with Liquid Membranes (LMs) Containing Ionic Liquids

Substances	Ionic Liquids	LMs	References
Phenol	$[C_4C_1Im][PF_6]$, $[NTf_2]$, [FAP]	BLM	[12]
Chlorophenol	$[C_4C_1Im][BF_4]$, $[PF_6]$, [Cl]	BLM	[13]
Chromium	$[C_4C_1Im][NTf_2]$	ELM	[14,15]
Phenols	$[C_4C_1Im][PF_6]$	ELM	[16,17]
Chromium	$[RNH_3][HSO_4]$	PEHFSD	[18]
Zinc	Cyphos® IL101	PIM	[19]
Zinc, cadmium, copper, iron	$[C_8C_8\ C_8C_1N][Cl]$	SILM	[20]
Vanadium	$[C_8C_8\ C_8C_1N][Cl]$	SILM	[21]
Bisphenol A	$[C_4C_4\ C_4C_4P][Cl]$, $[PF_6]$	SILM	[22]
Organic sulfur	$[C_4C_1Py][CH_3SO_4]$	SILM	[23]
Butanol	$[C_4C_1Im][PF_6]$	Gel-SLM	[24]
Phenol	$[C_4C_1Im][HSO_4]$	SILM	[25]
Hexane	$[C_4C_1Im][Cl]$, $[BF_4]$, $[PF_6]$, $[NTf_2]$,	SILM	[26]
Chromium	$[C_{14}C_6\ C_6C_6P][Cl]$, $[C_nC_1Im][BF_4]$, $[PF_6]$	PIM	[27,28]
Butanol	$[C_8C_8\ C_8C_1N][NTf_2]$, $[C_{14}C_{14}\ C_6C_6P][NTf_2]$, [DCN]	SILM	[29]
Saccharides	Aliquat 336®	SILM	[30]
Aromatics	$[C_8C_1Im]$ [Cl]	SILM	[31]
Lactic acid	Aliquat 336®	SILM	[32]
		PIM	[33]
Succinic acid	Aliquat 336®	SILM	[34]
		PIM	[35]
Lithium	$[C_3C_1\ Pi][TFSI]$	SILM	[36–38]

BLM, bulk liquid membrane; ELM, emulsion liquid membrane; SLM, supported liquid membrane; SILM, supported ionic liquid membrane; PEHFSD, pseudo-emulsion-based hollow fiber strip dispersion; PIM, polymer inclusion membrane.

studies, layered BLMs were used to clarify the transport mechanism and assess the effect of operational conditions on transport efficiency and selectivity. Layered BLMs, however, were not used for practical application in industries because of the thickness of the membrane and the small amount of species transported. In the presence of ILs, the membrane properties in BLMs were tunable, and it was concluded that the application of ILs increased the recovery efficiency of phenols and the stability of the membrane.

3. ELMs CONTAINING ILs

Processes utilizing ELMs are more complex because of the requirements of the emulsification step required before and the demulsification step required after the permeation. As a result, the application of ELMs is limited only to a laboratory scale. There are no studies of the use of the ILs as the membrane liquid in ELMs at the present time.

Goyal et al. [14,15] and Balasubramanian and Venkatesan [16,17] studied the permeations of chromium and phenols of ELMs containing ionic liquids. The presence of the IL $[C_4C_1Im][NTf_2]$ for chromium permeation and the IL $[C_4C_1Im][PF_6]$ for phenol permeation in kerosene as a membrane solvent were reported to enhance emulsion stability. The effect of IL on the permeation rate of both groups were opposing, likely as a result of the strong interaction between the carrier and the IL in the membrane phase.

4. SUPPORTED IL MEMBRANES

The use of SLM as a separation process is regarded as a promising technique because of the method's low energy consumption and high selectivity and a reduction in the amount of solvent used [39]. As described earlier, however, the stability of SLMs using conventional organic solvents has proven insufficient because of the decline in permeation flux over time. A number of attempts were made to improve the stability of membranes involving the application of ILs to the impregnated liquid. Table 1 lists the studies of SLMs using an IL as not the carrier, but the membrane solvent. Various combinations between permeated species and ILs were reported.

5. IL MEMBRANES FOR SUCCINIC ACID SEPARATION

With an increase in the demand for environmentally friendly material, the production of organic acids using fermentation is now of interest. The economics of the process, however, depend on the development of an effective recovery method for the organic acids from the broth. As an alternative to the conventional precipitation process, solvent extraction processes have been proposed as a promising recovery technique. Furthermore, in situ extractive fermentation involves the simultaneous recovery of organic acids using solvent extraction. This occurs during the production of the organic acids in the fermentation broth. Advantages regarding in situ processes include the suppression of the inhibition of the end product (acid) and enhanced reactor efficiency; disadvantages relate to the solvent extraction, which suffers from environmental problems inherent in the use of a volatile organic solvent.

The suitability of ILs as "green" solvents for chemical processes has recently been recognized extensively because of their lack of vapor pressure. This characteristic makes ILs ideal replacements for volatile conventional organic solvents. However, the use of a large amount of IL as a solvent in liquid–liquid extraction systems is expensive because of the extremely high cost of ILs in comparison to conventional organic solvents. Thus, use of an SLM, which is a porous solid structure impregnated with liquid, has been examined using ILs as the impregnating liquid.

As described earlier, although SLMs are subject to membrane instability, many articles characterize SILMs as more stable than SLMs. Using water as a feed/receiving phase,

Fortunato et al. [40] reported that the membrane pores were able to retain the $[C_4C_1Im]$ $[PF_6]$ for 230 h without leaking. Shortly after, PIMs containing ILs were proposed as an even more stable alternative to SILMs. PIMs are formed by casting a solution containing a carrier, a plasticizer, and a base polymer such as cellulose triacetate (CTA) or poly(vinyl chloride) (PVC) to form a thin, flexible, and stable film. PIMs have been reported to have superior stability over SLMs (not SILM) [41]. PVC and CTA have been the only two major polymers used for the PIM investigations conducted so far. Since CTA is susceptible to hydrolysis in acidic media, PVC was used in this study as the base polymer.

Following a brief statement as to why succinic acid should be separated, this section explains the permeation of succinic acid through SILMs and PIMs.

Succinic acid, which is a bio-based chemical produced that uses biomass conversion processes and is utilized in biorefineries, and its use in various applications have been studied in depth by various researchers. The future prospects of the fermentative production of succinic acid from renewable resources using *Actinobacillus succinogenes*, *Mannheimia succiniciproducens*, *Anaerobiospirillum succiniproducens*, and recombinant *Escherichia coli* is also being studied currently [42,43]. Assessment of raw material costs and estimation of the potential market size clearly suggest that petroleum-based succinic acid will be replaced by fermentative succinic acid. Bechtold et al. [43] explained that the typical fermentative process for the production of succinic acid includes seed cultivation, fermentation, product recovery, concentration, and purification. Among these, the purification step in particular has been studied intensively for cost-effectiveness and environmental safety. Methods such as precipitation, sorption, ion exchange, electrodialysis, and liquid—liquid extraction have been proposed to concentrate and purify succinic acid from the fermentation broth [44—47]. Cheng et al. [48] discussed the technological development of succinic acid purification, including direct crystallization, membrane separation, chromatography, and in situ separation. However, more economical and efficient recovery and purification processes for succinic acid are necessary.

With regard to the permeation experiment, the apparatus displayed in Figure 5 consists of feed and stripping phases. The SILM uses a hydrophilic polyvinylidene difluoride membrane for the supported membrane and the PIM uses a PVC membrane for the base polymer.

Figure 6(A) and (B) shows the typical experimental results for SILMs and PIMs, respectively. Aliquat 336 was used as an IL in the membrane phase, and hydrochloric acid was used as a stripping agent in the receiving phase. The concentration of succinic acid in the feed phase decreases with time, whereas the concentration in the stripping phase increases. The uphill transport of succinic acid has been observed in both cases. The succinate anion in the feed phase at pH 6.5 (the pKa_1 and pKa_2 values of succinic acid are 4.17 and 5.64) diffuses across the SILM or PIM into the hydrogen chloride—containing stripping phase. As a result, it reacts with hydrochloric acid to form undissociated

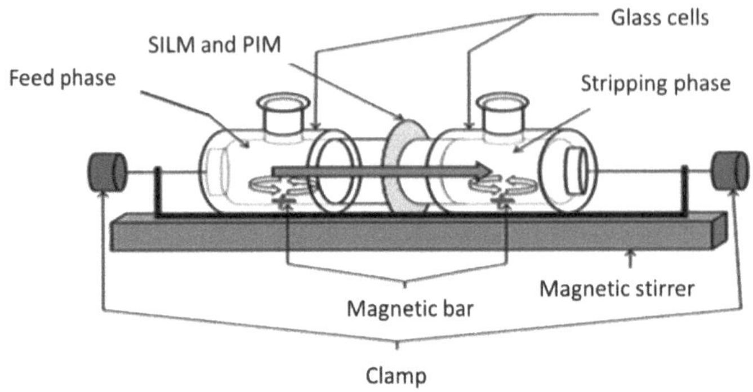

Figure 5 Experimental apparatus equipped with a flat membrane. PIM, polymer inclusion membrane; SILM, supported ionic liquid membrane.

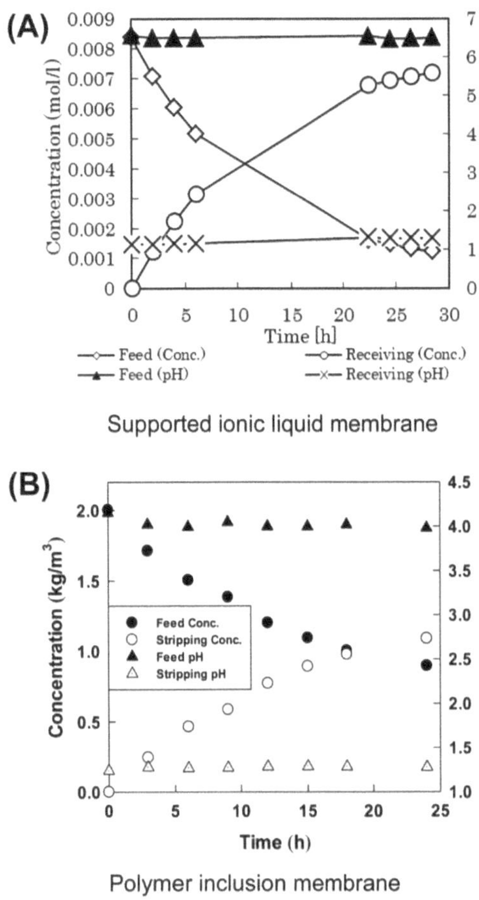

Figure 6 Typical experimental results for a supported ionic liquid membrane (A) and a polymer inclusion membrane (B).

succinic acid. Since this particular species is not soluble in the membrane phase, it becomes trapped in the stripping phase. Hence, permeation proceeds until the succinate concentration in the feed phase falls to zero. The permeation rate in PIM was observed to be lower than that of SILM, consistent with a previous report regarding diffusion coefficients, in which the PIM value was reported to be much lower than the corresponding SLM [41].

In general, the driving force for succinate transport is the gradient, which runs through the diffusional layers of the feed liquid film and the LM. This is a result of the assumption that the succinate concentration in the stripping phase is assumed to be zero, as described earlier. With this, the succinate flux can be expressed using the following equation [49]:

$$J = -\frac{V_f}{A}\frac{dC_f}{dt} = K_{OV} C_f \tag{1}$$

Here, K_{OV} is the overall mass transfer coefficient, V_f is the volume of the feed solution, A is the contact area between the feed and membrane phases, and C_f is the succinate concentration in the feed phase.

Thus,

$$\ln\left(\frac{C_f}{C_{f,0}}\right) = -a \cdot K_{OV} \cdot t \tag{2}$$

The subscript 0 denotes the initial state, and a ($=A/V_f = 12\ m^{-1}$) is the specific area. The K_{OV} values were obtained from the slopes of the straight-line plots based on Eqn (2).

The overall resistance to mass transfer, $1/K_{OV}$, is considered the sum of the membrane resistance, $1/K_m$, and the feed boundary layer resistance, $1/K_L$:

$$\frac{1}{K_{OV}} = \frac{1}{K_L} + \frac{1}{K_m} = \frac{1}{K_L} + \frac{\delta}{P} \tag{3}$$

Here, P is the permeability and δ is the membrane thickness. To evaluate the film mass transfer coefficient, the relationship between the overall mass transfer coefficient and the membrane thickness was investigated.

From the intercept shown in Figure 7, the feed boundary layer resistance, $1/K_L$, was determined. Hereafter, the permeability obtained from Eqn (3) is discussed, assuming that the feed boundary layer resistance is independent of the feed composition.

5.1 Supported IL Membrane

The effect of pH on permeability is presented in Figure 8. The maximum permeability was observed at pH 6.5, which was convenient because it was consistent with a previous study, in which the optimum pH in the fermentation broth was found to be between 6 and 7 [50].

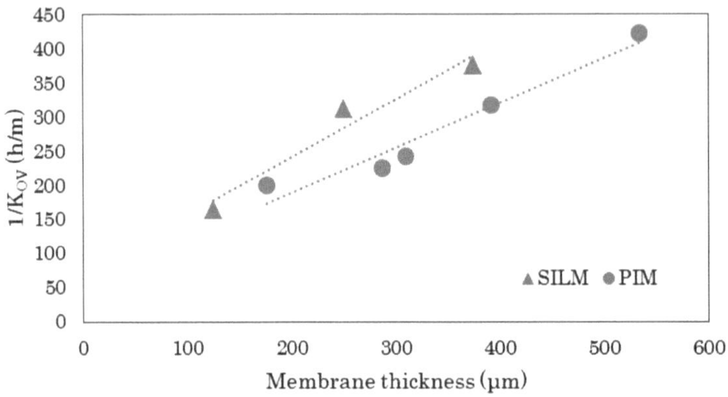

Figure 7 Effect of membrane thickness on overall mass transfer coefficient. $1/K_{OV}$, overall resistance to mass transfer; PIM, polymer inclusion membrane; SILM, supported ionic liquid membrane.

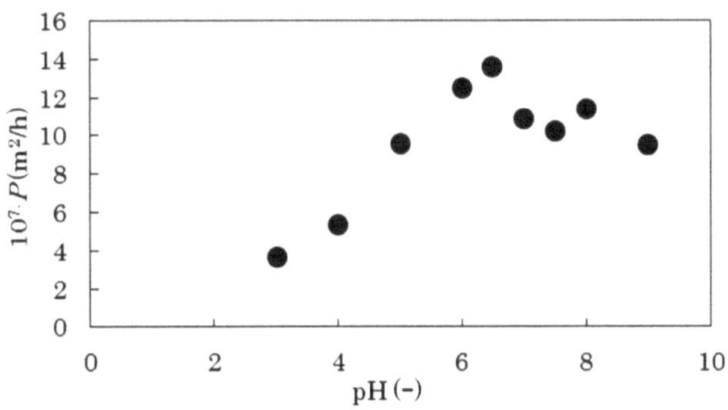

Figure 8 Effect of pH on permeability (10^7 *P*) in a supported ionic liquid membrane.

Based on the solution diffusion theory, permeability can be expressed using the following equation:

$$P = D_f \cdot K_D \tag{4}$$

Here, D_f denotes the diffusivity of the succinate complex; K_D is the succinate distribution ratio.

If the diffusivities of the succinate complex with Aliquat 336 are kept constant under the experimental conditions, the permeability is expected to be proportional to the distribution ratio. This relation is linear, as can be seen in Figure 9. The diffusivity obtained ($D_f = 1.2 \times 10^{-12}$ m^2/s) was similar to that of Aliquat 336-impregnated resin [51] and approximately 10 times greater than those in Aliquat 336/PVC membranes [33,52].

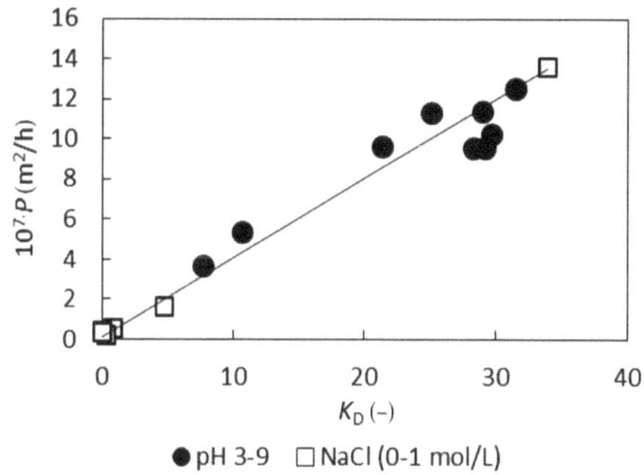

Figure 9 Relation between permeability (10^7 P) and distribution ratio (K_D) in a supported ionic liquid membrane. NaCl, sodium chloride.

The mechanism of permeation of succinic acid through the SILM consists of the following steps.

1. The succinate di-anion reacts with Aliquat 336 by an anion exchange reaction at the interface between the feed and membrane phases.
2. The Aliquat 336—succinate complex is diffused through the IL membrane.
3. The Aliquat 336—succinate complex reacts with hydrochloric acid to form undissociated succinic acid at the interface between the membrane and stripping phases.

The schematic permeation mechanism is shown in Figure 10.

5.2 Polymer Inclusion Membrane

As shown in Figure 6, the permeation rate of succinic acid through the PIM was considerably slower than that in the SILM when using hydrochloric acid as a stripping agent in the receiving phase. Therefore, sodium carbonate, rather than hydrochloric acid, was used as a stripping agent; these results are displayed in Figure 11. In doing so, the overall permeation rate of the succinate running through the PIM was able to reach a level almost the equivalent to the succinate in the SILM.

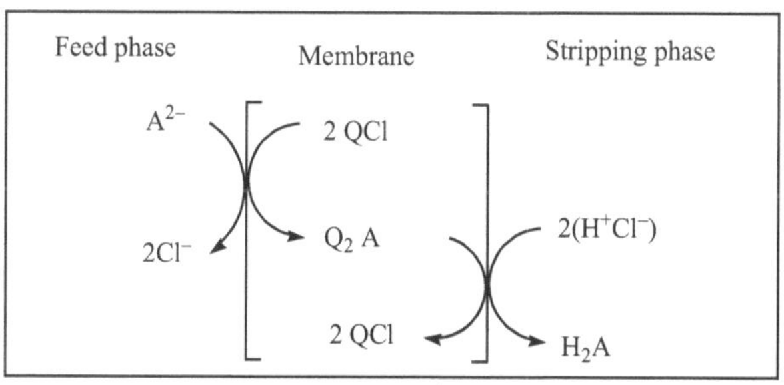

Figure 10 Transport mechanism for permeation of succinic acid by a supported ionic liquid membrane. H_2A, succinic acid; QCl, Aliquat 336.

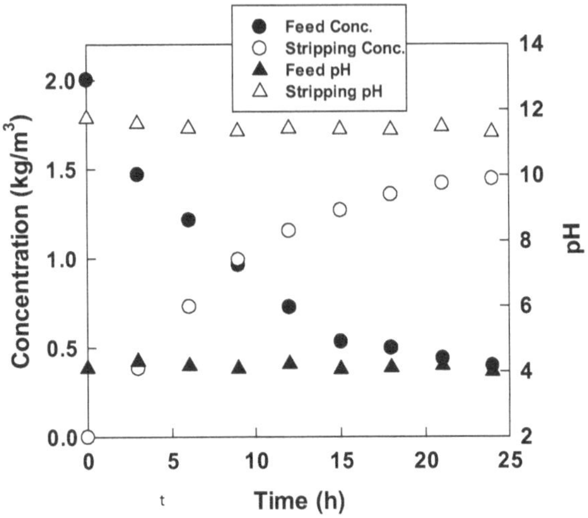

Figure 11 Permeation of succinic acid through a polymer inclusion membrane using sodium carbonate.

Although SILM has an optimum effect at a pH of 6.5, the effect of pH on the permeability observed under the experimental conditions occurred at pH 4, as presented in Figure 12. This discrepancy can likely be attributed to the difference in the permeation mechanisms of SILMs and PIMs, wherein the succinic acid species involved in each mechanism differ.

There are three types of succinic acid species: the undissociated, mono-anionic, and di-anionic forms. While the SILM is affected only by the di-anionic form, the PIM permeation is affected by all three species, as is evident from Figure 12; this is due to the pKa_1 and pKa_2 values of succinic acid (4.17 and 5.64, respectively). Figure 13 displays a

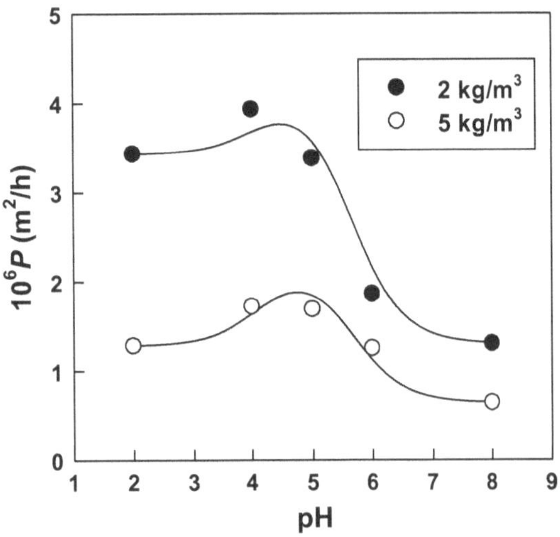

Figure 12 Effect of pH on permeability (10^6 *P*) in a polymer inclusion membrane.

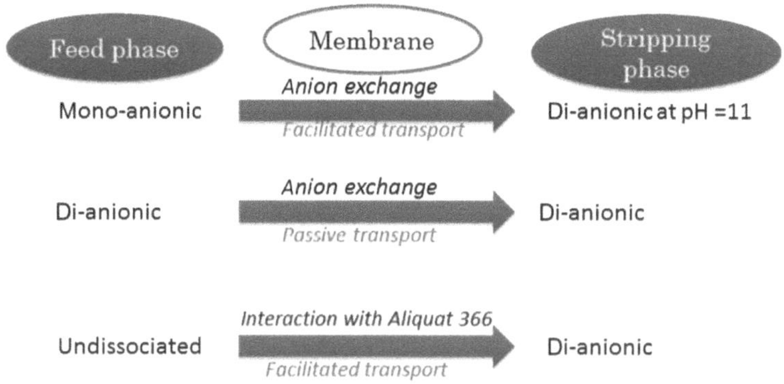

Figure 13 Transport mechanisms for permeation of succinic acid by a polymer inclusion membrane.

Table 2 Permeability of Acid Forms in a Polymer Inclusion Membrane

	Permeability (10^6 m^2/h)	
Form	**2 kg/m^3**	**5 kg/m^3**
Undissociated	3.4	1.3
Mono-anionic	4.1	2.2
Di-anionic	1.3	0.64

schematic explanation of the permeation mechanism of succinic acid through the PIM. The apparent permeability is expressed using the following equation:

$$P = P_{H2A} f_{H2A} + P_{HA-} f_{HA-} + P_{A2-} f_{A2-} \quad (5)$$

Here, *f* denotes the molar fraction of acid forms and is a function of pH. If the values of P_{H2A} and P_{A2-} can be assumed to be values at pH 2 and 8, respectively, the only parameter in Eqn (5) to be obtained is P_{HA-}.

Estimated values from the results displayed in Figure 12 are summarized in Table 2. Solid lines in Figure 12 are calculated from the values in Table 2, and the optimum pH is found to be 4.6 ~ 4.8. As expected from Figure 12, the permeability of the mono-anionic form of succinic acid is largest. This may be due to the smaller diffusivity of the bulky di-anionic complex with Aliquat 336 and the smaller distribution ratio of undissociated succinic acid to the membrane.

6. IL MEMBRANES FOR LACTIC ACID SEPARATION

In the previous section, the permeation of succinic acid through an SILM and PIM using a flat-sheet membrane was discussed. Next, hollow fiber membranes are examined. These membranes are considered a practical configuration because of their large specific

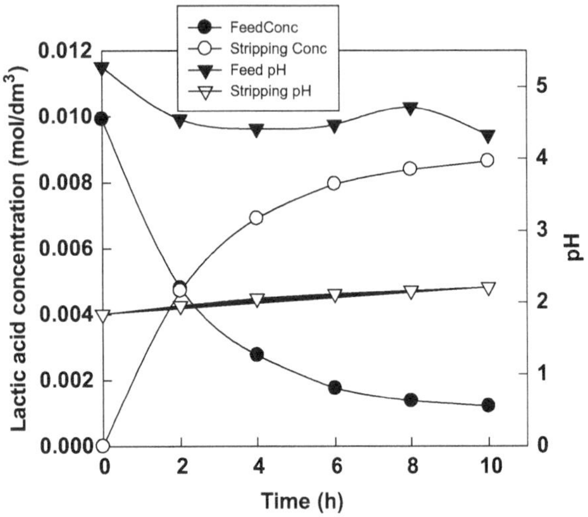

Figure 14 Permeation of lactic acid through Aliquat 336-impregnated hollow-fiber supported ionic liquid membrane.

area, which is necessary for the permeation of organic acid through the SILM. Here, the experiments use hollow fibers in the batch mode. In particular, the Aliquat 336-impregnated hollow fibers were immersed in the feed solution and the stripping aqueous solution was fed into the interior of the hollow fibers and circulated.

Figure 14 demonstrates the time course of permeation of lactic acid from an experiment using polypropylene (PP) hollow-fiber membranes impregnated with Aliquat 336. The observed permeation of lactic acid through the hollow fiber membranes was significantly faster what would be expected of a flat-sheet membrane. The recovery ratio of lactic acid in the hollow-fiber membrane system reached 87.1% after 10 h. The permeation fluxes in both cases, along with the membrane characteristics, are summarized in Table 3. The issue regarding the operation of the PP hollow-fiber membrane is the difficulty in adjusting the pressure difference necessary to keep the impregnated ILs in the membrane pore. This could reflect differences in the membrane microstructure, such as porosity and tortuosity, between polyvinylidene fluoride and PP membranes.

The superior stability of PIMs over SLMs (not SILMs) has been reported for membranes using Aliquat 336 as a carrier of chromate [53]; stable performance for

Table 3 Flux of Lactic Acid Through Supported Ionic Liquid Membranes with an Initial Lactic Acid Concentration of 0.01 mol/dm^3

	Flat-Sheet Membrane	Hollow-Fiber Membrane
Contact area (cm^2)	12.0	25.9
Membrane thickness (μm)	125	50
Permeation flux (mol/m^2/h)	3.71×10^{-6}	1.50×10^{-5}
Permeation flux (mol/m^2/h μm)	4.64×10^{-4}	7.50×10^{-4}

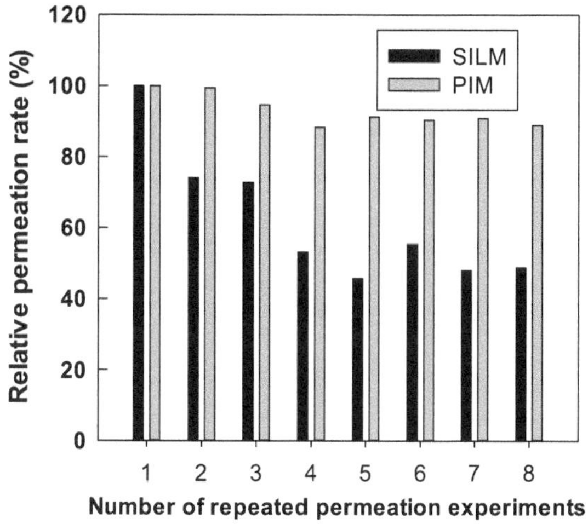

Figure 15 Comparison of stability of a supported ionic liquid membrane (SILM) and a polymer inclusion membrane (PIM) including Aliquat 336 as an ionic liquid.

30 days was recorded for PIMs. To now contrast the stability of SILMs and PIMs, a study of both was conducted by repeating a batch-wise transport experiment for each using Aliquat 336 as a carrier. The experiment was conducted 8 times for 24 h each, and the same membranes and fresh feed and stripping solutions were used. Figure 15 shows the relative permeation rates of PIMs and SILMs as a function of the number of runs. The permeation rates of the PIM after eight times remained at clearly >90% of the initial values, whereas those of SILM based on Aliquat 336 decreased with the number of experimental runs. This suggests that the dissolution rate of Aliquat 336 from the membrane pore of the SILM to water was not negligible, which differed from what was expected. Argiropoulos et al. [54] reported a negligible loss of Aliquat 336 in a hydrochloric acid solution and a considerable loss in distilled water from a PVC-based PIM. These behaviors were likely a kind of salting effect. Combining our results with those findings, the existence of salts was found to be essential to prevent the leakage of Aliquat 336 from the membranes. From the results, PIMs were proven to be highly resistant to leakage.

7. CONCLUSION

The application of ILs to LM systems has existed for approximately 15 years. We have seen recently a considerable number of articles published regarding gas separation in SILMs. Despite this, the practical applications of SLM processes have been limited because of membrane instability. In gas/vapor separation, the nonvolatile nature of ILs should allow for significant improvement of the current process and the development of new approaches to gas/vapor separation. However, significant effort will be required for ILs to become commercially viable [55].

This chapter gave a general description of LMs using ILs and liquid separation. In particular, it examined the recovery of fermented organic acids such as lactic and succinic acids using an SILM and PIM. Despite the considerable improvement of SILMs over SLMs (using a conventional organic solvent), however, the issues regarding membrane instability remain. This is mainly because of, in principal, the inevitability of the physical dissolution of ILs into adjacent liquid phases. In this case, because of higher membrane stability and a comparable permeation rate, PIMs have become an excellent alternative to SILM systems.

REFERENCES

[1] V.S. Kislik, Introduction, general description, definitions, and classification, in: V.S. Kislik (Ed.), Liquid Membranes, Elsevier, Amsterdam, The Netherlands, 2010, pp. 1–15.

[2] J.A. Jönsson, L. Mathiasson, Liquid membrane in analytical sample preparation: I. Principles, TrAC Trends Anal. Chem. 18 (1999) 318–325.

[3] V.S. Kislik, Bulk hybrid liquid membrane with organic water-immiscible carriers, in: V.S. Kislik (Ed.), Liquid Membranes, Elsevier, Amsterdam, The Netherlands, 2010, pp. 201–275.

[4] N.N. Li, Separating Hydrocarbons with Liquid Membranes. US Patent 3,410,794 (1968).

[5] N. Kouki, R. Tayeb, M. Dhahbi, A flat-sheet supported liquid membrane based on Aliquat 336® as carrier for the removal of salicylic acid from aqueous solution, Desalin. Water Treat. (2014), http://dx.doi.org/10.1080/19443994.2013, 814325.

[6] P. Dżygiel, P.P. Wieczorek, Supported liquid membranes and their modifications, in: V.S. Kislik (Ed.), Liquid Membranes, Elsevier, Amsterdam, The Netherlands, 2010, pp. 73–140.

[7] L.J. Lozano, C. Godinez, A.P. de los Rios, F.J. Hernandez-Fernandez, S. Sanchez-Segado, F.J. Alguacil, Recent advances in supported ionic liquid membrane technology, J. Membr. Sci. 376 (2011) 1–14.

[8] M.A. Malik, M.A. Hashim, F. Nabi, Ionic liquids in supported liquid membrane technology, Chem. Eng. J. 171 (2011) 242–254.

[9] P. Cserjési, K. Bélafi-Bakó, Application of ionic liquids in membrane separation processes, in: A. Kokorin (Ed.), Ionic Liquids: Application and Perspectives, InTech, Rijeka, Croatia, 2011, pp. 561–586.

[10] M. Matsumoto, Ionic liquid-based supported liquid membranes, in: K. Mohanty, M.K. Purkait (Eds.), Membrane Technology and Applications, CRC Press, Boca Raton, USA, 2012, pp. 305–316.

[11] A.P. de los Ríos, F.J. Hernández-Fernández, L.J. Lozano, C. Godínez, Supported ionic liquid membranes, in: A. Mohammad, Inamuddin (Eds.), Green Solvent II, Springer, Dordrecht, The Netherlands, 2012, pp. 275–288.

[12] Y.S. Ng, N.S. Jayakumar, M.A. Hashim, Behavior of hydrophobic ionic liquids as liquid membranes on phenol removal: experimental study and optimization, Desalination 278 (2011) 250–258.

[13] A.B. Lakshmi, S. Sindhu, S. Venkatesan, Performance of ionic liquid as bulk liquid membrane for chlorophenol removal, Int. J. ChemTech Res. 5 (2013) 1129–1137.

[14] R.K. Goyal, N.S. Jayakumar, M.A. Hashim, Chromium removal by emulsion liquid membrane using $[BMIM]^+[NTf_2]^-$ as stabilizer and TOMAC as extractant, Desalination 278 (2011) 50–56.

[15] R.K. Goyal, N.S. Jayakumar, M.A. Hashim, A comparative study of experimental optimization and response surface optimization of Cr removal by emulsion ionic liquid membrane, J. Hazard Mater. 195 (2011) 383–390.

[16] A. Balasubramanian, S. Venkatesan, Removal of phenolic compounds from aqueous solutions by emulsion liquid membrane containing ionic liquid $[BMIM]^+[PF_6]^-$ in tributyl phosphate, Desalination 289 (2012) 27–34.

[17] A. Balasubramanian, S. Venkatesan, Optimization of removal of phenol from aqueous solution by ionic liquid-based emulsion liquid membrane using response surface methodology, Clean 42 (2014) 64–70.

[18] F.J. Alguacil, I. Garcia-Diaz, F.A. López, Modeling of facilitated transport of Cr(III) using ($RNH_3^+HSO_4^-$) ionic liquid and pseudo-emulsion hollow fiber strip dispersion (PEHFSD) technology, J. Ind. Eng. Chem. 19 (2013) 1086–1093.

[19] D. Kogelnig, A. Regelsberger, A. Stojanovic, F. Jirsa, R. Krachler, B.K. Keppler, A polymer inclusion membrane based on the ionic liquid trihexyl(tetradecyl)phosphonium chloride and PVC for solid0liquid extraction of Zn(II) from hydrochloric acid solution, Monatsh Chem. 142 (2011) 769–772.

[20] A.P. de los Ríos, F.J. Hernández-Fernández, L.J. Lozano, S. Sánchez-Segado, A. Ginestá-Anzola, C. Godínez, F. Tomás-Alonso, J. Quesada-Medina, On the selective separation of metal ions from hydrochloride aqueous solution by pertraction through supported ionic liquid membranes, J. Membr. Sci. 444 (2013) 469–481.

[21] S. Nosrati, N.S. Jayakumar, M.A. Hashim, S. Mukhopadhyay, Performance evaluation of vanadium (IV) transport through supported ionic liquid membrane, J. Taiwan Inst. Chem. Eng. 44 (2013) 337–342.

[22] A. Panigrahi, S.R. Pilli, K. Mohanty, Selective separation of bisphenol A from aqueous solution using supported ionic liquid membrane, Sep. Purif. Technol. 107 (2013) 70–78.

[23] G.O. Yahaya, F. Hamad, A. Bahamdan, V.V.R. Tammana, E.Z. Hamad, Supported ionic liquid membrane and liquid-liquid extraction using membrane for removal of sulfur compounds from diesel/crude oil, Fuel Process. Technol. 113 (2013) 123–129.

[24] A. Plaza, G. Merlet, A. Hasanoglu, A. Isaacs, J. Sánchez, J. Romero, Separation of butanol from ABE mixtures by sweep gas pervaporation using a supported gelled ionic liquid membrane: analysis of transport phenomena and selectivity, J. Membr. Sci. 444 (2013) 201–212.

[25] S. Nosrati, N.S. Jayakumar, M.A. Hashim, Performance evaluation of supported ionic liquid membrane for removal of phenol, J. Hazard Mater. 192 (2011) 1283–1290.

[26] F.J. Hernández-Fernández, A.P. de los Ríos, F. Tomás-Alonso, J.M. Palacios, G. Víllora, Understanding the influence of the ionic liquid composition and the surrounding phase nature on the stability of supported ionic liquid membranes, AIChE J. 58 (2012) 583–590.

[27] L. Guo, Y. Liu, C. Zhang, J. Chen, Preparation of PVDF-based polymer inclusion membrane using ionic liquid plasticizer and Cyphos IL 104 carrier for Cr(VI) transport, J. Membr. Sci. 372 (2011) 314–321.

[28] L. Guo, J. Zhang, D. Zhang, Y. Liu, Y. Deng, J. Chen, Preparation of poly(vinylidene fluoride-co-tetrafluoroethylene)-based polymer inclusion membrane using bifunctional ionic liquid extractant for Cr(VI) transport, Ind. Eng. Chem. Res. 51 (2012) 2714–2722.

[29] H.R. Cascon, S.K. Choudhari, 1-Butanol pervaporation and intrinsic stability of phosphonium and ammonium ionic-liquid-based supported liquid membranes, J. Membr. Sci. 429 (2013) 214–224.

[30] M. Matsumoto, N. Oku, K. Kondo, Permeation of saccharides through supported ionic liquid membranes containing calixarenes as carriers, Solvent Extr. Res Dev. Jpn 20 (2013) 219–224.

[31] M. Chakraborty, D. Dobaria, P.A. Parikh, The separation of aromatic hydrocarbons through a supported liquid membrane, Petro. Sci. Technol. 30 (2012) 2504–2516.

[32] M. Matsumoto, A. Panigrahi, Y. Murakami, K. Kondo, Effect of ammonium- and phosphonium-based ionic liquids on the separation of lactic acid by supported ionic liquid membranes (SILMs), Membranes 1 (2011) 98–108.

[33] M. Matsumoto, Y. Murakami, K. Kondo, Separation of lactic acid through polymer inclusion membranes containing ionic liquids, Sep. Sci. Technol. 47 (2012) 354–359.

[34] A.I. Pratiwi, M. Matsumoto, K. Kondo, Permeation of succinic acid through supported ionic liquid membranes, J. Chem. Eng. Jpn 46 (2013) 383–388.

[35] A.I. Pratiwi, M. Matsumoto, K. Kondo, Permeation mechanism of succinic acid through polymer inclusion membranes with ionic liquid Aliquat 336, J. Chem. Eng. Jpn 47 (2014) 314–318.

[36] T. Hoshino, T. Terai, Basic technology for ^{6}Li enrichment using an ionic-liquid impregnated organic membrane, J. Nucl. Mater. 417 (2011) 696–699.

[37] T. Hoshino, T. Terai, High-efficiency technology for lithium isotope separation using an ionic-liquid impregnated organic membrane, Fusion Eng. Des. 86 (2011) 2168–2171.

[38] T. Hoshino, Preliminary studies of lithium recovery technology from seawater by electrodialysis using ionic liquid membrane, Desalination 317 (2013) 11–16.
[39] N.M. Kocherginsky, Q. Yang, L. Seelam, Recent advances in supported liquid membrane technology, Sep. Purif. Technol. 53 (2007) 171–177.
[40] R. Fortunato, C.A.M. Afonso, M.A.M. Reis, J.G. Crespo, Supported liquid membranes using ionic liquids: study of stability and transport mechanisms, J. Membr. Sci. 242 (2004) 197–209.
[41] M. Inês, G.S. Almeida, R.W. Cattrall, S.D. Kolev, Recent trends in extraction and transport of metal ions using polymer inclusion membranes (PIMs), J. Membr. Sci. 415 (2012) 9–23.
[42] H. Song, S.Y. Lee, Production of succinic acid by bacterial fermentation, Enzym. Microb. Technol. 39 (2006) 352–361.
[43] I. Bechtold, K. Bretz, S. Kabasci, R. Kopitzky, A. Springer, Succinic acid: a new platform chemical for biobased polymer from renewable resources, Chem. Eng. Technol. 31 (2008) 647–654.
[44] T. Kurzrock, D.W. Botz, Recovery of succinic acid from fermentation broth, Biotechnol. Lett. 32 (2010) 331–339.
[45] T. Kurzrock, D.W. Botz, New reactive extraction systems for separation of bio-succinic acid, Bioprocess Biosyst. Eng. 34 (2011) 779–787.
[46] M.D. Joshi, J.L. Anderson, Recent advances of ionic liquids in separation science and mass spectrometry, RSC Adv. 2 (2012) 5470–5484.
[47] F.S. Oliveira, J.M.M. Araujo, R. Ferreira, L.P.N. Rebelo, I.M. Marrucho, Extraction of l-lactic, l-malic, and succinic acid using phosphonium-based ionic liquids, Sep. Purif. Technol. 85 (2012) 137–146.
[48] K.K. Cheng, X.B. Zhao, J. Zeng, R.C. Wu, Y.Z. Xu, D.H. Liu, J.A. Zhang, Downstream processing of biotechnological produced succinic acid, Appl. Microbiol. Biotechnol. 95 (2012) 841–850.
[49] M.F. San Román, E. Bringas, R. Ibañez, I. Ortiz, Liquid membrane technology: fundamentals and review of its applications, J. Chem. Technol. Biotechnol. 85 (2009) 2–10.
[50] K.A. Berglund, S. Yedur, D.D. Dunuwila, Succinic Acid Production and Purification. US Patent, 5,958,744 (1999).
[51] B. Saha, R.J. Gill, D.G. Bailey, N. Kabay, M. Arda, Sorption of Cr (VI) from aqueous solution by Amberlite XAD-7 resin impregnated with Aliquat 336, React. Funct. Polym. 60 (2004) 223–244.
[52] L. Wang, R. Paimin, R.W. Cattral, W. Shen, S.D. Kolev, The extraction of cadmium (II) and copper (II) from hydrochloric acid using Aliquat 336/PVC membrane, J. Membr. Sci. 176 (2000) 105–111.
[53] Y.M. Scindia, A.K. Pandey, A.V.R. Reddy, Coupled-diffusion transport of Cr(VI) across anion-exchange membranes prepared by physical and chemical immobilization methods, J. Membr. Sci. 249 (2005) 143–152.
[54] G. Argiropoulos, R.W. Cattrall, I.C. Hamilton, S.D. Kolev, R. Paimin, The study of a membrane for extracting gold(III) from hydrochloric acid solution, J. Membr. Sci. 138 (1998) 279–285.
[55] P. Scovazzo, Determination of the upper limits, benchmarks, and critical properties for gas separations using stabilized room temperature ionic liquid membranes (SILMs) for the purpose of guiding future research, J. Membr. Sci. 343 (2009) 199–211.

CHAPTER 6

Separation of Proteins by Ionic Liquid-Based Three-Phase Partitioning

Enrique Alvarez-Guerra, Angel Irabien
Departamento de Ingenierías Química y Biomolecular, Universidad de Cantabria, Santander, Cantabria, Spain

1. INTRODUCTION

Proteins are components of living organisms that play critical roles in them and are useful for therapeutic and diagnostic applications, so the preparation of pure proteins in an efficient way is essential. Proteins are present in a natural way in biological fluids, or they are usually produced in media that are a highly complex mixture of other proteins, cell wall materials, and nucleic acids [1]. As a result, protein separation and purification is a very difficult task, so downstream processing usually represents between 50% and 80% of the total production costs of proteins [2]. For this reason, as the major manufacturing cost is the downstream processing, the competitive advantage in production of proteins will depend not only on innovations of basic biological areas, but also on innovation and optimization of separation and purification processes [3].

Traditional protein purification methods such as chromatography or electrophoresis are time and cost consuming [4,5]. Therefore, the demand for fast, cost-effective, and environmentally friendly downstream processes has encouraged the development of clean manufacturing methods and easily scaled-up, industrially relevant techniques [6]. In this way, a novel technique to recover proteins from waste streams and complex media called ionic liquid-based three-phase partitioning (ILTPP) is presented in this chapter. A study about the selection of the system used in the ILTPP technique and the influence of the main operating variables on the process has been performed. ILTPP can be considered a combination of two techniques widely described as alternatives to conventional protein separation methods: ionic liquid-based aqueous two-phase systems (ILATPS) and three-phase partitioning (TPP). The aim of ILTPP is the accumulation of the target protein at the liquid–liquid interface in the same way as TPP by means of the use of ILATPS. As a result, ILATPS and TPP are also described in the chapter in order to provide the context of ILTPP and the ideas that support this technique.

Ionic Liquids in Separation Technology
http://dx.doi.org/10.1016/B978-0-444-63257-9.00006-7
Copyright © 2014 Elsevier B.V.
All rights reserved.

2. PRECEDING TECHNIQUES OF IONIC LIQUID-BASED THREE-PHASE PARTITIONING

2.1 Ionic Liquid-Based Aqueous Two-Phase Systems

Aqueous two-phase systems (ATPS) consist of two liquid phases with high water content, which are formed by the addition of a sufficient amount of two different phase-forming components (PFC) to water. The typical combinations of PFC are usually constituted by polymer/polymer and polymer/salt; and these PFC must be hydrophilic and mutually incompatible so that the phase separation can take place [7]. The driving force of the phase separation in these systems is the enthalpy of the interactions between the polymer molecules, so the ATPS is formed when the interaction enthalpy is bigger than the loss in entropy according to the phase separation [8].

These aqueous biphasic systems were introduced by Beijernick in 1896, but they were rediscovered by Albertsson in the mid-1950s, when he proposed the use of ATPS as an alternative to traditional liquid—liquid extraction techniques, which commonly made use of volatile organic solvents [9,10]. ATPS have been applied extensively for the separation and purification of biomolecules such as amino acids and proteins because of the relevant advantages these systems present: high water content, biocompatibility, low interfacial tension, high capacity and yield, and ease of scale-up [11]. However, most of the polymers used in ATPS have high viscosity, form an opaque solution, sometimes have interference effects through analysis of various analytes, and present a limited range of polarity differences between the two phases that limits their applicability [12,13].

In 2003, Rogers and coworkers were the first to show that imidazolium-based ionic liquids were able to form ATPS in the presence of inorganic salts (K_3PO_4) [14]. Ionic liquid-based aqueous two-phase systems combine the advantages of ionic liquids and aqueous two-phase systems and show very interesting characteristics that make it possible to overcome the constraints associated with the use of polymers in conventional systems: low viscosity, short separation time, and, especially, the ability to control the phase polarities, since they can be adequately chosen taking into account the vast number of ionic liquids that can be obtained by means of cation and anion combination (at least 10^6 simple ionic liquids can be prepared) [4,13]. For these reasons, ILATPS are considered the common alternative to extracting biomolecules with ionic liquids and they have been widely used in the separation, concentration, and purification of proteins, amino acids, antibiotics, antioxidants, and alkaloids [2,15,16]. In the past few years, the number of publications in which new ILATPS are reported has been growing, so there is a high number of both ionic liquids and salts that are available to form ILATPS [17]. Furthermore, the use of carbohydrates, polymers, amino acids, or anionic surfactants as PFC in combination with ionic liquids to obtain aqueous biphasic systems has also been stated in the literature [18].

2.1.1 Formation of Ionic Liquid-Based Aqueous Two-Phase Systems

Table 1 shows the number of works in which each ionic liquid/inorganic salt system has been assessed, according to Freire et al. [9]. As can be seen, a high number of studies are based on ILATPS formed by phosphate salts and ionic liquids with $[Cl]^-$, $[Br]^-$, $[CF_3SO_3]^-$ (or TfO), or $[BF_4]^-$ as anions. Regarding the cation, most of the ILATPS reported in the literature are based on ionic liquids with an imidazolium-based cation. In this way, the influence of the alkyl side chain length of the cation of these ionic liquids has been widely studied [18–21]. It has been concluded that the higher the alkyl side chain length, the higher its hydrophobicity and the higher its phase-forming ability, which corresponds to binodal curves that are closer to the origin and limit bigger biphasic regions [22]. Nevertheless, this trend is only valid for alkyl side chains with a maximum of six carbon atoms, because for longer chains, the ability of the ionic liquid to self-aggregate leads to an increase of the homogeneous region as the alkyl side chain increases [20]. Furthermore, the introduction in the alkyl chain of functionalized groups that enhance the hydrophilic character of the ionic liquids (e.g., hydroxyl groups) reduces their ATPS-promoting ability [9].

With respect to the effect of the cation core, the quaternary phosphonium- and ammonium-based ionic liquids have considerably higher phase-forming abilities than pyridinium- and imidazolium-based ionic liquids [22,23]. Among cation cores containing nitrogen-based rings, larger cations such as pyridinium and piperidinium have a greater capacity to induce ATPS than the smaller imidazolium and pyrrolidinium [18]. With respect to the influence of the anion, the lower the hydrogen bond basicity of the anions composing the ionic liquid, the higher their ability to undergo liquid–liquid demixing [17,24]. As a result, all the studies lead to the conclusion that the triflate anion, $[CF_3SO_3]^-$, exhibits the highest ATPS-promoting ability [9]. Nevertheless, some of the most common ionic liquids possess some toxicity and are poorly biodegradable, so cholinium [25] or amino acid-based ionic liquids [26] have been proposed to overcome the mentioned drawbacks.

The ability to induce the formation of ATPS of the cation and anion of the salt follows the Hofmeister series, which is a series proposed to describe the ability of ions to precipitate proteins but that can also describe many other phenomena [27,28]. Therefore, Shahriari et al. [27] concluded that the molar entropy of hydration of the salt ions is the driving force for ATPS formation, in contrast with previous studies in which this driving force is related to ions' Gibbs free energies of hydration [23,29,30]. As a result, $[PO_4]^{3-}$ is the anion with the highest phase-forming ability, whereas $[Cl]^-$ presents the lowest capacity to form ATPS for a given ionic liquid; in a similar way, divalent cations have greater ability than monovalent cations to constitute biphasic systems [9]. Furthermore, the use of organic salts, such as citrates or tartrates, has also been reported to substitute inorganic salts in ILATPS because of the biodegradability and nontoxicity

Table 1 Number of Works in Which Each Ionic Liquid/Salt System Has Been Studied, According to Freire et al. [9]

Ionic Liquid	Salt																																	
	K_3PO_4	Na_3PO_4	K_2HPO_4	Na_2HPO_4	K_2HPO_4/KH_2PO_4	NaH_2PO_4	K_2CO_3	Na_2CO_3	$NaHCO_3$	Na_2SO_4	$NaHSO_4$	$(NH_4)_2SO_4$	Na_2SO_3	KCl	NaCl	$CaCl_2$	$MgCl_2$	KOH	NaOH	$K_3C_6H_5O_7$	$K_3C_6H_5O_7/C_6H_8O$	$Na_3C_6H_5O_7$	$(NH_4)_3C_6H_5O_7$	$NaKC_4H_4O_6$	$Na_2C_4H_4O_6$	$(NH_4)_2C_4H_4O_6$	$Na_2C_4H_4O_6$	KCH_3CO_2	$NaCH_3CO_2$	$Ca(CH_3CO_2)_2$	$Mg(CH_3CO_2)_2$	$Al_2(SO_4)_3$	$AlK(SO_4)_2$	SDBS
$[Im][Cl]$	■																																	
$[C_nIm][Cl]$	■																																	
$[C_nC_1Im][Cl]$	■		■		▒		▒	■		■								▒		■														
$[C_4C_1C_1Im][Cl]$	▒		■				■											■																
$[aC_1Im][Cl]$	▒		■				■																											
$[OHC_2C_1Im][Cl]$	■																																	
$[C_7H_7C_1Im][Cl]$	■				■			■		■																								
$[C_4Py][Cl]$	■																																	
$[C_4\text{-}2\text{-}C_1Py][Cl]$					■																													
$[C_4\text{-}3\text{-}C_1Py][Cl]$					■																													
$[C_4\text{-}4\text{-}C_1Py][Cl]$					■																													
$[C_4C_1Pip][Cl]$					■																													
$[C_4C_1Pyrr][Cl]$					■																													
$[C_4C_4C_4C_4P][Cl]$																																■		
$[C_nC_1Im][Br]$	░		░	■			▒			■								▒		▒	■	▒												
$[C_4C_1Im][HSO_4]$	■																																	
$[C_nC_1Im][C_1SO_4]$	▒				■			■		■																								
$[C_7H_7C_1Im][C_1SO_4]$								■																										
$[C_1Py][C_1SO_4]$								■																										
$[C_nC_1Im][C_2SO_4]$	░				■		■	■		■		■																						
$[C_7H_7C_1Im][C_2SO_4]$										■																						■		
$[C_nC_1Im][C_4SO_4]$	■						■	■				■																						
$[C_nC_1Im][C_6SO_4]$	■						■	■				■																						
$[C_nC_1Im][C_8SO_4]$	■						■	■		■		■																						
$[C_nC_1Im][C_1CO_2]$	░		▒		■		■																											
$[C_nC_1Im][C_1SO_3]$	■				■																													
$[C_4C_1Im][tos]$					■					■																						■		
$[C_4C_1Im][SCN]$					■					■																						■		
$[C_4C_1Im][DCA]$	▒		■		■					■																						■		
$[C_8Py][DCA]$					■					■																						■	■	
$[C_nC_1Im][CF_3SO_3]$	■	■		■	■	■		■	■	▒	■		■	■	■	■	■		■			■		■				■	■	■	■	■		
$[C_nC_1Im][BF_4]$		░		■		▒		░		░		■	▒		░							░	▒		▒	■	■		▒					■
$[C_4C_4C_4C_4P][CF_3CO_2]$	■									■																								
$[C_4C_4C_4C_4P][Br]$	■																															■	■	
$[C_4C_4C_4C_1P][C_1SO_4]$	■																															■		
$[C_4C_4C_4C_1P][TOS]$	■																															■		
$[C_4C_4C_4C_4N][Cl]$												■																						
$[C_2C_2C_1C_{(O)nOH}N][Cl]$					■																													

Notation: □ 0 works; ■ 1 work; ▒ 2 works; ░ 3/4 works; ▒ 5/6 works; ■ >6 works.

that these organic salts present, so they can be discharged into biological wastewater treatment plants [31–33]. However, as can be seen in Table 1, the use of organic salts has been very little studied when compared with inorganic salts.

Conventional inorganic salts can also be replaced by carbohydrates like glucose, fructose, or sucrose, which are characterized for being noncharged, biodegradable, nontoxic, and a renewable feedstock [34–36]. ILATPS based on carbohydrates can be used to recycle or concentrate hydrophilic ionic liquids from aqueous solutions, in contrast to conventional salts that are dissociated into ions in the solution by the ionic liquid, which complicates the recycling process [37]. Moreover, these systems are attractive to recover bioactive products from natural sources because of the advantage that the availability of carbon-based sources to cells represents in separations from fermentative media [34]. Similar advantages present the ILATPS formed by ionic liquids and amino acids, which also provide an environmentally benign and less aggressive system to extract biomolecules [15,38]. However, the number of ionic liquids that form aqueous biphasic systems in combination with carbohydrates or amino acids is reduced, due to the lower phase forming ability of these compounds with respect to inorganic salts [9].

2.1.2 Applications to Protein Extraction

The most attractive property of ILATPS is their gentle environment for biomaterials such as proteins because of the presence of water in both phases, so ILATPS are the most studied approach to separate proteins that involve the use of ionic liquids [2,22]. Table 2 shows the ILATPS with which protein extraction has been performed. Most of the studies use phosphate salt-based systems to carry out the extraction of proteins, but many different ionic liquids are assessed.

Despite the high complexity of the systems that comprise the separation of proteins by means of ILATPS, several works have assessed the driving forces that are involved in the partitioning of proteins. Dreyer et al. [51] studied the effect of temperature and protein properties (molecular weight, sphericity factor, volume, surface area, hydrophobicity, and charge) on the extraction of myoglobin, trypsin, lysozyme, and bovine serum albumim (BSA), selected as model proteins. A system based on the Ammoeng™ 110 ([Amm110]) and K_2HPO_4/KH_2PO_4 was used and extraction efficiencies higher than 80% were obtained for all the proteins (with the exception of myoglobin). It was concluded that the main driving force governing protein partitioning within ILATPS is electrostatic interaction. Therefore, pH, charge, and molecular weight of proteins are of major importance for the prediction of the partition coefficient, whereas it was demonstrated that hydrophobic interactions could not explain the experimental results. In a similar way, Du et al. [39] reported the extraction of BSA with $[C_4C_1Im][Cl]$ and K_2HPO_4, achieving extraction efficiencies greater than 90%. The partitioning behavior of the protein is influenced by electrostatic interactions in the interfacial area: attraction

Table 2 Proteins for Which Their Partitions in Each ILATPS Have Been Studied

Ionic Liquid	Salt/Carbohydrate					
	K_2HPO_4	K_2HPO_4/KH_2PO_4	$(NH_4)_2SO_4$	K_2CO_3	$K_3C_6H_5O_7$	D-Fructose
$[C_nC_1Im][Cl]$	BSA [39,40] Cytochrome c [41] Hemoglobin [41] Horseradish peroxidase [43] Myoglobin [41] Ovalbumin [41]	Lipase B [42] Lipase from *Bacillus* sp. ITP-001 [44]				
$[C_7H_7C_1Im][Cl]$		Lipase B [42]				
$[C_4C_1Pyrr][Cl]$		Lipase B [42]				
$[C_4C_1Py][Cl]$		Lipase B [42] Lipase from *Bacillus* sp. ITP-001 [44]				
$[C_4C_1Pip][Cl]$		Lipase B [42]				
$[C_nC_1Im][Br]$	BSA [1,40] Cytochrome c [1] Hemoglobin [40] Lysozyme [40] Trypsin [1] γ-globulin [1]				Cytochrome c [45]	

$[C_4C_1Im][CF_3SO_3]$		Lipase B [42]			
$[C_4C_1Im][DCA]$	BSA [46]	Lipase B [42] Lipase from *Bacillus* sp. ITP-001 [44]			
$[C_8Py][DCA]$		Lipase B [42]			
$[C_4C_1Im][CH_3SO_3]$		Lipase B [42]			
$[C_2C_1Im][C_2SO_4]$				Tl lipase [47]	
$[C_2C_1Im][C_4SO_4]$			Lipase A [48]		
$[C_4C_1Im][BF_4]$					BSA [49]
$[C_nC_mC_pC_qN][Gly]$	Cytochrome c [26]				
$[C_4C_4C_4C_4P][Gly]$	Cytochrome c [26]				
TMG-based ILs[a]	BSA [4] Bovine hemoglobin [4] Ovalbumin [4]				
[Amm110][Cl]		Alcohol dehydrogenases [50] BSA [51] Lysozyme [51] Myoglobin [51] Trypsin [51]			

[a]TMG-based ILs: ionic liquids based on 1,1,3,3-tetramethylguanidine cation and different types of anions (acrylate, methyl acrylate, acetate, lactate, maleate, itaconate sorbate, cinnamate).

between the positively charged $[C_4C_1Im]^+$ and the negatively charged protein in basic media, and repulsion between the protein and the $[HPO_4]^{2-}$ groups of the salt. The transfer of the BSA into the ionic liquid-rich phase is also dominated by the salting-out effect, which leads to a decrease in the solubility of the protein in the salt-rich phase as a result of competition from the salt ions (present at very high concentrations) with proteins for water molecules.

Nevertheless, other studies report very different mechanisms for protein partitioning in ILATPS. Pei et al. [1] obtained extraction efficiencies between 75% and 100% for the extraction of BSA, trypsin, cytochrome c, and γ-globulins using imidazolium-based ionic liquids and K_2HPO_4. In this work, the main driving force for the protein extraction is the hydrophobic interaction due to π-π interaction between the imidazolium cation and the aromatic residues of proteins, even though electrostatic interactions and salting-out effects were also important for the transfer of these compounds. As a result, it is concluded that the partition is a function of the characteristics of both the ILATPS (nature and concentration of the ionic liquid and salt, pH, and temperature) and the protein (molecular weight and surface hydrophobicity) [1]. Pei et al. [40] also obtained very similar results in terms of extraction efficiency when they assessed the extraction of BSA, lysozyme, and hemoglobin with imidazolium-based ionic liquids and K_2HPO_4. They also concluded that hydrophobic interactions are the main driving force of the extraction process. In addition, although the phosphate-based salt is replaced by potassium citrate, the extraction of cytochrome c is also dominated by hydrophobic interactions [45]. However, the importance of hydrophobic interactions in the extraction of proteins with ILATPS has also been reported for other types of ionic liquids. In this way, Wu et al. [26] used glycine ionic liquids, with ammonium and phosphonium-based cations, in the presence of K_2HPO_4 to enrich cytochrome c in the ionic liquid-rich phase. The main driving force of the process, as in the previous references, was the hydrophobic interactions between the cation of the ionic liquids and the cytochrome c, despite the fact that electrostatic interactions and the salting-out effect also contribute to the protein extraction.

Other works carried out with BSA and phosphate-based salts lead to the conclusion that the formation of ionic liquid aggregates and complex between these aggregates and proteins are responsible for the effective separation of BSA. These results are obtained with very different ionic liquids: guanidine acrylate-based [4] and imidazolium dicynamide-based ionic liquid [46]; and in both works a quantitative extraction of BSA can be obtained under the most favorable conditions. Notwithstanding the predominance of the effect of aggregates, the transfer of the protein into the ionic liquid-rich phase is also influenced by the hydrophobic interactions, salting-out effect, and hydrogen bonding interactions [4].

As a consequence, the very different conclusions reported in previous references demonstrate that the predominant driving forces for a specific protein separation by means of ILATPS vary depending on the ionic liquid, salt, target protein to be extracted, and the experimental conditions.

The performance of ILATPS for the separation of enzymes should involve the study of how the activity of these biomolecules is affected by the presence of the ionic liquids, besides their partitioning between the two aqueous phases [22]. In this way, many works have used imidazolium-based ionic liquids to study the recovery of different types of lipases [42,44,47,48]. Ventura et al. [42,44] achieved enzyme recoveries higher than 90% in combination with purification factors notably higher than 1 (values up to 51 are obtained), which means that the enzyme-specific activity is considerably greater after the extraction procedure. Dreyer and Kragl [50] also reported increases of the enzyme-specific activities in the ionic liquid-rich phase by factors between 2 and 4 when alcohol dehydrogenases are separated by means of the Ammoeng™ 110/(K_2HPO_4/KH_2PO_4) system. In other works related to the extraction of trypsin [1] and horseradish peroxidase [43] with ILATPS formed by imidazolium-based ionic liquids, even though the enzyme-specific activity cannot be enhanced in the process, it remains approximately constant because 90% of the activity is maintained, whereas enzyme recoveries are greater than 80%. In conclusion, ILATPS provide the opportunity to combine the extraction process with the performance of enzyme-catalyzed reactions, because the enzyme activity can be enhanced or maintained in these systems [50].

ILATPS have also been applied to extract proteins in different configurations. Novak et al. [49] carried out the continuous extraction of BSA in an ILATPS composed of [C_4C_1Im][BF_4] and D-fructose using a microchannel system that enables the separation of both phases at the exit. Although the extraction efficiencies achieved were moderate (53%), this approach makes it possible to use ILATPS in combination with the advantages of microfluidic separation techniques, such as continuous operation, better performance, and easier capacity by a numbering-up approach. ILATPS have also been used in countercurrent chromatography columns, where they were much easier to retain in the columns than conventional polymer-based ATPS [41]. Despite the very high partition coefficients obtained for the four proteins studied (cytochrome c, myoglobin, ovalbumin, and hemoglobin), it was not possible to separate them because they never eluted with the phosphate lower mobile phase, whereas when the ionic liquid-rich phase was the mobile phase, they were eluted all together in an unretained peak.

2.1.3 Drawbacks of Ionic Liquid-Based Aqueous Two-Phase Systems

Many studies have reported the use of ILATPS to separate proteins, with very promising results. However, as was previously stated, research has been focused on the influence of different inorganic salts and ionic liquids on the phase formation and on the distribution coefficient of the target compounds [7]. Therefore, there are some challenges related to

ILATPS that have been very little studied and have to be overcome. In the majority of the literature concerning extraction with ionic liquids, the back extraction of the biomolecules is not addressed [52], which may constitute the main obstacle to assure the feasibility of these processes [2]. Another concern of ILATPS that has to be assessed is the mutual ionic liquid—water solubility and the ion exchange between salt and ionic liquid, which may make the recycling of the ionic liquid difficult [15,53]. Eventually, high concentrations of the inorganic salts used in ILAPTS are not desirable in the effluent streams due to environmental problems [32], so its recovery should also be considered.

2.2 Three-Phase Partitioning

Three-phase partitioning (TPP) is an emerging technique for protein separation that consists of the accumulation of precipitated material at the interface between an organic liquid phase, usually *t*-butanol, and an aqueous solution of a salting-out salt, such as ammonium sulfate [54,55]. In this way, this technique is able to precipitate and collect proteins as a layer formed between the aqueous and the organic phase [56]. Even though water is miscible with *t*-butanol in any proportion, the presence of a certain amount of electrolyte (e.g., ammonium sulfate) leads to the separation of the mixture into two liquid phases. The upper phase contains mostly alcohol and some water with ammonium sulfate only in traces, whereas the lower phase is constituted by an aqueous solution of electrolyte with a low *t*-butanol content, as it is represented in Figure 1(A) [57,58]. Pigments, lipids, and enzyme inhibitors are concentrated in the upper phase, the lower aqueous phase is enriched with polar components like saccharides, and these two liquid

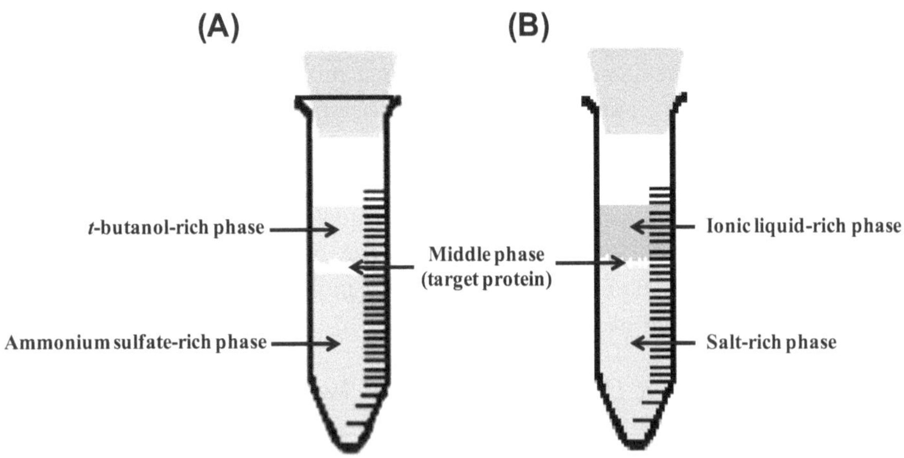

Figure 1 *Schematic Diagram of the Three Phase Partitioning, TPP (A) and Ionic Liquid-based Three Phase Partitioning, ILTPP (B) systems.*

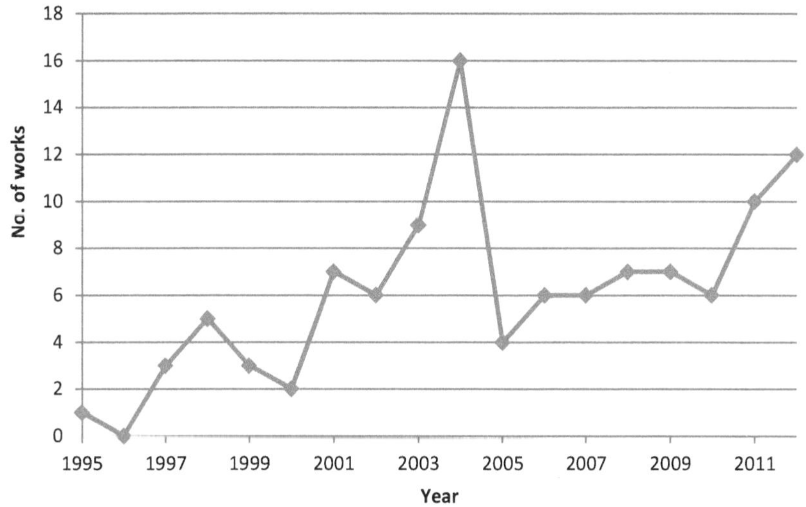

Figure 2 ***Number of published works related to TPP, per year.***

phases are separated by the intermediate protein-precipitated layer [59,60]. TPP systems can be formed with other organic solvents (e.g., *n*-butanol) and salts (e.g., $CaCl_2$) [55,61,62], but it is concluded that *t*-butanol and ammonium sulfate are the best organic solvent and salt, respectively, to carry out TPP [63].

The TPP technique was first used by Tan and Lovrien [64] in 1972 and it was described more in detail in 1987 by Lovrien et al. [65], who applied TPP to concentrate several proteins and enzymes [66,67]. Since then, the number of works based on TPP has increased notably, especially in the last decade, as can be seen in Figure 2, which shows the number of published works related to this technique per year. However, TPP has not attracted attention as much as other conventional techniques such as chromatography, because the industrial maturity of TPP is also lower than chromatographic-based processes or extraction with aqueous two phase systems [54,67].

The increasing interest in TPP may be attributed to the advantages that this technique presents. TPP can be defined as a simple, more efficient, and economical method for separation and purification of target proteins, with respect to the conventional techniques [68]. Therefore, whereas most of the conventional technologies (e.g., salting out, membrane filtration, ion exchange, gel permeation, or affinity chromatography) may involve a number of steps, are time and cost consuming, and require pretreatment and their scale-up may be difficult [58], TPP stands out because it is simple, inexpensive, scalable, a rapid procedure that can be used directly with crude suspensions, and works at room temperature, and the chemicals that it requires can be recycled [69]. For this reason, TPP seems to be an attractive prepurification or separation method [59].

2.2.1 Applications of Three-Phase Partitioning

Even though TPP was developed as an upstream technique, it has been used both for upstream and downstream protein purification processes and is also commonly used as a one-step purification method [68,70]. Furthermore, TPP also achieves the concentration of the target proteins, besides their purification, as well as the virtual elimination of viscosity and turbidity [71]. In this way, TPP usually generates greater than 100-fold purification, 70–85% of final purity, and at least a 50-fold decrease in volume [72]. The efficiency of a TPP process depends on both the properties of target protein (isoelectric point, molecular weight, hydrophobicity, solubility) and the operating conditions (pH, temperature, ammonium sulfate saturation, ratio of crude extract to *t*-butanol) at which TPP is performed [66,70]. These interesting results have made it possible to carry out some successful applications of TPP on a large scale [73].

The purification of many proteins has been successfully studied, such as BSA [59,62]; ovalbumin [59,62]; lysozyme [59,62]; gelatin [59]; β-lactoglobulin [62]; horseradish peroxidase [62]; *Ipomoea palmata* peroxidase [74]; α-galactosidase [55,58,75,76]; β-galactosidase [77]; laccase [66,78]; proteases from the viscera of farmed giant catfish [79], from papaya peels [80] and from *Calotropis procera* latex [81]; invertase [68,75,82]; amylase and trypsin inhibitors [83]; phospholipase D [84]; catalase [85]; and xylanase [86]. Furthermore, TPP has also been applied to the recovery of polysaccharides in the middle phase [87–89] and to the extraction of oil and fat in the *t*-butanol-rich phase [60,90].

Despite the high number of studies concerning the partitioning of proteins by means of TPP, the physicochemical principles underlying TPP are rather complex, so its mechanism is not completely understood [86,91]. In this way, Dennison and Lovrien [61] reported one of the most rigorous discussions on protein–ammonium sulfate interactions [67]. This work states that, on the one hand, *t*-butanol, which appears to be a kosmotrope and crowding agent, binds to TPP-precipitated proteins so protein/*t*-butanol co-precipitates are formed and float above the bottom phase, since bound *t*-butanol increases their buoyancy; while on the other hand, large concentrations of sulfate ions seem to exert themselves through their kosmotropic action, but electrostatic forces and the tendency for salt ions to bind and tighten protein molecule conformation must be considered to explain the sharp pH dependency of the TPP process. Furthermore, sulfate anions are densely hydrated, so, when they are in concentrations between 0.5 and 3 M, they tend to crowd or exclude proteins [61]. Nevertheless, the ideas stated by Dennison and Lovrien cannot explain a few phenomena, like the partitioning character of TPP, and they do not consider the fact that there are two liquid phases in the system [92]. Therefore, other authors consider that the interfacial adsorption of single proteins may be the initial step in the formation of the middle layer in TPP, which makes it possible to explain the constant partitioning ratio (independent of the total amount of protein) found in TPP that is characteristic of a partitioning process [59,62].

The middle phase formed in TPP is an emulsion gel, as it is deduced from the analysis of its rheological behavior [93]. This coherent interfacial phase is formed by 2–10% (w/w) gel-forming protein and about 90% (w/w) or more liquid with dissolved ammonium sulfate [57]. An important issue concerning the protein content at the interface is the possibility of suffering from structural changes. However, a recent work concluded that while the nature or extent of structural changes may vary, these changes are rather subtle and not drastic in nature, so TPP can be safely used for protein purification [63]. In fact, TPP can be used as a useful strategy for protein refolding, because *t*-butanol, which has a sufficient hydrophobic character, may bind to exposed hydrophobic sites of denatured proteins and interfere with aggregation, allowing ammonium sulfate to facilitate protein refolding [91]. Furthermore, TPP can refold proteins successfully with higher yields than obtained with conventional methods, so TPP can combine simultaneous protein purification and refolding [67].

Most of the proteins that have been purified by means of TPP are enzymes, as was previously stated. In many cases, TPP enhances the activity of various enzymes, resulting in apparent higher yields of 100–1000% [78], probably because of the subtle structural changes that occur in the process [63]. The enhancement of the enzyme activity with TPP has been reported for several enzymes, such as lipases [94,95], laccase [96], or α-chymotrypsin [97], and it represents an alternative application of TPP, besides the purification and refolding of proteins. TPP can also be applied in other applications related to the construction of biosensors [98], the extraction and purification of DNA [99,100], and the tuning permeabilization of microbial cells [101].

Eventually, it should be mentioned that TPP is the base of another technique called macroaffinity ligand-facilitated three-phase partitioning (MLFTPP). MLFTPP is an extension of TPP that uses, besides *t*-butanol and ammonium sulfate, a smart polymer that acts as a smart affinity macroligand. Therefore, in this case the middle phase contains the smart polymer and the protein that has affinity for the polymer [102,103]. This affinity complex can be dissociated like in any affinity-based separation process and the cognate protein recovered, so MLFTPP can be applied to concentrate, purify, and refold proteins [104]. The main advantage of MLFTPP is that it exhibits higher selectivity than simple TPP under the same or similar conditions [105].

2.2.2 Drawbacks of Three-Phase Partitioning

Even though *t*-butanol has a higher boiling point and is much less flammable than other conventional solvents used in extractions (e.g., hexane, methanol, or ethanol) [69], a concern with large-scale use of TPP is the use of process-scale quantities of this compound [54]. In this way, according to the Classification & Labeling Inventory of the

European Chemicals Agency, not only is *t*-butanol classified as flammable, but it also presents the following hazard classes: acute toxicity, eye irritation, and specific target organ toxicity—single exposure [106]. Consequently, TPP has the disadvantages that are associated with the use of volatile organic solvents.

3. PERFORMANCE OF IONIC LIQUID-BASED THREE-PHASE PARTITIONING

ILTPP is a novel technique for protein recovery that combines characteristics of ILATPS and TPP. On the one hand, ILTPP uses the same ionic liquid/salt systems of ILATPS; on the other hand, it promotes the partitioning of the target protein in the liquid—liquid interface in the same way as TPP. Therefore, the performance of ILTPP is very similar to TPP, but the *t*-butanol is replaced by the ionic liquid and other salts can be considered besides ammonium sulfate, as can be seen in Figure 1. As a result, the protein layer at the interface separates the ionic liquid-rich phase and the salt-rich phase, which are usually the top and bottom phases of the system, respectively (Figure 1(B)).

In this way, ILTPP can overcome the main drawbacks of both ILATPS and TPP techniques. The accumulation of the protein of interest at the interface makes its recovery easier, and the process is simplified because the back extraction step is not required, so the main disadvantage related to the use of ILATPS is avoided. Furthermore, ILTPP does not involve any volatile organic solvent such as *t*-butanol, because it is replaced by ionic liquids, which provide several advantages to the technique associated with their use due to the very interesting properties of ionic liquids [13,107,108]. As a consequence, ILTPP presents the advantages of ILATPS and TPP, but without the main disadvantages that may limit the application of these techniques, as Table 3 summarizes.

The study of the ILTPP technique has been focused on bovine whey proteins: BSA and, especially, lactoferrin (LF). Bovine whey is a by-product from cheese manufacturing that is composed mainly of water (93.5%) and lactose (4.5—5.0%), but it also contains

Table 3 Main Advantages and Disadvantages of ILATPS and TPP, and Characteristics of These Techniques That ILTPP Presents

	ILATPS	TPP	
Advantages	• Phase polarity control • Gentle biocompatible environment • Little emulsion formation • Low viscosity • Short separation time • Low energy requirements • Reliable scale-up	• Recovery of proteins at the interface • Simplicity • Easy of handling • Inexpensive • Short separation time • Low energy requirements • Reliable scale-up	ILTPP
Disadvantages	• Difficult biomolecule recovery	• Use of volatile organic solvents	

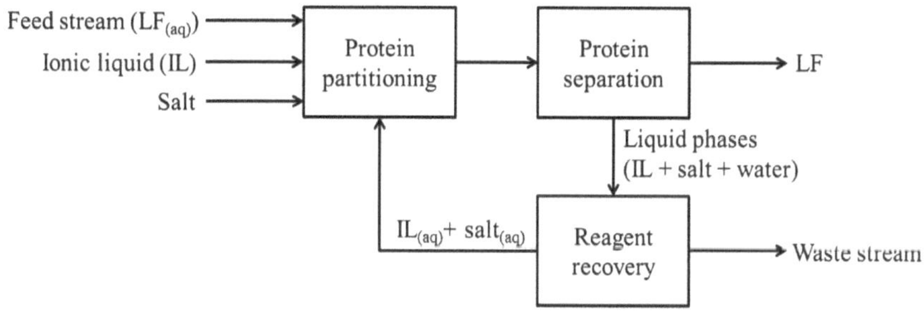

Figure 3 ***Block diagram of the ILTPP process for LF separation.***

valuable proteins (such as LF, BSA, or immunoglobulins) that stand out because of their nutritional, functional, and biological properties [109,110]. Nowadays, LF is separated from the bovine whey at industrial scale by means of cation-exchange chromatography, but this technique suffers from high costs and relatively low yields, whereas other alternative technologies like pressure-driven processes present relatively low selectivity and operational constraints associated with fouling [5]. Despite the promising results provided by liquid—liquid extraction with hydrophobic ionic liquids, LF could not be recovered because of the conformational changes of the target compound that prevent its back-extraction [2]. For this reason, ILTPP is presented as an alternative to the previous technologies to separate and recover LF at the liquid—liquid interface. Figure 3 shows the block diagram of the ILTPP process applied to the LF recovery. In this process, the feed stream containing the target protein (LF) is mixed with the ionic liquid and salt required to reach the desired composition of the mixture point at which the protein is partitioned. According to the definition of ILTPP, most of the LF should be accumulated at the liquid—liquid interface so that it can be recovered from the rest of the system. Once most of the target compound has been recovered, the maximum amount of the ionic liquid and salt contained in the liquid phases should be recycled to minimize both the net consumption of reagents and the waste stream. In the following sections, the selection of suitable ionic liquid/salt systems to perform ILTPP and the influence of the main operating variables on the protein recovery are stated.

3.1 Selection of the Ionic Liquid/Salt System

The ionic liquid/salt systems of ILTPP must present several properties so that they can be applied in this technique. As the conventional ILATPS, the ionic liquid and salt should be mutually incompatible to form two liquid phases in the region of compositions at which the process is performed. In addition, the salt should be completely dissolved; otherwise, the presence of precipitated salt will form a second solid phase than can make the recovery of the protein layer accumulated at the interface difficult. Another characteristic that the system should have is the capability to induce the formation of the protein layer,

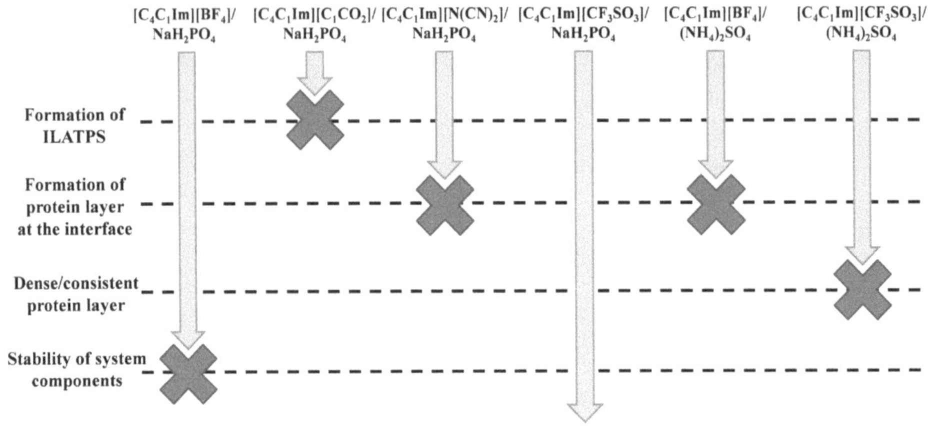

Figure 4 ***Assessment of several ionic liquid/salt systems to be used in the ILTPP technique [111].***

since it is the property that defines ILTPP and makes it different from the ILATPS technique. The protein layer should also be dense and coherent so that it is stable and can be easily separated from the liquid phases. Eventually, as in any process, the components of the system should be stable in the operating conditions, that is, they should not suffer from decomposition or degradation.

Figure 4 represents the assessment of different ionic liquid/salt systems to be used in the ILTPP technique, according to the previous criteria reported by Alvarez-Guerra and Irabien [111]. It should be mentioned that the ILTPP system that was first reported by the same group is based on the use of $[C_4C_1Im][BF_4]$ and NaH_2PO_4 [112]. This $[C_4C_1Im][BF_4]/NaH_2PO_4$ system was studied because it is the ILATPS for which several biomolecules have been successfully recovered from the ionic liquid-rich phase, which may solve the main constraint associated with ILATPS. However, these approaches involve the use of a second type of ionic liquid [53,113], volatile organic solvents [114], or extreme pH [115], with the consequent disadvantages. This ILATPS was applied for the extraction of bovine LF, but the interfacial accumulation of this protein was obtained [112]. Between 74% and 99% of the total LF can be recovered at the liquid–liquid interface with this system, whereas the other fraction of the protein remains in the salt-rich phase. However, the use of $[C_4C_1Im][BF_4]$ in contact with water is not advisable because it may suffer from hydrolysis [116], even though BF_4-based ionic liquids are one of the most studied, as stated in Table 1. As a result, despite the fact that the $[C_4C_1Im][BF_4]/NaH_2PO_4$ system fulfills all the requirements related to the ILATPS formation and protein accumulation at the interface, its lack of stability suggests that alternative ILTPP systems should be found.

For this reason, several ionic liquid/salt systems were assessed to replace the use of $[C_4C_1Im][BF_4]$ for the LF recovery [111]. In this way, $[C_4C_1Im][C_1CO_2]/NaH_2PO_4$

shows very poor results under the tested experimental conditions, because it cannot form two liquid phases or, for higher salt concentrations, this component cannot be completely dissolved and most of the protein is contained in the ionic liquid-rich phase. Regarding $[C_4C_1Im][N(CN)_2]/NaH_2PO_4$, the protein (LF) is mostly extracted in the ionic liquid-rich phase, so this system only provides promising results for the ILATPS technique. Therefore, this system cannot be used in ILTPP because protein is not accumulated at the interface.

The appropriate alternative to $[C_4C_1Im][BF_4]$ seems to be $[C_4C_1Im][CF_3SO_3]$, because the latter can form the two liquid phases characteristic of ILATPS but also the protein layer at the interface that defines the ILTPP technique. The experimental results are competitive with those obtained with $[C_4C_1Im][BF_4]$, because the fraction of LF present at the interface varies between 83% and 99% [111], which is slightly higher than the range of values reported for $[C_4C_1Im][BF_4]$. Moreover, $[C_4C_1Im][CF_3SO_3]$ does not suffer from hydrolysis when it is in contact with water, as happens with BF_4-based ionic liquids.

All the previous systems use NaH_2PO_4 as salt, whereas conventional TPP achieves the best results with $(NH_4)_2SO_4$. This salt was assessed with both $[C_4C_1Im][BF_4]$ and $[C_4C_1Im][CF_3SO_3]$, which are the ionic liquids with the most promising results, but the performance of ILTPP with the systems that include $(NH_4)_2SO_4$ clearly worsens. The protein layer obtained with $[C_4C_1Im][CF_3SO_3]/(NH_4)_2SO_4$ is neither dense nor coherent and it is mixed with the ionic liquid-rich phase, whereas the protein layer at the interface is not detected with the $[C_4C_1Im][BF_4]/(NH_4)_2SO_4$ system [111].

Apart from the previous studies, very few references report the interfacial partitioning of the target compound using ionic liquids. Ventura et al. [16] studied the isolation of red natural colorants from fermented broth using $[C_2C_2C_2C_2N][Br]$, $[C_4C_4C_4C_4N][Br]$, and $[C_4C_1Im][Cl]$ as ionic liquids and potassium citrate buffer solution as salt. For this purpose, although their target compounds were the colorants, these systems lead to the interfacial partition of the proteins present in the broth for all the systems, which improves their separation from the colorants. Nevertheless, the fraction of proteins that form the interfacial layer is relatively low, exhibiting a maximum of 45% under the best experimental conditions.

Some of the works related to the protein partition at the interface with ionic liquid-based systems are based on hydrophobic ionic liquids such as $[C_2C_1Im][NTf_2]$ [117] or $[C_8C_1Im][PF_6]$ [118], even in microinterfaces to form protein microcapsules [119]. This protein accumulation at the interface has also been reported for conventional polymer/salt ATPS [120,121], and the partitioning of biomolecules in this type of system has been modeled [122]. However, the cited works of these two different approaches do not usually have the protein recovery as a main objective and ionic liquid/salt systems are not used, so they cannot be classified as applications of the ILTPP technique.

3.2 Influence of Operating Variables

3.2.1 Ionic Liquid and Salt Concentration

The influence of this variable has been studied with the $[C_4C_1Im][BF_4]/NaH_2PO_4$ system to recover LF within the following ranges of mass fractions: 13.5–26.5% for the ionic liquid and 13.5–30.5% for the salt [112]. Both ionic liquid and salt concentrations have a significant effect on the fraction of protein that is accumulated at the liquid–liquid interface, and this effect is positive: the higher the concentration of any of these two compounds, the higher the protein that can be recovered at the interface. The effect of the salt is 64% higher than the effect of the ionic liquid, so changes in the salt concentration should be preferred to enhance the efficiency of the ILTPP process, because the cost of the salt is notably lower than the cost of ionic liquid but it has a greater influence on the performance of the process. Nevertheless, the effect of both variables on the protein recovery is greater at relatively high concentrations, as can be seen in Figure 5, where the experimental results reported by Alvarez-Guerra and Irabien [112] are plotted.

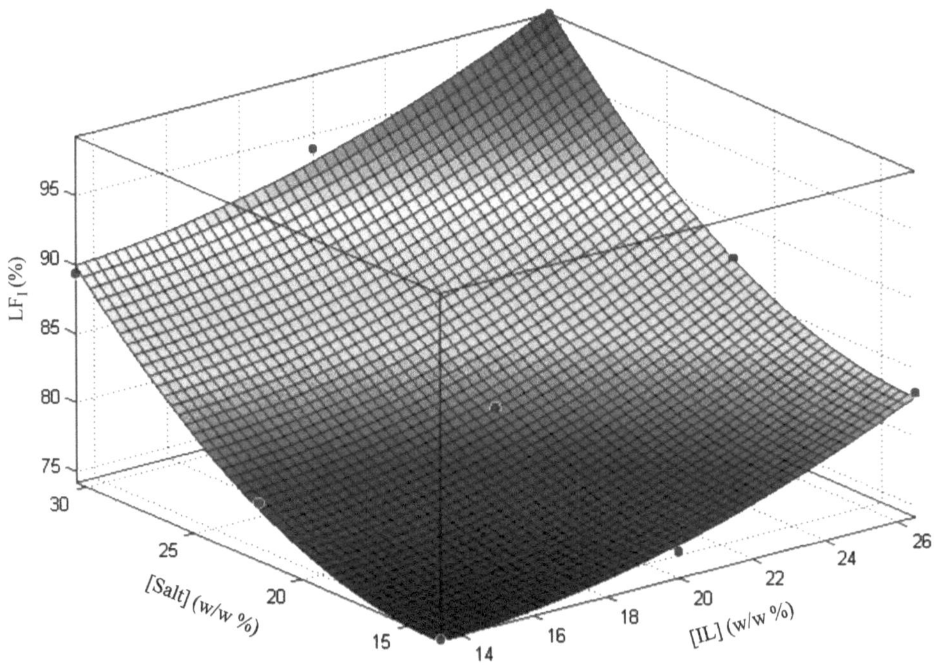

Figure 5 ***Fraction of the LF accumulated at the liquid interface, LF_I (%), as function of the mass fraction of salt, [Salt], and ionic liquid, [IL], expressed as percentage, for the system $[C_4C_1Im][BF_4]/NaH_2PO_4$.***

The influence of a second type of salt that is added to the $[C_4C_1Im][BF_4]/NaH_2PO_4$ system has also been assessed [112]. NaCl was selected as the second type of salt and its addition clearly enhances the protein recovery at the interface. When NaCl is added so that its mass fraction is equal to 6%, the amount of LF that is partitioned at the interface increases from 81% to 96%. It implies that the positive effect of NaCl on the protein recovery is higher than the effect of NaH_2PO_4, since the former achieves a greater increase in the fraction of protein present at the interface for the same change in the salt concentration.

Regarding the $[C_4C_1Im][CF_3SO_3]/NaH_2PO_4$ system, which has been identified as the most suitable to perform ILTPP according to Figure 4, it should be used in a narrow range of concentrations, around 23.3% for the ionic liquid and 26.3% for the salt. For lower concentrations of these components, a consistent layer of protein at the interface is not formed, whereas for higher concentrations, the salt cannot be completely dissolved [111]. For this reason, an analysis of the influence of the ionic liquid and salt concentration has not been carried out for the $[C_4C_1Im][CF_3SO_3]/NaH_2PO_4$ system.

3.2.2 Temperature

The dependence of the ILTPP technique on temperature has been reported for the recovery of LF with the $[C_4C_1Im][BF_4]/NaH_2PO_4$ system [112]. Slight increases of the protein contained at the interface are obtained for temperatures that are to a small extent higher than room temperature: between 25 and 35 °C the fraction of lactoferrin separated at the interface increases from 81% to 89%. Nevertheless, higher temperatures produce a drastic change in the previous trend and the fraction of protein recovered decreases 14 percentage points due to the increase of temperature to 40 °C, so the amount of LF that remains in the salt-rich phase increases in the same way. This behavior may be explained because of the conformational changes that LF may suffer at this temperature, since the denaturation of this protein at mild acidic conditions occurs at temperatures near 40 °C [112,123]. Therefore, from a process engineering point of view, the limited effect of this variable and the narrow range of temperatures that can be assessed to avoid protein denaturation imply that temperature adjustment should not be considered as one of the main ways to enhance the ILTPP technique in terms of protein recovery at the interface.

3.2.3 pH

The recovery of LF and BSA has been studied at pH equal to 3.1, 4.0, and 5.5, keeping constant the ionic strength of the salt solution [111]. For this purpose, the $[C_4C_1Im][CF_3SO_3]/NaH_2PO_4$ system is used, and the pH is adjusted by means of the addition of H_3PO_4 (in the case of pH = 3.1) or NaOH (pH = 5.5). BSA is considered besides LF because both are whey proteins with a similar molecular weight, which makes their separation with other technologies such as membranes more difficult [2].

The partition of these two proteins is very dissimilar: whereas most of LF is accumulated at the interface (higher than 83%), the fraction of BSA that can be located between the two liquid phases is always considerably lower (down to 32%). In addition, the influence of the pH on the protein partition is also very different. A total of 98% of the LF is accumulated at the interface at pH equal to 3.1 and 4.0, but this percentage decreases to 83% at pH = 5.5; in contrast, the lowest BSA recovery is obtained at pH = 3.1 (32%), whereas this variable increases to near 55% at pH = 4.0 and 5.5 [111]. As a result, the selectivity of the ILTPP process to LF increases from 1.5 at pH = 5.5 to 3.0 at pH = 3.1. However, at more acidic conditions (especially at pH = 3.1), the protein accumulated at the interface could not be redissolved with water, whereas at pH = 5.5, a complete redissolution is obtained. Consequently, it implies that protein may suffer from irreversible conformational changes that prevent the protein redissolution at pH = 3.1. As a result, despite the higher selectivity and higher lactoferrin recovery achieved at pH = 3.1, the performance of the ILTPP process at moderate pH such as 5.5 could be more interesting for the preservation of the protein properties.

The different experimental behavior exhibited by LF and BSA is attributed to their different isoelectric points (8–9 and 4.7–4.9, respectively) [111]. Therefore, LF is positively charged under all the tested experimental conditions, whereas BSA has a negative net charge at pH = 5.5 and a positive charge at pH = 3.1 and 4.0. This change in the charge of BSA between pH = 4.0 and 5.5 seems to be responsible for the different BSA partition. At pH = 5.5, the fraction of this protein that is not concentrated at the interface (45%) remains in the salt-rich phase; in contrast, at pH = 4.0, most of the LF that cannot be recovered at the interface (43%) is extracted into the ionic liquid-rich phase.

The influence of the pH has also been assessed between pH = 6 and 8 for the partition of proteins from fermented broth of *Penicillium purpurogenum* using the system composed of $[C_2C_2C_2C_2N][Br]$ and potassium citrate buffer [16]. The amount of protein present at the interface is not significantly affected by the pH variations, since in all cases it is around 40% and it does not show any trend with the pH. Nevertheless, it should be noted that the main aim of this work is the isolation of the natural colorants of this broth.

3.2.4 Protein Concentration

The influence of protein concentration on the performance of the ILTPP technique has been assessed for LF and BSA with systems based on $[C_4C_1Im][CF_3SO_3]$ and phosphate salts at pH = 4.0 and 5.5 within the range $0.4-1.6 \times 10^{-4}$ w/w [111]. This range is equivalent to concentrations between 80 and 320 mg/L in the water fraction of the ILTPP systems, which also belongs to the range of concentrations of LF in the bovine whey [124].

The effect of this variable on the process changes with pH. In this way, at pH = 4.0 the fraction of protein that is accumulated at the interface remains approximately constant with protein concentration: 95.9–99.5% for LF and 50.2–57.7% for BSA. As a result, the selectivity to LF does not exhibit any trend with protein concentration and its values are around 1.8. However, at pH = 5.5 the selectivity increases with protein concentration from slightly higher than 1 at the lowest protein concentration (4×10^{-5} w/w) to 1.5 (1.6×10^{-4} w/w). These results can be explained considering that the protein concentration decreases with the studied variable (from 94% to 83% and from 88% to 55% for LF and BSA, respectively), but this decrease is relatively more important for BSA, which produces the consequent increase in the selectivity [111].

4. POTENTIALITIES AND ONGOING RESEARCH ON IONIC LIQUID-BASED THREE-PHASE PARTITIONING

Study of the ILTPP technique has been focused on the recovery of LF and BSA. This approach should be considered as an example of the ILTPP potentiality to recover high-added-value proteins that are present in waste streams such as bovine whey. Furthermore, the possibility of applying the ILTPP technique to other media has been reported by Ventura et al. [16], since the interfacial partitioning of proteins from fermented broth of *Penicillium purpurogenum* is obtained using ILATPS, which is characteristic of the ILTPP process.

Moreover, this technique can be applied to concentrate and recover several types of proteins simultaneously or to separate a specific one selectively with the same ILTPP system, adjusting the experimental conditions. For instance, relatively low pH and/or high protein concentrations lead to moderate selectivities to LF with respect to BSA, whereas the selectivity is very low at relatively high pH and/or low protein concentrations.

Therefore, ILTPP shows a great potentiality to obtain, on the one hand, whey protein concentrates (WPC) or isolates (WPI) when the process is performed under the experimental conditions that lead to very low selectivity. In this case, ILTPP could be seen as a competitive alternative to the membrane processes that are used to concentrate whey proteins up to 10% or 20% w/w. WPC and WPI are mainly used for baby food, but also as dietary supplements [125].

On the other hand, ILTPP can also be considered as an alternative to separate LF from other whey proteins such as BSA, due to the relatively high selectivities (up to three) achieved. The importance of individual whey proteins is related to their own unique nutritional, functional, and biological characteristics that cannot be supplied with WPC or WPI [110]. Consequently, ILTPP can be competitive with other conventional techniques that are used to separate and purify specific proteins and are characterized for being time and cost consuming [2].

Despite the promising results for the recovery of individual proteins or groups of proteins by means of ILTPP, additional research on this technique is required to establish it in commercial processes. In this way, the main challenge that has to be overcome is the recyclability of the ionic liquid used in the process so that the economic and environmental impacts associated with the losses of this compound are minimized. For this purpose, several strategies can be used to reduce the net ionic liquid consumption, such as increasing the salt mass fraction after the protein recovery to reduce the concentration of the ionic liquid in the salt-rich phase [126,127] or removing water from the salt-rich phase by means of evaporation so that a higher fraction of this phase can be recycled [128]. The development of these strategies implies the thermodynamic characterization of ILTPP systems so that the concentration of the major components (ionic liquid, salt, and water) can be known.

In addition, the recovery of many other proteins from waste streams should be identified so that the applicability of the ILTPP technique could be broadened. Also, other ionic liquid/salt systems ought to be studied in order to determine other options with which the ILTPP process can be carried out. If a high number of available ILTPP systems are found, it will be easier to find one that enhances the selectivity to the target compounds in a given application.

5. CONCLUSIONS

ILTPP is a novel technique that leads to the accumulation of the target proteins at the liquid–liquid interface by means of the use of systems composed of ionic liquid and salt. ILTPP can be seen as a combination of ILATPS and TPP that makes it possible to overcome the constraints of these techniques while retaining their relevant advantages. The main potential applications of ILTPP are the concentration, separation, and recovery of proteins, especially from waste streams such as bovine whey, in which they are usually present in low concentrations and mixed with many other different types of compounds.

The system based on $[C_4C_1Im][CF_3SO_3]$ and phosphate salts is identified as the most suitable to perform this technique because it leads to the formation of the two liquid phases and a consistent protein layer at the interface without suffering from stability problems. Very high efficiencies with respect to the target proteins can be achieved, since under some experimental conditions (at high salt and ionic liquid mass fractions) almost 100% of the target protein is accumulated at the interface. Moreover, pH plays an important role in the selectivity of the process because it determines the net charge of proteins and it can be adjusted to enhance specifically the recovery of the target protein. Nevertheless, the simultaneous recovery of different types of proteins can also be sought in some applications, which can be fulfilled with ILTPP by means of the proper selection of the experimental variables such as pH or protein concentration.

Even though all the main studies about ILTPP have been carried out for LF recovery, it seems to have a great potential to valorize other high-added-value proteins that may be present in waste streams. In this way, further research in areas like the selection of other interesting ionic liquid/salt systems or the recyclability of these reagents (especially the ionic liquid) to minimize the associated economic and environmental impact is still required to carry on the development of this technique. For this purpose, additional work should be done to describe the thermodynamics of the ionic liquid/salt systems used in the ILTPP technique.

REFERENCES

[1] Y. Pei, J. Wang, K. Wu, X. Xuan, X. Lu, Ionic liquid-based aqueous two-phase extraction of selected proteins, Sep. Purif. Technol. 64 (2009) 288–295.

[2] E. Alvarez-Guerra, A. Irabien, Extraction of lactoferrin with hydrophobic ionic liquids, Sep. Purif. Technol. 98 (2012) 432–440.

[3] J.A. Asenjo, Selection of operations in separation processes, in: J.A. Asenjo (Ed.), Separation Processes in Biotechnology, Marcel Dekker, Inc., New York (USA), 1990, pp. 3–16.

[4] Q. Zeng, Y. Wang, N. Li, X. Huang, X. Ding, X. Lin, et al., Extraction of proteins with ionic liquid aqueous two-phase system based on guanidine ionic liquid, Talanta 116 (2013) 409–416.

[5] N. Ndiaye, Y. Pouliot, L. Saucier, L. Beaulieu, L. Bazinet, Electroseparation of bovine lactoferrin from model and whey solutions, Sep. Purif. Technol. 74 (2010) 93–99.

[6] K. Tonova, Separation of poly- and disaccharides by biphasic systems based on ionic liquids, Sep. Purif. Technol. 89 (2012) 57–65.

[7] A. Müller, A. Górak, Extraction of 1,3-propanediol from aqueous solutions using different ionic liquid-based aqueous two-phase systems, Sep. Purif. Technol. 97 (2012) 130–136.

[8] S. Oppermann, F. Stein, U. Kragl, Ionic liquids for two-phase systems and their application for purification, extraction and biocatalysis, Appl. Microbiol. Biotechnol. 89 (2011) 493–499.

[9] M.G. Freire, A.F.M. Cláudio, J.M.M. Araújo, J.A.P. Coutinho, I.M. Marrucho, J.N.C. Lopes, et al., Aqueous biphasic systems: a boost brought about by using ionic liquids, Chem. Soc. Rev. 41 (2012) 4966–4995.

[10] Y. Li, Q. Liu, M. Zhang, H. Su, Liquid-liquid equilibrium of the [4-MBP][BF_4]-NaCl-H_2O systems at T=293.15, 303.15, 313.15, and 323.15K: experimentation and correlation, Thermochim. Acta 565 (2013) 234–240.

[11] C. Wu, J. Wang, H. Wang, Y. Pei, Z. Li, Effect of anionic structure on the phase formation and hydrophobicity of amino acid ionic liquids aqueous two-phase systems, J. Chromatogr. A 1218 (2011) 8587–8593.

[12] A. Haghtalab, A. Paraj, B. Mokhtarani, [1-ethyl-3-methyl-imidazolium][ethylsulfate]-based aqueous two phase systems: new experimental data and new modelling, J. Chem. Thermodyn. 65 (2013) 83–90.

[13] J.F.B. Pereira, S.P.M. Ventura, F.A. E Silva, S. Shahriari, M.G. Freire, J.A.P. Coutinho, Aqueous biphasic systems composed of ionic liquids and polymers: a platform for the purification of biomolecules, Sep. Purif. Technol. 113 (2013) 83–89.

[14] K.E. Gutowski, G.A. Broker, H.D. Willauer, J.G. Huddleston, R.P. Swatloski, J.D. Holbrey, R.D. Rogers, Controlling the aqueous miscibility of ionic liquids: Aqueous biphasic systems of water-miscible ionic liquids and water-structuring salts for recycle, metathesis, and separations, J. Am. Chem. Soc. 125 (2003) 6632–6633.

[15] M. Domínguez-Pérez, L.I.N. Tomé, M.G. Freire, I.M. Marrucho, O. Cabeza, J.A.P. Coutinho, (Extraction of biomolecules using) aqueous biphasic systems formed by ionic liquids and aminoacids, Sep. Purif. Technol. 72 (2010) 85–91.

[16] S.P.M. Ventura, V.C. Santos-Ebinuma, J.F.B. Pereira, M.F.S. Teixeira, A. Pessoa, J.A.P. Coutinho, Isolation of natural red colorants from fermented broth using ionic liquid-based aqueous two-phase systems, J. Ind. Microbiol. Biotechnol. 40 (2013) 507–516.

[17] S.P.M. Ventura, S.G. Sousa, L.S. Serafim, A.S. Lima, M.G. Freire, J.A.P. Coutinho, Ionic-liquid-based aqueous biphasic systems with controlled pH: the ionic liquid anion effect, J. Chem. Eng. Data 57 (2012) 507–512.

[18] S.P.M. Ventura, S.G. Sousa, L.S. Serafim, A.S. Lima, M.G. Freire, J.A.P. Coutinho, Ionic liquid based aqueous biphasic systems with controlled pH: the ionic liquid cation effect, J. Chem. Eng. Data 56 (2011) 4253–4260.

[19] F.J. Deive, M.A. Rivas, A. Rodríguez, Sodium carbonate as phase promoter in aqueous solutions of imidazolium and pyridinium ionic liquids, J. Chem. Thermodyn. 43 (2011) 1153–1158.

[20] M.G. Freire, C.M.S.S. Neves, J.N. Canongia Lopes, I.M. Marrucho, J.A.P. Coutinho, L.P.N. Rebelo, Impact of self-aggregation on the formation of ionic-liquid-based aqueous biphasic systems, J. Phys. Chem. B 116 (2012) 7660–7668.

[21] C.M.S.S. Neves, S.P.M. Ventura, M.G. Freire, I.M. Marrucho, J.A.P. Coutinho, Evaluation of cation influence on the formation and extraction capability of ionic-liquid-based aqueous biphasic systems, J. Phys. Chem. B 113 (2009) 5194–5199.

[22] Z. Li, Y. Pei, H. Wang, J. Fan, J. Wang, Ionic liquid-based aqueous two-phase systems and their applications in green separation processes, Trac-Trends Anal. Chem. 29 (2010) 1336–1346.

[23] N.J. Bridges, K.E. Gutowski, R.D. Rogers, Investigation of aqueous biphasic systems formed from solutions of chaotropic salts with kosmotropic salts (salt-salt ABS), Green Chem. 9 (2007) 177–183.

[24] S.P.M. Ventura, C.M.S.S. Neves, M.G. Freire, I.M. Marrucho, J. Oliveira, J.A.P. Coutinho, Evaluation of anion influence on the formation and extraction capacity of ionic-liquid-based aqueous biphasic systems, J. Phys. Chem. B 113 (2009) 9304–9310.

[25] S. Shahriari, L.C. Tomé, J.M.M. Araújo, L.P.N. Rebelo, J.A.P. Coutinho, I.M. Marrucho, et al., Aqueous biphasic systems: a benign route using cholinium-based ionic liquids, RSC Adv. 3 (2013) 1835–1843.

[26] C. Wu, J. Wang, Z. Li, J. Jing, H. Wang, Relative hydrophobicity between the phases and partition of cytochrome-c in glycine ionic liquids aqueous two-phase systems, J. Chromatogr. A 1305 (2013) 1–6.

[27] S. Shahriari, C.M.S.S. Neves, M.G. Freire, J.A.P. Coutinho, Role of the Hofmeister series in the formation of ionic-liquid-based aqueous biphasic systems, J. Phys. Chem. B 116 (2012) 7252–7258.

[28] T. Mourão, A.F.M. Cláudio, I. Boal-Palheiros, M.G. Freire, J.A.P. Coutinho, Evaluation of the impact of phosphate salts on the formation of ionic-liquid-based aqueous biphasic systems, J. Chem. Thermodyn. 54 (2012) 398–405.

[29] J. Han, Y. Wang, W. Kang, C. Li, Y. Yan, J. Pan, et al., Phase equilibrium and macrolide antibiotics partitioning in real water samples using a two-phase system composed of the ionic liquid 1-butyl-3-methylimidazolium tetrafluoroborate and an aqueous solution of an inorganic salt, Microchim. Acta 169 (2010) 15–22.

[30] Y. Wang, X. Xu, Y. Yan, J. Han, Z. Zhang, Phase behavior for the [Bmim]BF_4 aqueous two-phase systems containing ammonium sulfate/sodium carbonate salts at different temperatures: experimental and correlation, Thermochim. Acta 501 (2010) 112–118.

[31] Y. Li, M. Zhang, Q. Liu, H. Su, Phase behaviour for aqueous two-phase systems containing the ionic liquid 1-butylpyridinium tetrafluoroborate/1-butyl-4-methylpyridinium tetrafluoroborate and organic salts (sodium tartrate/ammonium citrate/trisodium citrate) at different temperatures, J. Chem. Thermodyn. 66 (2013) 80–87.

[32] J. Han, Y. Wang, C.-L. Yu, Y.-S. Yan, X.-Q. Xie, Extraction and determination of chloramphenicol in feed water, milk, and honey samples using an ionic liquid/sodium citrate aqueous two-phase system coupled with high-performance liquid chromatography, Anal. Bioanal. Chem. 399 (2011) 1295–1304.

[33] J. Han, C. Yu, Y. Wang, X. Xie, Y. Yan, G. Yin, et al., Liquid-liquid equilibria of ionic liquid 1-butyl-3-methylimidazolium tetrafluoroborate and sodium citrate/tartrate/acetate aqueous two-phase systems at 298.15 K: experiment and correlation, Fluid Phase Equilib. 295 (2010) 98–103.

[34] M.G. Freire, C.L.S. Louros, L.P.N. Rebelo, J.A.P. Coutinho, Aqueous biphasic systems composed of a water-stable ionic liquid + carbohydrates and their applications, Green Chem. 13 (2011) 1536–1545.

[35] Y. Chen, Y. Meng, S. Zhang, Y. Zhang, X. Liu, J. Yang, Liquid-liquid equilibria of aqueous biphasic systems composed of 1-butyl-3-methyl imidazolium tetrafluoroborate + sucrose/maltose + water, J. Chem. Eng. Data 55 (2010) 3612–3616.

[36] B. Wu, Y.M. Zhang, H.P. Wang, Aqueous biphasic systems of hydrophilic ionic liquids + sucrose for separation, J. Chem. Eng. Data 53 (2008) 983–985.
[37] B. Wu, Y. Zhang, H. Wang, Phase behavior for ternary systems composed of ionic liquid + saccharides + water, J. Phys. Chem. B 112 (2008) 6426–6429.
[38] J. Zhang, Y. Zhang, Y. Chen, S. Zhang, Mutual coexistence curve measurement of aqueous biphasic systems composed of [bmim][BF_4] and glycine, L-serine, and L-proline, respectively, J. Chem. Eng. Data 52 (2007) 2488–2490.
[39] Z. Du, Y.-L. Yu, J.-H. Wang, Extraction of proteins from biological fluids by use of an ionic liquid/aqueous two-phase system, Chem-Eur. J. 13 (2007) 2130–2137.
[40] Y. Pei, L. Li, Z. Li, C. Wu, J. Wang, Partitioning behavior of wastewater proteins in some ionic liquids-based aqueous two-phase systems, Sep. Sci. Technol. 47 (2012) 277–283.
[41] M.J. Ruiz-Angel, V. Pino, S. Carda-Broch, A. Berthod, Solvent systems for countercurrent chromatography: an aqueous two phase liquid system based on a room temperature ionic liquid, J. Chromatogr. A 1151 (2007) 65–73.
[42] S.P.M. Ventura, S.G. Sousa, M.G. Freire, L.S. Serafim, Á.S. Lima, J.A.P. Coutinho, Design of ionic liquids for lipase purification, J. Chromatogr. B 879 (2011) 2679–2687.
[43] Q. Cao, L. Quan, C. He, N. Li, K. Li, F. Liu, Partition of horseradish peroxidase with maintained activity in aqueous biphasic system based on ionic liquid, Talanta 77 (2008) 160–165.
[44] S.P.M. Ventura, R.L.F. De Barros, J.M. De Pinho Barbosa, C.M.F. Soares, A.S. Lima, J.A.P. Coutinho, Production and purification of an extracellular lipolytic enzyme using ionic liquid-based aqueous two-phase systems, Green Chem. 14 (2012) 734–740.
[45] Y. Lu, W. Lu, W. Wang, Q. Guo, Y. Yang, Thermodynamic studies of partitioning behavior of cytochrome c in ionic liquid-based aqueous two-phase system, Talanta 85 (2011) 1621–1626.
[46] Y. Pei, Z. Li, L. Liu, J. Wang, H. Wang, Selective separation of protein and saccharides by ionic liquids aqueous two-phase systems, Sci. China Chem. 53 (2010) 1554–1560.
[47] F.J. Deive, A. Rodríguez, A.B. Pereiro, J.M.M. Araújo, M.A. Longo, M.A.Z. Coelho, et al., Ionic liquid-based aqueous biphasic system for lipase extraction, Green Chem. 13 (2011) 390–396.
[48] F.J. Deive, A. Rodríguez, L.P.N. Rebelo, I.M. Marrucho, Extraction of *Candida antarctica* lipase A from aqueous solutions using imidazolium-based ionic liquids, Sep. Purif. Technol. 97 (2012) 205–210.
[49] U. Novak, A. Pohar, I. Plazl, P. Žnidaršič-Plazl, Ionic liquid-based aqueous two-phase extraction within a microchannel system, Sep. Purif. Technol. 97 (2012) 172–178.
[50] S. Dreyer, U. Kragl, Ionic liquids for aqueous two-phase extraction and stabilization of enzymes, Biotechnol. Bioeng. 99 (2008) 1416–1424.
[51] S. Dreyer, P. Salim, U. Kragl, Driving forces of protein partitioning in an ionic liquid-based aqueous two-phase system, Biochem. Eng. J. 46 (2009) 176–185.
[52] F.S. Oliveira, J.M.M. Araújo, R. Ferreira, L.P.N. Rebelo, I.M. Marrucho, Extraction of l-lactic, l-malic, and succinic acids using phosphonium-based ionic liquids, Sep. Purif. Technol. 85 (2012) 137–146.
[53] Y. Jiang, H. Xia, C. Guo, I. Mahmood, H. Liu, Phenomena and mechanism for separation and recovery of penicillin in ionic liquids aqueous solution, Ind. Eng. Chem. Res. 46 (2007) 6303–6312.
[54] T.M. Przybycien, N.S. Pujar, L.M. Steele, Alternative bioseparation operations: life beyond packed-bed chromatography, Curr. Opin. Biotechnol. 15 (2004) 469–478.
[55] S.K. Dhananjay, V.H. Mulimani, Three-phase partitioning of α-galactosidase from fermented media of *Aspergillus oryzae* and comparison with conventional purification techniques, J. Ind. Microbiol. Biotechnol. 36 (2009) 123–128.
[56] J. Szamos, A. Jánosi, B. Tamás, É. Kiss, A novel partitioning method as a possible tool for investigating meat. I, Eur. Food Res. Technol. 206 (1998) 208–212.
[57] R. Borbás, S. Turza, J. Szamos, É. Kiss, Analysis of protein gels formed by interfacial partitioning, Colloid Polym. Sci. 279 (2001) 705–713.
[58] A. Sen, M. Eryilmaz, H. Bayraktar, S. Önal, Purification of α-galactosidase from pepino (*Solanum muricatum*) by three-phase partitioning, Sep. Purif. Technol. 83 (2011) 130–136.

[59] E. Kiss, J. Szamos, B. Tamás, R. Borbás, Interfacial behavior of proteins in three-phase partitioning using salt-containing water/tert-butanol systems, Colloid Surf. A-Physicochem. Eng. Aspects 142 (1998) 295–302.
[60] G.S. Vidhate, R.S. Singhal, Extraction of cocoa butter alternative from kokum (*Garcinia indica*) kernel by three phase partitioning, J. Food Eng. 117 (2013) 464–466.
[61] C. Dennison, R. Lovrien, Three phase partitioning: concentration and purification of proteins, Protein Expr. Purif. 11 (1997) 149–161.
[62] E. Kiss, R. Borbás, Protein adsorption at liquid/liquid interface with low interfacial tension, Colloid Surf. B-Biointerfaces 31 (2003) 169–176.
[63] G.M. Rather, M.N. Gupta, Three phase partitioning leads to subtle structural changes in proteins, Int. J. Biol. Macromol. 60 (2013) 134–140.
[64] K.H. Tan, R. Lovrien, Enzymology in aqueous-organic cosolvent binary mixtures, J. Biol. Chem. 247 (1972) 3278–3285.
[65] R.E. Lovrein, C. Goldensoph, P. Anderson, B. Odegard, Three phase partitioning (TPP) via t-butanol: enzyme separation from crudes, in: R. Burgess (Ed.), Protein Purification, Micro to Macro, Marcel Dekker Inc., New York (USA), 1987, pp. 131–148.
[66] V.V. Kumar, V. Sathyaselvabala, S.D. Kirupha, A. Murugesan, T. Vidyadevi, S. Sivanesan, Application of response surface methodology to optimize three phase partitioning for purification of laccase from *Pleurotus ostreatus*, Sep. Sci. Technol. 46 (2011) 1922–1930.
[67] S. Gautam, J. Mukherjee, I. Roy, M.N. Gupta, Emerging trends in designing short and efficient protein purification protocols, Am. J. Biochem. Biotechnol. 8 (2012) 230–254.
[68] E. Akardere, B. Özer, E.B. Çelem, S. Önal, Three-phase partitioning of invertase from Baker's yeast, Sep. Purif. Technol. 72 (2010) 335–339.
[69] S.M. Harde, R.S. Singhal, Extraction of forskolin from *Coleus forskohlii* roots using three phase partitioning, Sep. Purif. Technol. 96 (2012) 20–25.
[70] R.K. Wati, T. Theppakorn, S. Benjakul, S. Rawdkuen, Three-phase partitioning of trypsin inhibitor from legume seeds, Process Biochem. 44 (2009) 1307–1314.
[71] W.W. Ward, G. Swiatek, Protein purification, Curr. Anal. Chem. 5 (2009) 85–105.
[72] W.W. Ward, Three-phase partitioning for protein purification, Innovations Pharm. Technol. (2009) 28–34.
[73] V. Vinoth Kumar, M.P. Premkumar, V.K. Sathyaselvabala, S. Dineshkirupha, J. Nandagopal, S. Sivanesan, *Aspergillus niger* exo-inulinase purification by three phase partitioning, Eng. Life Sci. 11 (2011) 607–614.
[74] A.V. Narayan, M.C. Madhusudhan, K.S.M.S. Raghavarao, Extraction and purification of ipomoea peroxidase employing three-phase partitioning, Appl. Biochem. Biotechnol. 151 (2008) 263–272.
[75] S.K. Dhananjay, V.H. Mulimani, Purification of α-galactosidase and invertase by three-phase partitioning from crude extract of *Aspergillus oryzae*, Biotechnol. Lett. 30 (2008) 1565–1569.
[76] E. Çalci, T. Demir, E. Biçak Çelem, S. Önal, Purification of tomato (*Lycopersicon esculentum*) α-galactosidase by three-phase partitioning and its characterization, Sep. Purif. Technol. 70 (2009) 123–127.
[77] H.S. Choonia, S.S. Lele, Three phase partitioning of ß-galactosidase produced by an indigenous *Lactobacillus acidophilus* isolate, Sep. Purif. Technol. 110 (2013) 44–50.
[78] S. Rajeeva, S.S. Lele, Three-phase partitioning for concentration and purification of laccase produced by submerged cultures of *Ganoderma* sp. WR-1, Biochem. Eng. J. 54 (2011) 103–110.
[79] S. Rawdkuen, A. Vanabun, S. Benjakul, Recovery of proteases from the viscera of farmed giant catfish (*Pangasianodon gigas*) by three-phase partitioning, Process Biochem. 47 (2012) 2566–2569.
[80] P. Chaiwut, P. Pintathong, S. Rawdkuen, Extraction and three-phase partitioning behavior of proteases from papaya peels, Process Biochem. 45 (2010) 1172–1175.
[81] S. Rawdkuen, P. Chaiwut, P. Pintathong, S. Benjakul, Three-phase partitioning of protease from *Calotropis procera* latex, Biochem. Eng. J. 50 (2010) 145–149.
[82] B. Özer, E. Akardere, E.B. Çelem, S. Önal, Three-phase partitioning as a rapid and efficient method for purification of invertase from tomato, Biochem. Eng. J. 50 (2010) 110–115.
[83] L. Saxena, B.K. Iyer, L. Ananthanarayan, Three phase partitioning as a novel method for purification of ragi (*Eleusine coracana*) bifunctional amylase/protease inhibitor, Process Biochem. 42 (2007) 491–495.

[84] S. Sharma, M.N. Gupta, Purification of phospholipase D from *Dacus carota* by three-phase partitioning and its characterization, Protein Expr. Purif. 21 (2001) 310–316.
[85] Y.A. Duman, E. Kaya, Three-phase partitioning as a rapid and easy method for the purification and recovery of catalase from sweet potato tubers (*Solanum tuberosum*), Appl. Biochem. Biotechnol. 170 (2013) 1119–1126.
[86] I. Roy, A. Sharma, M.N. Gupta, Three-phase partitioning for simultaneous renaturation and partial purification of *Aspergillus niger* xylanase, BA-Proteins Proteomics 1698 (2004) 107–110.
[87] K. Mondal, A. Sharma, M.N. Gupta, Three phase partitioning of starch and its structural consequences, Carbohydr. Polym. 56 (2004) 355–359.
[88] A. Sharma, M.N. Gupta, Three phase partitioning of carbohydrate polymers: separation and purification of alginates, Carbohydr. Polym. 48 (2002) 391–395.
[89] C.G.d.O. Coimbra, C.E. Lopes, G.M.T. Calazans, Three-phase partitioning of hydrolyzed Levan, Bioresour. Technol. 101 (2010) 4725–4728.
[90] A. Sharma, S.K. Khare, M.N. Gupta, Three phase partitioning for extraction of oil from soybean, Bioresour. Technol. 85 (2002) 327–329.
[91] S. Gautam, P. Dubey, G.M. Rather, M.N. Gupta, Non-chromatographic strategies for protein refolding, Recent Pat. Biotechnol. 6 (2012) 57–68.
[92] R. Borbás, B.S. Murray, É. Kiss, Interfacial shear rheological behaviour of proteins in three-phase partitioning systems, Colloid Surf. A-Physicochem. Eng. Aspects 213 (2003) 93–103.
[93] R. Borbás, É. Kiss, M. Nagy, Elastic properties of protein gels obtained by three-phase partitioning, Prog. Colloid Polym. Sci. 117 (2001) 189–194.
[94] N. Saifuddin, A.Z. Raziah, Enhancement of lipase enzyme activity in non-aqueous media through a rapid three phase partitioning and microwave irradiation, E-J. Chem. 5 (2008) 864–871.
[95] I. Roy, M.N. Gupta, Enhancing reaction rate for transesterification reaction catalyzed by *Chromobacterium* lipase, Enzym. Microb. Technol. 36 (2005) 896–899.
[96] V.V. Kumar, V. Sathyaselvabala, M.P. Premkumar, T. Vidyadevi, S. Sivanesan, Biochemical characterization of three phase partitioned laccase and its application in decolorization and degradation of synthetic dyes, J. Molec. Catal. B 74 (2012) 63–72.
[97] G.M. Rather, J. Mukherjee, P.J. Halling, M.N. Gupta, Activation of alpha chymotrypsin by three phase partitioning is accompanied by aggregation, PLoS ONE 7 (2012) e49241.
[98] N. Adányi, J. Szamos, E.E. Szabó, M. Váradi, Interfacial enzyme partitioning as a tool for constructing biosensors, Acta Aliment. 28 (1999) 329–338.
[99] G. Ujhelyi, É. Gelencsér, J. Szamos, Rapid and economical DNA extraction method from meat matrices containing soybean flour, Acta Aliment. 40 (2011) 426–435.
[100] J. Szamos, E. Aubrecht, É. Gelencsér, Detection of wheat by adapted polymerase chain reaction (PCR) methodology, Acta Aliment. 27 (1998) 87–95.
[101] S. Raghava, M.N. Gupta, Tuning permeabilization of microbial cells by three-phase partitioning, Anal. Biochem. 385 (2009) 20–25.
[102] K. Mondal, A. Sharma, L. Lata, M.N. Gupta, Macroaffinity ligand-facilitated three-phase partitioning (MLFTPP) of α-amylases using a modified alginate, Biotechnol. Prog. 19 (2003) 493–494.
[103] K. Mondal, A. Sharma, M.N. Gupta, Macroaffinity ligand-facilitated three-phase partitioning for purification of glucoamylase and pullulanase using alginate, Protein Expr. Purif. 28 (2003) 190–195.
[104] I. Roy, K. Mondal, M.N. Gupta, Leveraging protein purification strategies in proteomics, J. Chromatogr. B 849 (2007) 32–42.
[105] A. Sharma, K. Mondal, M.N. Gupta, Separation of enzymes by sequential macroaffinity ligand-facilitated three-phase partitioning, J. Chromatogr. A 995 (2003) 127–134.
[106] ECHA (European Chemicals Agency), tert-butyl alcohol, 2013 (accessed 08. 2013.), http://clp-inventory.echa.europa.eu/SummaryOfClassAndLabelling.aspx?SubstanceID=97250&HarmOnly=no?fc=true&lang=es.
[107] D. Han, K.H. Row, Recent applications of ionic liquids in separation technology, Molecules 15 (2010) 2405–2426.
[108] M. Alvarez-Guerra, A. Irabien, Design of ionic liquids: an ecotoxicity (vibrio fischeri) discrimination approach, Green Chem. 13 (2011) 1507–1516.

[109] L. Chen, C. Guo, Y. Guan, H. Liu, Isolation of lactoferrin from acid whey by magnetic affinity separation, Sep. Purif. Technol. 56 (2007) 168–174.

[110] A.L. Zydney, Protein separations using membrane filtration: new opportunities for whey fractionation, Int. Dairy J. 8 (1998) 243–250.

[111] E. Alvarez-Guerra, A. Irabien, Ionic liquid-based three phase partitioning (ILTPP) systems for whey protein recovery: ionic liquid selection, J. Chem. Technol. Biotechnol. (2014) accepted, in press http://dx.doi.org/10.1002/jctb.4401.

[112] E. Alvarez-Guerra, A. Irabien, Ionic liquid-based three phase partitioning (ILTPP) for lactoferrin recovery, Sep. Sci. Technol. 49 (2014) 957–965.

[113] Y. Jiang, H. Xia, C. Guo, I. Mahmood, H. Liu, Enzymatic hydrolysis of penicillin in mixed ionic liquids/water two-phase system, Biotechnol. Prog. 23 (2007) 829–835.

[114] Z.-J. Tan, F.-F. Li, X.-L. Xu, J.-M. Xing, Simultaneous extraction and purification of aloe polysaccharides and proteins using ionic liquid based aqueous two-phase system coupled with dialysis membrane, Desalination 286 (2012) 389–393.

[115] Z. Tan, F. Li, X. Xu, Isolation and purification of aloe anthraquinones based on an ionic liquid/salt aqueous two-phase system, Sep. Purif. Technol. 98 (2012) 150–157.

[116] M.G. Freire, C.M.S.S. Neves, I.M. Marrucho, J.A.P. Coutinho, A.M. Fernandes, Hydrolysis of tetrafluoroborate and hexafluorophosphate counter ions in imidazolium-based ionic liquids, J. Phys. Chem. A 114 (2010) 3744–3749.

[117] T. Matsuyama, K. Domyoung, M. Umetsu, T. Ikawa, M. Yamanishi, N. Ishida, et al., Ionic liquid/water interfacial localization of a green fluorescent protein fused to a tryptophan-rich peptide, J. Biosci. Bioeng. 113 (2012) 160–165.

[118] W. Bi, M. Tian, K.H. Row, Ultrasonication-assisted extraction and preconcentration of medicinal products from herb by ionic liquids, Talanta 85 (2011) 701–706.

[119] M.-A. Morikawa, A. Takano, S. Tao, N. Kimizuka, Biopolymer-encapsulated protein microcapsules spontaneously formed at the ionic liquid-water interface, Biomacromolecules 13 (2012) 4075–4080.

[120] S.C. Ribeiro, G.A. Monteiro, J.M.S. Cabral, D.M.F. Prazeres, Isolation of plasmid DNA from cell lysates by aqueous two-phase systems, Biotechnol. Bioeng. 78 (2002) 376–384.

[121] K.A. Moreira, A.C. Chaves, E.T. Marques, D.M.F. Prazeres, W.M. De Azevedo, A.L.F. Porto, et al., Extraction of dengue 2 plasmid DNA vaccine (pD2) from cell lysates by aqueous two-phase systems, Biotechnology 6 (2007) 520–526.

[122] F. Luechau, T.C. Ling, A. Lyddiatt, A descriptive model and methods for up-scaled process routes for interfacial partition of bioparticles in aqueous two-phase systems, Biochem. Eng. J. 50 (2010) 122–130.

[123] A. Sreedhara, R. Flengsrud, V. Prakash, D. Krowarsch, T. Langsrud, P. Kaul, et al., A comparison of effects of pH on the thermal stability and conformation of caprine and bovine lactoferrin, Int. Dairy J. 20 (2010) 487–494.

[124] A. Heebøll-Nielsen, S.F.L. Justesen, T.J. Hobley, O.R.T. Thomas, Superparamagnetic cation-exchange adsorbents for bioproduct recovery from crude process liquors by high-gradient magnetic fishing, Sep. Sci. Technol. 39 (2004) 2891–2914.

[125] E. Eugster, E. Jakob, D. Wechsler, Cheese, Processed Cheese, and Whey, in: Ullmann's Encyclopedia of Industrial Chemistry, John Wiley & Sons, Inc., 2012.

[126] Y. Deng, T. Long, D. Zhang, J. Chen, S. Gan, Phase diagram of [Amim]Cl + salt aqueous biphasic systems and its application for [Amim]Cl recovery, J. Chem. Eng. Data 54 (2009) 2470–2473.

[127] C. Li, J. Han, Y. Wang, Y. Yan, J. Pan, X. Xu, et al., Phase behavior for the aqueous two-phase systems containing the ionic liquid 1-butyl-3-methylimidazolium tetrafluoroborate and kosmotropic salts, J. Chem. Eng. Data 55 (2010) 1087–1092.

[128] C.M.S.S. Neves, M.G. Freire, J.A.P. Coutinho, Improved recovery of ionic liquids from contaminated aqueous streams using aluminium-based salts, RSC Adv. 2 (2012) 10882–10890.

CHAPTER 7

The Role of Ionic Liquids in Enzyme-Membrane Integrated Systems

Laszlo Gubicza[1], Nándor Nemestóthy[1], Katalin Bélafi-Bakó[1], Zvjezdana Findrik[2]
[1]Research Institute on Bioengineering, Membrane Technology and Energetics, University of Pannonia, Veszprém, Hungary
[2]Faculty of Chemical Engineering and Technology, University of Zagreb, Croatia

1. ENZYME–MEMBRANE INTEGRATED SYSTEMS

1.1 Advantages of Integrated Systems

Enzymes are widely applied in various (bio)chemical transformations; however, the effectiveness of these processes should be enhanced to become more competitive in the market. To realize a bioprocess with high effectiveness, its features must be known in every detail. These features can be classified regarding the characteristics of the initial (substrate) and final (product) compounds, the reaction itself, and the behavior of the enzyme [1]; moreover the interactions between the materials present in the system [2,3] should be taken into account (Table 1).

1.1.1 Important Features of Biotransformations

To explain the characteristics of a particular enzymatic process reliable and complete, enzyme databases are currently available. Moreover, relevant literature can be collected from other sources. Still, it is advisable to perform preliminary experiments and carefully evaluate them. Among these measurements, kinetic investigations are extremely important (the effect of substrates on the reaction rate, analysis of progress curves, heat stability, activation energy as determined by the Arrhenius equation, enzyme activity, product/substrate inhibitions). In a real working system that manufactures products, however, it is not enough just to learn about the behavior of the enzyme process or the possible inhibition phenomena; we have to act to enhance the effectiveness, that is, to avoid inhibition and to eliminate the hindering factors. During these problem-solving procedures, experts have to have a strong engineering background, mainly related to reactor techniques and separation processes.

To study bioconversions from the aspect of *intensification* by applying or inserting a separation step, a special method that compares the optimal substrate and product concentrations may be useful [4]. It is important to emphasize that the *optimal* concentrations are of concern here, not the simple substrate or product concentrations.

Ionic Liquids in Separation Technology
http://dx.doi.org/10.1016/B978-0-444-63257-9.00007-9
Copyright © 2014 Elsevier B.V.
All rights reserved.

Table 1 Important Features of Biotransformations

Type and characteristics of the reaction	Equilibrium
	Formation/consumption of acid/alkali
	Formation/consumption of water
	Formation/consumption of gases
	Heat/compound/material balance
Features of substrates and products	Physical properties (melting point, boiling point, viscosity)
	Solubility
	Stability (pH, temperature)
Behavior of the enzyme	Single enzyme or complex
	Cofactor demand
	Location of the enzyme
	Structure of the active center
	Water demand
	Other ligands, activators
Interactions between the enzyme, substrates, and products	Kinetics
	enzyme-substrate complex
	Substrate/product inhibition
	Activity (pH, temperature dependence)
	Toxicity of substrate/product
	Stability

As can be seen in Figure 1, there are three operation regions:

- Case 1: the optimal substrate concentration is smaller than the optimal production concentration (P > S);
- Case 2: the two concentration values are similar; and
- Case 3: the optimal substrate concentration is higher than the optimal product concentration (S > P).

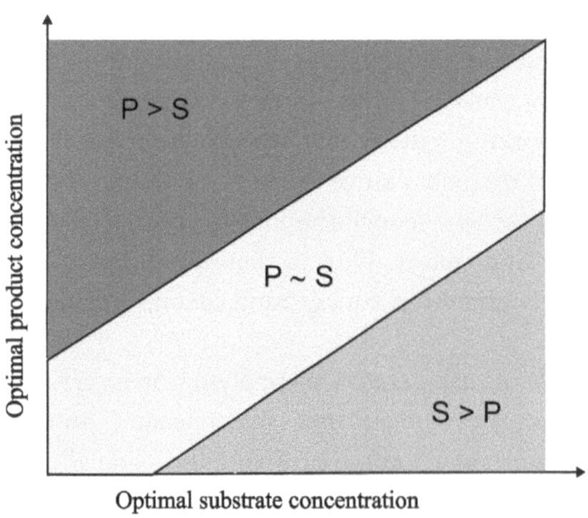

Figure 1 ***Comparison of the optimal substrate (S) and product (P) concentrations.***

In case 1 the optimal substrate concentration is significantly smaller than the product concentration, which often occurs when substrate conversion is fast and complete and/or increasing the initial substrate concentration is difficult (e.g., solubility problems, substrate inhibition). Here the gradual, controlled addition of substrate can direct the bioconversion toward an enhanced yield. In case 2 the concentrations of unreacted substrate and the product are similar; there is no big difference between the two concentration levels.

In case 3 the optimal substrate concentration is higher than the optimal product concentration, which is beneficial; it means that the equilibrium position can be shifted toward the product formation because of the excess of substrate present in the system. The product concentration, on the other hand, cannot be increased because of, for example, possible inhibition.

To shift the reaction equilibrium toward the synthesis, *product removal* can be applied (beyond the increased substrate concentration), as well. From the *separation* point of view, techniques with higher selectivity should be applied in cases 2 and 3 because the substrate is present in a larger quantity compared to that in case 1.

If the separation step is intended to combine simultaneously (*in situ*) with the bioconversion system, several other requirements should be satisfied. Although it is difficult to give general rules, some of them are listed here.

- The separation step should not disturb the bioconversion.
- The separation step must not poison the biocatalyst.
- The operation parameters of the system should be controlled properly.
- The reliable control of liquid level(s) is extremely important.

Depending on the compound, the substance to be separated, various separation processes can be applied in this way. Membrane separation processes, however, have numerous advantages over the other methods, such as low energy demand, variability, easy insertion, no addictives required, no hazardous material formed, and easy scale-up.

1.2 Membrane Bioreactors

Membrane bioreactors are special equipment in which a bioconversion step and a separation step take place simultaneously [5]. These reactors can be characterized by several aspects including basic features coming from either the bioconversion process or the membrane process (Table 2). For example, when describing a bioconversion it is important to give the reaction itself (together with the substrate and the product), the media, and the type and form of the biocatalyst. In membrane separation processes the most important points are the type of the method, membrane material, module type, and mode of operation (flow).

In Table 2 examples, opportunities for the application of membrane bioreactors are presented according to the aspects mentioned above. These features make it possible to classify the bioreactors, as well.

Table 2 Features of Membrane Bioreactors

Features	Description	References
Bioconversion		
Reaction	Hydrolysis of macromolecules and other compounds, redox reactions, esterification, enantioselective reactions	[6,7]
Media	Water, organic solvent, ionic liquids, supercritical fluids, two-phase systems, emulsions	[8]
Type of biocatalyst	Hydrolytic enzymes, lipases, esterases	[3]
Form of biocatalyst	Soluble, immobilized	[9]
Membrane Separation		
Type of process	Microfiltration, ultrafiltration, nanofiltration, dialysis, electrodialysis, pervaporation	[10]
Material of membrane	Polymer or inorganic; porous membrane, thick film membrane, ionselective membrane	[11,12]
Membrane module	Flat sheet, tubular, hollow fiber, spiral wound	[13]
Operation, flowing conditions	Batch (*dead end*), axial flow, cross flow	[14,15]

1.2.1 Features of Membrane Bioreactors

With regard to bioconversions, the vast majority of membrane bioreactors have been applied in hydrolysis of macromolecules such as polysaccharides and proteins, where hydrolytic enzymes in aqueous media are used. The macromolecules are degraded into small-molecular-weight compounds (glucose, amino acids), which often cause inhibition; therefore, it is advisable to remove them from the system. Separation can be carried out using porous membranes; the size difference between the substrate and the product is the principle for the separation. Moreover, the membrane also is able to retain the enzyme.

In membrane bioreactors other reactions can be carried out, too, although fewer results are reported in the literature. Beyond lipase and esterase reactions, special processes with the aim of cofactor regeneration can be used. In certain biocatalytic reactions, ionic (charged) compounds are formed (e.g., organic acids), which can be recovered by ion-selective membranes (dialysis, electrodialysis). Some processes can be accomplished in nonaqueous media, for example, organic solvent, solvent-free system, or, more recently, ionic liquids (ILs), and supercritical fluids. In a few applications even more special media are needed, for example, two- or multiphase systems and emulsions (micelles).

In most membrane bioreactors, the role of the membrane is to retain the enzyme (and other macromolecules). There are, however, other purposes identified: In enzymatic hydrolysis of triacylglycerols—unlike macromolecular hydrolysis—the membrane plays an important role in phase separation; moreover, it provides a surface for enzyme immobilization. In a couple of systems the membrane is considered simply as a selective barrier, and its selectivity can be exploited for special purposes, for example, separating charged solutes (ion-selective membranes) or removing water (pervaporation (PV) membrane).

To select the proper membrane separation technique and membrane material, it is important to clarify the exact role of the membrane in the bioreactor. In the case of macromolecular hydrolysis, porous ultrafiltration and microfiltration membranes are used. For other purposes more special techniques (PV, dialysis, electrodialysis) and membranes are available. To decide which membrane material to use (polymer, inorganic, or even liquid membrane), the reaction and the circumstances of operation should be taken into account. Polymer membranes are applied in aqueous systems, whereas inorganic, ceramic membranes—which have a higher chemical and physical stability—are suitable for special reaction media, as well. Thermal resistance of inorganic membranes is also better than that of polymeric ones; therefore it is suggested that they be used at higher temperatures. (Their extremely high price, however, strongly hinders their wider application.)

The selection of the membrane module also depends on the role of the membrane as well as the circumstances. For the set-up of laboratory-scale equipment, flat-sheet modules are suitable because of their simplicity. In higher-volume applications, possible fouling phenomena should be considered; if strong fouling or clogging is expected, tubular membranes are suggested because it is easier to clean them. In other cases, hollow-fiber membranes are used because of their high packing density, although some problems may arise because of sealing trouble, and the hollow fibers cannot be repaired.

In membrane bioreactor systems, bioconversion and membrane separation should operate in a simultaneous, synchronized mode; otherwise they will not work effectively. For this operation features of both systems have to be deeply learned, and the responses of the system to certain changes in operation conditions should be carefully studied.

2. BENEFITS OF USING ILs IN ENZYMATIC REACTIONS

In biocatalysis, the term *nonconventional media* refers to systems that use solvents other than water. Enzymes are well suited to their natural aqueous environment; however, biotransformations in industrial synthesis often involve organic molecules that are insoluble in water. More important, because of its high boiling point and high heat of vaporization, water is usually the least desired solvent for most organic reactions [16].

2.1 Enzymatic Reactions in Nonwater Media

Nonconventional media can be used for biocatalysis with either enzymes or whole cells. Compared to whole cells, isolated enzymes offer several benefits, including simpler reaction apparatus, higher productivity due to a higher catalyst concentration, and simpler product purification. Initial studies considered the addition of small quantities of water-miscible organic solvents (e.g., ethanol, acetone) to aqueous enzyme solutions, ensuring the availability of a large amount of water to retain the catalytic activity of enzymes. Then, biphasic mixtures were used: substrates from the organic phase diffuse to the aqueous phase, undergo enzymatic reaction, and the products diffuse back. The size of water droplets may be decreased to assist in mass transfer, resulting in the formation of microemulsions or reverse micelles. Their stabilization is achieved by adding surfactants [16,17]. Organic solvents were the first real nonwater media used. The use of nonaqueous organic solvents as reaction milieu is associated with advantages such as the improved solubility of substrates and products, enhanced thermostability of the enzyme, a shift in the thermodynamic equilibrium to favor synthesis over hydrolysis, the suppression of many water-dependent side reactions, and decreased microbial contamination, in addition to high levels of selectivity, specificity, and catalytic rates under mild conditions. For numerous molecules the synthetic routes based on enzyme catalysis have turned out to be competitive when compared with classic chemical as well as chemocatalytic synthetic approaches [18]. The application of enzymes in organic media is restricted because most enzymes are less active and stable in the presence of organic solvents. The most important disadvantages of using enzymes in organic solvents are (1) inactivation of enzymes, (2) labor- and cost-intensive preparation of biocatalysts in covalently modified systems, (3) mass-transfer limitations in the case of heterogeneous systems or viscous solvents, and (4) water activity control needed for processes involving condensation reactions [19]. However, the exploitation of such advantages often is limited by the low stability and activity of biocatalysts in these systems.

While water concentration itself seems to be of no significance for enzyme activity in nonwater media, the relative amount of water with respect to total water solubility is very important. Instead of using the concentration of the molar fraction of water, water activity (a_W) should be preferred because both enzyme activity and selectivity crucially depend on it [20]. If there is too little water the reaction rate will be low; if there is too much, undesirable hydrolysis reactions may occur, and the reaction rate often will decrease again [21]. The optimum a_W often does not change as other aspects of the system are altered, probably because it determines the amount of water bound to the enzyme. The addition of solid salt hydrates to the reaction system is a convenient method to control a_W.

Enzymes can express activity in supercritical fluids as well, such as carbon dioxide (CO_2); freons; hydrocarbons (ethane, ethene, propane); or inorganic compounds and ILs may be used as cosolvents to increase the effectiveness of the system. The most commonly used solvent is supercritical CO_2 ($scCO_2$), which is probably explained by the fact that its critical point of 73.8 bars and 31.1 °C makes equipment design and reaction set-up relatively simple. Supercritical fluids, which represent a state between the gaseous and liquid phases of the compound, exhibit properties similar to those of hydrophobic solvents such as hexane, so it is likely that the activity and stability of enzymes in these systems will be similar to those presented in hydrophobic solvents. Although the use of supercritical fluids is not restricted to hydrolases, the use of this class of enzymes, especially lipases, dominates [22]. Small changes in the temperature or pressure of a supercritical fluid may result in great changes in its viscosity and the diffusivity and solubility of compounds dissolved within it. Supercritical fluids, similar to ILs, are "designer solvents."

2.2 ILs and Enzymes

ILs, having no measurable vapor pressure, are an interesting class of tunable and designer solvents, and they have been used extensively in a wide range of applications including enzymatic biotransformations. In fact, ILs can be designed with different cation and anion combinations, which allow the possibility of tailoring reaction solvents with specific desired properties. These unconventional solvent properties of ILs provide an opportunity to carry out many important biocatalytic reactions that are impossible using traditional solvents [23]. The technological utility of enzymes can be enhanced greatly by their use in ILs rather than in conventional organic solvents because of their unusual solvent characteristics. Studies of enzymatic reactions in ILs revealed not only that ILs are environmentally friendly alternatives to volatile organic solvents but also that enzymes exhibit excellent selectivity, including substrate, regioselectivity, and enantioselectivity, in such a solvent. The first successful report of an enzyme-catalyzed reaction using an IL as a medium was published in 2000 by Erbeldinger et al. [24].

In organic solvents, the logarithm of the partition coefficient (log P value) of a solvent in an octanol/water mixture [25] is the most-used parameter and is considered a key determinant of enzyme activity. Experiments showed that solvents with a high log P (>3) are more hydrophobic and thus more favorable for enzymatic reactions than those with a low log P.

The esterification of 2-chloropropanoic acid was investigated in different organic solvents, and a good correlation was found between the activity of the enzyme and the log P value of the organic solvents used. High conversions were achieved using hydrophobic organic solvents such as n-hexane (log $P=3.5$) and n-octane (log $P=4.5$), whereas in hydrophilic organic solvents (e.g., tetrahydrofuran (log $P=0.5$)) the conversion was limited. The explanation of this phenomenon is simple: solvents with a low log P value strip water surrounding the enzyme and thus decrease its activity or inactivate it [26].

However, the experimentally measured log P values of ILs are extremely low, which seems to suggest that ILs are highly hydrophilic in nature and hence would likely inactivate enzymes. Investigating the esterification of 2-chloropropanoic acid in different ILs, we found that in solvents with log P values between -2 and -3, the same enzyme

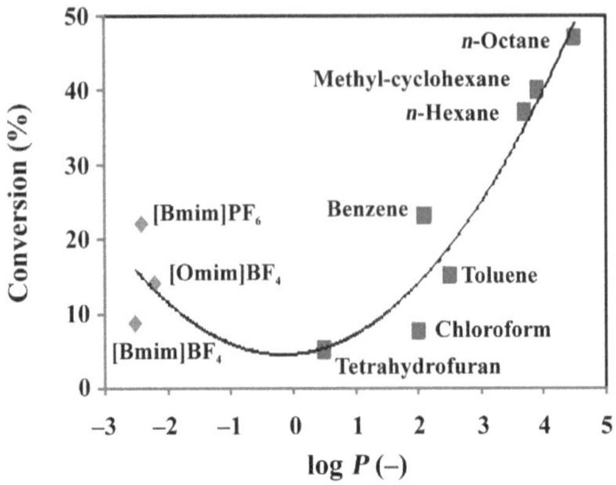

Figure 2 Correlation between the logarithm of the partition coefficient (log *P*) values of the solvents and the conversion (reaction tested: enzymatic enantioselective esterification of 2-chloropropanoic acid) ■: organic solvents, ♦: ionic liquids.

showed high activity (Figure 2). Therefore, log *P* does not seem to be able to act as a useful parameter for determining enzyme activity in ILs. On the other hand, one might argue that IL partitions in octanol/water mixtures an ion pair, and therefore its log *P* value may depend not only on the concentration of the cation but also on that of the anion and the extent of ion pairing in both phases [27].

ILs are not only suitable solvents for lignocellulosic biomass but they can assent to the separation of the biomass components. A key step in the conversion of lignocellulosic biomass to second-generation biofuels is the pretreatment of biomass to facilitate the enzymatic hydrolysis of polysaccharides into monomeric sugars, which can then be converted into fuel [28,29]. ILs were previously found to partially dissolve lignocellulosic biomass. It was reported that the biomass itself does not dissolve directly but that it is hydrolyzed first before the constituents (cellulose, hemicellulose, and lignin) dissolve into the IL [30]. In a new study, an environmentally friendly pretreatment process was developed to separate hemicelluloses and lignin from poplar wood; this process consists of IL pretreatment coupled with mild alkaline extraction. After the integrated pretreatment, cellulose was completely hydrolyzed to sugars for biofuel production, while residual lignin with a high purity was obtained [31].

2.3 Increasing Enzyme Selectivity Using ILs

The ability to carry out enantioselective biotransformations is possibly the most important factor that has motivated interest in biocatalytic processes. During recent years there have been an increasing number of publications and patents related to the enzymatic resolution of racemic mixtures or the enzymatic production of chiral compounds, stimulated by the demand for enantiopure compounds by the pharmaceutical industry. The total sales of chiral pharmaceuticals increased rapidly, and the market for enantiomerically pure or enantiomerically enriched organic compounds continues to expand. However, applications of stereoselective synthesis of chiral organic compounds are not limited to the pharmaceutical industry; they also involve the production of agrochemicals, cosmetics, and fine chemicals.

The enantioselectivity of biocatalytic reactions is normally expressed as the enantiomeric ratio, or the E value, a biochemical constant intrinsic to each enzyme that, contrary to enantiomeric excess, is independent of the extent of conversion. In an enzymatic resolution of a racemic substrate, the E value can be considered equal to the ratio of the rates of reaction for the two enantiomers when the conversion is close to zero. To be more precise, the E value is defined as the ratio of the specificity constants (k_{cat}/K_M) for the two enantiomers and can be obtained by determining the k_{cat} and K_M of a given enzyme for the two individual enantiomers [32].

In addition to solvent properties, the water content in the reaction medium is another important factor to be considered. The parameter that better describes the effects of water on enzyme properties such as K_M, V_{max}, stability, and conformation is the thermodynamic a_w. The effects of a_w on enzyme enantioselectivity are, however, controversial because an increase, decrease, or no variation in selectivity as a function of the water present in the reaction medium have all been reported. Therefore, when facing a resolution process it is advisable to perform the process at different a_w values [33,34].

Enzymes do not always show satisfying performance in terms of activity, stability, and, most important, enantioselectivity. These properties can be optimized by altering the substrates or the reaction system (medium engineering) or by using protein engineering techniques [35]. One of the most important influences an IL can have on a reaction is to change the selectivity of the enzymes. Enzymes in ILs (medium engineering) have been discovered in many cases to feature altered selectivity such as substrate specificity, enantioselectivity, prochiral selectivity, regioselectivity, and chemoselectivity [20].

As an example, enantioselective esterification of (±)-menthol was studied using *Candida rugosa* lipase (CRL) in the ILs 1-butyl-3-methyl-imidazolium hexafluorophosphate and 1-butyl-3-methyl-imidazolium tetraflouroborate, as well as in organic solvents with different hydrophobicities. Propionic anhydride was used as an acylating agent. Because the enzyme showed comparable conversion yield and enantioselectivity in $[C_4C_1Im][PF_6]$ and hexane in a 24-h reaction, more work focused on these two reaction media. The activity, stability, and enantioselectivity of CRL were compared by examining the effects of the mole ratio of substrates, temperature, incubation time, and enzyme recycling. It was found that temperature control was more crucial in the IL than in hexane to achieve high conversion and enantioselectivity. The IL system showed the advantage of using less acid anhydride to achieve a higher (±)-menthol conversion yield and better enantioselectivity. In addition, the enzyme showed the potential for recycled use in the IL. These advantages of the IL suggest that it could be used as a green alternative to organic solvents for the enantioselective esterification of (±)-menthol [36].

3. ESTERIFICATION IN ILs WITH INTEGRATED SYSTEMS

Integration of the reaction and separation steps has gained much attention in the past 10 years because of the necessity of building more efficient processes that will save time and money and reduce negative environmental effects. The advantages of such processes are enormous, in particular in the case of equilibrium reactions. By using an integrated process with *in situ* product removal, one can achieve an increased volumetric productivity, reduced substrate cost, reduced biocatalyst cost, reduced processing time, and higher reaction rates. In this manner, phenomena such as product inhibition or poor thermodynamic equilibrium can be overcome while the product is removed from the vicinity of the catalyst as soon as it is formed. In addition, product loss due to subsequent downstream processing can be reduced [4]. Integration of reaction and separation steps in esterification processes is gaining much attention and is progressing fast because it provides a convenient way to alleviate the limiting factors mentioned earlier [37].

The different configurations of integrated processes consisting of esterification and a separation step reported in the literature are described in this chapter. Such processes are of great interest in the case of esterifications because these are equilibrium reactions in which water is formed, which causes enzyme activity loss and promotes a reverse hydrolysis reaction. Therefore, water should be removed from the reaction system. The common nominator of all processes discussed in this chapter is the presence of IL, a green solvent that can overcome the disadvantages of organic solvents, can contribute to biocatalyst stability, or can act as a catalyst itself.

In general, downstream processing, for example, product separation steps, require energy, chemicals, and manpower. Hence, the possibility of using an integrated process with *in situ* product removal provides extensive savings and benefits.

3.1 Enzymatic Integrated Systems

3.1.1 Esterification in ILs Coupled with PV

PV is a membrane separation method that can be used in combination with esterification to remove one of the products from the reaction mixture and to shift the equilibrium toward the products [38]. Research involving ILs was started in a work by Schäfer et al. [39], who recovered different solutes from $[C_4C_1Im][PF_6]$ using PV. A year later the first research on an integrated system combining esterification in ILs with PV was published [40]. To date, four studies have been published and are presented in Table 3. These examples deal with simple esterifications [41,42] as well as the kinetic resolution of racemic organic acids [26,40]. In addition to product removal, PV used in these systems allowed the water content necessary for lipase activity to be kept constant. In all studied systems (Table 3), process integration enabled improved efficiency. In the first system studied [26,40] the authors investigated the synthesis of 1-butyl-2-chloro-propanoate. In both cases, the authors investigated both the efficiency of the integrated system as well as

Table 3 Integrated systems consisting of esterification and pervaporation.

Reaction	Ionic liquid	Separation method	Removed	X_{eq} / %	X_{per} / %	Membrane type	Membrane Properties	Reference
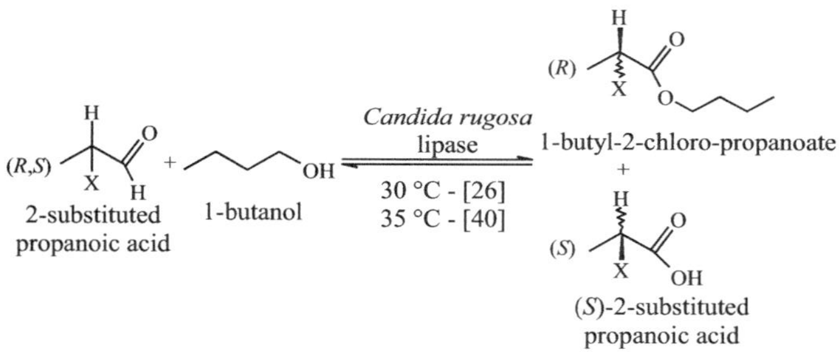 X: Cl-, Br-, MeO-, EtO-, PrO-, iPrO-, PheO-	$[C_4C_1Im][PF_6]$ $[C_8C_1Im][PF_6]$	pervaporation	water	37.6 23.8	48.2 31.8	Pervap 2201, Sulzer Chemtech GmbH	Hydrophilic composite membrane made from highly cross-linked PVA layer and a porous PAN support	[26]
	$[C_4C_1Im][PF_6]$	pervaporation	water	17.6	36.6	n.a.	Hydrophilic membrane	[40]
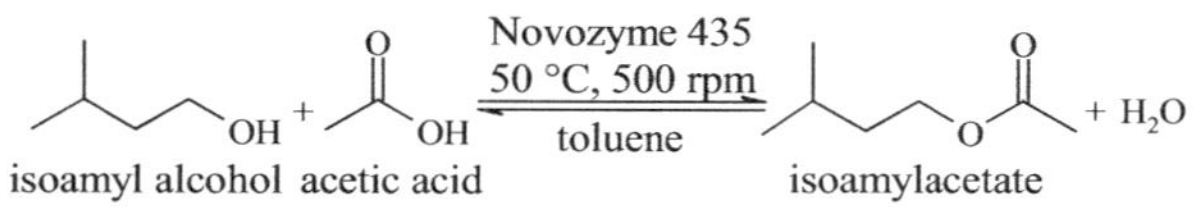	$[C_4C_1Im][PF_6]$	osmotic distillation pervaporation	water isoamyl acetate	n.a.	100	Pervap 2201/2205 Pervap 2256, Sulzer Chemtech GmbH	Hydrophilic membranes (Please see under Ulbert et al., 2004) Hydrophobic membrane – chemically stable non-woven fabric on porous support of polyacrylonitrile and separating layer of polymer	[41]
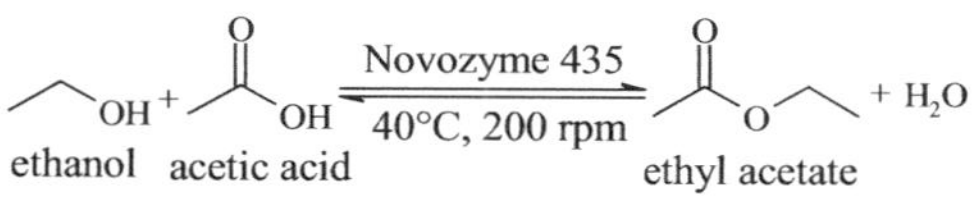	$[C_4C_1Im][PF_6]$	pervaporation	water ethyl acetate	n.a.	n.a.	Pervap 2001 Pervap 2255-50, Sulzer Chemtech GmbH	Hydrophilic membrane Hydrophobic membrane used for removal of methanol and ethanol from organics.	[42]

eq - equilibrium; N.A. - not available; PAN-polyacrylonitrile; per - with pervaporation; PVA – poly(vinyl alcohol); X-conversion.

enzyme enantioselectivity. Enantioselectivity was much better in ILs than in organic solvents.

In work reported by Fehér et al. [41], a semicontinuous system of isoamyl acetate synthesis was investigated. It consisted of a reaction system coupled with osmotic distillation for water removal and PV for isoamyl acetate removal. As shown for isoamyl acetate synthesis in this study, choosing the right membrane can achieve 100% selectivity toward the desired chemical. Gubicza et al. [42] moved a step further by combining two PV membranes in a cascade in a continuous system of ethyl acetate synthesis. The double PV system was used to remove both reaction products (i.e. water and ethyl acetate), and this is the first and only reported system of such type. The advantage of PV in these systems is the membrane selectivity, which is enabled by the great variety of membrane types and producers.

3.1.2 Esterifications with Integrated Supported Liquid Membranes

Supported IL membranes (SILMs) containing ILs can successfully be used for the separation of compounds such as alcohols, ketones, and amines by allowing selective transport through the membrane [43]. Their use could overcome the usual problems of conventional liquids because of their negligible vapor pressure and the possibility of minimizing the solubility of ILs in the surrounding phases by choosing their cation and anion [44]. In addition, the use of ILs enables a very stable supported liquid membrane (SLM) to be prepared without any significant loss of the IL to the atmosphere or other phases in which it comes in contact [45]. Separation methodology has been developed using lipases and esterification/hydrolysis in combination with SILMs. Esterification with integrated SILMs (Table 4) have demonstrated exciting potential in the development of new methodologies for the production of enantiomerically pure compounds [49]. The first report on the separation of racemic ibuprofen was published by Miyako et al. [51], who replaced the organic solvents [46] originally used in the membranes with ILs to increase the transport rate through the membrane and improve their stability. Organic solvents often cause protein damage, whereas ILs do not generally cause any damage to the enzyme and can even stabilize it. Therefore, this approach is of great interest. Transport through the membrane is facilitated by a lipase-catalyzed reaction (Figure 3), which enables the movement of the organic acid through the membrane in accordance with the following principle: Organic acid and alcohol react in the lipase-catalyzed esterification (feed phase) and form an ester, which can pass through the SILM because of its affinity toward the IL. Once it gets to the other side of the membrane (receiving phase), it hydrolyzes in the lipase-catalyzed (second) reaction and forms an organic acid and alcohol. Such a system has been shown to be of great benefit for kinetic resolutions. Miyako et al. proved the concept by using several organic solvents and ILs and showed that supported membranes containing ILs $[C_4C_1Im][PF_6]$ are more stabile (120 h) than a membrane containing isooctane (75 h). Coupling the SLM with

Table 4 Integrated systems of esterification and SILM.

Reaction	Ionic liquid	Removed	Membrane type	Membrane properties	Reference
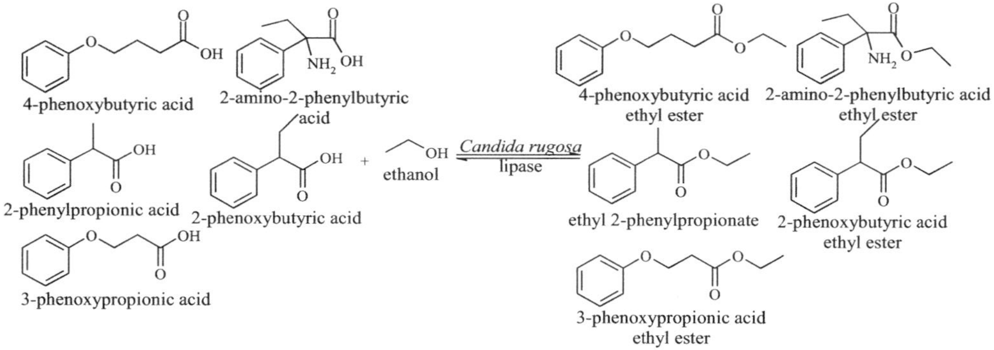	$[C_4C_1Im][PF_6]$ $[C_8C_1Im][PF_6]$	Transport of compounds through SILM was investigated	Polypropylene film Celgard 2500, Celanese Corporation	Hydrophobic 25 μm microporous monolayer membrane, uniform pore structure with high chemical and thermal stability	[46]
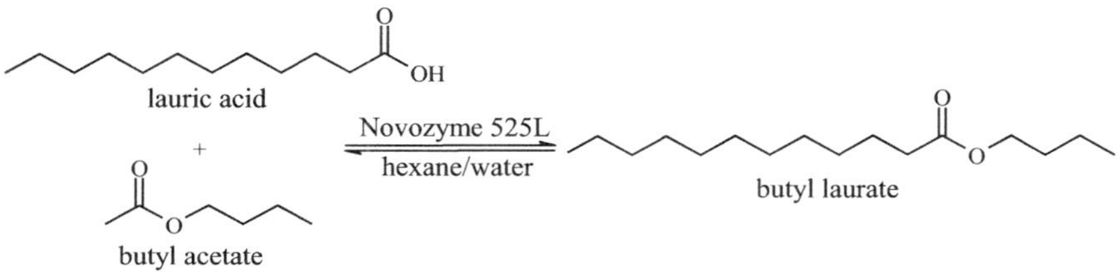	$[C_4C_1Im][PF_6]$ $[C_2C_1Im][NTf_2]$ $[C_4C_1Im][NTf_2]$	-	Ceramic tubular membrane	Lipase immobilized on the membrane and used for biocatalysis	[47]
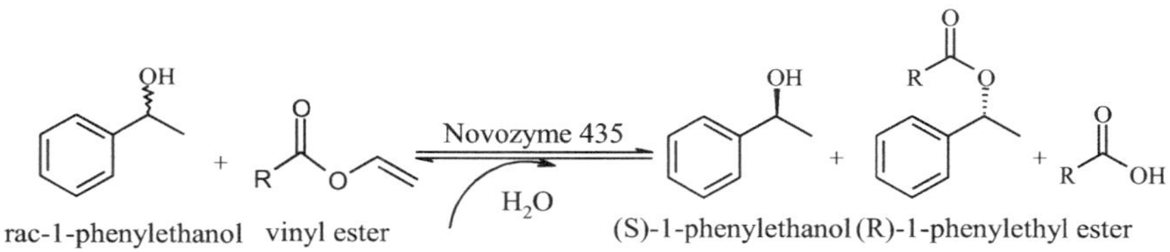 R= CH_3–, CH_3CH_2–, $CH_3(CH_2)_{10}$– [48]	$[C_4C_1Im][PF_6]$ $[C_8C_1Im][PF_6]$ $[C_4C_1Im][BF_4]$ $[C_8C_1Im][BF_4]$ $[C_4C_1Im][NTf_2]$ $[C_8C_1Im][NTf_2]$	(S)-1-phenylethanol	Nylon-polyamide, Millipore S.A.	Hydrophilic 0,45 μm pore size, 170 μm thickness	[48]
R= CH_3CH_2–, $CH_3CH_2CH_2$–, $CH_3(CH_2)_{10}$– [49]	$[C_4C_1Im][BF_4]$				[49]
R = CH_3CH_2–, $CH_3CH_2CH_2$ 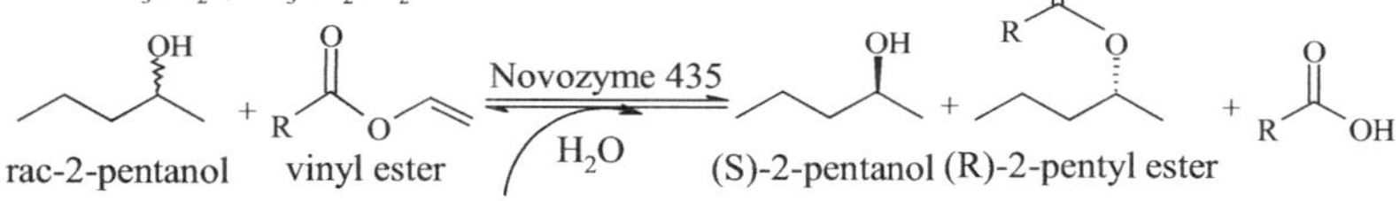	$[C_4C_1Im][PF_6]$ $[C_8C_1Im][PF_6]$ $[C_4C_1Im][BF_4]$ $[C_8C_1Im][BF_4]$ $[C_4C_1Im][NTf_2]$ $[C_8C_1Im][NTf_2]$	(S)-2-pentanol			[50]

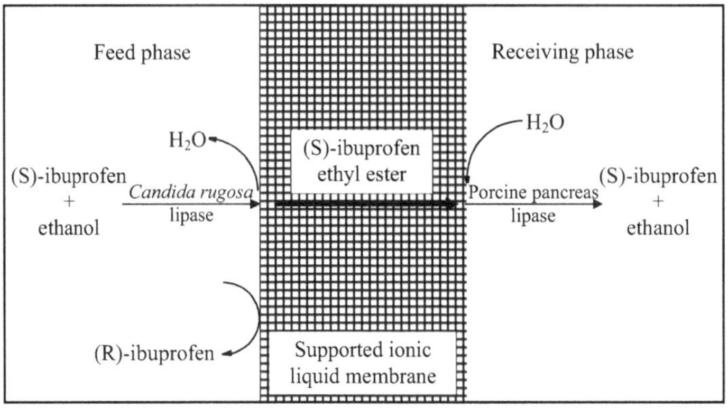

Figure 3 ***Illustrative example of a supported ionic liquid membrane used for the separation of racemic ibuprofen.***

the enzyme allows the permeation selectivity of the SILM system to be controlled [51]. The nature of the supporting membrane material strongly affects the transport phenomena [43]. The methodology of SILMs was successfully applied in other systems as well. Hernandez-Fernandez et al. [48,50] used it for the selective separation of rac-2-pentanol [50] and rac-1-phenylethanol [48] by transesterification with vinyl ester. Enantioselective esterification took place in the feed phase, while the nonreacted isomer of the alcohol diffuses through the membrane to the receiving phase [49]. Enzyme can also be immobilized on the membrane. Mori et al. [47] compared the performance of a regular membrane obtained by simple enzyme adsorption on a hydrophilic surface with a membrane composed of a hydrophobic IL phase with adsorbed lipase in a tubular membrane reactor. Even though the enzyme exhibited higher activity in the first case, it was more stable with time and maintained performance during consecutive uses over several months in the latter case. The membrane was used for the synthesis of butyl laurate (Table 4) in a hexane/water mixture.

3.1.3 Integrated Enzymatic Esterifications in Microreactor Devices with ILs

Microfluidic devices in the biotechnological applications have gained a lot of attention lately. Among the studied systems, integrated processes with reaction and separation in one step are of great importance. The application of ILs in these devices also is significant. Microfluidic devices have advantages in comparison to traditional macroreactors because of their large surface area—to—volume ratio in addition to effective heat and mass transfer, easier control of process parameters, and new production concepts [52]. Multiphase reactions present an exciting field of application of microstructured reactors for biocatalytic transformations. They enable process intensification and increase the reactor's overall performance [53]. Advantages of microreactors were successfully applied in esterification systems in combination with ILs (Table 5).

Table 5 Integrated Systems Consisting of Esterification and Microreactors

Reaction Scheme	Ionic Liquid	Separation Method	Removed	X/%	ee/%	References
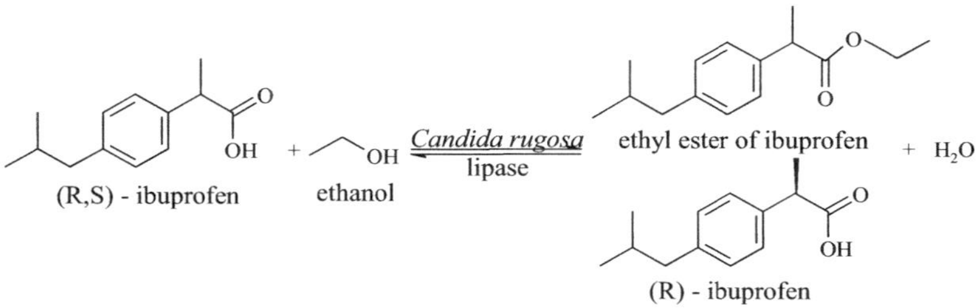 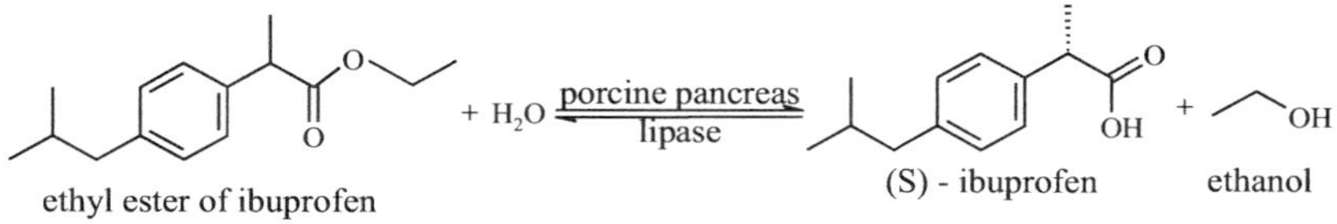	$[C_4C_1Im][PF_6]$	Ionic liquid flow (30–60 s)	(S)-ibuprofen	NA	77.67 In receiving phase	[54]
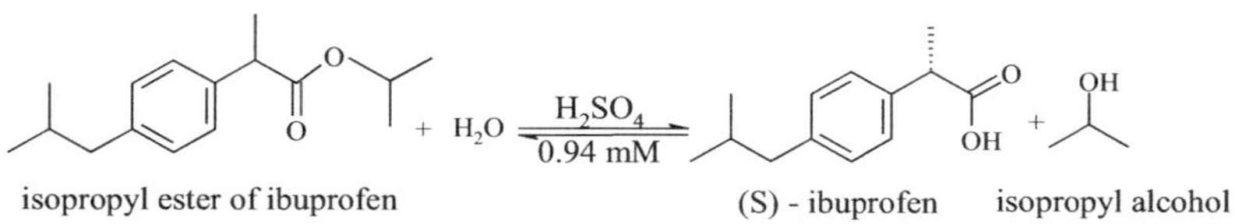	$[C_4C_1Im][PF_6]$			<50%	99.9 In receiving phase	[55]
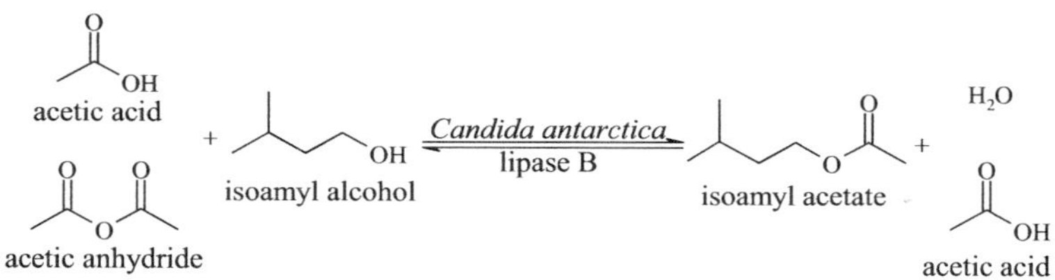	$[C_4C_1Pyrr]$ [DCA]	Extraction 77% Efficiency	Isoamyl acetate	NA	–	[56]
		Extraction in *n*-heptane	Isoamyl acetate	100	–	[57]
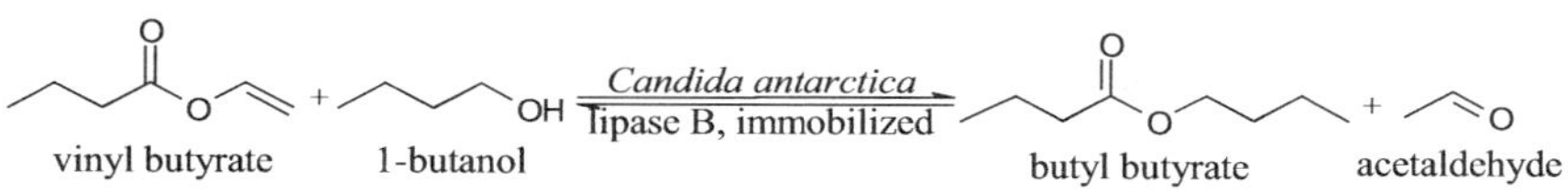	$[C_4C_1Py][NTf_2]$	Difference in boiling points, 91% efficiency	Butyl butyrate	100 (5 min)	–	[58]

NA—not available; X-conversion; ee-enantiomeric excess.

Huh et al. [54] used a microfluidic device to demonstrate their ability to replicate on a micro level a SILM with a somewhat changed configuration. They studied the separation of (S)-ibuprofen from a racemic mixture with the assistance of lipase and IL. In this case, the IL used was not supported in a traditional sense. A microreactor chip with three parallel inlets (shown in Figure 4(A)) was used. The IL enabled the transport of ester synthesized in the feed phase to the receiving phase, subsequent hydrolysis of ester in the receiving phase, and thus separation of (S)- and (R)-ibuprofen. An attractive feature of an IL membrane (or IL flow—ILF) is the ability to control its thickness by regulating its flow rate. Decreasing the ILF rate can decrease transportation resistance and improve transport. Another specialty of this system is fast separation (30–60 s) [54]. Yoon et al. [55] studied the separation of an ibuprofen racemic mixture as well. They used a similar microreactor chip (Figure 4(B)) and, unlike Huh et al., they combined enzymatic and chemical catalysis. The expensive enzymatic hydrolysis system mentioned earlier was replaced by the cheap aqueous acidic layer. As in the previous system, the conversion depended on the thickness of the IL layer. High enantiomeric excess in the receiving phase for both systems is presented in Table 5. Low conversion in the latter system might be improved by the change of IL.

Pohar et al. [56] studied lipase-catalyzed (free *Candida antarctica* lipase B) synthesis of isoamyl acetate synthesis in a $[C_4C_1Pyrr][DCA]/n$-heptane two-phase system in a microreactor. The choice of solvents in this work enabled good extraction of ester produced in the reaction (Table 5), while it dissolved in the organic phase. The reactants and the enzyme remained in the IL. A maximum isoamyl acetate yield of 150% was achieved after 33.8 min; this occurred because acetic anhydride was used as a donor and the acetic acid formed as a by-product during transesterification acts as an acyl donor in esterification reactions with isoamyl alcohol. Microreactor systems are known to potentially intensify processes, which can also be observed in this system by the tripled

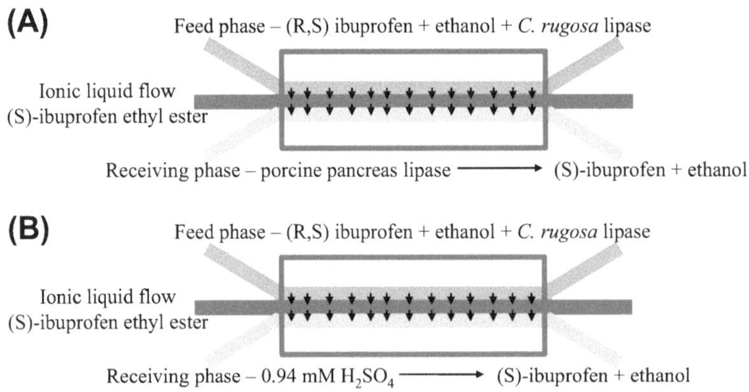

Figure 4 ***Illustrative example of ionic liquid flow in a microreactor as an alternative to a supported ionic liquid membrane in a macroreactor. A. porcine pancreas lipase is used as a catalyst in the receiving phase, B. 0.94mM sulphuric acid is used as a catalyst in the receiving phase.***

volume productivity in comparison to the batch reactor. This is enabled by the efficient mass transport and diffusion in the microchannel and large interface area. Enzymatic microreactors with IL/organic solvent two-phase systems are prospective tools for industrial esterifications, and their coupling with a microextraction unit can be implemented to enhance the separation of produced ester [56]. Žnidaršič-Plazl et al. [57] studied this system in microreactors with different geometries. In their most recent work, Pohar et al. [58] developed an integrated continuous microreactor consisting of a packed-bed microreactor and a miniaturized separator for continuous esterification in IL. The specifications of the system are presented in Table 5. Consecutive separation in the microseparator enabled product recovery with 91% efficiency. The separation was based on the difference in the boiling points between the products and the IL. This system successfully proves yet again process intensification using microdevices [58].

3.1.4 Special Cases of Esterification in Ionic Liquid in Integrated Systems

Among the integrated systems described in the literature, there are unconventional systems that do not include the conventional approach used in previous reports. These systems and their specifications are listed in Table 6. One such system combines the use of IL and $scCO_2$ [59] in the lipase-catalyzed kinetic resolution of chiral 1-phenylethanol. Reetz et al. [59] combined biocatalytic esterification with highly efficient separation in a continuous mode without the use of an organic solvent. The principle of the integrated system is as follows: substrates (racemic alcohol and an acylating agent) are transported into the reactor using $scCO_2$ as the mobile phase. One of the enantiomers is selectively esterified by lipase in the IL, and the product mixture is continuously extracted with the $scCO_2$ stream. Ester and unreacted alcohol consequently are separated by controlled density reduction by variation of temperature and/or pressure of CO_2. The choice of acylating agent determines the creation of a product with a certain desired solubility in CO_2, and efficient extractive separation can be achieved. Vinyl laurate was found to be an excellent acylating agent for this purpose [58]. Other excellent features of this system are high enantioselectivity and the long-term stability of lipase.

Ha et al. [60] developed an integrated system consisting of a continuous, double-layer, stirred tank reactor for the esterification of butyl alcohol and vinyl laurate in IL (Table 6). They successfully used the immiscibility of the product in IL for its separation, which consequently shifted the equilibrium. The systems' continuous mode of operation, which ran in steady state for 5 days, is its specialty.

An interesting new concept of switchable IL/solid phase $[C_{16}C_1C_1C_1N][NTf_2]$ was applied by Lozano et al. [44] in a reaction of alkyl carboxylic acid and flavor alcohol (Table 6). Considering that the melting point of this IL is 64 °C, this property was used for the reaction and consequent separation by centrifugation. After being cooled to room temperature, the reaction solution was in a monophasic solid phase. The solid solution could be separated into two phases by centrifugation: the upper phase contained no IL,

Table 6 Special Cases of Integrated Esterification Processes with Ionic Liquids (ILs)

Reaction	IL	Separation Method	Removed	X/%	References
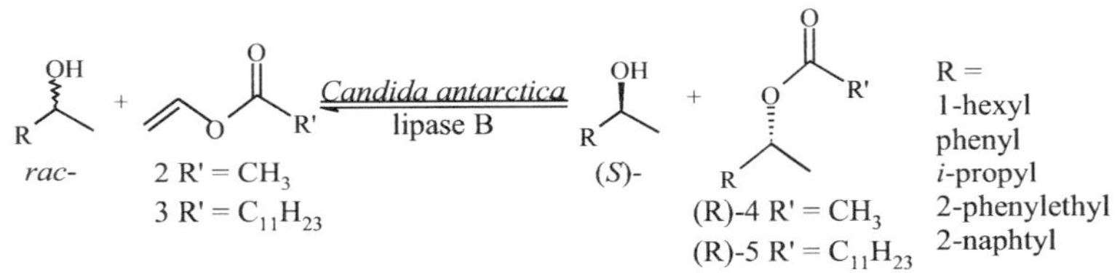	$[C_4C_1Im][NTf_2]$ $[C_4C_1Im][PF_6]$ $[C_4C_1Im][PF_4]$	$scCO_2$ extraction	Ester and unreacted alcohol	>90	[59]
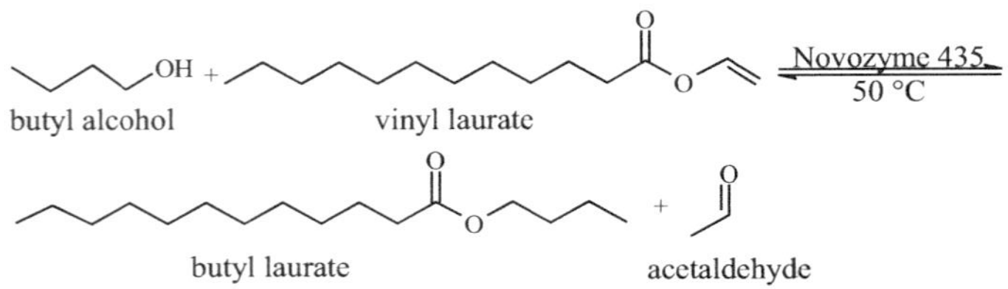	$[C_8C_1Im][TfO]$	Phase separation	Ester immiscible in IL	NA	[60]
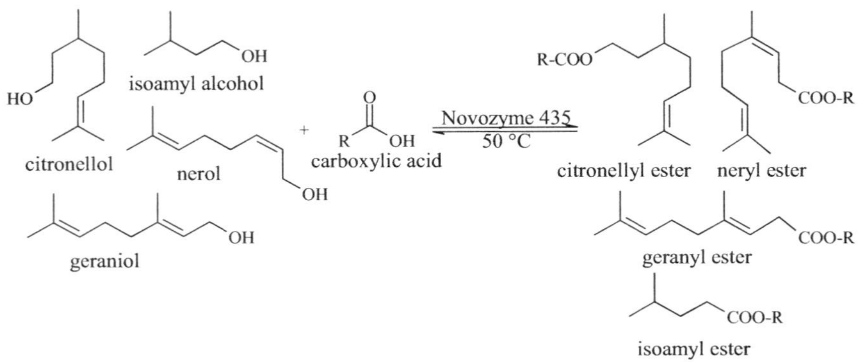 $R = -CH_3, -CH_2-CH_3, -(CH_2)_2-CH_3, -(CH_2)_3-CH_3$	$[C_{16}C_1C_1C_1N][NTf_2]$	Phase separation and centrifugation for 16 flavor esters produced	Liquid ester easily separated from solid IL	>90 Near 100% in most cases	[44]

NA—not available; X-conversion.

and the bottom solid contained the IL. The resulting liquid/solid biphasic system remained stable at room temperature over time. The proposed system is a "green" method for the efficient separation of hydrophobic compounds (flavor esters) from a homogeneous IL/organic compound mixture using a simple procedure.

3.2 Nonenzymatic Integrated Systems

Nonenzymatic esterifications in IL with integrated separation processes are rare in the literature. Examples we found are listed in Table 7. While in some cases the function of IL is limited to its solvent properties, in other cases it serves both as catalyst and cosolvent. The first reported system, investigated by Izák et al. [61], was published 3 years after the first integrated system of enzymatic esterification and PV was described [40]. Izák et al. coupled the esterification of acetic acid and (−)-borneol in IL with PV to remove the water produced and to shift the reaction equilibrium toward ester production. The IL does not permeate the hydrophilic membrane used in PV, and there is no loss to the environment [61]. By using PV reaction esterification, conversion can be doubled, similar to enzymatic systems (Table 7).

Uragami et al. [65] used PV and a membrane separation technique called evopomeation (EV) to separate water from a reaction mixture to shift the esterification equilibrium. This method combines the advantages of PV and circumvents its disadvantages. In this method, the feed solution is vaporized and subsequently fed to the membrane. Such an approach protects the membrane against swelling or shrinking, and using a water-permselective membrane, as in this case, allows water to be selectively removed from the system. Uragami et al. studied the esterification of *n*-butanol and acetic acid in $[C_4C_{3A}Im][NTf_2]$, which served as both catalyst and separating reagent in the reaction. Sulfuric acid also was used as a catalyst in this system (Table 7). The IL dissolves the reactants and the produced ester, and the water forms a separate phase, which simplifies its separation. These authors compared different integrated systems and found that PV and microwave heating improves the conversion. However, the best results were achieved when EV was used for water separation in combination with microwave heating (Table 7). Besides conversion, the reduction of reaction time, from 2 h to 10 min, also was improved.

In previously described cases, IL was used as a cosolvent and/or catalyst of esterification. In the following examples, ILs were used solely as esterification catalysts (Table 7). Tao et al. [62] performed batch reaction distillation with the addition of IL as the catalyst. Reactive distillation is a multifunctional reactor concept that combines distillative separation and chemical reaction in one apparatus. Hence, both the chemical reaction and distillation occur at the same time and place. The catalyst can be homogeneously or heterogeneously integrated into a column within or on internal packing. It was first invented in 1921 [66]; however, industrial application occurred in the 1980s [67]. Since that time, reactive distillation has been rapidly incorporated into more than 150 commercial processes in the petrochemical and chemical industries. These processes greatly

Table 7 Chemical Esterification Processes with In Situ Product Removal

Reaction	IL	Separation Method	Removed	X_{eq}/%	X_{IL}/%	Membrane Type	Membrane Properties	References
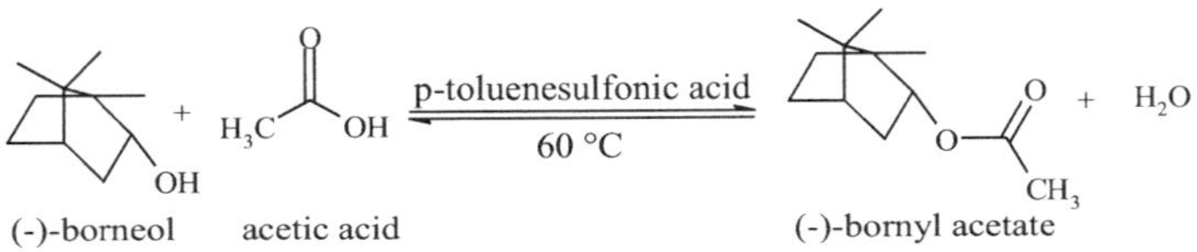	$[C_4C_1Im][PF_4]$	Pervaporation	Water	22	44	PVA membrane, PERVAP 2205, Sulzer, Germany	Suitable for contact with organic acids without any limitation	[61]
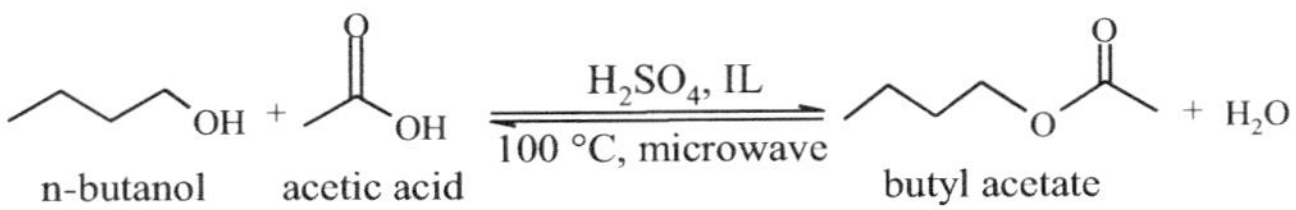	$[C_4C_{3A}Im]$ $[NTf_2]$ Solvent (20 wt%)	Evopomeation	Water	15.3	76.5	PVA PVA-TEOS self-made	**1.** Water permselective **2.** Hybrid water permselective membrane	[65]
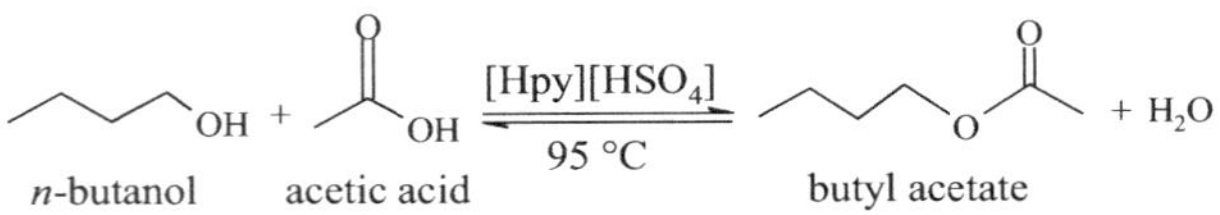	$[Py][HSO_4]$ Catalyst, 25 wt%	Reactive distillation	Water	48	91	–	–	[62]
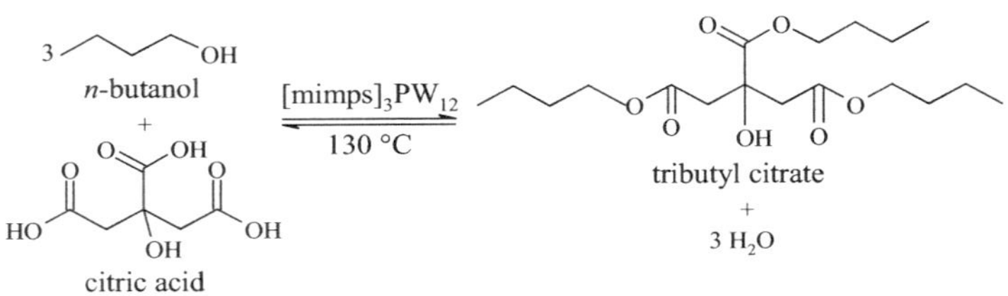	$[MIMPS]_3$ $[PW_{12}O_{40}]$ Catalyst	Reaction induced self-separation catalyst	Water	62.5 Without IL	95.4 With IL	–	–	[63]
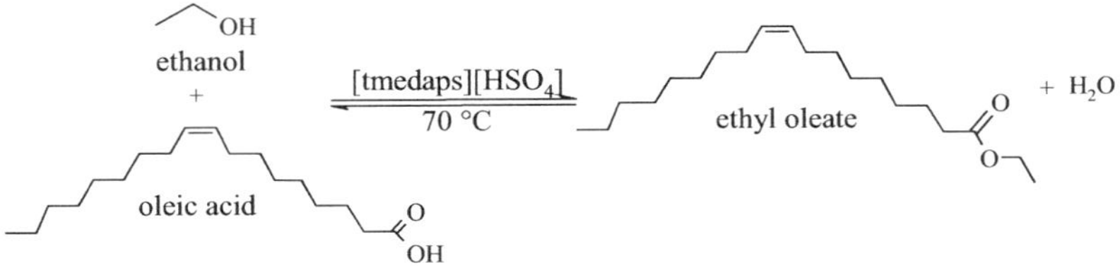	[TMEDAPS] $[HSO_4]$ catalyst	Liquid/liquid biphase separation at room temperature	Ethyl oleate is not soluble and forms a separate phase	Without catalyst: traces of product, With sulfuric acid as catalyst 61%	96	–	–	[64]

IL—ionic liquid; PVA—poly(vinyl alcohol); TEOS—tetraethoxysilane; X-conversion.

reduce capital and energy costs, lower environmental effects, provide safe and reliable operation [68], and often have better conversions and selectivities [69]. The use of reactive distillation in combination with enzyme esterification processes is limited because of the increased temperature, which can hurt the enzyme activity. Still, there are several examples of this [70-72], but they do not include the application of ILs and are therefore not discussed in this chapter. Park and Kazlauskas [73] showed that PV is more effective than distillation at recovering solutes from IL because of its milder process conditions.

Tao et al. [62] were able to replace the corrosive and toxic esterification catalyst sulfuric acid (which is also difficult to separate and recycle in traditional processes) with ILs. They synthesized the so called Brönsted acidic ILs composed of HSO_4 anions and different cations. Among the tested IL catalysts, [Py][HSO_4] was the most successful in achieving high conversion (equimolar concentrations) after 2 h (Table 7). The catalytic performance of [Py][HSO_4] is very similar to a sulfuric acid-catalyzed process without water removal. Therefore, this shows how a combination of reactive distillation with a noncorrosive catalyst can improve the reaction equilibrium and lead to the completion of esterification.

Another example of an integrated esterification process was described by Leng et al. [63], whose process is based on special IL catalysts (reaction-induced self-separation catalysts); a novel material made by combining propane sulfonate functionalized organic cations and heteropolyanions (or polyoxometalate anions) called heteropolyacids. They were investigated as a solid acid catalyst, and their melting point, which is above 100 °C, differs from that of the usual ILs. These catalysts are well soluble in reaction media (polycarboxylic acid or polyol) but are immiscible with ester product and therefore precipitate at the end of a reaction and ease their separation. They combine the advantages of homogeneous and heterogeneous catalysis. So, this is not a case of a usual integrated system in which reaction and separation are combined, but rather a case of using IL properties to separate product. Researchers have used various esterification reactions to test the synthesized catalysts and the separation possibilities. Most of these reactions achieved high yields and selectivities, as well as phase separation at the end of the process. The most thoroughly investigated reaction was the esterification of citric acid and *n*-butanol with $[MIMPS]_3[PW_{12}O_{40}]$ as the catalyst (reaction conditions and outcomes are presented in Table 7). The reaction was performed under reflux and vigorous stirring and achieved a 95.4% yield of tributyl citrate and a selectivity of 98%. The catalyst could be reused several times with moderate activity loss.

Fang et al. [64] studied the chemical esterification catalyzed by a di-cationic IL, [TMEDAPS][HSO_4], which also is used as a solvent and dissolves both reactants and water upon heating. The details of the reaction can be found in Table 7. Ester solubility in IL is very low, and it forms a separate layer (upper phase) in the reactor, which leads to its easy separation upon generation and cooling. In this manner, the reaction equilibrium is immediately shifted to the product side. Researchers have shown that di-cationic ILs are promising novel catalysts for the synthesis of biodiesel, with satisfactory yields. Hence, from the last two examples one can conclude that IL offers numerous advantages and potential in esterification reactions with *in situ* product removal.

4. CONCLUSION

The use of ILs in esterification and integrated systems offers numerous advantages and possibilities, which have been illustrated by the literature discussed here. They enable reactions with improved enantioselectivity and better enzyme stability and allow easier product separation, which is usually based on differences in the solubilities of reactant(s) and product(s) in the chosen IL. By combining cations and anions in IL, one can design an IL with desired properties. These properties can be successfully applied in integrated systems such as esterification with SILMs or ILF in microreactors. ILs offer new possibilities for the selective removal of different compounds from reaction mixtures, which is an industrially important area, especially if racemate resolution is concerned. We showed that esterifications in IL can successfully be combined with different separation processes such as PV, EV, reactive distillation, and extraction and form excellent integrated systems for the production of important compounds. Hence, we can expect that their use will be significantly broadened in the future.

ABBREVIATIONS

a_w water activity
CRL *Candida rugosa* lipase
EV evopomeation
IL ionic liquid
ILF ionic liquid flow
PV pervaporation
PVA poly(vinyl alcohol)
PVA-TEOS poly(vinyl alcohol) tetraethoxysilane
SILM supported ionic liquid membrane
$scCO_2$ supercritical carbon dioxide

ACKNOWLEDGMENT

This work was supported by the project TÁMOP-4.2.2.A. – 1/11/KONV-2012-0071.

REFERENCES

[1] M. Dixon, E.C. Webb, Enzymes, Academic Press, New York, 1964.
[2] J.E. Bailey, D.F. Ollis, Biochemical Engineering Fundamentals, McGraw Hill Co, Singapore, 1986.
[3] J.M.S. Cabral, D. Best, L. Boross, J. Tramper (Eds.), Applied Biocatalysis, Harwood Academic Publishers, Chur, Switzerland, 1994.

[4] G.J. Lye, J.M. Woodley, Application of *in situ* product removal techniques to biocatalytic processes, Trends Biotechnol. 17 (1999) 395–402.
[5] K. Bélafi-Bakó, Effect of operation conditions on the efficiency of membrane bioreactors, Desalination 200 (2006) 511–513.
[6] A. Guadix, F. Camacho, E.M. Guadix, Production of whey protein hydrolysate with reduced allergenicity in a stable membrane reactor, J. Food Eng. 72 (2006) 398–405.
[7] K. Nakamura, M. Aizava, O. Miyawaki, Electro-enzymology; Co-enzyme Regeneration, Springer-Verlag, Berlin, 1988.
[8] L. Giorno, Membrane bioreactors, in: Integration of Membrane Processes into Bioconversions, Kluwer Academic, London, 2000.
[9] W.M. Fogarty, Microbial Enzymes and Biotechnology, Applied Science Publishers, London; New York, 1983.
[10] E. Drioli, L. Giorno, Biocatalytic Membrane Reactors, Taylor and Francis Group, London, 1999.
[11] G. Belfort, Membranes and bioreactors: a technical challenge in biotechnology, Biotechnol. Bioeng. 33 (1989) 1047–1066.
[12] M. Mulder, Basic Principles of Membrane Technology, Kluwer Academic Publishers, Dordrecht, 1996.
[13] G.M. Rios, M.P. Beleville, D. Paolucci, J. Sanchez, Progress in enzymatic membrane reactors – a review, J. Membr. Sci. 242 (2004) 189–196.
[14] D.M.F. Prazeres, J.M.S. Cabral, Enzymatic membrane bioreactors and their applications, Enzyme Microb. Technol. 16 (1994) 738–750.
[15] S.C. Cheison, Z. Wang, S.Y. Xu, Hydrolysis of whey protein isolate in a tangential flow filter membrane reactor I.: Characterisation of permeate flux and product recovery by multivariate data analysis, J. Membr. Sci. 283 (2006) 45–56.
[16] S.J. Krishna, Developments and trends in enzyme catalysis in nonconventional media, Biotechnol. Adv. 20 (2002) 239–267.
[17] N. Krieger, T. Bhatnagar, J.C. Baratti, A.M. Baron, V.M. de Lima, D. Mitchell, Non-aqueous biocatalysis in heterogeneous solvent systems, Food Technol. Biotechnol. 42 (2004) 279–286.
[18] K. Drauz, H. Groger, O. May (Eds.), Enzyme Catalysis in Organic Synthesis, third ed., Wiley-VCH Verlag GmbH & Co. KGaA, Weinheim, 2012.
[19] N. Doukyu, H. Ogino, Organic solvent-tolerant enzymes, Biochem. Eng. J. 48 (2010) 270–282.
[20] A.S. Bommarius, B.R. Riebel, Biocatalysis, Wiley VCH Verlag GmbH &Co. KGaA, Weinheim, 2004.
[21] P.J. Halling, Salt hydrates for water activity control with biocatalysts in organic media, Biotechnol. Technol. 6 (1992) 271–276.
[22] T. Matsuda, Recent progress in biocatalysis using supercritical carbon dioxide, J. Biosci. Bioeng. 115 (2013) 233–241.
[23] M. Moniruzzaman, K. Nakashima, N. Kamiya, M. Goto, Recent advances of enzymatic reactions in ionic liquids, Biochem. Eng. J. 48 (2010) 295–314.
[24] M. Erbeldinger, A.J. Mesiano, A.J. Russel, Enzymatic catalysis of formation of Z-aspartame in ionic liquid: an alternative to enzymatic catalysis in organic solvents, Biotechnol. Prog. 16 (2000) 1129–1131.
[25] E.G. Kellogg, D.J. Abraham, Hydrophobicity: is logP(o/w) more than the sum of its parts? Eur. J. Med. Chem. 35 (2000) 651–661.
[26] O. Ulbert, T. Fráter, K. Bélafi-Bakó, L. Gubicza, Enhanced enantioselectivity of *Candida rugosa* lipase in ionic liquid as compared to organic solvents, J. Mol. Catal. B Enzym. 31 (2004) 39–45.
[27] Z. Yang, W. Pan, Ionic liquids: green solvents for nonaqueous biocatalysis, Enzym. Microb. Technol. 37 (2005) 19–28.
[28] R.P. Swatloski, S.K. Spear, J.D. Holbrey, R.D. Rogers, Dissolution of cellose with ionic liquids, J. Am. Chem. Soc. 124 (2002) 4974–4975.
[29] K. Shill, K. Miller, S. Douglas, D.S. Clark, W. Harvey, H.W. Blanch, A model for optimizing the enzymatic hydrolysis of ionic liquid-pretreated lignocellulose, Bioresour. Technol. 126 (2012) 290–297.

[30] J. van Spronsen, A. Miguel, M.A. Tavares Cardoso, G.-J. Geert-Jan Witkamp, W. de Jong, M.C. Kroon, Separation and recovery of the constituents from lignocellulosic biomass by using ionic liquids and acetic acid as co-solvents for mild hydrolysis, Chem. Eng. Proc. 50 (2011) 196–199.
[31] T.-Q. Yuan, W. Wang, F. Xu, R.-C. Sun, Synergistic benefits of ionic liquid and alkaline pretreatments of poplar wood. Part 1: effect of integrated pretreatment on enzymatic hydrolysis, Bioresour. Technol. 144 (2013) 429–434.
[32] V. Gotor, I. Alfonso, E. Garc (Eds.), Asymmetric Organic Synthesis with Enzymes, WILEY-VCH Verlag GmbH & Co. KGaA, Weinheim, 2008.
[33] U.T. Bornscheuer, Methods to increase enantioselectivity of lipases and esterases, Curr. Opin. Biotechnol. 13 (2002) 543–547.
[34] E.N. Vulfson, P.J. Halling, H.L. Holland (Eds.), Enzymes in Nonaqueous Solvents: Methods and Protocols, Humana Press Inc, 2001.
[35] J. Durand, E. Teuma, M. Gómez, Ionic liquids as a medium for enantioselective catalysis, C. R. Chim. 10 (2007) 152–177.
[36] Y. Yuan, S. Bai, Y. Sun, Comparison of lipase-catalyzed enantioselective esterification of (±)-menthol in ionic liquids and organic solvents, Food Chem. 97 (2006) 324–333.
[37] A.L. Pavia, F.X. Malcata, Integration and separation with lipases: an overview, J. Mol. Cat. B Enzym. 3 (1–4) (1997) 99–109.
[38] Z. Findrik, G. Németh, Đ. Vasić-Rački, K. Bélafi-Bakó, Z. Csanádi, L. Gubicza, Pervaporation-aided enzymatic esterifications in non-conventional media, Proc. Biochem. 47 (12) (2012) 1715–1722.
[39] T. Schäfer, C.M. Rodrigues, C.A.M. Afonso, J.G. Crespo, Selective recovery of solutes from ionic liquids pervaporation – a novel approach for purification and green processing, Chem. Commun. 17 (2001) 1622–1623.
[40] K. Bélafi-Bakó, N. Dörmő, O. Ulbert, L. Gubicza, Application of pervaporation for removal of water produced during enzymatic esterification in ionic liquids, Desalination 149 (2002) 267–268.
[41] E. Fehér, B. Major, K. Bélafi-Bakó, L. Gubicza, Semi-continuous enzymatic production and membrane assisted separation of isoamyl acetate in alcohol-ionic liquid biphasic system, Desalination 241 (2009) 8–13.
[42] L. Gubicza, K. Bélafi-Bakó, E. Fehér, T. Fráter, Waste-free process for continuous flow enzymatic esterification using a double pervaporation system, Green Chem. 10 (2008) 1284–1287.
[43] A. Riisager, R. Fehrmann, M. Haumann, P. Wasserscheid, Supported ionic liquids: versatile reaction and separation media, Top. Catal. 40 (2006) 91–102.
[44] P. Lozano, J.M. Bernal, A. Navarro, Clean enzymatic process for producing flavour esters by direct esterification in switchable ionic liquid/solid phases, Green Chem. 14 (2012) 3026–3033.
[45] L.C. Branco, J.G. Crespo, C.A.M. Afonso, Highly selective transport of organic compounds by using supported liquid membranes based on ionic liquids, Angew. Chem. Int. Ed. 41 (2002) 2771–2773.
[46] E. Miyako, T. Maruyama, N. Kamiya, M. Goto, A supported liquid membrane encapsulating a surfactant-lipase complex for the selective separation of organic acids, Chem. Eur. J. 11 (2005) 1163–1170.
[47] M. Mori, R. Gomez Garcia, M.P. Belleville, D. Paolucci-Jeanjean, J. Sanchez, P. Lozano, M. Vaultier, G. Rios, A new way to conduct enzymatic synthesis in an active membrane using ionic liquids as catalyst support, Catal. Today 104 (2005) 313–317.
[48] F.J. Hernandez-Fernandez, A.P. de los Rios, F. Tomas-Alonso, D. Gomez, M. Rubio, G. Villora, Integrated reaction/separation processes for the kinetic resolution of rac-1-phenylethanol using supported liquid membranes based on ionic liquids, Chem. Eng. Proc. 46 (2007) 818–824.
[49] F.J. Hernandez-Fernandez, A.P. de los Rios, L.J. Lozano-Blanco, C. Godinez, Biocatalytic ester synthesis in ionic liquid media, J. Chem. Technol. Biotechnol. 85 (2010) 1423–1435.
[50] F.J. Hernandez-Fernandez, A.P. de los Rios, F. Tomas-Alonso, D. Gomez, G. Villora, On the development of an integrated membrane process with ionic liquids for the kinetic resolution of rac-2-pentanol, J. Memb. Sci. 314 (2008) 238–246.
[51] E. Miyako, T. Maruyama, N. Kamiya, M. Goto, Enzyme-facilitated enantioselective transport of (S)-ibuprofen through a lipase-facilitated supported liquid membrane based on ionic liquid, Chem. Commun. 23 (2003) 2926–2927.

[52] W. Ehrfeld, V. Hessel, H. Löwe, Microreactors: New Technology for Modern Chemistry, Wiley VCH, Weinheim, 2000.
[53] J.M. Bolivar, J. Weisbauer, B. Nidetzky, Biotransformations in microstructured reactors: more than flowing with the stream? Trends Biotechnol. 29 (7) (2011) 333–342.
[54] Y.S. Huh, Y.-S. Jun, Y.K. Hong, W.H. Hong, D.H. Kim, Microfluidic separation of (S)-ibuprofen using enzymatic reaction, J. Mol. Cat. B Enzym. 43 (2006) 96–101.
[55] T.-H. Yoon, L.-Y. Hong, D.-P. Kim, Chiral separation by a pseudo membrane in a triple-laminar flow with a microfluidic contactor, Chem. Asian J. 6 (2011) 1015–1018.
[56] A. Pohar, I. Plazl, P. Žnidaršič-Plazl, Lipase-catalyzed synthesis of isoamylacetate in an ionic liquid/ n-heptane two-phase system at the microreactor scale, Lab Chip 9 (2009) 3385–3390.
[57] P. Žnidaršič-Plazl, A. Pohar, I. Plazl, Theoretical and experimental studies of enzyme-catalyzed isoamyl acetate synthesis with ionic liquid at the microreactor scale, Chem. Eng. Trans. 17 (2009) 1077–1082.
[58] A. Pohar, P. Žnidaršič-Plazl, I. Plazl, Integrated system of a microbioreactor and a miniaturized continuous separator for enzyme catalyzed reactions, Chem. Eng. J. 189–190 (2012) 376–382.
[59] M.T. Reetz, W. Wiesenhöfer, G. Francio, W. Leitner, Continuous flow enzymatic kinetic resolution and enantiomer separation using ionic liquid/supercritical carbon dioxide media, Adv. Synth. Catal. 345 (2003) 1221–1228.
[60] S.H. Ha, M.N. Lan, Y.-M. Koo, Continuous production and *in situ* separation of fatty acid ester in ionic liquids, Enzym. Microb. Technol. 47 (2010) 6–10.
[61] P. Izák, N.M.M. Mateus, C.A.M. Alfonso, J.G. Crespo, Enhanced esterification conversion in a room temperature ionic liquid by integrated water removal with pervaporation, Sep. Purif. Technol. 41 (2005) 141–145.
[62] D.-J. Tao, X.-M. Lu, J.-F. Lu, K. Huang, Z. Zhou, Y.-T. Wu, Noncorrosive ionic liquids composed of [H_2SO_4] as esterification catalysts, Chem. Eng. J. 171 (2011) 133–1339.
[63] Y. Leng, J. Wang, D. Zhu, X. Ren, H. Ge, L. Shen, Heteropolyanion-based ionic liquids: reaction induced self-separation catalyst for esterification, Angew. Chem. Int. Ed. 48 (2009) 168–171.
[64] D. Fang, J. Yang, C. Jiao, Dicationic ionic liquids as environmentally benign catalysts for biodiesel synthesis, ASC Catal. 1 (2011) 42–47.
[65] T. Uragami, J. Kishimoto, T. Miyata, Membrane reactor for acceleration of esterification using a special ionic liquid with reaction and separation and microwave heating, Catal. Today 193 (2012) 57–63.
[66] A.A. Backhaus, Continuous Process for the Manufacture of Esters. US Patent 1,400,849, 1921.
[67] V.H. Agreda, L.R. Partin, W.H. Heise, High-purity methyl acetate via reactive distillation, Chem. Eng. Prog. 86 (1990) 40–46.
[68] G.J. Harmsen, Reactive distillation: the front-runner of industrial process intensification. A full review of commercial application research, scale-up and operation, Chem. Eng. Proc. 46 (2007) 774–780.
[69] Z. Švandova, J. Markoš, Theoretical study on transesterification in a combined process consisting of a reactive distillation column and a pervaporation unit, Chem. Pap. 65 (2) (2011) 167–176.
[70] G.A. Hills, A.R. Macrae, R.R. Poulina, Ester Preparation. Eur. Patent 0 383 405 A1, February 14, 1990.
[71] G. Lin, H.-C. Liu, S.H. Liu, J. Chin. Chem. Soc. 42 (1995) 957–961.
[72] R. Heils, A. Sont, P. Bubenheim, A. Liese, I. Smirnova, Integration of enzymatic catalysts in a reactive distillation column with structured packings, Ind. Eng. Chem. Res. 51 (35) (2012) 11482–11489.
[73] S. Park, R.J. Kazlauskas, Biocatalysis in ionic liquids-advantages beyond green technology, Curr. Opin. Biotechnol. 14 (2003) 432–437.

CHAPTER 8

Separation of Gases Using Membranes Containing Ionic Liquids

Katalin Bélafi-Bakó, Nándor Nemestóthy, Péter Bakonyi
Research Institute on Bioengineering, Membrane Technology and Energetics, University of Pannonia, Veszprém, Hungary

1. INTRODUCTION

Ionic liquids (ILs) are multipurpose, fluid organic salts with large structural variability and are used in various fields such as chemical engineering and biotechnology. For example, ILs may contribute to the successful pretreatment of lignocellulosic biomass to achieve high sugar yields from such complex feedstock [1,2]. Thereafter, they can serve as reaction media for the bioconversion of fermentable sugars into valuable products [3–5]. Furthermore, ILs are potential candidates to recover these compounds of interest for enhanced product quality [6–8]. In summary, ILs are likely able to play a vital role in supporting the upstream ("product formation") as well as the downstream ("product separation") side of production, which is within the scope of this book. The purification of the target molecule(s) can be done either in the liquid or the gaseous phase and represents a noticeable issue for overall process efficiency. Because of the specificity of this chapter, it will address the separation of gases using membranes rely on ILs.

Approaches to enriching gases cover several well-established absorptive, adsorptive, and cryogenic applications that have routinely been used in the past. These options, however, possess the disadvantages of significant energy demand, complexity, and an intense need for chemicals, raising economic and even environmental concerns. Therefore, engineers and scientists are proposing new alternatives by means of membranes, for example [9–11].

Gas upgrading by membranes is a promising method and is definitely in the spotlight because of its relatively low energy consumption, gentle operating conditions, simplicity, and ecofriendly characteristics. In general, both the laboratory-scale and commercial applications of membrane gas separation are dominated by membranes made of artificial polymers. These materials have been widely used during the last 30–40 years, and therefore remarkable knowledge has been gained. One experience is that traditional polymeric membranes face performance limitations [12–14]. As a consequence, scientists are still pursuing ongoing research to develop novel polymers with improved characteristics. Although many advanced materials have been found, typically they are not commercialized. Because of the performance issues of conventional polymer

Ionic Liquids in Separation Technology
http://dx.doi.org/10.1016/B978-0-444-63257-9.00008-0
Copyright © 2014 Elsevier B.V.
All rights reserved.

membranes, a new direction has recently been established using unconventional chemicals: the ILs. Membranes prepared with ILs, such as supported IL membranes (SILMs), became a particular and promising field of interest since, based on research findings, they are potentially able to concentrate gases from multicompound mixtures [7,15–21]. The purification of certain gases is crucial from two equally important aspects: first, to avoid technological issues related to gas delivery, transportation, and utilization, and second, to satisfy certain environmental standards [16].

According to these requirements, the most common need is the removal of carbon dioxide (CO_2), nitrogen, aggressive/acidic gases (hydrogen sulfide (H_2S) and sulfur dioxide (SO_2)), and water vapor (dehumidification); major separation topics include capture after combustion (flue gas purification), natural sweetening of gas, and anaerobic fermentation gas (e.g., biohydrogen) conditioning [15,16,22–29].

2. GAS TRANSPORT IN SILMs

SILMs (Figure 1) are structurally described as simple combinations of porous, usually polymeric membrane supports, with ILs filling in the pores [30].

Thus, the product is a nonporous phase barrier that acts in a way similarly to the conventional dense polymer membranes from a gas transport point of view. Therefore, the solution–diffusion mechanism is applied to describe gas permeation across SLIMs [7,15,17,20]. The steady-state flux (J_i) can be expressed by considering Fick's law, which explicitly indicates that the process is a concentration-driven one:

$$J_i = D_i\left(c_{i,F} - c_{i,P}\right)/l \tag{1}$$

where D_i is the diffusion coefficient of the permeating gas compound (i), l is the membrane thickness, and $c_{i,F}$ and $c_{i,P}$ are the concentration of gas compound (i) on the feed and the permeate sides, respectively.

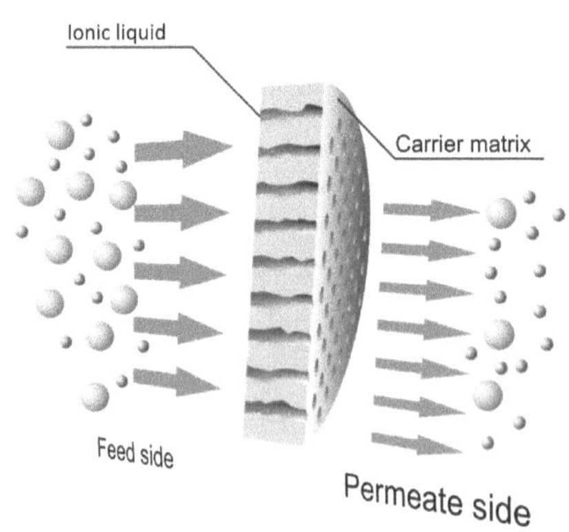

Figure 1 Structure of conventional supported ionic liquid membranes.

Nevertheless, taking into account Henry's law, one can conclude that the partial pressure difference between the feed and permeate phases is a key factor governing the mass transport rate [31,32]:

$$J_i = D_i S_i \left(p_{i,F} - p_{i,P} \right) / l \tag{2}$$

where S_i is the solubility coefficient of the permeating gas compound (i) and $p_{i,F}$ and $p_{i,P}$ are its partial pressures on the feed and the permeate sides, respectively.

The term $D_i S_i$ usually identifies the permeability of the given gas species (P_i), which is one of the most important data in evaluating membrane performance:

$$P_i = D_i S_i \tag{3}$$

In addition to permeability, which gives information about the transfer rate of gas across a membrane, the theoretical selectivity ($\alpha_{i/j}$) is another remarkable trait and is calculated as a ratio of the permeabilities of the gases (i,j):

$$\alpha_{i/j} = P_i / P_j = \left(D_i / D_j \right) \left(S_i / S_j \right) \tag{4}$$

where the D_i / D_j term is the diffusivity selectivity and S_i / S_j is the solubility selectivity.

From Eqn (4) it is seen that the theoretical selectivity is determined by the diffusivity and solubility differences of the gas components in the membrane or, to be more precise, in the IL occupying the pores of the support material. The gas transport through the solid membrane support itself is ideally negligible in comparison with the permeation taking place in the IL pores.

There are reliable methods to tentatively measure the sorption and diffusivity of certain gases in ILs [33–37]; however, computational, predictive methods and molecular modeling (e.g., COSMO-RS, UNI-FAC, Camper Molar Volume Model, Kilaru Viscosity Model) are also popular nowadays because of their time- and resource-saving characteristics [7,16,20,38].

Studies revealed that the specific properties of ILs (such as molar volume and viscosity) and the solute size (molar volume) are apparently crucial parameters for achieving good process efficiency. It also has been shown that the quality of separation for SILMs is almost solely affected by the solubility selectivity, whereas the gas transfer rates are likely influenced by the diffusivity of the migrating gases [19,20,39,40]. The fact that solubility plays a primary role in separation permits the ideal and real selectivities to be close values, representing a huge advantage of SILMs since the successful recovery of gases by conventional polymeric membranes can be strongly dependent on the composition of the gas mixture. Moreover, effective gas purification using SLIMs is usually carried out under gentle operating conditions in terms of pressure and temperature, increasing their attractiveness and economic viability.

It is important to mention that operating conditions maintained during the gas separation process are substantial elements in the feasibility of SILMs since they have a large effect on gas solubility, diffusivity, and the inherent properties of ILs, such as viscosity [15]. Although IL membranes are able to compete or even exceed the performance of conventional, polymeric gas separation membranes because of their favorable properties, further research is presumably needed to develop ILs with extraordinary permeability and selectivity so that they can take up post in the attractive region of Robeson's plot (above the current upper limit) [14].

More often than not, the characterization of gas separation membranes starts with single gas experiments, where the individual permeation rates/permeabilities and theoretical selectivities can be determined. These measurements provide basic and important data about the behavior of the membrane. Nevertheless, in real cases the task is the separation of gas mixtures, and hence the pure gas tests should be followed by subjecting the membrane to mixed gases. The results obtained under such circumstances allow one to extract more realistic information regarding the intended membrane, and subsequently its feasibility can be more precisely evaluated. The next section presents applications of the IL-based gas separation membranes.

3. APPLICATION OF ILS IN GAS SEPARATION

Removal of CO_2 is the most widely examined issue in SILM applications related to gas separation and therefore claims considerable background in the scientific literature [41]. Recent research and review articles on the topic have summarized that CO_2 has unique solubility properties in ILs, which can be enhanced by varying the constituents—the anions and cations—building up the molecule and/or changing their substituents. This is because the nature and strength/weakness of the interactions between an IL and a given solute is determined not only by the quality of the gas species but also by the chemical structure of the IL standing as the solvent. In addition, it has been observed that the structure of the support matrix can influence the permeation characteristic of the IL applied [42,43].

Upon revealing a better understanding about CO_2 sorption in ILs, new opportunities for improving the CO_2 separation performance of SILMs and so-called functionalized ILs have been developed and can be used. These ILs comprise adequate functional groups designed to selectively interact with certain gas molecules and hence facilitate their passage through the membrane. This approach provides the chance to fabricate membranes that are not strictly limited by solubility and diffusion so that the process can be enhanced through the higher affinity of the IL to the respective gaseous substance to be separated [7,19]. Some authors have reported on the modulation of SILM permeability by external magnetic field [44].

Among the ILs tested for CO_2 removal, the imidazolium (Figure 2(A)), ammonium (Figure 2(B)), and phosphonium (Figure 2(C)) types and their functionalized derivates containing fluoroalkyl, sulfonyl, nitrile, and alkyne groups as well as oligo- and

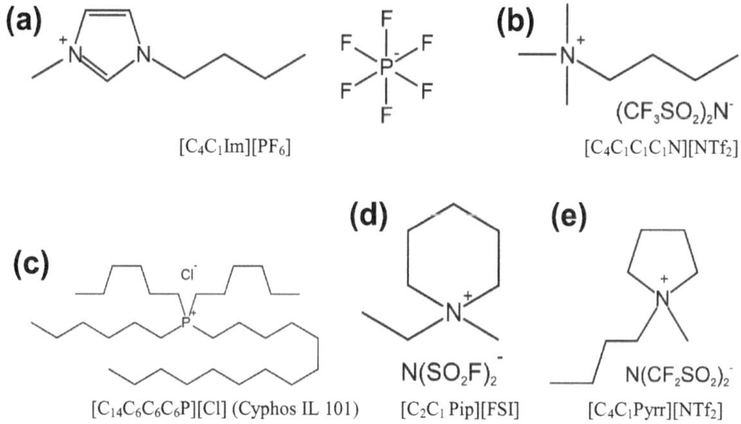

Figure 2 Several types of ionic liquids with various cations and anions. (a) imidazolium type IL; (b) ammonium type IL; (c) phosphonium type IL; (d) piperidinium type IL; (e) pyrrolidinium type IL.

polyethylene glycol, with their polar ether linkages, and others (e.g., those with ring-opened, piperidinium (Figure 2(D)), and pyrrolidinium (Figure 2(E)) cations) can be found [20,41,45–47].

Furthermore, it is also possible to incorporate ILs with amine groups, leading to highly CO_2-selective, tailor-made SILMs. This task-specific solution depicts a combination of classic amine-based CO_2 scrubbing (chemical sorption) and novel, prosperous supported IL based technology. Apart from the well-known solvent, amino acids might also be used as carrier molecules for such purposes [48]. Although amine-facilitated SILMs are appealing they can be characterized by some drawbacks, for example their increased viscosity after forming the complex with CO_2 presumably causes a reduced gas transport (diffusion) rate [16]. As a suggested alternative, ILs might be mixed with traditional amine solutions [89].

ILs can be functionalized not only by linking them with certain pendant polar groups for enhanced CO_2 separation performance but also through their polymerization. This is attractive from various points of views. Basically, the stability of SILMs is dependent on the adhesion between the IL and the support membrane matrix, which is determined by the capillary forces. However, these forces are often not strong enough to ensure as high mechanical endurance as required to attain the desired separation efficiency under increased transmembrane pressures. It is attributed to the fact that ILs could be squeezed from the pores in such circumstances and, subsequently, their long-term applicability is questionable.

This bottleneck can be overcome by synthesizing the polymerized form of ILs since solid ILs show improved stability even at elevated pressures [19,23,49,50]. The integration and gelation of ILs with certain artificial and natural polymers are also options, as reported previously [51–56]. In addition, common SILMs are relatively thick; this thickness hinders the gas flux, which may be recovered by means of poly-ILs. Although the ideal and mixed gas selectivities in IL membranes seem to be more or less identical, it is worthwhile to mention that polymerized IL membranes may be sensitive to the threat of plasticization, similar to conventional dense polymeric membranes [57]. Like normal SILMs, the membranes made of poly-ILs can also be functionalized by attaching a polar group(s) to the IL's cation part to achieve better gas separation performance [58,59]. In addition, varying the anion in the polymerized IL can change the gas separation behavior [60].

ILs have recently started to be used in attractive mixed matrix membranes, demonstrating a possible way ahead in IL research [61–63].

Beside the polymerizability of ILs, some other approaches also have been proposed to engineer more stable SILMs. As mentioned, capillary binding forces play an important role in SILM stability since it is their responsibility to suitably entrap the molten salts in the solid phases. Therefore, the pore diameter of the support membrane is a significant factor associated with the membrane integrity. From this aspect, it was demonstrated that replacing the commonly used microfiltration carriers by nanofiltration membranes made it possible to get stronger capillary holding forces resulting in better stabilized SILMs [64,65]. It also has been reported that the preparation method of the SILM itself can affect the operational stability of the SILMs constructed [43,66]. Finally, covering the SILM surface with a thin layer of polymer film has been shown to be an appropriate technique to retain the ILs in the pores of the carrier matrices [67].

Although membrane stability over time is an issue to be taken into account, only a few related publications can be referenced [20,56,68–70]. To our knowledge, the most extensive experiments on this topic were run for approximately 260 days with a CO_2/methane mixture [71].

Water content of the various gas streams can have a remarkable effect on the separation efficiency. First, facilitated transport membranes made of task-specific, amine-linked ILs may require humidity for sufficient performance [20]. Second, the reliable and efficient operation of SILM—in terms of membrane consistency, long-term usability, and, again, separation performance—could be influenced by the presence of humidity since it is able to modify the IL's characteristics and behavior. This fact gains particular importance during purification tasks when the gas to be handled carries water or is saturated with it (e.g., when it is produced in a fermentative bioprocess) [15,28,48,72–74]. In the aspect of fermentative gas enrichment, there is a special interest in biohydrogen as a future energy carrier, and SILMs have demonstrated a high potential to finish this work [15,72,75,76].

Thermal stability is also a beneficial property of ILs, leading to the opportunity for use in high-temperature applications (200–300 °C) aiming to get rid of CO_2 [77,78]. Nevertheless, in such cases the support material is of concern, and therefore the sustainable integrity of the SILM and its long-term performance needs proper, temperature resistant carriers.

Table 1 Studies Using Gas Mixtures to Test the Performance of Ionic Liquids for Carbon Dioxide (CO_2)/Methane (CH_4) and CO_2/Nitrogen (N_2) Separation

		Operational Circumstances			Selectivity		
Ionic Liquid	**Support**	**Gas Composition**	**p_F**	**Temperature (°C)**	**CO_2/N_2**	**CO_2/CH_4**	**Reference**
Block copolymer ion gels with $[C_4C_1Im][NTf_2]$	PVDF	50% CO_2:50% N_2; 50% CO_2:50% CH_4	2 atm	Room temperature	22–39	12–21	[86]
$[C_4C_1Im][NTf_2]$, $[C_3NH_2C_1Im][NTf_2]$, $[C_3NH_2C_1Im][CF_3SO_3]$	PTFE	CO_2, CH_4 (various partial pressure ratios)	1 atm	15–70	—	10–120	[71]
$[C_4C_1Im][PF_6]$, $[C_8C_1Im][PF_6]$, $[C_4C_1Im][BF_4]$, $[C_{10}C_1Im][BF_4]$, $[C_4C_1Im][NTf_2]$	PVDF	50% CO_2:50% N_2; 50% CO_2:50% CH_4	0.7 bar	30	20–32	98–200	[72]
polyRTIL membrane with styrene-based IL monomers	PES	50% CO_2:50% CH_4	10–40 bar	10–40	—	~8–29	[57]
$[C_2C_1Im][BF_4]$, $[C_2C_1Im][DCA]$, $[C_2C_1Im][CF_3SO_3]$, $[C_2C_1Im][NTf_2]$, $[C_6C_1Im][NTf_2]$, $[C_4C_1Im][BETI]$	PES, PVDF	Various CO_2/CH_4 binary mixtures	207–307 kPa	30	—	10–27	[20]
$[C_4C_1Im][PF_6]$, $[C_4C_1Im][NTf_2]$	PVDF	30–50% CO_2; 70–50%N_2	1.05–1.21 atm	40	10–22	—	[87]
$[C_{2v}C_4Im][NTf_2]$/ $[C_2C_1Im][BF_4]$, $[C_2C_1Im][NTf_2]$, $[C_2C_1Im][B(CN)_4]$/ZIF-8	polyRTIL	50% CO_2/50% N_2; 50% CO_2/50% CH_4	3.5 bar	35	20.9	11.6	[61]

IL: ionic liquids; p_F: feed pressure; PVDF: Polyvinylidene fluoride; PTFE: Polytetrafluoroethylene; PES: Polyethersulfone; polyRTIL: polymerized Room Temperature Ionic Liquid.

Table 2 Studies Using Gas Mixtures to Test the Performance of Ionic Liquids for Separation Tasks other than Carbon Dioxide (CO_2)/Methane and CO_2/Nitrogen (N_2)

Ionic Liquid	Support	Operational Circumstances			Selectivity			Reference
		Gas Composition	p_F	Temperature	H_2/CO	CO_2/H_2	CO/N_2	
$[C_4C_1Im][NTf_2]$, $[C_{10}C_1Im][NTf_2]$, $[C_8C_8C_8C_1N][NTf_2]$, $[C_8Py][NTf_2]$	Nanofiltration membrane	NS	3–7 bar	20 °C	4.3	–	–	[64]
$[H_2NC_3C_1Im][NTf_2]$, $[C_6C_1Im][NTf_2]$	Nylon 66	20% CO_2, 20% H_2, Ar to balance	108 kPa	37–300 °C	–	~9–15	–	[78]
$[C_4C_1Im][PF_6]$	–	55–50% CO_2: 45–50% H_2	1.25–5 Mpa	253–308 K	–	30–300	–	[88]
$[C_6C_1Im][Cl]/CuCl$	PVDF	50% CO:50% N_2	150–250 kPa	30–50 °C	–	–	2.3–3	[69]
$[C_2C_1Im][C_2SO_4]$	α-Al_2O_3	3.3% SO_2, 0–10% CO_2, air to balance	0.11–0.125 bar	15 °C	–	–	–	[84]

NS: not specified; p_F: feed pressure.

The selective removal of sulphur-containing, aggressive gases such as H_2S and SO_2 is an emerging field in IL-assisted separation. H_2S is a contaminant that usually accompanies natural gas, and it also is expectedly formed during anaerobic biological processes, for example, in biohydrogen fermentation. IL-based applications have recently started to claim a laboratory-scale niche for the sweetening of the gaseous streams instead of using the traditional amine-based solutions. Although studies report their potential for the objectives mentioned, more feedback and relevant continuous investigation are required to reveal their real applicability [15,16,79,80].

SO_2 is likely present in postcombustion (flue) gases and is considered to be a regional air pollutant with high environmental and health risks. Therefore, its emission must be prevented by efficient separation for multiple reasons. ILs have recently achieved footing on this specific area as well. Some interesting results have been published, assigning ground for future research [81−85].

Several particular examples in which certain ILs were subjected to real (mixed gas) separation experiments are listed in Tables 1 and 2, along with the operational conditions and the performances reached in terms of selectivity. One drawback is that the relevant literature generally deals with single-gas permeation studies; the performances of IL membranes in real cases still remain unclear.

4. CONCLUSIONS

ILs undoubtedly have claimed a segment in gas separation and proven themselves for various applications, as demonstrated thoroughly in this chapter; however, to date this has occurred only at the laboratory scale. One of the main advantages of ILs is the tunability by altering the structural anions and cations, providing the ultimate chance to engineer case-specific membranes. Impressive progress was made in recent years and, as a result, many research directions (e.g., IL polymerization) have been developed, aiming at more attractive membranes based on ILs. In addition, this research has revealed the major theoretical basics of gas transport through IL membranes and the factors influencing it. Nevertheless, continuing research is required to make IL membranes more reliable by utilizing their exceptional properties. Furthermore, evaluation studies are advised under real gas separation conditions to get a better picture of their (long-term) feasibility [15]. Moreover, addressing the scale-up issue also is encouraging for future research [19].

ACKNOWLEDGMENTS

Péter Bakonyi thanks for the support of the TÁMOP 4.2.4.A/2-11-1-2012-0001 'National Excellence Program' by the European Union and the State of Hungary, co-financed by the European Social Fund. Nándor Nemestóthy acknowledges the János Bolyai Research Scholarship of the Hungarian Academy of Sciences. The research infrastructure was supported by TÁMOP 4.2.2/A-11/1/KONV-2012-0071 project financed by the European Union and the European Social Fund.

REFERENCES

[1] M. Mora-Pale, L. Meli, T.V. Doherty, R.J. Linhardt, J.S. Dordick, Room temperature ionic liquids as emerging solvents for the pretreatment of lignocellulosic biomass, Biotechnol. Bioeng. 108 (2011) 1229–1245.

[2] H. Tadesse, R. Lugue, Advances on biomass pretreatment using ionic liquids: an overview, Energy Environ. Sci. 4 (2011) 3913–3929.

[3] S. Park, L.J. Kazlauskas, Biocatalysis in ionic liquids – advantages beyond green technology, Curr. Opin. Biotech. 14 (2003) 432–437.

[4] Z. Yang, W. Pan, Ionic liquids: green solvents for nonaqueous biocatalysis, Enzyme Microb. Technol. 37 (2005) 19–28.

[5] F.V. Rantwijk, R.A. Sheldon, Biocatalysis in ionic liquids, Chem. Rev. 107 (2007) 2757–2785.

[6] D. Han, K.H. Row, Recent applications of ionic liquids in separation technology, Molecules 15 (2010) 2405–2426.

[7] L.J. Lozano, C. Godinez, A.P. de los Rios, F.J. Hernandez-Fernandez, S. Sanchez-Segado, F.J. Alguacil, Recent advances in supported ionic liquid membrane technology, J. Membr. Sci. 376 (2011) 1–14.

[8] H. Zhao, S. Xia, P. Ma, Use of ionic liquids as 'green' solvents for extractions, J. Chem. Technol. Biotechnol. 80 (2005) 1089–1096.

[9] A. Brunetti, F. Scura, G. Barbieri, E. Drioli, Membrane technologies for CO_2 separation, J. Membr. Sci. 359 (2010) 115–125.

[10] C.A. Scholes, S.E. Kentish, G.W. Stevens, Carbon dioxide separation through polymeric membrane systems for flue gas applications, Recent Pat Chem. Eng. 1 (2008) 52–66.

[11] S. Sridhar, B. Smitha, T.M. Aminabhavi, Separation of carbon dioxide from natural gas mixtures through polymeric membranes – a review, Sep. Purif. Rev. 36 (2007) 113–174.

[12] B.D. Freeman, Basis of permeability/selectivity tradeoff relations in polymeric gas separation membranes, Macromolecules 32 (1999) 375–380.

[13] L.M. Robeson, Correlation of separation factor versus permeability for polymeric membranes, J. Membr. Sci. 62 (1991) 165–185.

[14] L.M. Robeson, The upper bound revisited, J. Membr. Sci. 320 (2008) 390–400.

[15] P. Bakonyi, N. Nemestóthy, K. Bélafi-Bakó, Biohydrogen purification by membranes: an overview on the operational conditions affecting the performance of non-porous, polymeric and ionic liquid based gas separation membranes, Int. J. Hydrogen Energy 38 (2013) 9673–9687.

[16] F. Karadas, M. Atilhan, S. Aparicio, Review on the use of ionic liquids (ILs) as alternative fluids for CO_2 capture and natural gas sweetening, Energy Fuels 24 (2010) 5817–5828.

[17] F.F. Krull, C. Fritzmann, T. Melin, Liquid membranes for gas/vapor separations, J. Membr. Sci. 325 (2008) 509–519.

[18] M.A. Malik, M.A. Hashim, F. Nabi, Ionic liquids in supported liquid membrane technology, Chem. Eng. J. 171 (2011) 242–254.

[19] R.D. Noble, D.L. Gin, Perspective on ionic liquids and ionic liquid membranes, J. Membr. Sci. 369 (2011) 1–4.

[20] P. Scovazzo, Determination of the upper limits, benchmarks, and critical properties for gas separations using stabilized room temperature ionic liquid membranes (SILMs) for the purpose of guiding future research, J. Membr. Sci. 343 (2009) 199–211.

[21] P. Scovazzo, D. Havard, M. McShea, S. Mixon, D. Morgan, Long-term, continuous mixed-gas dry fed CO_2/CH_4 and CO_2/N_2 separation performance and selectivities for room temperature ionic liquid membranes, J. Membr. Sci. 327 (2009) 41–48.

[22] R.E. Baltus, R.M. Counce, B.H. Culbertson, H. Luo, D.W. DePaoli, S. Dai, D.C. Duckworth, Examination of the potential of ionic liquids for gas separations, Sep. Sci. Technol. 40 (2005) 525–541.

[23] J.E. Bara, D.E. Camper, D.L. Gin, R.D. Noble, Room-temperature ionic liquids and composite materials: platform technologies for CO_2 capture, Acc. Chem. Res. 43 (2010) 152–159.

[24] E.D. Bates, R.D. Mayton, I. Ntai, J.H. Davis, CO_2 capture by task-specific ionic liquids, J. Am. Chem. Soc. 124 (2002) 926–927.

[25] J.F. Brennecke, B.E. Gurkan, Ionic liquids for CO_2 capture and emission reduction, J. Phys. Chem. Lett. 1 (2010) 3459–3464.
[26] J.D. Figueroa, T. Fout, S. Plasynski, H. McIlvried, R.D. Srivastava, Advances in CO_2 capture technology – the U.S. Department of Energy's Carbon Sequestration Program, Int. J. Green house Gas Control 2 (2008) 9–20.
[27] M. Hasib-ur-Rahman, M. Siaj, F. Larachi, Ionic liquids for CO_2 capture – Development and progress, Chem. Eng. Process. 49 (2010) 313–322.
[28] P. Scovazzo, Testing and evaluation of room temperature ionic liquid (RTIL) membranes for gas dehumidification, J. Membr. Sci. 355 (2010) 7–17.
[29] C. Wang, S.M. Mahurin, H. Luo, G.A. Baker, H. Li, S. Dai, Reversible and robust CO_2 capture by equimolar task-specific ionic liquid–superbase mixtures, Green Chem. 12 (2010) 870–874.
[30] X. Han, D.W. Armstrong, Ionic liquids in separations, Acc. Chem. Res. 40 (2007) 1079–1086.
[31] W.J. Koros, K.G. Fleming, Membrane-based gas separation, J. Membr. Sci. 83 (1993) 1–80.
[32] M.H.V. Mulder, Basic Principles of Membrane Technology, Kluwer Academic Publishers, Dordrecht, 1996.
[33] D. Camper, C. Becker, C. Koval, R. Noble, Diffusion and solubility measurements in room temperature ionic liquids, Ind. Eng. Chem. Res. 45 (2006) 445–450.
[34] R. Condemarin, P. Scovazzo, Gas permeabilities, solubilities, diffusivities, and diffusivity correlations for ammonium-based room temperature ionic liquids with comparison to imidazolium and phosphonium RTIL data, Chem. Eng. J. 147 (2009) 51–57.
[35] A. Finotello, J.E. Bara, S. Narayan, D. Camper, R.D. Noble, Ideal gas solubilities and solubility selectivities in a binary mixture of room-temperature ionic liquids, J. Phys. Chem. B 112 (2008) 2335–2339.
[36] Y. Hou, R.E. Baltus, Experimental measurement of hte solubility and diffusivity of CO_2 in room temperature ionic liquids using a transient thin-liquid-film method, Ind. Eng. Chem. Res. 46 (2007) 8166–8175.
[37] D. Morgan, L. Ferguson, P. Scovazzo, Diffusivities of gases in room-temperature ionic liquids: data and correlations obtained using a lag-time technique, Ind. Eng. Chem. Res. 44 (2005) 4815–4823.
[38] C. Cadena, J.L. Anthony, J.K. Shah, T.I. Morrow, J.F. Brennecke, E.J. Maginn, Why is CO_2 so soluble in imidazolium-based ionic liquids, J. Am. Chem. Soc. 126 (2004) 5300–5308.
[39] D. Camper, J. Bara, C. Koval, R.D. Noble, Bulk-fluid solubility and membrane feasibility of Rmim-based room-temperature ionic liquids, Ind. Eng. Chem. Res. 45 (2006) 6279–6283.
[40] A. Finotello, J.E. Bara, D. Camper, R.D. Noble, Room-temperature ionic liquids: temperature dependence of gas solubility selectivity, Ind. Eng. Chem. Res. 47 (2008) 3453–3459.
[41] J.E. Bara, T.K. Carlisle, C.J. Gabriel, D. Camper, A. Finotello, D.L. Gin, R.D. Noble, Guide to CO_2 separations in imidazolium-based room-temperature ionic liquids, Ind. Eng. Chem. Res. 48 (2009) 2739–2751.
[42] J.J. Close, K. Farmer, S.S. Moganty, R.E. Baltus, CO_2/N_2 separations using nanoporous alumina-supported ionic liquid membranes: effect of the support on separation performance, J. Membr. Sci. (2012) 390–391, 201–210.
[43] D.H. Kim, I.H. Baek, S.U. Hong, H.K. Lee, Study on immobilized liquid membrane using ionic liquid and PVDF hollow fiber as a support for CO_2/N_2 separation, J. Membr. Sci. 372 (2011) 346–354.
[44] E. Santos, J. Albo, C.I. Daniel, C.A.M. Portugal, J.G. Crespo, A. Irabien, Permeability modulation of supported magnetic ionic liquid membranes (SMILMs) by an external magnetic field, J. Membr. Sci. 430 (2013) 56–61.
[45] J.E. Bara, C.J. Gabriel, S. Lessmann, T.K. Carlisle, A. Finotello, D.L. Gin, R.D. Noble, Enhanced CO_2 separation selectivity in oligo(ethylene glycol)functionalized room-temperature ionic liquids, Ind. Eng. Chem. Res. 46 (2007) 5380–5386.
[46] S.M. Mahurin, J.S. Yeary, S.N. Baker, D. Jiang, S. Dai, G.A. Baker, Ring-opened heterocycles: promising ionic liquids for gas separation and capture, J. Membr. Sci. 401–402 (2012) 61–67.
[47] L.G. Sánchez, J.R. Espel, F. Onink, G.W. Meindersma, A.B. de Haan, Density, viscosity, and surface tension of synthesis grade imidazolium, pyridinium, and pyrrolidinium based room temperature ionic liquids, J. Chem. Eng. Data 54 (2009) 2803–2812.

[48] S. Kasahara, E. Kamio, T. Ishigami, H. Matsuyama, Effect of water in ionic liquids on CO_2 permeability in amino acid ionic liquid-based facilitated transport membranes, J. Membr. Sci. 415−416 (2012) 168−175.
[49] O. Green, S. Grubjesic, S. Lee, M.A. Firestone, The design of polymeric ionic liquids for the preparation of functional materials, Polym. Rev. 49 (2009) 339−360.
[50] J. Tang, H. Tang, W. Sun, M. Radosz, Y. Shen, Poly(ionic liquid)s as new materials for CO_2 absorption, J. Polym. Sci., Part A: Polym. Chem. 43 (2005) 5477−5489.
[51] P. Bernardo, J.C. Jansen, F. Bazzarelli, F. Tasselli, A. Fuoco, K. Friess, P. Izák, V. Jarmarova, M. Kacírková, G. Clarizia, Gas transport properties of Pebax®/room temperature ionic liquid gel membranes, Sep. Purif. Technol. 97 (2012) 73−82.
[52] R.M. Couto, T. Carvalho, L.A. Neves, R.M. Ruivo, P. Vidinha, A. Paiva, I.M. Coelhoso, S. Barreiros, P.C. Simões, Development of Ion-Jelly® membranes, Sep. Purif. Technol. 106 (2013) 22−31.
[53] K. Friess, J.C. Jansen, F. Bazzarelli, P. Izák, V. Jarmarová, M. Kacírková, J. Schauer, G. Clarizia, P. Bernardo, High ionic liquid content polymeric gel membranes: correlation of membrane structure with gas and vapour transport properties, J. Membr. Sci. 415−416 (2012) 801−809.
[54] J.C. Jansen, K. Friess, G. Clarizia, J. Schauer, P. Izák, High ionic liquid content polymeric gel membranes: preparation and performance, Macromolecules 44 (2011) 39−45.
[55] B.A. Voss, J.E. Bara, D.L. Gin, R.D. Noble, Physically gelled ionic liquids: solid membrane materials with liquidlike CO_2 gas transport, Chem. Mater. 21 (2009) 3027−3029.
[56] I.N. Yoon, S. Yoo, S.J. Park, J. Won, CO_2 separation membranes using ion gels by self-assembly of a triblock copolymer in ionic liquids, Chem. Eng. J. 172 (2011) 237−242.
[57] K. Simons, K. Nijmeijer, J.E. Bara, R.D. Noble, M. Wessling, How do polymerized room- temperature ionic liquid membranes plasticize during high pressure CO_2 permeation? J. Membr. Sci. 360 (2010) 202−209.
[58] J.E. Bara, C.J. Gabriel, E.S. Hatakeyama, T.K. Carlisle, S. Lessmann, R.D. Noble, D.L. Gin, Improving CO_2 selectivity in polymerized room-temperature ionic liquid gas separation membranes through incorporation of polar substituents, J. Membr. Sci. 321 (2008) 3−7.
[59] T.K. Carlisle, G.D. Nicodemus, D.L. Gin, R.D. Noble, CO_2/light gas separation performance of cross-linked poly(vinylimidazolium) gel membranes as a function of ionic liquid loading and cross-linker content, J. Membr. Sci. 397−398 (2012) 24−37.
[60] R.S. Bhavsar, S.C. Kumbharkar, U.K. Kharul, Polymeric ionic liquids (PILs): effect of anion variation on their CO_2 sorption, J. Membr. Sci. 389 (2012) 305−315.
[61] L. Hao, P. Li, T. Yang, T.S. Chung, Room temperature ionic liquid/ZIF-8 mixed-matrix membranes for natural gas sweetening and post-combustion CO_2 capture, J. Membr. Sci. 436 (2013) 221−231.
[62] Y.C. Hudiono, T.K. Carlisle, J.E. Bara, Y. Zhang, D.L. Gin, R.D. Noble, A three-component mixed-matrix membrane with enhanced CO_2 separation properties based on zeolites and ionic liquid materials, J. Membr. Sci. 350 (2010) 117−123.
[63] Y.C. Hudiono, T.K. Carlisle, A.L. LaFrate, D.L. Gin, R.D. Noble, Novel mixed matrix membranes based on polymerizable room-temperature ionic liquids and SAPO-34 particles to improve CO_2 separation, J. Membr. Sci. 370 (2011) 141−148.
[64] Q. Gan, D. Rooney, M. Xue, G. Thompson, Y. Zou, An experimental study of gas transport and separation properties of ionic liquids supported on nanofiltration membranes, J. Membr. Sci. 280 (2006) 948−956.
[65] S.U. Hong, D. Park, Y. Ko, I. Baek, Polymer-ionic liquid gels for enhanced gas transport, Chem. Commun. 46 (2009) 7227−7229.
[66] F.J. Hernández-Fernández, A.P. de los Ríos, F. Tomás-Alonso, J.M. Palacios, G. Víllora, Preparation of supported ionic liquid membranes: influence of the ionic liquid immobilization method on their operational stability, J. Membr. Sci. 341 (2009) 172−177.
[67] P. Izák, W. Ruth, Z. Fei, P.J. Dyson, U. Kragl, Selective removal of acetone and butan-1-ol from water with supported ionic liquidepolydimethylsiloxane membrane by pervaporation, Chem. Eng. J. 139 (2008) 318−321.

[68] S. Yoo, J. Won, S.W. Kang, Y.S. Kang, S. Nagase, CO_2 separation membranes using ionic liquids in a Nafion matrix, J. Membr. Sci. 363 (2010) 72–79.
[69] G. Zarca, I. Ortiz, A. Urtiaga, Copper(I)-containing supported ionic liquid membranes for carbon monoxide/nitrogen separation, J. Membr. Sci. 438 (2013) 38–45.
[70] W. Zhao, G. He, F. Nie, L. Zhang, H. Feng, H. Liu, Membrane liquid loss mechanism of supported ionic liquid membrane for gas separation, J. Membr. Sci. 411–412 (2012) 73–80.
[71] S. Hanioka, T. Maruyama, T. Sotani, M. Teramoto, H. Matsuyama, K. Nakashima, M. Hanaki, F. Kubota, G. Masahiro, CO_2 separation facilitated by task-specific ionic liquids using a supported liquid membrane, J. Membr. Sci. 314 (2008) 1–4.
[72] L.A. Neves, J.G. Crespo, I.M. Coelhoso, Gas permeation studies in supported ionic liquid membranes, J. Membr. Sci. 357 (2010) 160–170.
[73] P. Scovazzo, J. Kieft, D.A. Finan, C. Koval, D. DuBois, R.D. Noble, Gas separations using non-hexafluorophosphate $[PF_6]^-$ anion supported ionic liquid membranes, J. Membr. Sci. 238 (2004) 57–63.
[74] W. Zhao, G. He, L. Zhang, J. Ju, H. Dou, F. Nie, C. Li, H. Liu, Effect of water in ionic liquid on the separation performance of supported ionic liquid membrane for CO_2/N_2, J. Membr. Sci. 350 (2010) 279–285.
[75] P. Cserjési, N. Nemestóthy, K. Bélafi-Bakó, Gas separation properties of supported liquid membranes prepared with unconventional ionic liquids, J. Membr. Sci. 349 (2010) 6–11.
[76] L.A. Neves, N. Nemestóthy, V.D. Alves, P. Cserjési, K. Bélafi-Bakó, I.M. Coelhoso, Separation of biohydrogen by supported ionic liquid membranes, Desalination 240 (2009) 311–315.
[77] J. Ilconich, C. Myers, H. Pennline, D. Luebke, Experimental investigation of the permeability and selectivity of supported ionic liquid membranes for CO_2/He separation at temperatures up to 125 °C, J. Membr. Sci. 298 (2007) 41–47.
[78] C. Myers, H. Pennline, D. Luebke, J. Ilconich, J.K. Dixon, E.J. Maginn, J.F. Brennecke, High temperature separation of carbon dioxide/hydrogen mixtures using facilitated supported ionic liquid membranes, J. Membr. Sci. 322 (2008) 28–31.
[79] S.H. Lee, B.K. Kim, E.W. Lee, Y.I. Park, J.M. Lee, The removal of acid gases from crude natural gas by using novel supported liquid membranes, Desalination 200 (2006) 21–22.
[80] Y.I. Park, B.S. Kim, Y.H. Byun, S.H. Lee, E.W. Lee, J.M. Lee, Preparation of supported ionic liquid membranes (SILMs) for the removal of acidic gases from crude natural gas, Desalination 236 (2009) 342–348.
[81] J.L. Anderson, J.K. Dixon, E.J. Maginn, J.F. Brennecke, Measurement of SO_2 solubility in ionic liquids, J. Phys. Chem. B 110 (2006) 15059–15062.
[82] Y.Y. Jiang, Z. Zhou, Z. Jiao, L. Li, Y.T. Wu, Z.B. Zhang, SO_2 gas separation using supported ionic liquid membranes, J. Phys. Chem. B 111 (2007) 5058–5061.
[83] P. Luis, L.A. Neves, C.A.M. Afonso, I.M. Coelhoso, J.G. Crespo, A. Garea, A. Irabien, Facilitated transport of CO_2 and SO_2 through supported ionic liquid membranes (SILMs), Desalination 245 (2009) 485–493.
[84] P. Luis, A. Garea, A. Irabien, Zero solvent emission process for sulfur dioxide recovery using a membrane contactor and ionic liquids, J. Membr. Sci. 330 (2009) 80–89.
[85] A. Yokozeki, M.B. Shiflett, Separation of carbon dioxide and sulfur dioxide gases using room-temperature ionic liquid [hmim][Tf_2N], Energy Fuels 23 (2009) 4701–4708.
[86] Y. Gu, E.L. Cussler, T.P. Lodge, ABA-triblock copolymer ion gels for CO_2 separation applications, J. Membr. Sci. 423–424 (2012) 20–26.
[87] P. Jindaratsamee, A. Ito, S. Komuro, Y. Shimoyama, Separation of CO_2 from the CO_2/N_2 mixed gas through ionic liquid membranes at the high feed concentration, J. Membr. Sci. 423–424 (2012) 27–32.
[88] A. Yokozeki, M.B. Shiflett, Hydrogen purification using room-temperature ionic liquids, Appl. Energ 84 (2007) 351–361.
[89] D. Camper, J.E. Bara, D.L. Gin, R.D. Noble, Room temperature ionic liquid-amine solutions: tunable solvents for efficient and reversible capture of CO_2, Ind. Eng. Chem. Res. 47 (2008) 8496–8498.

CHAPTER 9

Ion-Selective Electrodes Based on Ionic Liquids

Joaquín A. Ortuño[1], Francisca Tomás-Alonso[2], Aurora M. Rubio[2]
[1]Department of Analytical Chemistry, University of Murcia, Murcia, Spain
[2]Department of Chemical Engineering, Regional Campus of International Excellence "Campus Mare Nostrum," University of Murcia, Campus de Espinardo, Murcia, Spain

1. INTRODUCTION

There is a fundamental relationship between ion separation with membranes and the potentiometric response of ion-selective electrodes based on these membranes toward the corresponding ions. The more selective the ion-exchange processes for a particular ion at the sample/membrane interface, the more selective the potentiometric response to that ion. This relation justifies the inclusion of a chapter on ion-selective electrodes based on ionic liquids (IL-ISEs) in this edition.

An ion-selective electrode (ISE) is defined as an electrochemical sensor based on thin films or selective membranes as recognition elements, akin to an electrochemical half-cell needing a reference electrode (RE) to complete the full electrochemical cell. The potential differences measured between the electrodes, the ISE and the RE, immersed in a given solution, are linearly dependent on the logarithm of the activity of the ions [1].

To date, around 25,000 reports on ISEs have been published, not only in analytical chemistry but also in other fields such as chemistry, engineering, medicine, pharmacology, biochemistry, food science, and technology, among others.

The first ISE was reported in 1906, when pH-sensitive glass was discovered by Max Cremer [2], leading to the production of the first commercial pH glass electrodes in the 1930s [3]. The introduction of crystalline compounds as sensing materials in the early 1960s considerably increased the number of ion-selective electrodes. Later, the limited number of ions of interest measurable with ISEs based on glass and crystalline materials gave rise to the development of ISEs based on ionophores. The first electrode of this type was reported by Stefanac and Simon [4], who doped a thin film of water-immiscible organic solvents with antibiotics. This triggered the development of new synthetic ionophores for use in ISEs [5]. Although they were still unknown as such, ionic liquids (ILs) were first used in 1972 by James et al. [6] as part of a coated wire ISE consisting of a liquid membrane based on anionic association complexes with a quaternary ammonium ion, Aliquat 336. Knowledge of ionic liquids was still scant and the importance given to them was also low. On the other hand, the rapid growth of research into ISEs made it necessary

Ionic Liquids in Separation Technology
http://dx.doi.org/10.1016/B978-0-444-63257-9.00009-2
Copyright © 2014 Elsevier B.V.
All rights reserved.

to standardize the related nomenclature, which was carried out by the International Union of Pure and Applied Chemistry (IUPAC) in 1976 [7] and subsequently updated [1] and supplemented with technical reports [8−10].

To the best of our knowledge, the first report on the use of a modern ionic liquid in the construction of ion-selective electrodes was signed by Coll et al. [11]. These authors used the ionic liquid 1-butyl-3-methylimidazolium hexafluorophosphate $[C_4C_1Im][PF_6]$ to promote the potentiometric response of a plasticized polymeric membrane containing a suitable ionophore toward sulfate, and explained its behavior by the increase in the dielectric constant of the membrane due to the presence of the ionic liquid.

For several years now, room-temperature ionic liquids (RTILs) have shown very good features for constructing the ion-sensing layer of ISEs, not only because of their well-known ion-exchange properties, but also because of their ability to improve the selectivity in ion sensors as a result of their larger dielectric constants. This can be seen in the growing number of ionic liquids that are used in ISEs today. As an example, Figure 1(A)−(D) shows the high number of ionic liquids that are included in this chapter.

2. IONIC LIQUIDS AS PLASTICIZERS AND ION EXCHANGERS IN ION-SELECTIVE ELECTRODE MEMBRANES

Shvedene et al. [12] proposed the use of ISE membranes composed only of an ionic liquid and a polymer. They reasoned the well-known ion-exchange [13,14] and polymer plasticizing properties of ILs [15−17] made them suitable for use as both ion exchangers and plasticizers in ion-selective electrode membranes. They found that imidazolium-based ionic liquids were capable of plasticizing poly(methyl methacrylate) (PMMA) but not poly(vinyl chloride) (PVC), while a phosphonium-based ionic liquid plasticized both PMMA and PVC. These authors studied the potential response of several IL-polymer membranes to hydrophobic ions. The membranes composed of $[C_4C_1Im][PF_6]$ and PMMA exhibited only a weak potentiometric response toward hydrophobic cations such as protonated iso-leucine tert-butyl ester, for which a sub-Nernstian response (18 mV/dec) in a narrow concentration range was obtained for the best composition. The fact that the electrode lost sensitivity to these ions within a few days was attributed to the gradual leaching of the IL from the membrane due to the substantial solubility of $[C_4C_1Im][PF_6]$ in water. The authors obtained a higher response when they used a less water-soluble IL, $[C_4C_1C_1Im][NTf_2]$, for which the calibration slope become almost Nernstian toward some cationic surfactants such as cetylpyridinium. No response was obtained for the anionic surfactant dodecyl sulfate. With regard to membranes constructed with the IL $[C_{12}Phe_2C_2P][NTf_2]$, the membranes with PMMA were found to be insensitive to both cations and anions, while those with PVC membranes responded to cationic surfactants. This means that the polymer used may affect the membrane potentiometric response. After preliminary conditioning in a proper

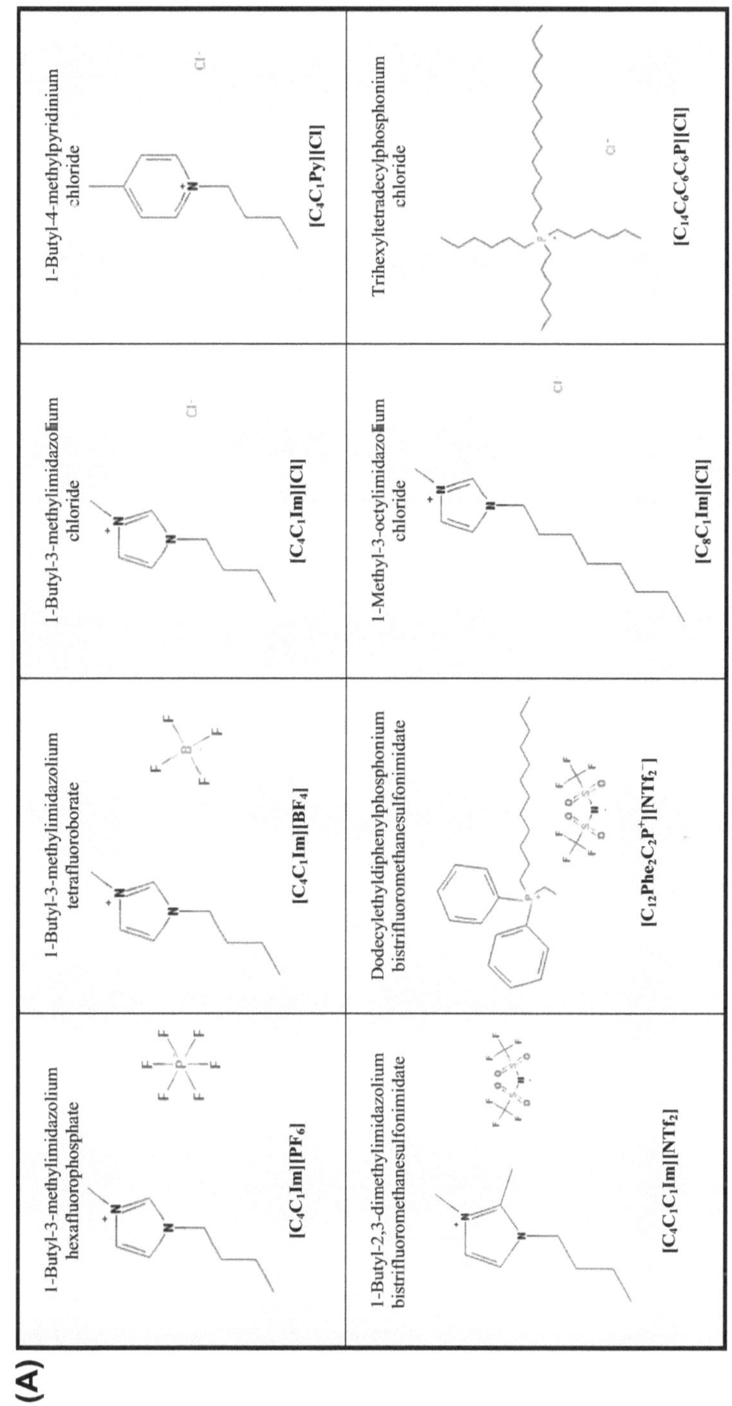

Figure 1 Name and Structure of the Ionic Liquids Used in Ion-Selective Electrode Membranes.

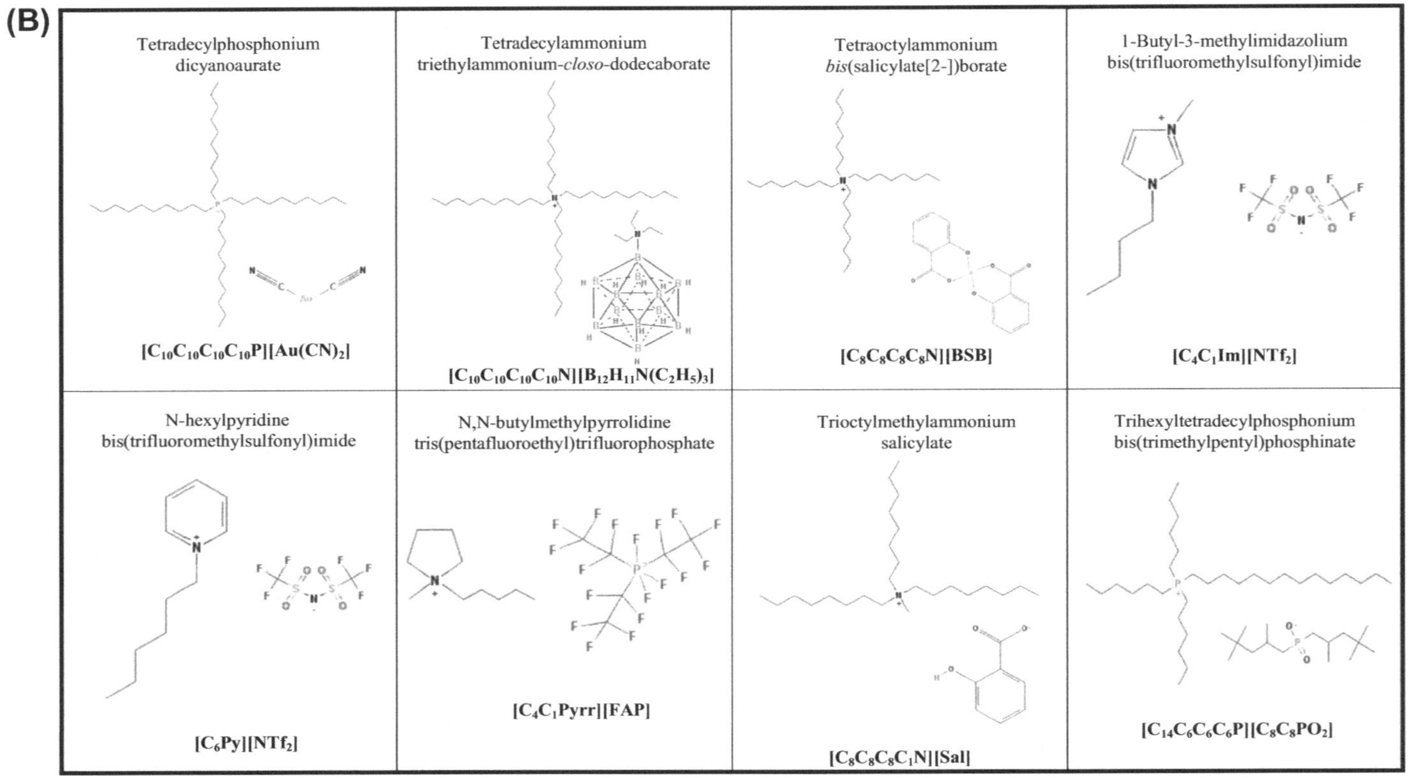

Figure 1 (*continued*).

(C)

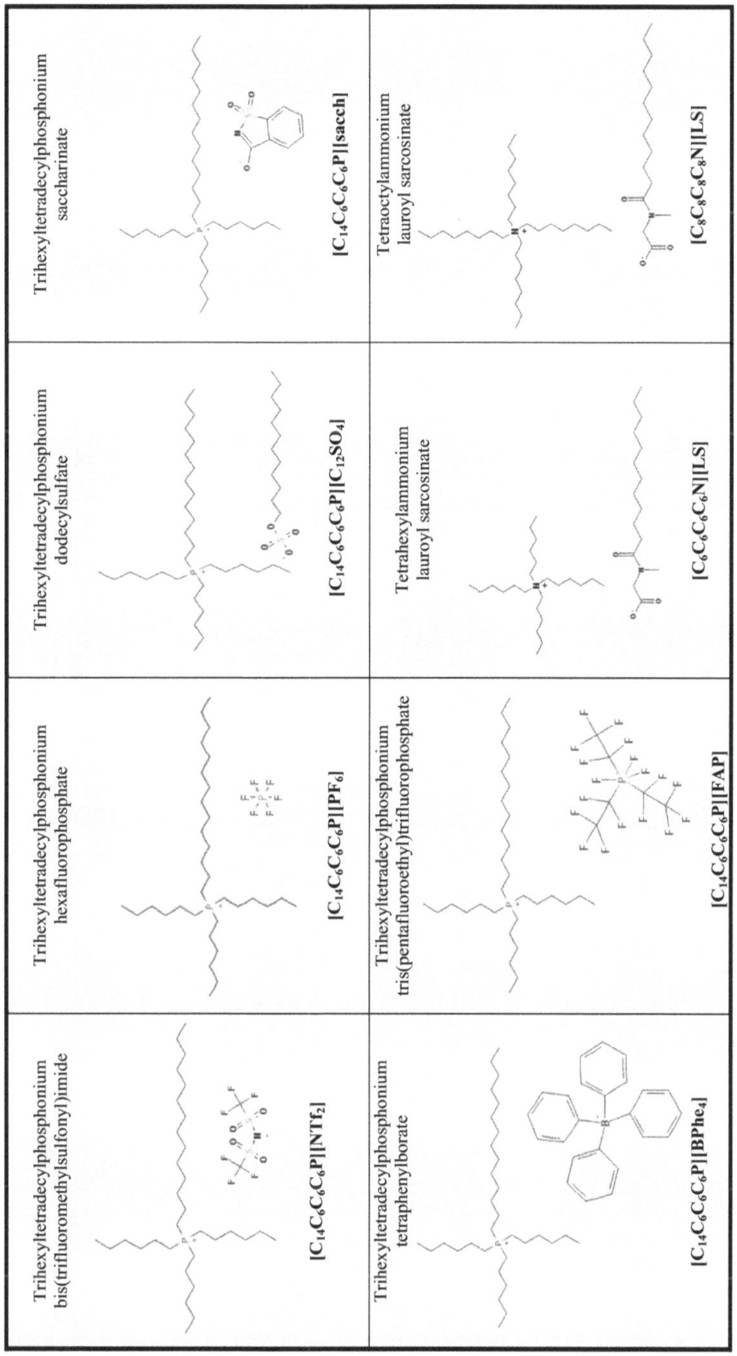

Figure 1 *(continued)*.

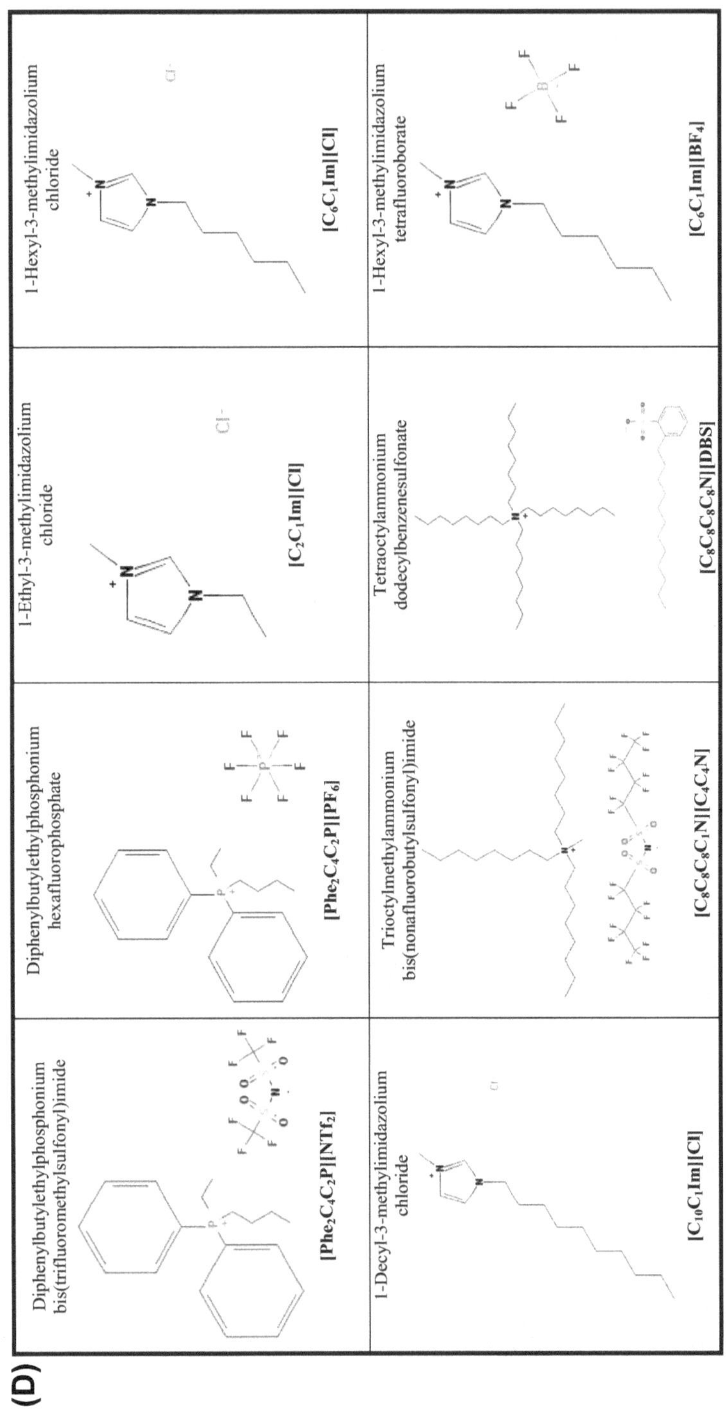

Figure 1 *(continued)*.

solution, the membrane showed sensitivity to several anions, such as dodecyl sulfate. The slope of the electrode response to other anions decreased in accordance with the Hofmeister series, as observed using conventional anion-selective electrodes based on anion exchangers. As is well known, the Hofmeister series shows the following order for monovalent anions: $ClO_4^- > SCN^- > I^- > NO_3^- > Br^- > Cl^- > HCO_3^- > CH_3COO^- > F^- > H_2PO_4^-$ [18]. This order is usually associated with the lipophilicity of the anions since the higher the affinity of the anion to the membrane, the better the potentiometric response of the ISE to the given anion. Thus, the highly hydrophilic anions located at the end of the series, such as sulfate and phosphate, are harder to measure than the more lipophilic anions located at the beginning of the series, such as perchlorate or thiocyanate.

With regard to the potentiometric response mechanism, the same authors suggested that the IL acts as ion exchanger. Their results opened the way for the construction of new IL-ISEs, which has been followed by other groups. The composition and performance characteristics of IL-ISEs reported on this or similar membrane categories are collated in Table 1.

A key paper for interpreting the potential response and selectivity of IL-ISEs was that of Kakiuchi [26], which included a hydrophobicity scale for IL ions. This scale is based on the experimental values of the standard ion-transfer potential for various ions between nitrobenzene and water. The essential extra-thermodynamic assumption underlying this scale is that the Gibbs transfer energy of tetraphenylarsonium is equal to that of tetraphenylborate. The Kakiuchi scale can be used to predict the viability and extension of the ion-exchange process between the target ion present in the aqueous sample and the corresponding IL ion of the membrane, and, therefore, to interpret the corresponding potentiometric response.

Two years later the Pletnev group [27] obtained some improvement in the detection limits and operational stability by including a conventional plasticizer (2-nitrophenyl octyl ether, NPOE) in the IL-ISE membrane composition. Later the same group [23] continued using highly hydrophobic liquids composed of substituted imidazole, pyridine, and pyrrolidine cations and hydrophobic anions together with PVC and NPOE. These membranes showed a response to the hydrophobic organic cations and the slope of the calibration for cationic surfactants was close to the Nernstian value. The potentiometric selectivity coefficients for cetyltrimethylammonium (CTMA) as primary ion with respect to some organic ions (such as tyrosine methyl ester, benzylamine, tetramethylamine, and dibutylamine cations) and over hydrophilic inorganic cations (such as sodium and potassium) were within a very narrow range (log K from -1.05 to -1.35). This narrowing of the interval of the selectivity coefficient values seems to be a feature of membranes based on IL.

In the same paper the authors also described a salicylate-selective electrode using the ionic liquid $[C_8C_8C_8C_1N][Sal]$. The fact that some membranes constructed with other

Table 1 Composition and Performance Characteristics of Ion-Selective Electrodes Based on Ionic Liquids as Plasticizers and Ion Exchangers

Ionic Liquid	Membrane Composition	Analytes	Response Characteristics	References
$[C_4C_1C_1Im][NTf_2]$	1:1 IL:PVC	Iso-leucine tert-butyl ester protonated form	$S = 25$ mV/dec, $C_{min} = 3 \times 10^{-5}$ M	[12]
		Cetylpiridinium	$S = 56$ mV/dec, $\Delta pC = 3.3 - 5$, $C_{min} = 5 \times 10^{-6}$ M	
		Cetyltrimethylammonium	$S = 53$ mV/dec, $\Delta pC = 3.3 - 5.3$, $C_{min} = 3 \times 10^{-6}$ M	
$[C_{12}Phe_2C_2P][NTf_2]$	2:1 IL:PMMA	Dodecylpyridinium	$S = 48$ mV/dec, $C_{min} = 3.5 \times 10^{-5}$ M	[12]
		Cetylpyridinium	$S = 57$ mV/dec, $C_{min} = 4.5 \times 10^{-6}$ M	
		Cetyltrimethylammonium	$S = 59$ mV/dec, $C_{min} = 8.5 \times 10^{-6}$ M	
		Dodecyl sulfate[a]	$S = -52$ mV/dec, $C_{min} = 1.2 \times 10^{-5}$ M	
		Salicylate[a]	$S = -44$ mV/dec, $C_{min} = 5.4 \times 10^{-5}$ M	
		Nitrate[a]	$S = -23$ mV/dec, $C_{min} = 5.5 \times 10^{-4}$ M	
$[C_{10}C_{10}C_{10}C_{10}P][Au(CN)_2]$	45% IL 55% PVC	Dicyanoaurate	$S = 58$ mV/dec, $C_{min} = 6.2 \times 10^{-7}$ M	[19]
$[C_{14}C_6C_6C_6P][Cl]$	20% IL 80% PVC	Sulfate[a]	$S = -30.2$ mV/dec, $\log K^{pot}_{Cl^-,SO_4^{2-}} = -1.53$	[20]
		Chloride[a]	$S = -58.7$ mV/dec	
		Salicylate[a]	$S = -62.0$ mV/dec, $\log K^{pot}_{Cl^-,Sal^-} = 2.57$	
		Perchlorate[a]	$S = -61.4$ mV/dec, $\log K^{pot}_{Cl^-,ClO_4^-} = 3.79$	
	20%IL 80%PVC	Ibuprofen	$S = -52.4$ mV/dec, $C_{min} = 1.8 \times 10^{-5}$ M Linear range $= 2.9 \times 10^{-4} - 1.8 \times 10^{-2}$ M.	[21]
$[C_8C_1Im][Cl]$	20% IL 80% PVC	Sulfate[a]	$S = -28.9$ mV/dec	[20]
		Chloride[a]	$S = -27.5$ mV/dec, $\log K^{pot}_{SO_4^{2-},Cl^-} = 1.88$	
		Salicylate[a]	$S = -33.4$ mV/dec, $\log K^{pot}_{SO_4^{2-},Sal^-} = 1.66$	
		Thiocyanate[a]	$S = -35.8$ mV/dec, $\log K^{pot}_{SO_4^{2-},SCN^-} = 1.86$	
		Perchlorate[a]	$S = -40.0$ mV/dec, $\log K^{pot}_{SO_4^{2-},ClO_4^-} = 1.91$	

$[C_4C_1Im][PF_6]$	1% IL 33% PVC 66% NPOE	$[Bmim^+][PF_6^-]$	$S = 58.8$ mV/dec, $C_{min} = 1.1 \times 10^{-6}$ M Response range $= 1.8 \times 10^{-6} - 1.0 \times 10^{-2}$ M	[22]
$[C_4C_1Im][NTf_2]$	5% IL 32% PVC 63% NPOE	Benzylammonium Cetyltrimethylammonium	$S = 49.5$ mV/dec, $C_{min} = 5.5 \times 10^{-4}$ M $S = 58.1$ mV/dec, $C_{min} = 3.1 \times 10^{-6}$ M	[23]
$[C_6Py][NTf_2]$	5% IL 32% PVC 63% NPOE	Benzylammonium Cetyltrimethylammonium	$S = 30.8$ mV/dec, $C_{min} = 6.8 \times 10^{-4}$ M $S = 56.2$ mV/dec, $C_{min} = 2.2 \times 10^{-6}$ M	[23]
$[C_4C_1Pyrr][FAP]$	5% IL 32% PVC 63% NPOE	Benzylammonium Cetyltrimethylammonium Tetradecyltrimethylammonium	$S = 25.8$ mV/dec, $C_{min} = 3.5 \times 10^{-4}$ M $S = 57.9$ mV/dec, $C_{min} = 1.6 \times 10^{-6}$ M $S = 53.2$ mV/dec, $C_{min} = 2.8 \times 10^{-6}$ M	[23]
$[C_8C_8C_8C_1N][Sal]$	0.5% IL –	Salicylate	$S = 57.0$ mV/dec, $C_{min} = 5.2 \times 10^{-5}$ M	[23]
$[C_{14}C_6C_6C_6P]$ $[C_8C_8PO_2]$	20% IL 80% PVC	Monohydrogenphosphate Sulfate Chloride Bromide Nitrate Salicylate Thiocyanate Perchlorate	$\log K^{pot}_{ClO_4^-,HPO_4^{2-}} < -5.00$ $\log K^{pot}_{ClO_4^-,SO_4^{2-}} < -5.00$ $S = -37.6$ mV/dec, $\log K^{pot}_{ClO_4^-,Cl^-} = -3.22$ $S = -46.9$ mV/dec, $\log K^{pot}_{ClO_4^-,Br^-} = -2.13$ $S = -50.9$ mV/dec, $\log K^{pot}_{ClO_4^-,NO_8^-} = -1.64$ $S = -55.5$ mV/dec, $\log K^{pot}_{ClO_4^-,Sal^-} = -0.83$ $S = -54.8$ mV/dec, $\log K^{pot}_{ClO_4^-,SCN^-} = -0.47$ $S = -55.8$ mV/dec	[24]
$[C_{10}C_{10}C_{10}C_{10}N]$ $[B_{12}H_{11}N(C_2H_5)_3]$	65–75% IL 35–25% PVC	Triethylammonium- closo-dodecaborate	Linear rang $= 1.0 \times 10^{-6} - 3.2 \times 10^{-3}$ M $K_{[B_{12}H_{11}N(C_2H_5)_3]^-,NO_8^-} = 8.1 \times 10^{-5}$ M $K_{[B_{12}H_{11}N(C_2H_5)_3]^-,Cl^-} = 1.3 \times 10^{-5}$ M $K_{[B_{12}H_{11}N(C_2H_5)_3]^-,F^-} = 1.2 \times 10^{-6}$ M	[25]

[a]After proper conditioning.

ILs did not give a response to salicylate was attributed to the higher hydrophobicity of the anions forming these ILs compared with the salicylate ion, which significantly complicated the anion exchange. Interestingly, the detection limit was decreased substantially when the IL in the membrane composition was reduced. Moreover, the selectivity coefficient range, which followed the Hofmeister series, widened considerably.

Because some popular ILs based on the cations of substituted imidazolium with fluorinated anions do not easily decompose, accumulate in the environment, and may be dangerous for living organisms, Shvedene et al. [28] studied the use of ILs based on hydrophobic cations of tetraalkylammonium and the anion of a substituted amino acid, a component of an industrially produced anionic surfactant, lauroyl sarcosinate, for the construction of an IL-ISE. The selected electrode permitted the potentiometric determination of lauroyl sarcosinate.

As regards the results of other groups, Peng et al. [20] made a comparative study of the response of various membranes constituted by PVC and several ionic liquids, based on [$C_4C_1Im^+$], [$C_8C_1Im^+$] and [$C_{14}C_6C_6C_6P^+$] cations, with and without added plasticizer (NPOE). As shown before, ILs based on [$C_4C_1Im^+$] do not plasticize PVC, so the authors used a conventional plasticizer, NPOE, to give mechanical stability to the membranes based on the ionic liquids [C_4C_1Im][PF_6], [C_4C_1Im][BF_4], and [C_4C_1Im][Cl]. On the other hand, ionic liquids based on [$C_8C_1Im^+$] and [$C_{14}C_6C_6C_6P^+$] were able to plasticize PVC without additional plasticizers. [C_4C_1Py][Cl], even with additional plasticizers, was difficult to combine with PVC to make stable membranes.

A $C_4C_1ImPF_6$-NPOE-PVC (10:60:30, wt%) electrode conditioned in a NaCl solution showed no response to the different anions and cations tested. After conditioning in $NaClO_4$ solution, it showed some response to benzoate, salicylate, and saccharin, with very sub-Nernstian slope values, but still did not respond to hydrophilic anions such as SO_4^{2-}, Cl^- and NO_3^-. Similar results were obtained with a $C_4C_1ImBF_4$-NPOE-PVC (10:60:30) membrane. The authors then tested ionic liquids containing a more exchangeable anion as chloride to improve the response. A membrane made with C_4C_1ImCl-NPOE-PVC (10:60:30) was tested but the high solubility of the ionic liquid [C_4C_1Im][Cl] made its response very unstable. Interestingly, a membrane constructed with 1-methyl-3-octylimidazolium chloride [C_8C_1Im][Cl] and PVC (20:80) exhibited a Nernstian response to sulfate in a wide concentration range with good selectivity toward SO_4^{2-} over Cl^-, ClO_4^-, NO_3^-, and SCN^-. Membranes constructed with the ionic liquid trihexyltetradecylphosphonium chloride [$C_{14}C_6C_6C_6P$][Cl] and PVC (20:80) and $C_{14}C_6C_6C_6PCl$-NPOE-PVC (10:60:30) showed a Nernstian response to SO_4^{2-}, Cl^-, ClO_4^-, and salicylate in a similar way. The response range was similar to that of a traditional PVC-NPOE membrane containing a conventional anion exchanger, tridodecylmethylammonium chloride ([$C_{12}C_{12}C_{12}C_1N$][Cl]), but the selectivity for chloride over lipophilic anions such as ClO_4^- and salicylate was better. In view of these results, the authors concluded that an ISE composed exclusively of $C_{14}C_6C_6C_6PCl$-PVC is a suitable alternative to a membrane constructed with the traditional ([$C_{12}C_{12}C_{12}C_1N$][Cl]) anion exchanger.

The versatility that ionic liquids possess, which allows customization of their ionic composition, means that other ion exchangers are unnecessary. Additionally, RTILs have also shown very good performance as plasticizers and increase the stability of membranes compared with conventional plasticizers. Thus, Kopytin et al. [19] considered that the smaller the number of membrane components, the more straightforward and more reliable the prediction of the membrane's properties. On this basis, they first studied the simplest system, composed of an ionic liquid and PVC, using the IL [$C_{10}C_{10}C_{10}C_{10}P$][$Au(CN)_2$], which acted simultaneously as plasticizer and anion exchanger. For comparative purposes, these researchers prepared other membranes with an additional conventional plasticizer, dibutyl phthalate, noting that the detection limit for these last membranes was worse than that of membranes without the additional plasticizer. The electrode constructed with 55 wt% PVC and 45 wt% ionic liquid showed the widest linearity range and lowest detection limit of all the electrodes assayed. The selectivity of this electrode for the dicyanoaurate(I) anion in the presence of several anions was high. Later [25] they tested a different ionic liquid, [$C_{10}C_{10}C_{10}C_{10}N$][$B_{12}H_{11}N(C_2H_5)_3$], in an electrode with two components (PVC and ionic liquid in different proportions) and another with three components (PVC, ionic liquid, and one conventional plasticizer, NPOE). These authors studied the membranes composed of PVC (30 wt%) and ionic liquid (70 wt%) in detail and obtained a slight improvement in the detection limit with respect to the membranes containing NPOE. This last electrode showed a high selectivity in the presence of inorganic anions, and so the authors proposed it for the potentiometric titration of the closo-dodecaborate anion. The authors concluded that when membranes for ion-selective electrodes are made with viscous ionic liquids, the percentage of ionic liquid can be considerably increased because it is easy to mix with the PVC while the presence of the external plasticizer does not provide additional benefits to the potentiometric response of the electrode.

Ortuño et al. [22] studied the potentiometric response of membranes based on $C_4C_1ImNTf_2$-NPOE-PVC, $C_4C_1ImNTf_2$-FFNE-PVC, and $C_4C_1ImPF_6$-NPOE-PVC toward [$C_4C_1Im^+$] and other ionic liquid cations. The membrane plasticized with 2-fluorophenyl-2-nitrophenyl ether (FFNE), which had a higher dielectric constant than NPOE, provided lower response than the corresponding membrane plasticized with NPOE. Also, the IL [C_4C_1Im][PF_6] provided a higher response than [C_4C_1Im][NTf_2], probably due to the higher hydrophobicity of the first ionic liquid. With regard to the concentration of [C_4C_1Im][PF_6] in the membrane composition, it was observed that 1 wt % [C_4C_1Im][PF_6] produced a better response than 5 wt% When the IL concentration was further lowered (0.3 wt% [C_4C_1Im][PF_6]), membrane lifetime was shortened, which was explained by IL leaching. The electrode showed good selectivity for [$C_4C_1Im^+$] with respect to some common inorganic ions. The response toward other ionic liquid cations,

1-ethyl-3-methylimidazolium [$C_2C_1Im^+$], 1-benzyl-3-methylimidazolium [BzC_1Im^+], 1-hexyl-3-methylimidazolium [$C_6C_1Im^+$], and 1-methyl-3-octylimidazolium [$C_8C_1Im^+$], in the form of chlorides and toward tetraethylamonium [$C_2C_2C_2C_2N^+$] was also studied by making independent calibrations. From the results the selectivity coefficients were calculated. Partition coefficients of the imidazolium cations assayed between water and membrane were obtained from the corresponding selectivity coefficient and the ionic partition coefficient (P) of [$C_2C_2C_2C_2N^+$]. A relationship between the P values and the toxicity of the corresponding imidazolium chlorides was found.

Gourishetty et al. [24] analyzed 20 ion-selective electrodes based on PVC membranes, of which all but one were composed of an ionic liquid with the [$C_{14}C_6C_6C_6P^+$] cation and different anions. The aim was to study the influence of the hydrophobic/hydrophilic character of the anion on the potentiometric response of the electrode. Another ISE, which did not include ionic liquid in its composition but a conventional plasticizer (NPOE) and a conventional anion exchanger ([$C_{12}C_{12}C_{12}C_1N$][Cl]), was used for comparative purposes. The response slopes of the electrodes prepared with ILs composed of hydrophilic anions, such as [$sacch^-$], [$C_8C_8PO_2^-$], and [Cl^-], were similar to those of the ISE without ionic liquid toward salicylate, thiocyanate, and perchlorate, but did not respond to hydrophilic analyte anions, such as chloride or nitrate. The electrodes containing the most hydrophobic anions [FAP^-], [NTf_2^-], and [$C_{12}SO_4^-$] did not respond to the most hydrophilic anions. However, when [$C_{12}C_{12}C_{12}C_1N$][Cl] was incorporated in the membrane composition the slopes improved considerably. These authors also found that all of the electrodes based on 66 wt% IL exhibited regions of super-Nernstian response in the calibrations of some anions, such as thiocyanate and nitrate, which grew with the hydrophilicity of the anion in the membrane. The authors used the standard ion-transfer potentials of the constituent IL anions reported by Kakiuchi [26] to quantify anion hydrophilicity, finding that, in general, the addition of [$C_{12}C_{12}C_{12}C_1N$][Cl] to the membrane produced higher super-Nernstian responses and reduced the linear response range.

Comparing the performance of different electrodes constructed with different ionic liquid percentages in the membrane, the above authors concluded that a reduction in the amount of ionic liquid leads to a reduction of the super-Nernstian response region. In line with the presented arguments, these observations can be attributed to the reduction in the ion-exchange capacity and the reduction of the membrane ion diffusivity when the percentage of ionic liquid decreases. The authors finally concluded that the electrode composed of [$C_{14}C_6C_6C_6P$][$C_8C_8PO_2$] (20 wt%) provided the best combination of slopes and linear ranges of all ISEs studied.

Tomás-Alonso et al. [21] studied the potentiometric response of ISEs constructed with PVC membranes prepared with the ionic liquids [C_4C_1Im][PF_6] and [$C_{14}C_6C_6C_6P$][Cl]. They also characterized the membranes by scanning-electron microscopy (SEM) combined with energy-dispersive x-ray (EDX) analysis.

The membranes were prepared with 20, 67, and 80 wt% of $[C_4C_1Im][PF_6]$ and 20 wt% of $[C_{14}C_6C_6C_6P][Cl]$. The membrane with 80% $[C_4C_1Im][PF_6]$ was extremely fragile and could not be used in potentiometric assays. When 20% $[C_4C_1Im][PF_6]$ membrane was exposed to NO_3^-, Cl^-, and SCN^- anions, its response was highly sub-Nernstian and followed the Hofmeister series. This low response can be attributed to the lipophilicity of the anion $[PF_6^-]$, which is higher than the lipophilicity of the analyte anions, hindering anion exchange. Additionally, from the EDX spectra the authors concluded that the IL leached from the first layer (c. 1 μm) of the membrane (the P and F peaks disappear in the spectrum of the membrane exposed to the studied anions).

In view of these results, the authors decided to use the ionic liquid $[C_{14}C_6C_6C_6P]$ [Cl] with its more lipophilic cation, which prevents the ionic liquid from leaching out of the membrane, and very hydrophilic anion, which facilitates anion exchange. The 20% $[C_{14}C_6C_6C_6P][Cl]$ membrane provided high responses with a total potential change of 240 mV in response to 0.1 M thiocyanate. Due to this high response, the membrane was selected for further response studies toward an anionic drug, ibuprofen. The electrode showed an excellent response to the ibuprofen anion, with good repeatability and reproducibility between days. From the EDX spectra, before and after exposure to sample solutions, it was concluded that there was no loss of $[C_{14}C_6C_6C_6P^+]$ cation in the outer side of the membrane and that anion exchange between chloride and ibuprofen anion occurred.

Generally, the IL-ISEs mentioned above have a high tolerance to pH variations.

3. ION-SELECTIVE ELECTRODES BASED ON IONIC LIQUIDS WITH ADDED IONOPHORE

Ionophores are lipophilic complexing agents capable of reversibly binding ions. In their uncomplexed form, they may be charged or electrically neutral [29]. In many cases, a number of different ionophores can be used to measure the same ion [30]. Indeed, selectivity, not binding strength, is the most important characteristic of these ion carriers [5], since the potentiometric selectivity depends on the selective extraction of ions.

The selective behavior of an ionophore-based ISE may be partly influenced by choosing the appropriate membrane components. A list of the latest reports on ISEs made with ionophore-based membranes of ionic liquids is shown in Table 2, according to the target ion.

4. ION-SELECTIVE ELECTRODES WITH SOLID CONTACT

Conventional ISEs are based on polymeric membranes in contact, on their inner side, with a solution that usually contains the primary ion and an inner reference

Table 2 Ion-Selective Electrodes Based on Ionic Liquids with Added Ionophore

Analytes	Ionophore	Membrane Composition	References
Anion Analytes			
Sulfate	Polyazacycloalkane I	7.11% ionophore 25.94% $[C_4C_1Im][PF_6]$ 41.84% PVC 25.11% NPOE	[11]
	Hydroxyl Schiff base	200 mg ionophore 5 mL $[C_4C_1Im][PF_6]$ 100 mg PVC 50 g of CTAB 2 mL DOP	[31]
Chromate	Europium β-diketone complex	10% ionophore 15% $[C_4C_1Im][BF_4]$ 67% graphite 5% MWCNTs 3% nano-silica	[32]
Permanganate	Modified SBA-15	20% ionophore 20% $[C_4C_1Im][BF_4]$ 55% graphite powder 5% MWCNTs	[33]
Dichromate	Modified SBA-15	20% ionophore 20% $[C_4C_1Im][BF_4]$ 55% graphite powder 5% MWCNTs	[34]
Monohydrogen phosphate	Cerium acetylacetonate complex	15% ionophore 15% $[C_4C_1Im][BF_4]$ 65% graphite powder 5% MWCNTs	[35]
Cation Analytes			
Potassium	Dicyclohexano-18-crown-6	180 mmol/kg DCH18C6 $[C_8C_8C_8C_1N][C_4C_4N]$ 27 mmol/kg $[K][C_4C_4N]$ 4.2% ionophore 60.8% $[C_8C_8C_8C_1N][C_4C_4N]$ 3.5% $[K][C_4C_4N]$ 31.5% P(VDF-HFP)	[36]
Copper (II)	*N′*-(1-oxoacenaphthylen-2(1H)-ylidene) furan-2-carbohydrazide	20% ionophore 25% $[C_4C_1Im][BF_4]$ 49% graphite 5% MWCNTs 1% nanosilica	[37]

Table 2 Ion-Selective Electrodes Based on Ionic Liquids with Added Ionophore—cont'd

Analytes	Ionophore	Membrane Composition	References
	N,*N*,*N*′,*N*′-tetracyclohexyl-2,2′-thiodiacetamide	0.5% ionophore 3% $[C_2C_1Im][Cl]$ 33% PVC 63.5% NPOE	[38]
Lead (II)	Tetr-butylcalix[4]arene-tetracis(*N*,*N*-dimethylthioacetamide)	1% ionophore 3% $[C_4C_1Im][Cl]$ 33% PVC 31.5% NPOE 31.5% BBPA	[39]
	Dithiodibenzoic acid	1.41% ionophore 1.41% $[C_4C_1Im][PF_6]$ 36.60% PVC 60.58% DOP	[40]
Cadmium (II)	Bismarck brown R or [5-amino-4-[5-(2-amino-4-azaniumyl-5-methyl phenyl)diazenyl-2-methylphenyl]diazenyl-2-methyl phenyl] azanium dichloride	17% ionophore 20% $[C_4C_1Im][PF_6]$ 50% graphite powder 10% MWCNTs 3% silica nanoparticles	[41]
	N,*N*,*N*′,*N*′-tetrabutyl-3,6-dioxaoctanedi (thioamide)	1% ionophore 3% $[C_2C_1Im][Cl]$ 33% PVC 63% NPOE	[42]
Mercury (II)	1-(2-ethoxyphenyl)-3-(3-nitrophenyl) triazene	8% ionophore 30% $[C_4C_1Im][BF_4]$ 42% graphite powder 20% MWCNTs	[43]
	Ion imprinted polymer	15% ionophore 10% $[C_4C_1Im][BF_4]$ 69% graphite powder 5% MWCNTs 1% nanosilica	[44]
Zinc (II)	Dibenzo-24-crown-8	3.2% ionophore 0.64% $[C_8C_8C_8N][BDS]$ 64.08% PVC 32.08% DOP	[45]
Lanthanum (III)	1-[({2-[2-(2-hydroxy-1-naphtyl)-3-(2-{[(E)-1-(2-hydroxy-1-naphtyl) methylidene] amino}ethyl)-1-imidazolidinyl]ethyl} imino) methyl]-2-naphthol	10% ionophore 20% $[C_4C_1Im][BF_4]$ 60% graphite powder 10% MWCNT-COOH	[46]

(Continued)

Table 2 Ion-Selective Electrodes Based on Ionic Liquids with Added Ionophore—cont'd

Analytes	Ionophore	Membrane Composition	References
Cerium (III)	N-[(2-hydroxyphenyl) methylidene]-2-furohydrazide	16% ionophore 25% $[C_4C_1Im][BF_4]$ 44% graphite powder 15% MWCNTs	[47]
	(Z)-2-((1h-1,2,4-triazol-3-ylimino)methyl)phenol	16% ionophore 20% $[C_4C_1Pyrr][NTf_2]$ 51% graphite powder 10% MWCNTs 3% nanosilica	[48]
Praseodymium (III)	*N′*-(pyridin-2-ylmethylene) benzohydrazide	17% ionophore 30% $[C_4C_1Im][BF_4]$ 38% graphite powder 15% MWCNTs	[49]
	N,N-bis(a-methylsalicylidene) diethylenetriamine	25% ionophore 25% $[C_4C_1Im][BF_4]$ 45% graphite powder 5% MWCNTs	[50]
Neodymium (III)	2-{[(6-aminopyridin-2-yl) imino] methyl}-phenol	20% ionophore 20% $[C_4C_1Im][BF_4]$ 53% graphite powder 5% MWCNT-NH_2 2% nanosilica	[51]
Europium (III)	Bis(thiophen)buyane 2,3-dihydrazone	15% ionophore 15% $[C_4C_1Im][BF_4]$ 63% graphite powder 5% MWCNTs 2% nanosilica	[52]
Terbium (III)	*N′*-(1-oxoacenaphthylen-2(1H)-ylidene) furan-2-carbohydrazide	15% ionophore 15% $[C_4C_1Im][BF_4]$ 60% graphite powder 10% MWCNTs	[53]
Erbium (III)	[5-(dimethylamino) naphthalene-1-sulfonyl 4-phenylsemicarbazide]	20% ionophore 20% $[C_4C_1Im][BF_4]$ 45% graphite powder 15% MWCNTs	[54]
Thulium (III)	4-methyl-1,2-bis(2-pyridine carboxamido) benzene	20% ionophore 25% $[C_4C_1Im][BF_4]$ 50% graphite 5% nanographene	[55]
	2,2′-dianilinedisulfide	15% ionophore 20% $[C_4C_1Im][BF_4]$ 57% graphite 2% nanosilica 6% nanographene	[56]

Table 2 Ion-Selective Electrodes Based on Ionic Liquids with Added Ionophore—cont'd

Analytes	Ionophore	Membrane Composition	References
Ytterbium (III)	*N'*-(1-oxoacenaphthylen-2(1H)-ylidene) furan-2-carbohydrazide	25% ionophore 25% $[C_4C_1Im][BF_4]$ 45% graphite powder 5% MWCNTs	[57]
Lutetium (III)	*N*-[(Z)-1-(2-thienyl) methylidene]-*N*-[4-(4-{[(Z)-1-(2-thienyl) methylidene] amino} benzyl)phenyl] amine	25% ionophore 20% $[C_4C_1Im][BF_4]$ 49% graphite powder 5% MWCNT-NH_2 1% nanosilica	[58]

DOP: dioctylphthalate; CTAB: cetyltrimethylammonium bromide; SBA-15: mesoporous silica; P(VDF-HFP): poly(vinylidene fluoride-co-hexafluoropropylene); BBPA: bis(1-butylpenthyl) adipate

electrode. On the other hand, solid-contact ion-selective electrodes (SC-ISEs) do not contain internal solution and the electrical conductor is in direct contact with the membrane.

The advantages of SC-ISEs over conventional ISEs include the possibility of miniaturization, which makes it possible to work with small sample volumes, lowers the cost of production, avoids the problems inherent to inner filling solutions (drying out of the inner solution, maintenance, etc.), and increases the ability to work in environments where conventional ISEs may be damaged.

Even though new solid contact materials are contributing to improving SC-ISE properties, providing better potential stability and less susceptibility to interferences [59], SC-ISEs have not yet replaced conventional ISEs. Nevertheless, their many advantages promise a new generation of potentiometric ion sensors [60].

Ionic liquids can act as useful components of SC-ISEs because of their capacity to lower the membrane resistance and reduce interference from co-ions. Additionally, they improve the analytical parameters of the solid-contact sensor, including the detection limit, measuring range, working pH range, and selectivity [38].

The beginning of the development of solid-contact ISEs can be traced to the early 1970s with the construction of the coated-wire electrode [61], but it was not until 2008 [62] that SC-ISEs based on ionic liquids were reported for the first time.

4.1 Screen-Printed Solid-Contact Ion-Selective Electrodes

Screen-printed electrodes are constructed by printing multiple successive layers on a plastic or ceramic solid support. Their low cost and easy mass production, the flexibility in design, and the possibility of using small volumes make screen-printed electrodes attractive for practical applications [63].

Chernyshov et al. [62] made a study of the design and properties of SC-ISEs made from commercially available screen-printed electrodes (SPEs) covered with PVC or PMMA polymeric membranes containing the ionic liquid [$C_{12}Phe_2C_2P$][NTf_2] or [$C_4C_1C_1Im$][NTf_2]. In some cases, NPOE was added as an additional plasticizer. The response of these electrodes to cetylpyridinium bromide, a hydrophobic cation, and to sodium dodecyl sulfate, a hydrophobic anion, was studied. Later, the same group [27] used this type of sensor to study low-melting ionic solids (LMISs) without PVC as sensing material for the potentiometric detection of several anions.

4.2 Carbon Paste Electrodes

Carbon pastes are easy to obtain at minimal cost, and their composition can be modified simply to create new sensors with the required properties, which makes them a very useful material for constructing analytical electrometric sensors [64].

In recent years a wide number of papers with ionic liquid-based carbon paste electrodes (CPEs) have been reported for potentiometric determinations of a wide range of ions (Table 2). Carbon pastes were prepared by mixing graphite powder combined, in some cases, with other solid materials, such as multi-walled carbon nanotubes (MWCNTs), nanographene, and nanosilica with IL ([C_4C_1Im][BF_4]) and a suitable ionophore. Carbon nanotubes (CNTs) are used in the construction of carbon paste electrodes due to their interesting physicochemical properties, such as ordered structure, ultra-light weight, high mechanical strength, high electrical and thermal conductivity, metallic or semi-metallic behavior, and high surface area [54]. Ionic liquids are used as binders because of their chemical stability, low vapor pressure, low toxicity, low melting temperature, high ionic conductivity, and good electrochemical and thermal stability [57]. Usually carbon paste electrodes based on MWCNTs and RTILs show improved performance over traditional CPEs. Adding MWCNT can improve analytical parameters such as the detection limit or the response time, while using RTILs instead of paraffin oil can also increase the sensitivity of the sensor.

4.3 Other Solid-Contact Electrodes

Another type of solid-contact electrode with ionic liquids has been reported by Wardak and associates [38,39,42]. These authors prepared a two-layer membrane, where the inner layer contained PVC, ionic liquid, and additional plasticizer and the outer layer contained the same components plus an ionophore. They found that the use of an ionic liquid instead of a conventional lipophilic additive decreased membrane resistance and shortened the response time. Furthermore, the ionic liquid used [C_4C_1Im][Cl] allowed a constant concentration of chloride ions to be maintained in the membrane phase, lending potential stability to the internal contact Ag/AgCl/membrane and avoiding potential drifts [38,42].

5. OTHER TYPES OF IONIC LIQUID-BASED ION-SELECTIVE ELECTRODES

As mentioned before, Chernyshov et al. [27] used low-melting ionic solids on screen-printed electrodes for the potentiometric detection of several anions in aqueous solutions. LMISs form ion-sensing layers without the need for polymer or plasticizer, and can be easily doped and liquefied, thus simplifying the manufacture of ion-sensing devices. The procedure followed involved melting the chosen LMIS on the working electrode surface. After cooling to room temperature, a homogeneous solid membrane with strong adhesion properties was obtained.

Of three potential LMISs ($[C_8C_8C_8C_8N][BSB]$, $[C_8C_8C_8C_1N][BSB]$, or $[C_6C_6C_6C_6N][BSB]$) synthesized and characterized, $[C_8C_8C_8C_8N][BSB]$ (Figure 1(B)) was selected for covering the screen-printed electrode surface because it had a lower melting point than the other LMISs. The authors compared the potentiometric response and selectivity between the above mentioned SPE and a liquid-contact electrode made with a membrane of 5% $[C_8C_8C_8C_8N][BSB]$, 63% NPOE, and 32% PVC. Neither electrode gave a response toward cations, but they did respond to hydrophilic and hydrophobic anions. The former had a lower selectivity than the latter. However, the potentiometric response of the SPE was reproducible and reversible and showed a rapid response for relatively dilute solutions.

Another variation on ISE components is that used by Yan et al. [65], who synthesized new ionic associates formed of the ionic liquid $[C_4C_1Im][PF_6]$ and the cationic dyes malachite green and methylene blue. These were used as novel ionophores for $[PF_6^-]$ selective electrodes. Of the membrane compositions, NPOE was found to be the most effective plasticizer, and it was found that the addition of $[C_{12}C_{12}C_{12}C_1N][Cl]$ led to Nernstian response slopes. As regards selectivity, common anions give rise to no interference, so the electrodes proposed can be used for the determination of $[PF_6^-]$ and also for $[BF_4^-]$.

Ganjali's group developed IL-ISEs for some cationic drugs, e.g., clonidine [66], clomiphene [67], and cyclizine [68], using as sensing materials the ion pairs between the corresponding cationic drug and tetraphenylborate in a PVC membrane plasticized with a conventional plasticizer. A small amount of the ionic liquid $[C_4C_1Im][BF_4]$ was incorporated in the membrane composition to improve the potentiometric response of the sensor. For clonidine-ISE, the best sensor response was obtained with a membrane composition of 2% ionic liquid, 6% ion pair, 62% dibutyl phthalate, and 30% PVC.

6. APPLICATIONS OF ION-SELECTIVE ELECTRODES BASED ON IONIC LIQUIDS IN THE ANALYSIS OF REAL SAMPLES

Several applications of IL-ISEs in the analysis of real samples have been reported. Below, the applications found in the literature are classified according to the kind of sample.

6.1 Analysis of Waters (Drinking Water, Well Water, River Water, and Industrial Waste Water)

In this section a number of IL-ISEs used to determine ions in water and waste water will be discussed.

The determination of sulfate in drinking water using an ion-selective electrode based on a C_8C_1ImCl-PVC membrane was proposed by Peng et al. [20], who used water samples with known added sulfate amounts and found good recoveries. A monohydrogen phosphate sensor was used by Norouzi et al. [35] for the determination of phosphate ions in wastewater samples from a monocalcium phosphate factory. They found monohydrogen phosphate concentrations in the range of 53–65 ppm, similar to those found using a spectrophotometric method. Ganjali et al. [32] determined the chromate concentration in wastewater samples of chromium electroplating samples, finding values ranging between 2×10^{-4} and 6×10^{-4} M, which agreed with those found by inductively coupled plasma-atomic emission spectrometry (ICP-AES).

Some authors have used IL-ISEs for the determination of heavy metal ions in different types of waters. Thus, Wardak used an IL-ISE [42] for the determination of Cd^{2+} in tap water, river water, and wastewater. Their results were similar to those obtained by anodic stripping voltammetry. Afkhami et al. [41] also determined trace amounts of cadmium in tap water and wastewater at concentrations as low as 1.6×10^{-8} M, obtaining similar results to those obtained with ICP-AES. Wardak [69] proposed an IL-ISE for the direct determination of Pb^{2+} ions in real water samples. Khani et al. [43] determined Hg^{2+} in sea water, lab water, and river water with added concentrations of Hg^{2+} (5×10^{-8} M and 5×10^{-5} M), obtaining good recoveries. Ganjali et al. determined concentrations of around 5×10^{-6} M mercury [44] and around 5×10^{-4} M copper [37] in wastewater, which were similar to the results obtained with ICP-AES.

With regard to lanthanide ions, Afkhami et al. [48] determined Ce^{3+} in different environmental samples (well water, river water, and industrial wastewater), to which different Ce^{3+} amounts (c. $10^{-7} - 5 \times 10^{-6}$ M) were added and a synthesized Ce^{3+}-polluted water sample. They found good recoveries in all cases. Lastly, Ganjali et al. [53] determined Tb^{3+} in waters.

In the case of organic ions, Ortuño et al. [22] applied an IL-ISE to the determination of $[C_4C_1Im^+]$ in spiked waters to monitor the biosorption of $[C_4C_1Im^+]$ by dry biomass. They also used the electrode for the determination of ion partition coefficients of the imidazolium cations between water and the membrane.

6.2 Detergents

Shvedene et al. [12] applied a PVC-plasticized ISE based on $[C_{12}Phe_2C_2P][NTf_2]$ to the determination of sodium dodecyl sulfate in detergents.

6.3 Alloys

Faridbod et al. [54] determined the erbium content of two erbium alloys ($Mg_{95}Al_3Er_2$ and $ErDy_4Al_{10}$). Khani et al. [43] determined the mercury content in a dental amalgam.

6.4 Pharmaceutical Samples

Some of the IL-ISEs described for pharmaceutical analyses are presented below. Ganjali's group developed a set of selective electrodes for the determination of clonidine (*N*-[2,6-dichlorophenyl]-4,5-dihydro-1H-imidazol-2-amine) [66], clomiphene (2-[4-(2-chloro-1,2-diphenylethenyl)phenoxy]-*N*,*N*-diethylethanamine) [67], and ciclyzine (1-[diphenylmethyl]-4-methylpiperazine or 1-benzhydryl-4-methylpiperazine) [68], in some pharmaceutical commercial formulations, obtaining results that were in satisfactory agreement with the labeled amounts.

6.5 Biological Samples

Interestingly, Afkhami et al. [41] used an IL-ISE for the determination of cadmium in several biological samples: rice, tobacco, and human hair. The obtained results agreed well with those obtained by ICP-AES.

7. CONCLUSIONS

Ionic liquids are useful as membrane materials for use in ion-selective electrodes since they act as suitable plasticizers and ion exchangers. Some membranes composed only of an ionic liquid and a polymer respond to inorganic anions, generally following the Hofmeister series, and also to some organic anions and cations. The incorporation of a conventional plasticizer together with the ionic liquid in the membrane can improve some performance characteristics of IL-ISE. The incorporation of a suitable ionophore in the membrane permits IL-ISEs to be used for specific metal ions and also increases the selectivity for hydrophilic anions. Ionic liquids alone or combined with nanomaterials allow the development of solid-contact ISEs. As we have seen, ion-selective electrodes based on ionic liquids are applicable to the analysis of real samples in many situations. In this review we have looked at many of these, and it only remains for us to apologize for any omission.

ACKNOWLEDGMENT

The authors gratefully acknowledge the financial support of the Ministerio de Economía y Competitividad, Spain (Project CTQ2011-27049).

REFERENCES

[1] R.P. Buck, E. Lindner, Recommendations for nomenclature of ion-selective electrodes, Pure Appl. Chem. 66 (1994) 2527–2536.
[2] M. Dole, The early history of the development of the glass electrode for pH measurements, J. Chem. Educ. 57 (1980) 134.
[3] W.G. Hines, R. de Levie, The early development of electronic pH meters, J. Chem. Educ. 87 (2010) 1119–1280.
[4] Z. Stefanac, W. Simon, In-Vitro-verhalten von makrotetroliden in membranen als grundlage fur hochselektive kationenspezifische elektrodensysteme chimia, Chimia 20 (1966) 436.
[5] P. Bühlmann, L.D. Chen, Ion-selective electrodes with ionophore-doped sensing membranes, in: P.A. Gale, J.W. Steed (Eds.), Supramolecular Chemistry: From Molecules to Nanomaterials, John Wiley & Sons, Ltd, 2012, pp. 2539–2579.
[6] H. James, G. Carmack, H. Freiser, Coated wire ion selective electrodes, Anal. Chem. 44 (1972) 856–857.
[7] Commission on Analytical Nomenclature, Recommendations for nomenclature of ion-selective electrodes (RECOMMENDATIONS 1975), Pure Appl. Chem. 48 (1976) 127–132.
[8] Y. Umezawa, P. Bühlmann, K. Umezawa, K. Tohda, S. Amemiya, Potentiometric selectivity coefficients of ion-selective electrodes part I. Inorganic cations (technical report), Pure Appl. Chem. 72 (2000) 1851–2082.
[9] Y. Umezawa, K. Umezawa, P. Bühlmann, N. Hamada, H. Aoki, J. Nakanishi, M. Sato, K.P. Xiao, Y. Nishimura, Potentiometric selectivity coefficients of ion-selective electrodes part II. Inorganic anions (IUPAC technical report), Pure Appl. Chem. 74 (2002) 923–994.
[10] Y. Umezawa, P. Bühlmann, K. Umezawa, N. Hamada, Potentiometric selectivity coefficients of ion-selective electrodes part III. Organic ions (IUPAC technical report), Pure Appl. Chem. 74 (2002) 995–1099.
[11] C. Coll, R.H. Labrador, R. Martínez-Máñez, J. Soto, F. Sancenón, M.J. Seguí, E. Sánchez, Ionic liquids promote selective responses towards the highly hydrophilic anion sulfate in PVC membrane ion-selective electrodes, Chem. Commun. 24 (2005) 3033–3035.
[12] N.V. Shvedene, D.V. Chernyshov, M.G. Khrenova, A.A. Formanovsky, V.E. Baulin, I.V. Pletnev, Ionic liquids plasticize and bring ion-sensing ability to polymer membranes of selective electrodes, Electroanalysis 18 (2006) 1416–1421.
[13] M.L. Dietz, J.A. Dzielawa, Ion-exchange as a mode of cation transfer into room-temperature ionic liquids containing crown ethers: implications for the 'greenness' of ionic liquids as diluents in liquid-liquid extraction, Chem. Commun. 20 (2001) 2124–2125.
[14] S.V. Smirnova, I.I. Torocheshnikova, A.A. Formanovsky, I.V. Pletnev, Solvent extraction of amino acids into a room temperature ionic liquid with dicyclohexano-18-Crown-6, Anal. Bioanal. Chem. 378 (2004) 1369–1375.
[15] J. Fuller, A.C. Breda, R.T. Carlin, Ionic liquid-polymer gel electrolytes, J. Electrochem. Soc. 144 (1997) 67–70.
[16] A. Noda, M. Watanabe, Highly conductive polymer electrolytes prepared by in situ polymerization of vinyl monomers in room temperature molten salts, Electrochim. Acta 45 (2000) 1265–1270.
[17] M.P. Scott, M. Rahman, C.S. Brazel, Application of ionic liquids as low-volatility plasticizers for PMMA, Eur. Polym. J. 39 (2003) 1947–1953.
[18] K. Wojciechowski, M. Kucharek, W. Wróblewski, P. Warszynski, On the origin of the hofmeister effect in anion-selective potentiometric electrodes with tetraalkylammonium salts, J. Electroanal. Chem. 638 (2010) 204–211.
[19] A.V. Kopytin, E.N. Pyatova, A.F. Zhukov, YuA. Politov, K.E. German, A.Yu Tsivadze, Ion-selective sensors for determining $Au(CN)_2^-$ with membranes based on the ionic liquid tetradecylphosphonium dicyanoaurate, J. Anal. Chem. 63 (2008) 888–890.
[20] B. Peng, J. Zhu, X. Liu, Y. Qin, Potentiometric response of ion-selective membranes with ionic liquids as ion-exchanger and plasticizer, Sensor Actuat. B Chem. 133 (2008) 308–314.

[21] F. Tomás-Alonso, A.M. Rubio, R. Álvarez, J.A. Ortuño, Dynamic potential response and SEM-EDX studies of polymeric inclusion membranes based on ionic liquids, Int. J. Electrochem. Sci. 8 (2013) 4955–4969.

[22] J.A. Ortuño, M. Cuartero, M.S. García, M.I. Albero, Response of an ion-selective electrode to butylmethylimidazolium and other ionic liquid cations. Applications in toxicological and bioremediation studies, Electrochim. Acta 55 (2010) 5598–5603.

[23] N.V. Shvedene, D.V. Chernyshev, Yu. P. Gromova, M.Yu Nemilova, I.V. Pletnev, Hydrophobic ionic liquids in plasticized membranes of ion-selective electrodes, J. Anal. Chem. 65 (2010) 861–865.

[24] R. Gourishetty, A.M. Crabtree, W.M. Sanderson, R.D. Johnson, Anion-selective electrodes based on ionic liquid membranes: effect of ionic liquid anion on observed response, Anal. Bioanal. Chem. 400 (2011) 3025–3033.

[25] A.V. Kopytin, K.Yu Zhizhina, Yu. I. Urusovb, V.N. Mustyatsaa, Yu. V. Kokunova, N.T. Kuznetsova, Potentiometric sensors with membranes based on ionic liquid tetradecylammonium triethylammonio-closo-dodecaborate, J. Anal. Chem. 67 (2012) 168–171.

[26] T. Kakiuchi, Electrochemical aspects of ionic-liquid|water two-phase systems, Anal. Chem. 79 (2007) 6442–6449.

[27] D.V. Chernyshov, V.M. Egorov, N.V. Shvedene, I.V. Pletnev, Low-melting ionic solids: versatile materials for ion-sensing devices, ACS Appl. Mater. Interfaces 1 (2009) 2055–2059.

[28] N.V. Shvedene, S.A. Krasnov, M.Yu Nemilova, A.V. Grigoreva, K.M. Sotnezova, I.V. Pletnev, Ionic liquids with an anion of n-lauroyl sarcosinate in membranes of ion-selective electrode, J. Anal. Chem. 67 (2012) 834–841.

[29] E. Bakker, P. Bülhmann, E. Pretsch, Carrier-based ion-selective electrodes and bulk optodes. 1. General characteristics, Chem. Rev. 97 (1997) 3083–3132.

[30] P. Bülhmann, E. Pretsch, E. Bakker, Carrier-based ion-selective electrodes and bulk optodes. 2. Ionophores for potentiometric and optical sensors, Chem. Rev. 98 (1998) 1593–1688.

[31] A. Sathyapalan, A. Zhou, T. Kar, F. Zhou, H. Su, A novel approach for the design of a highly selective sulfate-ion-selective electrode, Chem. Commun. 3 (2009) 325–327.

[32] M.R. Ganjali, M.H. Eshraghi, S. Ghadimi, S.M. Moosavi, M. Hosseini, H. Haji-Hashemi, P. Norouzi, Novel chromate sensor based on MWCNTs/nanosilica/ionic liquid/Eu complex/graphite as a new nano-composite and its application for determination of chromate ion concentration in waste water of chromium electroplating, Int. J. Electrochem. Sci. 6 (2011) 739–748.

[33] F. Faridbod, M.R. Ganjali, M. Hosseini, P. Norouzi, Permanganate selective nano-composite electrode, Int. J. Electrochem. Sci. 7 (2012) 1927–1936.

[34] M.R. Ganjali, Z. Rafiei-Sarmazdeh, T. Poursaberi, S.J. Shahtaheri, P. Norouzi, Dichromate ion-selective sensor based on functionalized SBA-15/ionic liquid/MWCNTs/graphite, Int. J. Electrochem. Sci. 7 (2012) 1908–1916.

[35] P. Norouzi, M.R. Ganjali, F. Faridbod, S.J. Shahtaheri, H.A. Zamani, Electrochemical anion sensor for monohydrogen phosphate based on nano-composite carbon paste, Int. J. Electrochem. Sci. 7 (2012) 2633–2642.

[36] N. Nishi, H. Murakami, Y. Yasui, T. Kakiuchi, Use of highly hydrophobic ionic liquids for ion-selective electrodes of the liquid membrane type, Anal. Sci. 24 (2008) 1315–1320.

[37] M.R. Ganjali, T. Poursaberi, M. Khoobi, A. Shafiee, M. Adibi, M. Pirali-Hamedani, P. Norouzi, Copper nano-composite potentiometric sensor, Int. J. Electrochem. Sci. 6 (2011) 717–726.

[38] C. Wardak, J. Lenik, Application of ionic liquid to the construction of Cu(II) ion-selective electrode with solid contact, Sensor Actuat. B Chem. 189 (2013) 52–59.

[39] C. Wardak, Ionic liquids as new lipophilic additives to the membrane of lead ion-selective electrodes with solid contact, Int. J. Environ. Anal. Chem. 89 (2009) 735–748.

[40] M.F. Al-Asousi, A.H. Bu-Olayan, A.F. Shoukry, Atomic force microscopy and X-ray photoelectron spectroscopy of lead(II) membrane electrode based on dithiodibenzoic acid as ionophore, Electroanalysis 24 (2012) 224–228.

[41] A. Afkhami, H. Bagheri, A. Shirzadmehr, H. Khoshsafar, P. Hashemia, A potentiometric sensor for Cd^{2+} based on carbon nanotube paste electrode constructed from room temperature ionic liquid, ionophore and silica nanoparticles, Electroanalysis 24 (2012) 2176–2185.

[42] C. Wardak, A comparative study of cadmium ion-selective electrodes with solid and liquid inner contact, Electroanalysis 24 (2012) 85–90.
[43] H. Khani, M.K. Rofouei, P. Arab, V.K. Gupta, Z. Vafaei, Multi-walled carbon nanotubes-ionic liquid-carbon paste electrode as a super selectivity sensor: application to potentiometric monitoring of mercury ion(II), J. Hazard. Mater. 183 (2010) 402–409.
[44] M.R. Ganjali, T. Alizadeh, F. Azimi, B. Larjani, F. Faridbod, P. Norouzi, Bio-mimetic ion imprinted polymer based potentiometric mercury sensor composed of nano-materials, Int. J. Electrochem. Sci. 6 (2011) 5200–5208.
[45] M.F. Al-Asousi, A.F. Shoukry, A.H. Bu-Olayan, Halogen-free ionic liquid as an additive in zinc(II)-selective electrode: surface analyses as correlated to the membrane activity, Talanta 94 (2012) 184–188.
[46] M.R. Ganjali, S. Aghabalazadeh, M. Rezapour, M. Hosseini, P. Norouzi, La^{3+} carbon nanotube (CNT) based electrode using a new (1-[({2-[2-(2-hydroxy-1-naphtyl)-3-(2-{[(E)-1-(2-hydroxy-1-naphtyl)methylidene]amino}ethyl)-1-imidazolidinyl]ethyl} imino)methyl]-2-naphthol), Int. J. Electrochem. Sci. 5 (2010) 1743–1753.
[47] M.R. Ganjali, H. Khoshsafar, A. Shirzadmehr, M. Javanbakht, F. Faridbod, Improvement of carbon paste ion selective electrode response by using room temperature ionic liquids (RTILs) and multi-walled carbon nanotubes (MWCNTs), Int. J. Electrochem. Sci. 4 (2009) 435–443.
[48] A. Afkhami, T. Madrakian, A. Shirzadmehr, M. Tabatabaee, H. Bagheri, New Schiff base-carbon nanotube-nanosilica-ionic liquid as a high performance sensing material of a potentiometric sensor for nanomolar determination of cerium(III) ions, Sensor Actuat. B Chem. 174 (2012) 237–244.
[49] M.R. Ganjali, H. Khoshsafar, F. Faridbod, A. Shirzadmehr, M. Javanbakht, P. Norouzi, Room temperature ionic liquids (RTILs) and multiwalled carbon nanotubes (MWCNTs) as modifiers for improvement of carbon paste ion selective electrode response; a comparison study with PVC membrane, Electroanalysis 21 (19) (October 2009) 2175–2178.
[50] F. Faridbod, H.A. Zamani, M. Hosseini, M. Pirali-Hamedani, M.R. Ganjali, P. Norouzi, Praseodymium selective carbon paste electrode based on carbon nanotubes and ionic liquids, Int. J. Electrochem. Sci. 6 (2011) 3694–3703.
[51] M.R. Ganjali, M. Hosseini, M. Pirali-Hamedani, H.L. Zamani, MWCNT-NH_2/nanosilica/RTIL/ionophore nanocomposite carbon paste electrode for Nd(III), Int. J. Electrochem. Sci. 6 (2011) 2808–2817.
[52] P. Norouzi, M. Hosseini, M.R. Ganjali, M. Rezapour, M. Adibi, A nano-composite modified carbon paste europium(iii) sensor, Int. J. Electrochem. Sci. 6 (2011) 2012–2021.
[53] M.R. Ganjali, H. Ganjali, M. Hosseini, P. Norouzi, A novel nano-composite Tb^{3+} carbon paste electrode, Int. J. Electrochem. Sci. 5 (2010) 967–977.
[54] F. Faridbod, M.R. Ganjali, B. Larijani, P. Norouzi, Multi-walled carbon nanotubes (MWCNTs) and room temperature ionic liquids (RTILs) carbon paste Er(III) sensor based on a new derivative of dansyl chloride, Electrochim. Acta 55 (2009) 234–239.
[55] M.R. Ganjali, S.O. Ranaei-Siadat, H. Rashedi, M. Rezapour, P. Norouzi, Thulium selective sensor based on nanographene/RTIL/ionophore/graphite, Int. J. Electrochem. Sci. 6 (2011) 3684–3693.
[56] M.R. Ganjali, M. Rezapour, S.K. Torkestani, H. Rashedi, P. Norouzi, Long-term stable fabrication of a nanocomposite Tm(III) sensor containing nanographene/nanosilica/RTIL/ionophore, Int. J. Electrochem. Sci. 6 (2011) 2323–2332.
[57] F. Faridbod, M.R. Ganjali, M. Pirali-Hamedani, P. Norouzi, MWCNTs-ionic liquids-ionophore-graphite nanocomposite based sensor for selective determination of ytterbium(III) ion, Int. J. Electrochem. Sci. 5 (2010) 1103–1112.
[58] P. Norouzi, M. Pirali-Hamedani, S.O. Ranaei-Siadat, M.R. Ganjali, Nanocomposite carbon paste sensor for selective determination of Lu(III) ion, Int. J. Electrochem. Sci. 6 (2011) 3704–3713.
[59] E. Lindner, R.E. Gyurcsányi, Quality control criteria for solid-contact, solvent polymeric membrane ion-selective electrodes, J. Solid State Electrochem. 13 (2009) 51–68.
[60] M. Zhou, S. Gan, B. Cai, F. Li, W. Ma, D. Han, L. Niu, Effective solid contact for ion-selective electrodes: tetrakis(4-chlorophenyl)borate (TB−) anions doped nanocluster films, Anal. Chem. 84 (2012) 3480–3483.

[61] R.W. Cattrall, H. Freiser, Coated wire ion-selective electrodes, Anal. Chem. 43 (1971) 1905–1906.
[62] D.V. Chernyshov, M.G. Khrenova, I.V. Pletnev, N.V. Shvedene, Screen-printed ion-selective electrodes covered with membranes containing ionic liquids, Mend. Comm. 18 (2008) 88–89.
[63] C. Cristea, E. Bodoki, V. Sima, R. Sandulescu, Modified screen printed electrodes for the development of biosensors, in: S. Vlad, R.V. Ciupa, A.I. Nicu (Eds.), International Conference on Advancements of Medicine and Health Care through Technology, IFMBE Proceedings, vol. 26, 2009, pp. 89–92 (Springer Berlin Heidelberg).
[64] Z. Stanic, S. Girousi, Carbon paste electrodes in potentiometry: the state of the art and applications in modern electroanalysis (A review), Sens. Electroanal. 6 (2011) 89–128.
[65] Z. Yan, Y. Pei, J. Fan, S. Wang, J. Wang, Selective electrodes for $[PF_6]^-$ and $[BF_4]^-$ anions based on the associates formed by ionic liquid and cationic dyes, Mat. Sci. Eng. C 33 (2013) 356–361.
[66] M.R. Ganjali, A. Karimi, S.J. Shahtaheri, P. Norouzi, Determination of clonidine by potentiometry using PVC membrane electrode, Int. J. Electrochem. Sci. 8 (2013) 1999–2008.
[67] F. Faridbod, M. Hosseini, M.R. Ganjali, P. Norouzi, Potentiometric sensor for determination of clomiphene, Int. J. Electrochem. Sci. 8 (2013) 1976–1985.
[68] M.R. Ganjali, B. Larijani, F. Faridbod, P. Norouzi, Potentiometric determination of cyclizine by a PVC membrane sensor, Int. J. Electrochem. Sci. 8 (2013) 10487–10497.
[69] C. Wardak, 1-Hexyl-3-methylimidazolium hexafluorophosphate as new component of polymeric membrane of lead ion-selective electrode, Desalin. Water Treat. 51 (2013) 658–664.

CHAPTER 10

Use of Ionic Liquids in Electrochromic Devices

Sofiane Hassab, Javier Padilla
Department of Applied Physics, Technical University of Cartagena (UPCT), Cartagena, Spain

1. INTRODUCTION

1.1 Color is Everywhere

Color changes have always been a source of information, and we are continuously receiving color stimuli as electromagnetic radiation, originally from the sun and afterward through interaction with a myriad of objects or living bodies around us. From the slow and basic nature changes that inform us of the season or the daily darkening and bleaching of the sky, to the highly sophisticated signs that you are using to interpret these words, we are always surrounded by color.

As with other phenomena in nature, humanity has learned to control color. From the static and permanent ways of generating color images or codes (painting, printing), technology has evolved toward dynamic, nonpermanent methods of showing images that flood everyday life through myriad electronic devices (mobile phones, television, cinema, advertising spots, etc.) and different color technologies, among them liquid crystal displays (LCDs) and light-emitting diodes (LEDs) being probably the most consolidated ones. In this chapter we will explore some aspects of an emerging alternative that tries to find its space among these mature technologies.

New or improved applications on different fields, like architectural glazing windows for energy control and glare reduction in building industry, antiglare rearview mirrors and sun roofs for cars, or dynamic eyewear, among others, are the ones in which electrochromic devices, which will be defined in this text, are showing promising results.

1.2 Electrochromism: A Way of Controlling Color

Electrochromism refers to the property of certain materials to modify their color by means of an electrical current. Additionally, the color change is expected to be reversible and gradually controlled. Although the term can be used for different physical or chemical processes (e.g., electrophoretic displays), most of the time it involves redox processes in an electrolytic medium, that is, an electrochemical process. An electrochromic device is an electrochemical cell composed of two electrodes (electronic conductors) embedded in an electrolyte (an ionic conductor, usually a salt and a solvent).

Ionic Liquids in Separation Technology
http://dx.doi.org/10.1016/B978-0-444-63257-9.00010-9
Copyright © 2014 Elsevier B.V.
All rights reserved.

As we will see later, the electrochromic material can be deposited over the electrodes as a thin film or be part of the electrolyte. In any case, the application of an electric potential promotes the oxidation (extraction of electrons) at one side and the reduction (donation of electrons) on the other side (understanding "side" as solid material or in solution species), and this results in a color change.

To have an initial understand of how this process results in a color change, we must bear in mind a simple but practical definition of color as a result of interaction of light with materials, and the concept that electrons inside matter are organized in different energy levels: when visible light (understood as electromagnetic radiation of different wavelengths) reaches a material, the incident photon energy is able to excite an electron toward an upper level, and this could be resulting in an absorbed photon, among other processes. The color we perceive of a material is the resulting nonabsorbed wavelengths. The oxidation or reduction of a material creates or eliminates new energy levels, therefore promoting different transitions and absorptions, creating a different color. This is the process that we can control gradually and reversibly by means of an applied electrical current.

1.3 Electrochromic Materials and Devices

Many different materials are electrochromic; we can, roughly speaking, divide them into two general groups: organic and inorganic materials. Among the organic materials, conjugated polymers and viologens are well-known ones, whereas we usually refer to transition metal oxides when talking about inorganic ones.

Conjugated or conducting polymers are a special type of polymers with semi-conducting to metal properties [1]. From a chemical point of view, the alternation of double and single bonds along their structure is the main characteristic, this being also the origin of the possibility of electrical conduction. These materials undergo reversible redox reactions that promote color changing from the visible to infrared (IR) wavelengths.

Viologens are bipyridinium derivatives. These bipyridinium ions undergo a reversible one-electron reduction to switch from a dicationic to a monocationic radical state, promoting color changes, usually in the visible region, as the redox chemistry occurs.

Transition metal oxides are also able to reversibly change their color by a process of ion intercalation. Among them, WO_3 is the most studied one, but other oxides like V_2O_5, NiO, CeO_2, and MoO_3 have been also investigated.

The more common electrochromic architectures include two transparent electrodes: one electrochromic redox system, which has high transmittance in the reduced form and low transmittance in the oxidized form, or vice versa; and a second redox system, which can have opposite electrochromism to the first, or is highly transparent in all oxidation levels. This configuration refers to transmission modulation applications, which are the most frequent. Exceptions to this configuration occur for reflecting devices, in which additional materials are included in the electrodes or in solution to reflect light (metallic for mirrors or IR modulators, white-diffusing for displays) or in both-colored state devices. In any of the possible configurations, an electrochemical reaction occurs, meaning that oxidizing species at one electrode and reducing species at the other must be present simultaneously. Some electrochromic materials are colored when oxidized (anodically coloring materials), others when neutral or reduced (cathodically coloring materials). Some device configurations take advantage of this situation using two complementary coloring materials to construct the device.

According to Rauh [2], there are three different electrochromic configurations: the "solution type," where both redox systems exist together in solution; the "hybrid type," where one redox system is immobilized as a thin film on a transparent electrode and the other one is in solution; and the "battery type," where both redox systems form thin layers on a conductive substrate. In the case of a layer as redox system, ions are also inserted into or extracted from the layer, simultaneously with the electrons, to balance the charge. Usually the battery type is considered bi-stable (meaning that, ideally, both systems retain their color states without continuous potential application), whereas the solution type and the hybrid type require a continuous current to maintain the color.

When quantitatively describing an electrochromic device, as their main function is a color change, optical magnitudes are, obviously, the most important. A color state can be described by its transmittance spectrum, which measures the amount of light that the material transmits for a defined wavelength range. Transmittance is expressed as a ratio of intensities $T = \frac{I}{I_0}$ and is usually written as a percentage $\%T$. Transmittance and absorbance (A), a directly related magnitude, have an exponential relation: $T = 10^{(2-A)}$. Optical contrast is defined as the percentage transmittance change between two color states, usually at a certain wavelength, when the maximum variation occurs. Another relevant magnitude, especially when talking of reflective devices, is the reflectance ($\%R$), which measures the amount (more correctly, a ratio, therefore without units) of light reflected.

Coloration efficiency refers to the amount of charge consumed to promote a color change, and it is usually expressed as $\mathrm{CE} = \frac{\Delta A}{Q}(\mathrm{cm^2/C})$. Certain studies also include colorimetric parameters, which define more precisely the color of a material. There are several ways for defining a color, based on different color spaces that rely on standard parameters defined by the CIE (Commission Internationale de l'éclairage). All of them use three coordinates to specify a given color.

Other usual magnitudes defining a device performance are switching time (sometimes ambiguously termed switching speed) and lifetime. The former is the time consumed to promote a certain color change. Because of the difficulty of exactly defining this term, usually the reference value is taken as the time to promote 90% of the total transmittance change. The latter is the number of cycles (color changes) that a device is able to stand without detrimental performance. This is also a difficult parameter to define, and usually is reported as the decrease in transmittance change after a certain number of cycles.

Finally, optical memory is a relevant magnitude, especially concerning device energy consumption. It measures the time that a device is able to retain a given color in the absence of any potential applied.

1.4 Relevance of Ionic Liquids in Electrochemical Applications

As mentioned before, electrochemical devices need an electrolyte to operate; specifically, ions are needed to allow current flow between both electrodes. Usually these ions (cations and anions) come from a salt dissolved in a solvent. The main problem, in terms of device operation, is the gradual evaporation of the solvent, which finally hinders the overall electrochemical reaction, together with possible leakage. Ionic liquids, defined as salts with low melting temperature (by convention, below 100 °C), act as both solvent and salt: one of their most important properties is a negligible vapor pressure (and generally thermal stability and nonflammability), so evaporation is greatly reduced or eliminated.

The leakage problem is not avoided, however; but we will see different possibilities to overcome this issue, presenting solid or semi-solid electrolytes including ionic liquids.

Other important characteristics, from an electrochemical point of view, are their high ionic conductivity (allowing fast processes) and their large electrochemical window (e.g., −2.5 to 2.5 V versus Ag wire quasi RE [3]) that allows a broad range of electrochromic reactions without side decomposition reactions that often occur in conventional solvents.

1.5 Chapter Outline

From a user's point of view, electrochromic devices appearance is not complex. However, they are usually composed of several layers of different materials. From the synthesis and deposition of each material to the final assembly and durability, construction of these devices is more complex than might be thought.

This chapter aims to show that complexity, starting from the single electrochromic materials, through the key role that electrolytes play in the device performance, to finally reach the complete devices' operation and characterization.

2. USE OF IONIC LIQUIDS TO SYNTHESIZE ELECTROCHROMIC MATERIALS

Either organic or inorganic materials are synthesized using many different techniques. "Dry" techniques, like sputtering, chemical vapor deposition, spray pyrolysis, or pulsed laser deposition, are more often used for inorganic oxides, while "wet," especially electrochemical, methods are employed for conducting polymers.

Electropolymerization is probably the most used technique in academics for synthesizing conducting polymers. Ionic liquids can act as an electrolytic medium, which is always needed in these processes. The need for broad-potential windows for high-oxidation-potential monomers, or structural or morphologic changes arising from electropolymerization in ionic liquids, are the key for the interest shown by the scientific community. Additional benefits of polymerization processes in ionic liquids are reviewed by Kubisa [4].

Although there are a vast number of conducting polymers, three main families (attending to their primary monomer) stand out over the rest: polyanilines (PANI), polypyrroles (PPy), and polythiophenes (PT). Historically, the simple monomers PANI and PPy or PT have been studied, but the need for specific tuning of color, stability, or processability has displaced them with a vast family of more complex monomers (but usually containing these basic monomers in their structures). The case of thiophene is particularly relevant, for instance, as different substituted poly-alkyl-thiophenes are able to cover the whole color gamut [5]. More recently, alkyl-substituted pyrroles have been used [6]. Figure 1 shows the chemical structures of some of these monomers.

2.1 Polyanilines

There are several reports concerning the electropolymerization of anilines in ionic liquids for different applications [7–11]. Research toward electrochromic applications

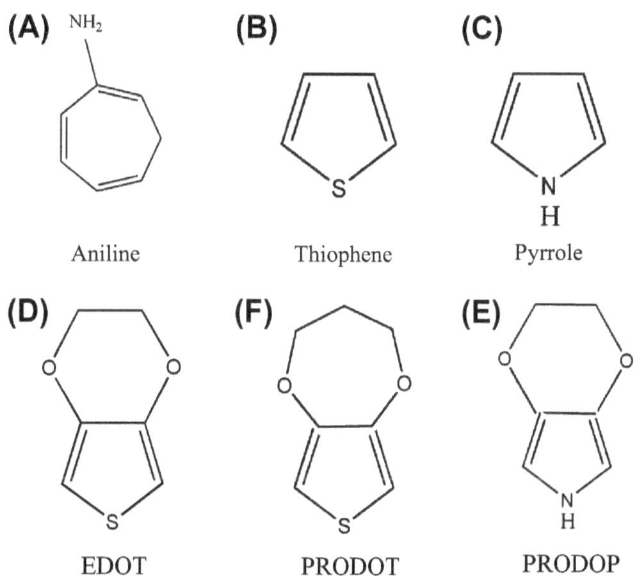

Figure 1 Chemical structures of some conjugated monomers.

is more restricted, though. The work carried out by Lu et al. [3,12,13] was pioneering in successfully including ionic liquids as substitutes of water in the polymerization of PANI (this work also represented a breakthrough in other aspects, increasing the lifetime of devices up to 1,000,000 cycles. This will be further explained later in this chapter).

PANI is usually polymerized in aqueous electrolytes, and due to the reduced potential windows, this process is specially affected by parallel degradation processes. Also, PANI is able to undergo two different redox processes: leucoemeraldine to emeraldine and emeraldine to pernigraniline at higher potentials. If the electrolyte used does not have a broad enough potential window, the electrochromic process of PANI is limited and cannot be fully obtained. Lu et al. focused on air-stable, hydrophobic ionic liquid such as $[C_4C_1Im][BF_4]$ as polymerization electrolyte, allowing the process to be in ambient conditions (absence of glove box). Due to the necessity of acid conditions for aniline to polymerize, an organic acid (CF_3COOH) was used, avoiding again the use of water.

Mu [14] explored the polymerization of PANI in highly acidic ionic liquid containing electrolyte and the redox response of the resulting films in variable pH medium, showing that films polymerized in the presence of ionic liquid show redox activity in higher pH mediums than conventionally polymerized PANI.

It is also interesting to mention the work of Fernandes et al. [15], who synthesized PANI in deep eutectic solvents (DES) as an alternative to ionic liquids. DES are mixtures of two or more components, forming an eutectic with melting point lower than those of their constituting components. The lower toxicity and cost of DES, plus, in certain cases, the good biodegradability of their components, may allow seeing them as inexpensive alternatives to ionic liquids.

2.2 Polypyrroles

Electropolymerization of pyrrole from ionic liquids has been made since the 1980s [16–18], although the synthesis was made in non-air-stable chloroaluminate-based ionic liquids and required the use of inert atmosphere. We can again consider the work from Lu et al. as a turning point for easier ambient condition polymerizations [3,12,13]. Different studies have been carried out since then [19–22], although, as in the case of polyaniline, fewer are directly intended for electrochromic purposes. Color transitions of polypyrrole (yellow to dark green) are not especially interesting for most electrochromic applications, in which at least one transparent state is desired. Ahmad et al. [23] have explored possible color tuning by synthesizing of N-substituted pyrroles and subsequent electropolymerization in 1-ethyl-3-methylimidazolium bis(pentafluoroethylsulfonyl) imide ($[C_3C_1Im][PFSI]$) and also compared [24] electrochromic response of PPy films grown in different ionic liquids or acetonitrile.

2.3 Polythiophenes

Polythiophene derivatives are probably the most successfully applied electrochromic conducting polymers, from the well-known PEDOT (polyethylenedioxythiophene) to a whole family of substituted thiophenes, ethylenedioxy or propylenedioxy based [25].

There are some reports of propylenedioxythiophene electropolymerization in ionic liquids, although the final application was not electrochromism [26,27]. It is also worth mentioning the synthesis of an unsymmetrical fused thiophene made by Inagi et al. [28]: in this case ionic liquid allowed electropolymerization of the monomer, which was not possible to make in organic common solvents. Other thiophene derivatives, like chlorothiophene, which requires high anodic potential to be polymerized, have been successfully synthesized [29] due to the broad electrochemical window of, in that case, $[C_4C_1Im][PF_6]$.

Although 3-substituted derivatives like poly-3-hexyl-thiophene (P3HT) or poly-3-octyl-thiophene (P3OT) are more important in the field of organic solar cells, they also have been used for electrochromism. Pang et al. [30,31] were able to synthesize them, together with poly-3-methyl-thiophene (P3MT), obtaining good contrast values of 45%, 39%, and 46%, respectively. Other research teams also synthesized P3MT, in the same ionic liquid [32] ($[C_4C_1Im][PF_6]$). Song et al. explored the electropolymerization of another derivative, 3-methoxy-thiophene, in a different ionic liquid, $[C_4C_1Im][BF_4]$ [33].

Some efforts have been made also with the omnipresent PEDOT [3,32,34], on direct electropolymerization from ionic liquids. However, as in many other applications, it is the PEDOT: Polystyrenesulfonate (PEDOT:PSS) dispersion, due to its processability and superior performance, that attracts more interest. Dobbelin et al. [35] made a comprehensive review of different approaches toward new PEDOT dispersions, including ionic liquids. The first successful approach has been to substitute PSS, used as a stabilizer for chemical oxidative polymerization of EDOT, for polymeric ionic liquids (PIL) [36]. The role that PIL play in the polymerization of PEDOT has also been studied by Kim et al. [37].

Another approach pointed out on that review is the use of ionic liquids as additives in PEDOT:PSS dispersions. By adding an excess of ionic liquids to an aqueous dispersion of PEDOT:PSS, the hydrogen counter-cations of PSS were exchanged by the corresponding cation coming from the ionic liquid, resulting in a precipitate that can be redispersed in some organic solvents [38].

Finally, they report a further step beyond inclusion of ionic liquid in the electrochromic polymer. Ionic liquid functionalized EDOT monomers were synthesized and polymerized, either chemically [39] or electrochemically [40]. Although the work is more focused on obtaining well-ordered structures of conducting polymers, Lee et al. [41] synthesized and electropolymerized an ionic liquid monomer containing a thiophene moiety together with a vinyl group, that showed electrochromic behavior.

2.4 Other Conducting Polymers

Finally, other conducting polymers not directly related to the families mentioned above have also been studied. Dong et al. [42] reported the electrochemical polymerization of 1,2 methylenedioxybenzene (MDOB) as analogue to other successful electrochromic polymers like 3,4-alkylenedioxythiophene and 3,4-alkylenedioxypyrrole. Although an inert atmosphere was needed to obtain good-quality films, Zhuang et al. [43] explored the electropolymerization of poly-carbazole, an anodically coloring polymer.

It is worth noting that literature cited in this section only considers electropolymerization in ionic liquids. Scarcity of these works may give a wrong impression about the scientific efforts made toward electrochromic polymers, which is not the case. Extensive work has been made on every aspect treated here, although the review of every one falls out of the scope of this chapter. As an example, the color tuning of different conducting polymers: an introduction to this vast field can be found in the two extensive reviews from Beaujuge et al. or Walczak et al. [6,25].

2.5 Composites

The synthesis of electrochromic materials in modified surface substrates or in the presence of other particles or chemicals, resulting in composites (understood as a combination of two or more materials with different, improved properties, compared with the single-material ones), is a useful tool previously used with the aim of, among others, increasing adherence, redox activity, color switching speed, or optical properties of the electrochromic material.

The interesting optoelectronic properties of other carbon-based materials (carbon nanotubes or graphene) have attracted enormous attention in many fields [44—47]. The good values of transparency and electrical conductivity make them also interesting in the electrochromic field for its use as electrodes. It has been also proposed that they can increase the adherence of the electrochromic materials to the substrate and therefore the lifetime of the corresponding device. Ahmad et al. [48] explored that possibility through electropolymerization of poly-methylpyrrole (PMPy) in the presence of carbon nanotubes.

Reddy et al. [49] synthesized poly-propylenedioxypyrrole (PRODOP) using reduced graphene oxide (RGO) and 1-ethyl-3-methylimidazolium bis(trifluoromethylsylfonyl) imide [EMIM][TFSI] ionic liquid (IL), resulting in PRODOP-RGO-IL, with increased conductivity and coloration efficiency versus PRODOP-IL films. Inclusion of the reduced graphene also modified the optical properties in the near-infrared region. The same authors were able also to synthesize nanocomposites of PEDOT-ionic liquid functionalized graphene (ILFG) [50] and compare them with PEDOT-RGO films. The ILFG films showed superior electrochemical activity and coloration efficiency.

A different approach is the inclusion of metallic nanoparticles in the structure of the electrochromic material. Bhandari et al. [51] synthesized PEDOT films in the presence of an ionic liquid, with subsequent inclusion of Au nanoparticles and CdSe quantum dots. The resulting films showed higher values of optical density change, coloration efficiency, and switching speed compared with PEDOT control films without nanoparticles. Similar work was done by the same research group including Ag or Au nanoparticles in Polyethylenedioxypyrrole (PEDOP) [52]: again, increasing switching speeds were obtained with the nanocomposited film.

Mesoporous TiO_2 substrates with highly increased available surfaces (compared to "flat" substrates) have been successfully applied in the development of dye-sensitized solar cells [53]. Increasing availability of redox sites boosts the light collection and final efficiency of the device. The same mesoporous substrates have been applied to develop fast electrochromic displays [54,55]. Additional benefits of the direct electropolymerization of conducting polymers onto the TiO_2 surface are stronger adhesion and therefore possibility of increasing the lifetime of devices by avoiding cracking or delamination. Pang et al. [56] obtained better stability of poly-chlorothiophene (PClT) films deposited over a nanostructured TiO_2 surface. Similar stability increase results were obtained by Ma et al. [57] with PEDOT/TiO_2 films.

Finally, an interesting approach from an optical point of view is to synthesize composites of two electrochromic materials. The idea behind this is to obtain complementary colors or reinforce the coloration of one of the materials. Song et al. [58] obtained combined dark red and dark purple coloration in PBrT/WO_3 films (electropolymerized poly-bromothiophene over WO_3 substrate) with logically increased redox activity compared with the single-material films. The same research group followed similar methodology to obtain poly-methoxythiophene/WO_3 [59] and poly-methoxythiophene/NiO (nickel oxide) films [60].

2.6 Viologens

The fact that viologens are in-solution electrochromic materials (that is, the species are present dissolved in an electrolyte rather than as a solid thin film) allows thought of interesting approaches combined with ionic liquids, which act as both electrolyte and solvent.

Kavanagh et al. [61] incorporated a viologen into an ionogel, with subsequent photocuring to obtain a solid electrolyte. Ionogels are hybrid materials in which an ionic liquid is confined, chemically or physically, into a polymer matrix. The use of ionogels as solid electrolytes for electrochemical devices is being explored, as we will see in more detail later on in this chapter. More recently, the same research group has gone further in this approach [62], synthesizing a viologen-ionic liquid compound in which both materials are chemically bonded, with subsequent photopolymerization (using **2,2-Dimethoxy-2-phenylacetophenone** (DMPA)) to obtain a solid electrochromic electrolyte.

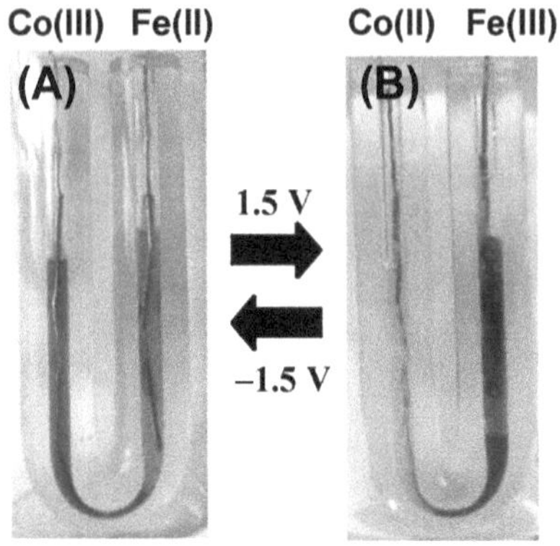

Figure 2 Electrochromic cell designed in a U tube showing electrochromic behavior of two immiscible magnetic ionic liquids. *(Reproduced from reference [68] with permission of The Royal Society of Chemistry).*

Ionic liquidized DNA (deoxyribonucleic acid) has been used as a host polymer to incorporate viologen molecules in the work carried out by Kakibe et al. [63,64]. One of the aims of this approach is to use DNA to prevent dimerization of the viologens [65], an irreversible process that limits the electrochromic response of these materials.

Synthesis of ionic liquids with transition metal oxide anions leads to interesting magnetic properties [66,67], but also electrochromism [68] when subjected to redox processes (Figure 2).

2.7 Inorganic Materials

The only work that can be found related to electrochemical deposition of electrochromic inorganic oxides in ionic liquid mediums is from Cai et al. [69]. In this work NiO was deposited from a choline chloride-based deep eutectic solvent. They explored the different electrochromic response of the resulting film as a function of process temperature, and reported differences as large as 40% in contrast (27% compared to 67%) in films deposited at 70 and 90 °C, respectively.

3. IONIC LIQUIDS AS ELECTROLYTES

Due to its intrinsic nature, ionic liquids can be considered pure electrolytes. Additional properties like a broad electrochemical window make them suitable candidates for characterization of electrochromic materials. Most of the materials referred to in the previous section, synthesized in ionic liquids, were also optoelectrochemically characterized in the same or similar liquid mediums [3,14,22,23,29,31–34,42,43,48,50, 56,58–60,70,71]. Other studies in which electrochromic materials are characterized in ionic liquids can be easily found. Fujimoto et al. [72] were able to optoelectrochemically characterize a new molecule, octathio[8]circulene, in ionic liquids, otherwise not possible because of the insolubility in aqueous and organic solvents. On the inorganic materials side, Tebby et al. [73] deposited TiO_2 films by doctor blade technique, and afterward irradiated them with ultraviolet light, obtaining electrochromic films successfully tested in lithium-containing ionic liquid, in a low-temperature process that could be applied to plastic substrates. Surca Vuk et al. [74] synthesized a specially designed ionic liquid, from the sol–gel precursor 1-[3-(trimethoxy-λ^4-silyl)propyl] imidazole, for its use in electrochromic devices of WO_3 and different counter-electrode oxides ($CeVO_4$, V_2O_5, Ti/V-oxide). Polyhedral oligomeric silsesquioxanes (POSS) were used by Colovic et al. [75] to synthesize solid (melting point 150–200 °C) $I^-/I^-{}_3$ ionic liquids, although for practical reasons they finally used room temperature liquid derivatives in WO_3 hybrid electrochromic devices.

Additionally, use of pure ionic liquids (no gel or solid compositions) as electrolytes can be found in patent literature. Some researchers have designed or proposed specific ionic liquids [76,77], electrochromism being just one of the possible applications among other electrochemical devices. McEwen and Koch proposed, in 2000 [78], ionic liquids as nonflammable electrolytes for electrochemical devices, including electrochromic. Abundant electrochromic device architectures have been proposed suggesting ionic liquids as electrolytes [79–98]. Ionic liquid mixed with other materials, but still in liquid final form, can also be found [99,100].

Although ionic liquids show different advantages to be used as electrolytes in electrochromic devices, it is desirable, for practical reasons, that they may be used in a partially solid form. This will prevent leakage or gravity deformation and increase security in eyewear and other applications. For that reason, in the next paragraphs we would like to focus in the efforts made toward that objective. As in previous sections, it must be remarked that research toward solid or quasi-solid electrolytes is not restricted to electrochromic applications. On the contrary, in electrolyte development science, electrochromism could be considered a marginal topic that has seen considerable benefit from advances in other applications, especially batteries. There is, nevertheless, enough literature concerning strategies to obtain ionic liquid-based quasi solid electrolytes for electrochromic applications.

3.1 Polymer Electrolytes Containing Ionic Liquids

Polymers combined with salts, showing adequate ionic conductivity, have been extensively used as electrolytes in electrochemical devices, polyethylene oxide (PEO) being the first and more studied one [101]. Improved properties, especially increasing ionic conductivity, can be obtained with the inclusion of a plasticizer, usually organic solvents like ethylene carbonate, propylene carbonate, dimethyl sulfoxide, dimethyl carbonate, or diethyl carbonate. This allows better ion mobility inside the electrolyte, but it is still subjected to evaporation processes. Ionic liquids [102] have been proven as an alternative, nonvolatile component for the same purposes.

Desai et al. [103] compared PEO-based polymer electrolytes, using lithium perchlorate and lithium trifluoromethanesulfonylimide salts and plasticized with ionic liquid or conventional organic solvents. Electrochromic devices were made and tested using both options, and the ones using ionic liquid as plasticizer showed superior stability in a 10,000 switching cycle stability test.

Many other analogue polymers have been studied for polymer electrolyte purposes. Bircan et al. [104] studied the effect of ionic liquid concentration in poly(ethyleneglycol) ethyl ether methacrylate (PEG-MA) and poly(ethyleneglycol) diacrylate (PEG-DA) polymer electrolytes, with lithium trifluoromethanesulfonate (LITRIF) as salt, in PEDOT/PBENMCZ (poly[3,6-bis(2-(3,4-ethylenedioxy)thienyl)-*N*-methylcarbazole]) electrochromic devices.

Polymethylmethacrylate (PMMA) is another often studied polymer for electrolytes, due to its high degree of transparency (especially relevant for electrochromic applications). In a recent work by Duluard et al. [105], different weight ratio mixtures of PMMA and two ionic liquids (1-butyl-3-methylimidazolium hexafluorophosphate ($[C_4C_1Im][PF_6]$) and 1-butyl-3-methylimidazolium bis(trifluoromethanesulfonyl)imide ($[C_4C_1Im][NTf_2]$), together with variable concentrations of Lithium Bistrifluoromethanesulfonyllmide (LiTFSI) salt, resulted in free-standing membranes that the authors characterized thermally and optically (Figure 3). Ionic conductivity was also

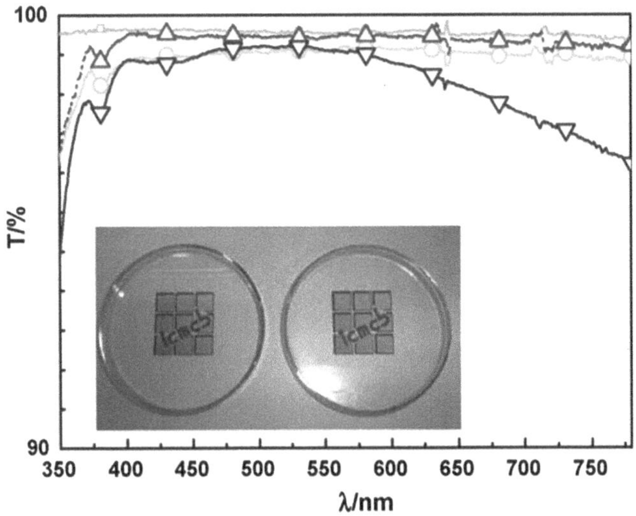

Figure 3 Ultraviolet–visible spectra of PMMA-ionic liquid-based membranes. Inset photographs showing achieved transparency (left-hand side with no deposited membrane and right-hand side with a membrane). *(Reprinted from Ref. [105], Copyright (2010), with permission from Elsevier).*

studied for the different combinations. These values were logically higher for the pure ionic liquids, but mixtures with 30% PMMA were used as electrolytes in PEDOT/Prussian blue (PB) devices showing appropriate electrochromic behavior [106]. Some organic solvents (dichloromethane, acetone, or butanone) were used for adding in the dissolution processes, although they were evaporated prior to device assembly or polymer electrolyte characterization.

Other general compositions resulting in polymer electrolytes, including any of the above mentioned polymers (PEG, PMMA, etc.), can be found in patent literature [107–117].

Ionic liquids have the chemical versatility to be produced as polymer structures, termed polymeric ionic liquids. This has opened new ways of using these materials [118–121]. Du et al. [122] have recently synthesized a polymeric ionic liquid, poly(-methyl 2-(3-vinylimidazolidin-1-yl)acetate bis(trifluoromethanesulfonyl)imide) ([PMVIm][NTf_2]) and developed polymer electrolytes composed of [C_4C_1Im][NTf_2], lithium bis(trifluoromethanesulfonyl)imide (Li[NTf_2]), and poly(methylmethacrylate co-vinyl acetate) (P(MMA-VAc)). The thermal stability and ionic conductivity were found to increase considerably compared to the electrolyte without the polymeric ionic liquid. Similar studies were conducted with another polymeric ionic liquid [123] synthesized by the same group, poly-1-methyl-3-[2-(methacryloyloxy)ethyl]imidazolium bis(trifluoromethane sulfonylimide) ([PMMEIm][NTf_2]).

It is also this chemical versatility that allows new, fascinating compositions in which a polymeric ionic liquid is dissolved in its own analogue pure ionic liquid, resulting in functional quasi-solid electrolytes, as in the work by Marcilla et al. [124]. The dual function of ionic liquid, as salt and solvent, makes it somehow difficult to classify the different ionic liquid-containing electrolyte compositions in a classical way. Adding this new polymer functionality make this differences more subtle.

At this point, we believe it is useful to explain the criteria, which of course are arguable, that we have followed to classify the different types of quasi-solid electrolytes in this section: polymer electrolytes and ionogels (these are to be discussed in the next subsection). From the simplest definition of a polymer electrolyte, it is composed of a polymer with certain ionic conductivity and a dissolved salt. From the viewpoint of functionality of electrochromic devices, the ions from the salt are the ones participating in the electrochromic material redox processes, usually compensating a charge excess or defect. In this context, added ionic liquid to this composition acts as a plasticizer, usually increasing conductivity and mechanical flexibility.

We therefore distinguish three components and functions: polymer as the main component and solvent for salt, salt as ion source for redox processes, and ionic liquid as plasticizer to improve mechanical properties and ionic conductivities.

On the other hand, ionogels are hybrid materials in which an ionic liquid is embedded or trapped inside a structure (usually a polymer matrix, although inorganic gelators, for instance, can also be used). In this case, there are only two basic components, polymer matrix and ionic liquid. The ionic liquid acts as a salt now, providing the ions necessary to complete the redox processes, and the polymer matrix provides mechanical integrity to obtain a quasi-solid electrolyte.

Considering the desired properties of an electrolyte, solid ones usually have low conductivity, but liquid ones are prone to leakage or evaporation. A balance between these two limits can be obtained from two directions. In the case of polymer electrolytes with low ionic conductivities, ionic liquids help by plasticizing the mixture. Ionic liquids, however, can act as electrolytes by themselves, but are liquid. In this case, another compound, a polymer matrix, helps in making the mixture more solid (an ionogel).

Based on these criteria, we have included the above mentioned electrolytes developed by Marcilla et al. [124] as polymer electrolytes. In their work, the ionic liquids 1-butyl-3-methylimidazolium bis(trifluoromethanesulfonimide), $[C_4C_1Im][NTf_2]$, 1-butyl-3-methylimidazolium tetrafluoroborate, $[C_4C_1Im][BF_4]$, and 1-butyl-3-methylimidazolium bromide $[C_4C_1Im][Br]$ were mixed with their analogues, polymeric ionic liquids poly[1-vinyl-ethyl-imidazolium] bearing similar counter-anions $[NTf_2]$, $[BF_4]$, and [Br]. We considered that the polymeric ionic liquid will act as polymer structure and salt, and the added ionic liquids as plasticizers. In this case, the same anion is present in the polymer structure and in the ionic liquid, hence the difficulty of classifying the type of electrolyte based on the above mentioned criteria. Some of the compounds developed are shown in Figure 4.

The last compound we want to comment on is also difficult to clearly classify. Colovic et al. [125] synthesized a complex ionic liquid monomer in which an inorganic phase (silica) and a PEO-based functionalization were included. The resulting monomer structure is shown in Figure 5. The study focused on the process of hydrolysis and condensation of the electrolyte, but they also tested it in a WO_3-based hybrid electrochromic device.

3.2 Ionogels

As explained in the previous section, the simplest way of defining an ionogel is a combined material in which an ionic liquid is confined, usually with the aim of obtaining a solid or quasi-solid platform to be used as electrolyte. More detailed insight on the different materials and possibilities used to obtain these materials can be found in two detailed reviews by Ueki [126] and Le Bideau [127]: in this last one, ionogels are presented as "hybrid materials, in which the properties of the IL are hybridized with those of another component, which may be organic (low molecular weight gelator, (bio) polymer), inorganic (e.g. carbon nanotubes, silica, etc.) or hybrid organic–inorganic (e.g. polymer and inorganic fillers)".

Regarding electrochromic applications, some of these options can be found in literature. One interesting approach is the use of abundant and inexpensive biopolymers, like gelatin or agarose, as gelling materials. Vidinha et al. [128] used gelatin dissolved in several ionic liquids to form what they termed "ion-jelly," showing acceptable

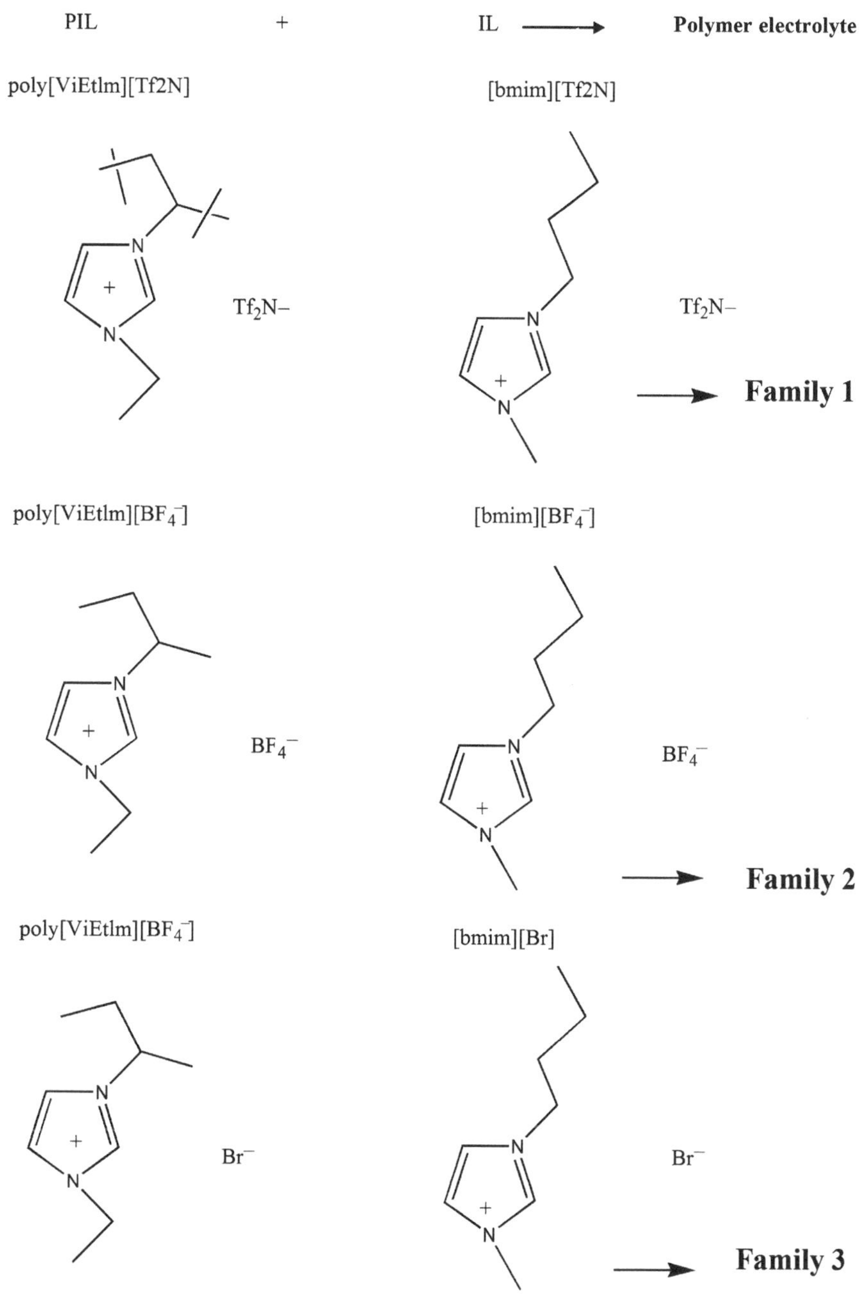

Figure 4 Different ionic liquid-based polymer electrolytes, as developed in reference [124].

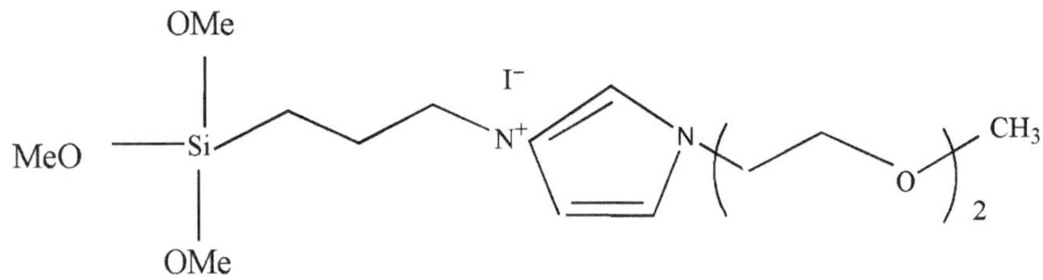

Figure 5 The structure of a single end-capped ionic liquid 1-(2-(2-methoxyethoxy)ethoxy)ethyl)-3-(3-(trimethoxysilyl)propylimidazolium iodide [EO_2-$TMSC_3$Im][I], for short. *(Structure drawn according to reference [125]).*

ionic and electronic conductivities, for operation of an electrochromic device, or even a light-emitting diode, respectively. Some of these researchers reported recently the electrospinning of these "ion-jelly" fibers [129] (Figure 6).

Leones et al. [130] explored similar gels, but also included europium salts in the search for additional luminescence properties that could broaden the application fields of these gels beyond electrochromism. The physical appearance of one of these membranes is shown in Figure 7. The same researchers recently reported an analogue work (without any salt in this case) on obtaining agarose-based ionogels [131]. Additional research on agarose-based ionogels has been developed by Trivedi et al. [132]. Another interesting approach is the use of cellulose, as proposed by Yang et al. [133].

Regarding more traditionally used polymer matrices, based on polyethylene oxides or analogues, Deepa et al. [134] used a poly-vinyl alcohol-based formulation in PEDOT/PANI and PEDOT/PB devices. The same author participated in another study in which ionogels of PMMA and ionic liquid were obtained [135]. The interesting approach of this work is that they studied two different ways of entrapping the ionic liquid inside the polymer matrix. In the first one, they used the conventional mixing, stirring, and drying method before clipping the device. In the second one they mixed the ionic liquid, 1-butyl-1-methyl pyrrolidinium bis(trifluoromethylsulfonyl) imide in this case, with MMA monomer and then subjected the mixture to thermal polymerization. They assembled the device with the liquid mixture and then gradually polymerized it in contact with the electrochromic films. This approach improved the ions' accessibility and the interfacial contact between electrolyte and films, which resulted in better operation of the device in terms of contrast and switching speed. Jia et al. [136], dealing with the topic of thermal stability of electrolytes for electrochromic devices, developed poly(vinylidene fluoride) (PVDF)-ionic liquid membranes. They compared ionic liquid-loaded PVDF-hexafluoropropylene gel membranes with supercritical CO_2-treated electrospun membranes, showing great improvement in the optical stability of the latter after extensive thermal exposure.

Analogue approaches, based on Polyvinyl alcohol (PVA), PMMA, etc. (polymers or monomers that can be thermally or photo-polymerized) are proposed by several authors [137–143].

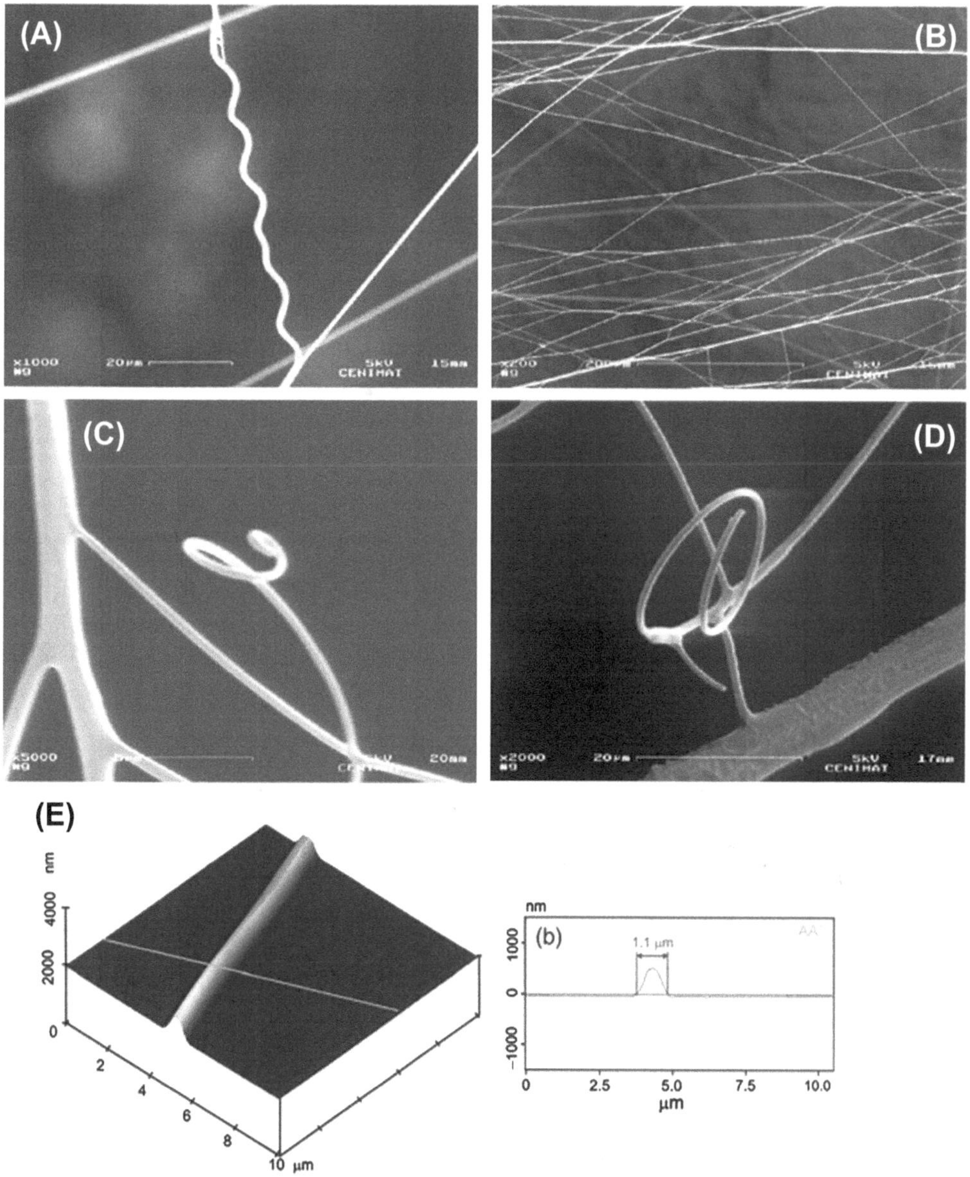

Figure 6 Scanning electron microscope micrographs of ion-jelly fibers (A,B,C,D) and atomic force microscopy topography of a single fiber (E. (b) depicts a lateral view showing thickness of a single fiber). *(Reprinted from Ref. [129], copyright (2012), with permission from Elsevier).*

Figure 7 Gelatin-ionic liquid-based membrane. *(Reprinted from Ref. [130], copyright (2012), with permission from Elsevier).*

Another option to obtain immobilizing matrices for the ionic liquids, as mentioned before, is the use of inorganic materials. Ahmad and Deepa [144] used tetraethyl orthosilicate, formic acid, and lithium bis(perfluoroethylsulfonylimide) Li[PFSI] salt, together with the ionic liquid 1-ethyl-3-methylimidazolium bis(perfluoroethylsulfonyl) imide $[C_2C_1Im][PFSI]$, to form an ionogel.

4. ELECTROCHROMIC DEVICES CONTAINING IONIC LIQUIDS

In this last section of the chapter we will explore the final result of almost all the previous works cited: functional electrochromic devices. The criterion we have followed to include any research work in this section is the presence of ionic liquids in the final structure of the device. Again, the picture presented here of electrochromic device materials and configurations is restricted and should not be interpreted as a complete view. For instance, after reading this section one could interpret that inorganic materials are less used than organic materials, which is not the case.

There are different options to classify electrochromic devices. Sometimes they are classified according to the type of materials used as color changing material (organic or inorganic), other times by their configuration (battery type, hybrid type, etc.): for more detailed information, it is always worthwhile to check the classic books of Monk et al. [145,146]. In our classification, we have included three traditional subsections, conducting polymer, inorganic oxide, and viologen-based devices, but also, due to extensive work found, subsections on hybrid and IR modulation devices. It is also worth mentioning that most of the patent literature concerning electrochromic devices has been included in the previous section on electrolytes. Due to their invention protection nature, patents tend to cover as many options as possible, and therefore, specific data become less important. Also, ionic liquid use as electrolyte in devices is generally proposed as just one option among other possible ones. In the case that ionic liquids are part of the invention disclosure, they also cover many general options. For these reasons, we chose then to show more specific device studies in the next section. Table 1 summarizes the main performance values of some of the devices commented on in the next sections.

4.1 Conducting Polymer-Based Devices

When talking about electrochromic devices and ionic liquids, the seminal work by Lu et al. [12], and subsequent ones [3,13] (Figure 8), are probably the most relevant ones. Ionic liquids played a key role, first as ionic medium to synthesize the conducting polymers used on the EC devices (PEDOT and PANI), and afterwards as electrolytes for the devices. Devices constructed showed 63% contrast and fast speeds (100 ms for 21% change in transmittance). PANI films cycled in an ionic liquid were electrochemically stable for up to one million cycles. Although unfortunately this stability test was not made for the assembled devices, it was the proof that the use of ionic liquids could dramatically increase the lifetime of electrochromic (and other electrochemical) devices. It is interesting to note that, although this aspect is frequently pointed out in electrochromic research using ionic liquids, comparative studies with other solvents and long-term lifetime tests are not so often found. We will pay special attention to this aspect (lifetime) together with optical modulation (% contrast obtained), switching time, and operating potential.

Abundant literature of conducting polymer-based electrochromic devices using an ionic liquid as electrolyte (or being part of it) can be found after the above mentioned works. Most of them use a dual configuration, where two electrochromic materials are present in the device, usually changing color in a complementary way. Despite the vast amount of electrochromic polymers available, curiously one of them, PEDOT, is present in most of the resulting devices. PEDOT-based devices are well known and thoroughly studied, so it is reasonable to think that researchers trying to include novelties in other aspects of electrochromic devices (i.e., a novel electrolyte) choose them as a standard testing system. Prussian blue is the most frequently chosen material to act as a counter-electrode. For instance, Bhandari et al. [71] studied a viologen-doped PEDOT//Prussian blue system, obtaining modest contrast values (20%), but showing the feasibility of this dual electrochrome approach. Vidinha et al. [128] also chose the PEDOT//PB system to prove the viability of their "ion-jelly" as electrolyte in electrochromic devices (however, no data on the device performance is shown). Deepa et al. [134] studied the same system devices, with a different ion liquid-based gel electrolyte, obtaining approximately 30% contrast at 633 nm. However, when subjected to ± 1 V steps for 2500 cycles, the devices lost 74% of optical modulation. The same research group tested different PMMA-based electrolytes [135], obtaining approximately 48% contrast at 585 nm for ± 2 V applied potential. No stability test was performed in this case. In an effort toward up-scaling, Duluard et al. [106] also worked with PMMA-based

Table 1 Performance of Several Electrochromic Devices

References	Device Configuration (Working electrode// Counter electrode)	Applied Voltage (V)	Contrast (%)	Reported Wavelength (nm)	Switching Time (s)		Number of Cycles	Degradation (Electrochemical or Optical)
					Bleaching	Coloring		
Conducting Polymer								
[3,12,13]	PANI//PEDOT-P3OT	(−1.5,+1.5)	63	560	0.1	0.1	1,000,000	Stable
[134]	PEDOT//PB	(−1,+1)	30[a]	633	5	2.7	2500	74%
[135]	PEDOT//PB	(−2,+2)	48[a]	585	2.2	2	—	—
[106]	PEDOT//PB	(−1,+1)	33[a]	640	2	7	1000	15%
[124]	PEDOT-viologen//PB	(−1.5,+1.5)	20	568	6	15	—	—
[32]	PEDOT//P3MT	(−1.2,+0.8)	26	665	1.2[b]	0.8	5000	35%
[104]	PEDOT//PBENMCZ	(−1,+1.4)	35–40	555	0.4	0.5	—	—
[147]	PEDOT//PEDOT	(0,+2)	51	Visible range	3.8	2.7	—	—
[49]	PEDOT:RGO-[EMIM][TFSI]//PB	(−3,+3)	<10[a]	500	>10	>10	—	—
[52]	PEDOT-Ag//PB	(−2,+2)	—	633	7	6.5	—	—
	PEDOT-Au//PB	(−2,+2)	—	633	6	6	—	—
[57]	PEDOT-TiO_2// Indium tin oxide (ITO)	(−1.1,+0.8)	30	570	0.8	0.8	10,000	5%
[56]	PClT-TiO_2//ITO	(−1.8,+2.2)	18	770	5.6	13.4	1000	1.30%
[31]	P3HT//ITO	(−1.8,+1.8)	—	440–780	—	—	1000	1.50%
[48]	(PMPy–CNT)// Fluorine tin oxide (FTO)	(−2,+2)	25	550	—	—	—	—

Inorganic Oxide								
[144]	WO_3//PB	(−1.5,+1.5)	46	633	0.8	2	1000	Stable
[135]	WO_3//PB	(−2,+2)	40[a]	685	1.7	2.1	—	—
[130]	WO_3//CeO_2-TiO_2	(−2.5,+2)	20	550	2	15	400	—
[148]	WO_3//V_2O_5	(−2,+2)	50	650	100	100	2000	Stable
Viologen								
[62]	[PV,8,8,8][3X] + DMPA	(−0.8,0)	—	610	221	221	—	—
[64]	DNA-viologen//TMPD	(−1,+1)	64[a]	600	6[a]	6[a]	100	Stable
Hybrid								
[125]	WO_3//Pt	(−1.5,+1.7)	54	634	8	40	4240	36%
[75]	WO_3//Pt	(−1,+1.7)	25	634	—	—	1000	—
[149]	WO_3//Pt	(0,+2)	62	Vis	From 2 s to 2 min		—	—
IR Modulation								
[147]	PEDOT//PEDOT	(0,+2)	44	1971	3.8	2.7	—	—
[150]	PEDOT//PEDOT	(−1.2,+1.2)	40[c]	10000−25000	—	—	20,000	20%

[a]Approximated values, calculated from plots. Not directly reported by the authors. (TMPD): N,N,N′,N′-tetramethyl-p-phenylenediamine
[b]Switching from red to blue colors, no bleached state.
[c]Values of reflectance, not transmittance.

Figure 8 Photograph of a PANI/[C_4C_1Im][BF_4]/PEDOT−P3OT display (6 × 6 cm) showing the number 8. *(Reproduced from Ref. [3] by permission of ECS – The Electrochemical Society).*

electrolytes, obtaining approximately 33% contrast at 640 nm and a 15% loss in absorbance variation after 1000 cycles. They also performed a longer test for 12 × 15 cm^2 devices, up to 21,000 cycles, after which the contrast variation was substantially decreased. This work also deals with one important topic for electrochromic devices: the influence of the electrical properties of the transparent conducting oxide (TCO) used as substrate in the device performance. Conductivities of these materials are limited, and this is reflected in lower switching speeds or higher operating potential of electrochromic devices when surface increases. The reader must pay special attention to this fact when comparing switching speeds from different devices.

Other materials used for dual combinations with PEDOT include an anodically coloring polymer like PBEDOT-NMCz [104], a red coloring polymer P3MT [32], or even also PEDOT in a symmetrical configuration proposed by Pozo-Gonzalo et al. [147]. In the first study, the main focus is on the thermal stability of the ionic liquid gel compositions: 30 cm^2 devices were constructed and tested, showing 35% contrast (at 555 nm) and fast switching speeds of 0.5 s for −1 to +1.4 V applied potential. No stability tests were performed. In the second case, the aim was to show a dual coloring device (red and blue, as part of the three red-green-blue primary colors) with no transmissive state. Devices using pure ionic liquid as electrolyte (with subsequent sealing to prevent leaking) showed 26% contrast at 665 nm and were able to retain 65% of that contrast after 5000 cycles (applied potential +0.8 and −1.2 V). Switching speeds were 1.2 for red coloring (PEDOT bleaching) and 0.8 s for blue coloring (PEDOT coloring). Finally, using a symmetrical configuration, in which the same material is used as electrochromic film at one electrode and as ion storage layer film at the counter-electrode, contrast values up to 50.5% in the visible region were obtained. Switching speeds were 3.8 s for bleaching and 2.7 s for coloring, for 0–2 V applied potentials.

Although, as mentioned previously, in most of the studies PEDOT is used as active material for the resulting devices, other polymers can be found. Reddy [49] and Kharkwal et al. [52] used PRODOP as the main component of several composites (containing reduced graphene oxide or Au (or Ag) nanoparticles). They found interesting improvements in electrical properties of the composites compared to the single polymers. From the electrochromic point of view, the more relevant result is the substantial decrease of switching time, from 31 to 6 s, when using these composites in devices.

As mentioned before, when constructing devices, in most of the cases two redox materials are included, due to the necessity of completing the reaction in the two electrodes. Both of them can be electrochromic, or one of them can be optically passive, acting as an ion storage layer. Use of just one material (leaving the bare TCO as counter-electrode) is not as usual because this frequently implies higher operating potentials and poorer electrochromic performance. However, several functional devices with this architecture can be found. Pang et al. [31,56], used P3HT and P3ClT to construct devices that did not appreciably decrease their performance (less than 2%) over 1000 cycles. Ahmad and Singh [48] used the same architecture to show carbon nanotube (CNTs) functionalized poly(methylpyrrole) devices. Ma et al. [57] obtained remarkable lifetime (10,000 cycles with only 5% contrast loss) and contrast (30% at 570 nm) with PEDOT/TiO_2 composite devices.

4.2 Inorganic Oxide-Based Devices

If PEDOT was the material mainly chosen for organic-based devices, WO_3 is the one on the inorganic side. Prussian blue was the counter-electrode material chosen by Ahmad et al. [144] and Sydam et al. [135]. In the first case, devices showing 46% contrast at 633 nm for 1000 cycles were obtained. In the second study, WO_3/PB devices were constructed (together with PEDOT/PB devices, as mentioned in previous paragraphs) obtaining almost 40% contrast at 685 nm. In both studies, switching speeds (coloring and bleaching) were below 2 s. Leones et al. [130] chose a CeO_2-TiO_2 transparent ion storage layer to test their gelatin-based gels. Brazier et al. [148] used the well-known vanadium oxide V_2O_5 as complementary material to obtaining 2000-cycle stable devices with 50% contrast at 650 nm.

4.3 Viologen-Based Devices

Viologens, when used as part of an electrochromic device, are usually dissolved in the electrolyte. As ionic liquids can act as solvent and/or salt, mixing of both materials appears a natural and straightforward way to obtain functional devices. Kavanagh et al. [61] followed this approach with their ionogels. They also studied an interesting route for fast and easy device assembly, by ultraviolet photocuring and photopatterning. The same research group has recently reported [62] a more advanced approach, in which the viologen moiety was chemically attached to an ionic liquid, obtaining improved switching times compared to analogue previous efforts on electrochromic ionic liquids [68]. Agrawal et al. have reported, in several patents [149–151], a number of different options for viologen-ionic liquid-containing devices, including liquid or gel electrolytes.

In another interesting approach, reported by Kakibe et al. [63,64], the authors take advantage of the chemical affinity of DNA with viologens to immobilize the latter. With the formation of these stable complexes, the aim of this research is to prevent dimerization of the viologen, which can shorten its lifetime. They obtained a remarkable 63% contrast at 600 mn, with approximately 6 s switching speeds.

Finally, Pozo-Gonzalo et al. [152] used their previously mentioned polymeric ionic liquid approach, including viologen compounds, to develop all-plastic electrochromic devices.

4.4 Hybrid Devices

Hybrid devices, as mentioned previously, are a type of electrochromic device containing a redox pair in solution that completes the reaction occurring in the electrochromic electrode. All of the published research found related to this configuration (and using ionic liquids) deals with inorganic oxide systems, more precisely WO_3//Pt. However, we have not included them as part of the previous "inorganic oxide devices" subsection because these hybrid systems are not restricted to inorganic or organic materials. Georg and Georg [153] studied this system, using a gelled ionic liquid containing $\frac{I^-}{I_3^-}$ They reported devices showing 62% contrast in the visible range, and different switching speeds for different device sizes (ranging from 2 min to 2 s), up to 30 × 30 cm^2(Figure 9). Using the same electrode materials and redox system, Colovic et al. [75,125] explored more sophisticated approaches (synthesizing iodide-containing ionic liquids as commented in the previous "electrolytes" section), obtaining up to 50% contrast at 634 nm.

Figure 9 A 30 × 30 cm^2 device in its bleached and dark state. *(Reprinted from Ref. [153], copyright (2009), with permission from Elsevier).*

4.5 Infrared Modulation

Modulation of optical transmission out of the visible range, especially the IR region, has been a well-known and, although not studied by a lot of researchers, interesting topic in electrochromism during the last years. The main application of these devices is in this case thermal control. Apart from specific applications like spacecraft thermal modulation or military night vision camouflage, modulation of the near-infrared region is gaining more importance in the field of sustainable buildings. Realizing that a great percentage of building heating through windows is caused by near-infrared radiation, efforts toward enlarging modulation from visible to IR are becoming more frequent [154]. As for hybrid devices, the construction of this special type of electrochromic device is not restricted to organic or inorganic materials; therefore we do not assign them to any of those previous sections. In this case, contrary to hybrid devices, conducting polymer-based devices are the only ones present in the literature.

Chandrasekhar has carried out extensive research on this type of device [155,156], especially for spacecraft applications. In the last years he has included ionic liquids in the electrolytes used [157—159], although there is not detailed information about the exact compositions used. One of these devices is shown in Figure 10.

Figure 10 Variable emittance device array. A 4-device array is shown, each device being 7.5 cm^2. *(Reprinted with permission from Ref. [159], copyright (2010), AIP Publishing LLC).*

Chevrot's group has recently introduced ionic liquids in the structure of their extensively studied concept of self-supported semi-interpenetrating polymer network electrochemical devices. They have successfully applied this concept for visible electrochromic devices or actuators. Recent reports focus on spacecraft thermal control applications [160–164], reporting a 20,000 cycle test with a 20% decrease in optical modulation (no decrease for the first 5000 cycles) [161]. In all the cases, the electrochromic material used was PEDOT. Pozo-Gonzalo et al. [147] showed that their symmetrical PEDOT–PEDOT configuration, previously mentioned in this chapter, can also be used for remarkable IR modulation (44% transmittance modulation).

Finally, a new concept has been recently reported by Wang et al. [165]. In this case, in the absence of redox reactions, potential applied to single-walled nanotubes in the presence of ionic liquid promotes a strong density variation of holes, resulting in electro-optical modulation, especially in the IR region (1.8 μm).

5. CONCLUSIONS

The different uses that ionic liquids can have in electrochromic devices have been reviewed, from the synthesis of electrochromic materials to the many solutions proposed to be used as electrolytes. Finally, performance of the resulting devices has been compared.

Ionic liquids are shown to be a convenient medium for the synthesis of electrochromic conducting polymers, sometimes resulting in better performance materials. It can be concluded that they show a broad range of chemical or physical procedures to obtain high-performance solid, semi-solid, or liquid electrolytes. Some problems like evaporation of the electrolyte, or narrow electrochemical windows, are minimized, resulting in long-lasting devices.

In summary, although inclusion of ionic liquids in electrochromic devices is relatively recent, the scientific community has enthusiastically embraced them as a significant opportunity to further develop this technology. Maybe these fascinating materials could represent the final impulse toward a massive further commercialization of electrochromic devices, which has been elusive for a long time.

ACKNOWLEDGMENTS

S. Hassab acknowledges a Ph.D. grant from the Erasmus-Mundus (Campus Mare Nostrum) program. The authors acknowledge funding from Fundacion Seneca-CARM (through project 11955/PI/09).

REFERENCES

[1] T.A. Skotheim, Handbook of Conducting Polymers, M. Dekker, 1986.
[2] R.D. Rauh, Electrochromic windows: an overview, Electrochim. Acta 44 (1999) 3165–3176.

[3] W. Lu, A.G. Fadeev, B. Qi, B.R. Mattes, Fabricating conducting polymer electrochromic devices using ionic liquids, J. Electrochem. Soc. 151 (2004) H33—H39.
[4] P. Kubisa, Application of ionic liquids as solvents for polymerization processes, Prog. Polym. Sci. 29 (2004) 3—12.
[5] L. Groenendaal, G. Zotti, P.H. Aubert, S.M. Waybright, J.R. Reynolds, Electrochemistry of poly(3,4-alkylenedioxythiophene) derivatives, Adv. Mater. 15 (2003) 855—879.
[6] R.M. Walczak, J.R. Reynolds, Poly (3,4-alkylenedioxypyrroles): the PXDOPS as versatile yet underutilized electroactive and conducting polymers, Adv. Mater. 18 (2006) 1121—1131.
[7] N. BiÇak, B.F. Senkal, E. Sezer, Preparation of organo-soluble polyanilines in ionic liquid, Synth. Met. 155 (2005) 105—109.
[8] H. Gao, T. Jiang, B. Han, Y. Wang, J. Du, Z. Liu, J. Zhang, Aqueous/ionic liquid interfacial polymerization for preparing polyaniline nanoparticles, Polymer 45 (2004) 3017—3019.
[9] P.C. Innis, J. Mazurkiewicz, T. Nguyen, G.G. Wallace, D. MacFarlane, Enhanced electrochemical stability of polyaniline in ionic liquids, Curr. Appl. Phys. 4 (2004) 389—393.
[10] D. Wei, C. Kvarnstrom, T. Lindfors, A. Ivaska, Polyaniline nanotubules obtained in room-temperature ionic liquids, Electrochem. Commun. 8 (2006) 1563—1566.
[11] M.C. Li, C.A. Ma, B.Y. Liu, Z.M. Jin, A novel electrolyte 1-ethylimidazolium trifluoroacetate used for electropolymerization of aniline, Electrochem. Commun. 7 (2005) 209—212.
[12] W. Lu, A.G. Fadeev, B. Qi, E. Smela, B.R. Mattes, J. Ding, G.M. Spinks, J. Mazurkiewicz, D. Zhou, G.G. Wallace, D.R. MacFarlane, S.A. Forsyth, M. Forsyth, Use of ionic liquids for pi-conjugated polymer electrochemical devices, Science 297 (2002) 983—987.
[13] W. Lu, A.G. Fadeev, B. Qi, B.R. Mattes, Stable conducting polymer electrochemical devices incorporating ionic liquids, Synth. Met. 135-136 (2003) 139—140.
[14] S. Mu, Pronounced effect of the ionic liquid on the electrochromic property of the polyaniline film: color changes in the wide wavelength range, Electrochim. Acta 52 (2007) 7827—7834.
[15] P.M.V. Fernandes, J.M. Campiña, C.M. Pereira, F. Silva, Electrosynthesis of polyaniline from choline-based deep eutectic solvents: morphology, stability and electrochromism, J. Electrochem. Soc. 159 (2012) G97—G105.
[16] P.G. Pickup, R.A. Osteryoung, Electrochemical polymerization of pyrrole and electrochemistry of polypyrrole films in ambient temperature molten salts, J. Am. Chem. Soc. 106 (1984) 2294—2299.
[17] P.G. Pickup, R.A. Osteryoung, Charging and discharging rate studies of polypyrrole films in AlCl3: 1-methyl-(3-ethyl)-imidazolium chloride molten salts and in CH3CN, J. Electroanal. Chem. 195 (1985) 271—288.
[18] J. Zawodzinski, L. Janiszewska, R.A. Osteryoung, On the chemistry of pyrrole in room-temperature chloroaluminate melts, J. Electroanal. Chem. 255 (1988) 111—117.
[19] J.M. Pringle, M. Forsyth, G.G. Wallace, D.R. MacFarlane, Solution-surface electropolymerization: a route to morphologically novel poly(pyrrole) using an ionic liquid, Macromolecules 39 (2006) 7193—7195.
[20] K. Sekiguchi, M. Atobe, T. Fuchigami, Electropolymerization of pyrrole in 1-ethyl-3-methylimidazolium trifluoromethanesulfonate room temperature ionic liquid, Electrochem. Commun. 4 (2002) 881—885.
[21] A.M. Fenelon, C.B. Breslin, The formation of polypyrrole at iron from 1-Butyl-3-methylimidazolium hexafluorophosphate, J. Electrochem. Soc. 152 (2005) D6—D11.
[22] J.M. Pringle, J. Efthimiadis, P.C. Howlett, J. Efthimiadis, D.R. MacFarlane, A.B. Chaplin, S.B. Hall, D.L. Officer, G.G. Wallace, M. Forsyth, Electrochemical synthesis of polypyrrole in ionic liquids, Polymer 45 (2004) 1447—1453.
[23] S. Ahmad, S. Sen Gursoy, S. Kazim, A. Uygun, Growth of N-substituted polypyrrole layers in ionic liquids: synthesis and its electrochromic properties, Sol. Energy Mater. Sol. Cells 99 (2012) 95—100.
[24] M. Deepa, S. Ahmad, Polypyrrole films electropolymerized from ionic liquids and in a traditional liquid electrolyte: a comparison of morphology and electro-optical properties, Eur. Polym. J. 44 (2008) 3288—3299.
[25] P.M. Beaujuge, J.R. Reynolds, Color control in pi-conjugated organic polymers for use in electrochromic devices, Chem. Rev. 110 (2010) 268—320.

[26] B. Gaihre, S. Ashraf, G.M. Spinks, P.C. Innis, G.G. Wallace, Comparative displacement study of bilayer actuators comprising of conducting polymers, fabricated from polypyrrole, poly(3,4-ethylenedioxythiophene) or poly(3,4-propylenedioxythiophene), Sensor. Actuat. A-Phys. 193 (2013) 48–53.
[27] S. Ahmad, J.H. Yum, H.J. Butt, M.K. Nazeeruddin, M. Graetzel, Efficient platinum-free counter electrodes for dye-sensitized solar cell applications, ChemPhysChem 11 (2010) 2814–2819.
[28] S. Inagi, T. Fuchigami, Electronic property and reactivity of novel fused thiophene, Synth. Met. 158 (2008) 782–784.
[29] Y. Pang, H. Xu, X. Li, H. Ding, Y. Cheng, G. Shi, L. Jin, Electrochemical synthesis, characterization, and electrochromic properties of poly(3-chlorothiophene) and its copolymer with 3-methylthiophene in a room temperature ionic liquid, Electrochem. Commun. 8 (2006) 1757–1763.
[30] Y. Pang, X. Li, H. Ding, G. Shi, L. Jin, Electropolymerization of high quality electrochromic poly(3-alkyl-thiophene)s via a room temperature ionic liquid, Electrochim. Acta 52 (2007) 6172–6177.
[31] X.Y. Li, Y.H. Pang, G.Y. Shi, J.R. Zhu, F. Wang, L.T. Jin, Synthesis and electrochromic properties of poly(3-hexyl-thiophene) in a room temperature ionic liquid and its application to an electrochromic device, Chin. J. Chem. 26 (2008) 677–680.
[32] L. Ma, Y. Li, X. Yu, Q. Yang, C.H. Noh, Fabricating red–blue-switching dual polymer electrochromic devices using room temperature ionic liquid, Sol. Energy Mater. Sol. Cells 93 (2009) 564–570.
[33] X.C. Song, X. Wang, Y.F. Zheng, H.Y. Yin, Electrochemical synthesis and electrochromic properties of nano-poly(3-methoxythiophene) in a room temperature ionic liquid, Curr. Nanosci. 8 (2012) 410–413.
[34] P. Damlin, C.K. varnstrom, A. Ivaska, Electrochemical synthesis and in situ spectroelectrochemical characterization of poly(3,4-ethylenedioxythiophene) (PEDOT) in room temperature ionic liquids, J. Electroanal. Chem. 570 (2004) 113–122.
[35] M. Dobbelin, R. Marcilla, C. Pozo-Gonzalo, D. Mecerreyes, Innovative materials and applications based on poly(3,4-ethylenedioxythiophene) and ionic liquids, J. Mater. Chem. 20 (2010) 7613–7622.
[36] C. Pozo-Gonzalo, R. Marcilla, M. Salsamendi, D. Mecerreyes, J.A. Pomposo, J. Rodriguez, H.J. Bolink, PEDOT: poly(1-vinyl-3-ethylimidazolium) dispersions as alternative materials for optoelectronic devices, J. Polym. Sci. A1 46 (2008) 3150–3154.
[37] T.Y. Kim, T.H. Lee, J.E. Kim, R.M. Kasi, C.S.P. Sung, K.S. Suh, Organic solvent dispersion of poly(3,4-ethylenedioxythiophene) with the use of polymeric ionic liquid, J. Polym. Sci. A1 46 (2008) 6872–6879.
[38] M. Dobbelin, R. Marcilla, C. Tollan, J.A. Pomposo, J.R. Sarasua, D. Mecerreyes, A new approach to hydrophobic and water-resistant poly(3,4-ethylenedioxythiophene):poly(styrenesulfonate) films using ionic liquids, J. Mater. Chem. 18 (2008) 5354–5358.
[39] M. Dobbelin, R. Tena-Zaera, R. Morcilla, J. Iturri, S. Moya, J.A. Pomposo, D. Mecerreyes, Multiresponsive PEDOT-ionic liquid materials for the design of surfaces with switchable wettability, Adv. Funct. Mater. 19 (2009) 3326–3333.
[40] M. Dobbelin, C. Pozo-Gonzalo, R. Marcilla, R. Blanco, J.L. Segura, J.A. Pomposo, D. Mecerreyes, Electrochemical snthesis of PEDOT derivatives bearing imidazolium-Ionic liquid moieties, J. Polym. Sci. A1 47 (2009) 3010–3021.
[41] S. Lee, G.A. Becht, B. Lee, C.T. Burns, M.A. Firestone, Electropolymerization of a bifunctional ionic liquid monomer yields an electroactive liquid-crystalline polymer, Adv. Funct. Mater. 20 (2010) 2063–2070.
[42] B. Dong, L. Zheng, J. Xu, H. Liu, S. Pu, Electropolymerization of 1,2-methylenedioxybenzene in 1-butyl-3-methylimidazolium hexafluorophosphate room temperature ionic liquid, Polymer 48 (2007) 5548–5555.
[43] D.X. Zhuang, P.Y. Chen, Electrochemical formation of polycarbazole films in air- and water-stable room-temperature ionic liquids, J. Electroanal. Chem. 626 (2009) 197–200.

[44] A.K. Geim, K.S. Novoselov, The rise of graphene, Nat. Mater. 6 (2007) 183–191.

[45] A.H. Castro Neto, F. Guinea, N.M.R. Peres, K.S. Novoselov, A.K. Geim, The electronic properties of graphene, Rev. Mod. Phys. 81 (2009) 109–162.

[46] R.H. Baughman, A.A. Zakhidov, W.A. De Heer, Carbon nanotubes – the route toward applications, Science 297 (2002) 787–792.

[47] A. Thess, R. Lee, P. Nikolaev, H. Dai, P. Petit, J. Robert, C. Xu, Y.H. Lee, S.G. Kim, A.G. Rinzler, D.T. Colbert, G.E. Scuseria, D. Tomanek, J.E. Fischer, R.E. Smalley, Crystalline ropes of metallic carbon nanotubes, Science 273 (1996) 483–487.

[48] S. Ahmad, S. Singh, Electrochromic device based on carbon nanotubes functionalized poly (methyl pyrrole) synthesized in hydrophobic ionic liquid medium, Electrochem. Commun. 10 (2008) 895–898.

[49] B.N. Reddy, M. Deepa, A.G. Joshi, A.K. Srivastava, Poly(3,4-Ethylenedioxypyrrole) enwrapped by reduced graphene oxide: how conduction behavior at nanolevel leads to increased electrochemical activity, J. Phys. Chem. C 115 (2011) 18354–18365.

[50] A.P. Saxena, M. Deepa, A.G. Joshi, S. Bhandari, A.K. Srivastava, Poly(3,4-ethylenedioxythiophene)-ionic liquid functionalized graphene/reduced graphene oxide nanostructures: improved conduction and electrochromism, ACS Appl. Mater. Interfaces 3 (2011) 1115–1126.

[51] S. Bhandari, M. Deepa, S.N. Sharma, A.G. Joshi, A.K. Srivastava, R. Kant, Charge transport and electrochromism in novel nanocomposite films of poly(3,4-ethylenedioxythiophene)-Au nanoparticles-CdSe quantum dots, J. Phys. Chem. C 114 (2010) 14606–14613.

[52] A. Kharkwal, D. Melepurath, A.G. Joshi, A.K. Srivastava, Red to blue high electrochromic contrast and rapid switching poly(3,4-ethylenedioxypyrrole)-Au/Ag nanocomposite devices for smart windows, ChemPhysChem 12 (2011) 1176–1188.

[53] U. Bach, D. Lupo, P. Comte, J.E. Moser, F. Weissortel, J. Salbeck, H. Spreitzer, M. Gratzel, Solid-state dye-sensitized mesoporous TiO2 solar cells with high photon-to-electron conversion efficiencies, Nature 395 (1998) 583–585.

[54] M. Gratzel, Materials science – ultrafast colour displays, Nature 409 (2001) 575–576.

[55] P. Bonhote, E. Gogniat, F. Campus, L. Walder, M. Gratzel, Nanocrystalline electrochromic displays, Displays 20 (1999) 137–144.

[56] Y. Pang, X. Li, G. Shi, F. Wang, L. Jin, Electrochromic properties of poly(3-chlorothiophene) film electrodeposited on a nanoporous TiO2 surface via a room temperature ionic liquid and its application in an electrochromic device, Thin Solid Films 516 (2008) 6512–6516.

[57] L. Ma, Y. Li, X. Yu, Q. Yang, C.H. Noh, Using room temperature ionic liquid to fabricate PEDOT/TiO2 nanocomposite electrode-based electrochromic devices with enhanced long-term stability, Sol. Energy. Mater. Sol. Cells 92 (2008) 1253–1259.

[58] X.C. Song, Y.F. Zheng, R. Ma, H.Y. Yin, Electrochemical polymerisation of poly(3-bromothiophene)/WO 3 nanocomposite films, Micro Nano Lett. 6 (2011) 273–276.

[59] H.Y. Yin, X.C. Song, Y.F. Zheng, X. Wang, Z.A. Yang, R. Ma, Organic/inorganic nanocomposite films based on poly(3-methoxythiophene) and WO 3, Mater. Sci. Eng. B-Solid 176 (2011) 684–687.

[60] W.T. Li, Y.T. Wang, X.C. Song, Electrochemical polymerization and electrochromic properties of poly(3-methoxythiophene)/NiO composite films, Appl. Mech. Mater. 268 (2013) 45–48.

[61] A. Kavanagh, R. Copperwhite, M. Oubaha, J. Owens, C. McDonagh, D. Diamond, R. Byrne, Photo-patternable hybrid ionogels for electrochromic applications, J. Mater. Chem. 21 (2011) 8687–8693.

[62] A. Kavanagh, K.J. Fraser, R. Byrne, D. Diamond, An electrochromic ionic liquid: design, characterization, and performance in a solid-state platform, ACS Appl. Mater. Interfaces 5 (2013) 55–62.

[63] T. Kakibe, R. Kurihara, H. Ohno, Multistep coloring and bleaching of viologens monomolecularly incorporated in DNA, Chem. Lett. 38 (2009) 494–495.

[64] T. Kakibe, H. Ohno, Quasi-reversible electrochromic behavior of alkyl viologens dispersed monomolecularly in double-stranded DNA chains, J. Mater. Chem. 19 (2009) 4960–4964.

[65] T. Kakibe, H. Ohno, Immobilization of heptyl viologens in DNA strands both to inhibit dimerization and to accelerate quasi-reversible electron transfer reaction, Chem. Commun. (2008) 377–379.

[66] R.E. Del Sesto, T.M. McCleskey, A.K. Burrell, G.A. Baker, J.D. Thompson, B.L. Scott, J.S. Wilkes, P. Williams, Structure and magnetic behavior of transition metal based ionic liquids, Chem. Commun. 8 (2008) 447–449.
[67] I. De Pedro, D.P. Rojas, J.A. Blanco, J.R. Fernandez, Antiferromagnetic ordering in magnetic ionic liquid Emim[FeCl4], J. Magn. Magn. Mater. 323 (2011) 1254–1257.
[68] A. Branco, L.C. Branco, F. Pina, Electrochromic and magnetic ionic liquids, Chem. Commun. 47 (2011) 2300–2302.
[69] G.F. Cai, J.P. Tu, C.D. Gu, J.H. Zhang, J. Chen, D. Zhou, S.J. Shi, X.L. Wang, One-step fabrication of nanostructured NiO films from deep eutectic solvent with enhanced electrochromic performance, J. Mater. Chem. A1 (2013) 4286–4292.
[70] S. Duluard, B. Ouvrard, A. Celik-Cochet, G. Campet, U. Posset, G. Schottner, M.H. Delville, Comparison of PEDOT films obtained via three different routes through spectroelectrochemistry and the differential cyclic voltabsorptometry method (DCVA), J. Phys. Chem. B 114 (2010) 7445–7451.
[71] S. Bhandari, M. Deepa, S. Pahal, A.G. Joshi, A.K. Srivastava, R. Kant, A dual electrochrome of poly-(3,4-ethylenedioxythiophene) doped by N,N'-bis(3-sulfonatopropyl)-4-4'-bipyridinium-redox chemistry and electrochromism in flexible devices, ChemSusChem 3 (2010) 97–105.
[72] T. Fujimoto, M.M. Matsushita, H. Yoshikawa, K. Awaga, Electrochemical and electrochromic properties of octathio[8]circulene thin films in ionic liquids, J. Am. Chem. Soc. 130 (2008) 15790–15791.
[73] Z. Tebby, O. Babot, T. Toupance, D.H. Park, G. Campet, M.H. Delville, Low-temperature UV-processing of nanocrystalline nanoporous thin TiO 2 films: an original route toward plastic electrochromic systems, Chem. Mater. 20 (2008) 7260–7267.
[74] A. Surca Vuk, V. Jovanovski, A. Pollet-Villard, I. Jerman, B. Orel, Imidazolium-based ionic liquid derivatives for application in electrochromic devices, Sol. Energy Mater. Sol. Cells 92 (2008) 126–135.
[75] M. Colovic, I. Jerman, M. Gaberscek, B. Orel, POSS based ionic liquid as an electrolyte for hybrid electrochromic devices, Sol. Energy Mater. Sol. Cells 95 (2011) 3472–3481.
[76] A. Kawada, T. Kumagai, JP2008239514 (A).
[77] N.M. Ignatyev, K. Kawata, M. Schulte, H. Yoshizaki, WO2012041437 (A3).
[78] A.B. McEwen, V.R. Koch, WO0193363 (A3).
[79] C. Xu, J. Zhang, J. Zheng, CN103123431 (A).
[80] T. Shibuya, S. Hirano, S. Uchida, JP2006267833 (A).
[81] N. Leyland, D. Corr, M. Cassidy, F. Pichot, N. Vlachopoulos, M. Ryan, WO2007082067 (A2).
[82] J.J. Finley, I.G. Schwendeman, A.A. Polcyn, C.S. Harris, WO2010025119 (A1).
[83] I.G. Schwendeman, J.J. Finley, A.A. Polcyn, C.M. Boykin, US2010060971 (A1).
[84] T. Miyaura, I. Doi, A. Ikegawa, JP2005099249 (A).
[85] W. Lu, B.R. Mattes, A.G. Fadeev, B. Qi, US 6667825 B2.
[86] K.Y. Park, WO2006052067 (A1).
[87] J.R. Reynolds, A.L. Dyer, US2011062860 (A1).
[88] C.H. Noh, R.R. Das, J.W. Kim, S.J. Jeon, US2013033735 (A1).
[89] M. Graetzel, D. Kuang, P. Wang, S.M. Zakeerudin, WO2007093961 (A1).
[90] H. Lifka, A.M.B. Van Mol, P. Van De Weijer, EP2244317 (A1).
[91] A.G. Fadeev, W. Lu, B.R. Mattes, WO02063073 (A1).
[92] A. Agrawal, J.P. Cronin, R.S. Lecompte, J.C.L. Tonazzi, WO2004085567 (A3).
[93] K.Y. Park, WO2005121882 (A1).
[94] C. Michot, WO2007068822 (A8).
[95] I.G. Schwendeman, A.A. Polcyn, J.J. Finley, C.M. Boykin, J.M. Knowles, US7907322 (B2).
[96] C. Peng, P.Y. Liu, TWI354917 (B).
[97] U. Bruhnke, K. Wilfinger, H.C. Steffel, EP2080648 (A1).
[98] J.S. Oh, B.B. Lee, J.D. Park, S.J. Mun, WO2006033545 (A1).
[99] N. Kobayashi, JP2006058618 (A).
[100] S.A. Miller, US2008209876 (A1).

[101] D.E. Fenton, J.M. Parker, P.V. Wright, Complexes of alkali metal ions with poly(ethylene oxide), Polymer 14 (1973) 589.
[102] J.H. Shin, W.A. Henderson, S. Passerini, Ionic liquids to the rescue? Overcoming the ionic conductivity limitations of polymer electrolytes, Electrochem. Commun. 5 (2003) 1016–1020.
[103] S. Desai, R.L. Shepherd, P.C. Innis, P. Murphy, C. Hall, R. Fabretto, G.G. Wallace, Gel electrolytes with ionic liquid plasticiser for electrochromic devices, Electrochim. Acta 56 (2011) 4408–4413.
[104] H. Bircan, V. Seshadri, J. Padilla, M. Invernale, T.F. Otero, G.A. Sotzing, Use of polymer/ionic liquid plasticizers as gel electrolytes in electrochromic devices, J. Phys. Conf. Ser. 127 (2008).
[105] S. Duluard, I. Litas, A.J. Bhattacharyya, F. Mauvy, G. Campet, M.H. Delville, Soft matter electrolytes based on polymethylmetacrylate dispersions in lithium bis(trifluoromethanesulfonyl) imide/1-butyl-3-methylimidazolium bis(trifluoromethanesulfonyl)imide ionic liquids, Electrochim. Acta 55 (2010) 8839–8846.
[106] S. Duluard, A. Celik-Cochet, I. Saadeddin, A. Labouret, G. Campet, G. Schottner, U. Posset, M.H. Delville, Electrochromic devices based on in situ polymerised EDOT and Prussian Blue: Influence of transparent conducting oxide and electrolyte composition – towards up-scaling, New J. Chem. 35 (2011) 2314–2321.
[107] K.H. Lee, K.W. Yoo, S.J. Sung, S.T. Kim, KR20120000733 (A).
[108] K.H. Lee, K.W. Yoo, S.J. Sung, S.T. Kim, KR20120000735 (A).
[109] J.S. Oh, J.D. Park, B.B. Lee, W.J. Kwon, S.H. Kim, S.J. Mun, US2005231785 (A1).
[110] P. Gould Nigel, F. Abbassi, A.J.H. Batting, M.O. Sandberg, A. Sawatdee, L.B. Hennerdal, WO2008062149 (A1).
[111] S.H. Kim, W.J. Kwon, B.B. Lee, S.J. Mun, J.S. Oh, J.D. Park, WO2005116161 (A1).
[112] H. Miaoliang, L. Zhang, W. Jianming, W. Jihuai, CN101719425 (A).
[113] B. Ramard, G.A. Sotzing, WO2006094052 (A3).
[114] B. Ramard, G.A. Sotzing, US7626748 (B2).
[115] S.H. Kim, J.S. Yang, S.T. Kim, KR20130013991 (A).
[116] K.W. Yoo, K.H. Lee, S.J. Sung, S.T. Kim, KR20120032827 (A).
[117] U. Posset, B. Herbig, G. Schottner, K. Zaghib, J. Labrecque, A. Guerfi, M. Perrier, R. Ruffo, WO2013041562 (A1).
[118] D. Mecerreyes, Polymeric ionic liquids: broadening the properties and applications of polyelectrolytes, Prog. Polym. Sci. 36 (2011) 1629–1648.
[119] O. Green, S. Grubjesic, S. Lee, M.A. Firestone, The design of polymeric ionic liquids for the preparation of functional materials, Polym. Rev. 49 (2009) 339–360.
[120] A.S. Best, S. Viale, S.J. Picken, WO2005117188 (A1).
[121] S. Janietz, W. Meyer, D. Grothe, DE102011102438 (A1).
[122] Q. Du, X. Fu, S. Liu, L. Niu, G. Wang, X. Zou, Polymer electrolytes based on polymeric ionic liquid Poly(methyl 2-(3-vinylimidazolidin-1-yl)acetate bis(trifluoromethane sulfonyl)imide), J. Inorg. Organomet. Polym. 22 (2012) 316–323.
[123] Q. Du, X. Fu, S. Liu, L. Niu, The effect of ionic liquid fragment on the performance of polymer electrolytes, Polym. Int. 61 (2012) 222–227.
[124] R. Marcilla, F. Alcaide, H. Sardon, J.A. Pomposo, C. Pozo-Gonzalo, D. Mecerreyes, Tailor-made polymer electrolytes based upon ionic liquids and their application in all-plastic electrochromic devices, Electrochem. Commun. 8 (2006) 482–488.
[125] M. Colovic, A. Surca Vuk, M. Hajzeri, B. Orel, Structural studies of alkoxysilyl-functionalised ionic liquid and its application in a hybrid electrochromic device, Mater. Res. Bull. 47 (2012) 3160–3169.
[126] T. Ueki, M. Watanabe, Macromolecules in ionic liquids: progress, challenges, and opportunities, Macromolecules 41 (2008) 3739–3749.
[127] J. Le Bideau, L. Viau, A. Vioux, Ionogels, ionic liquid based hybrid materials, Chem. Soc. Rev. 40 (2011) 907–925.
[128] P. Vidinha, N.M.T. Lourenco, C. Pinheiro, A.R. Bras, T. Carvalho, T. Santos-Silva, A. Mukhopadhyay, M.J. Romao, J. Parola, M. Dionisio, J.M.S. Cabral, C.A.M. Afonso, S. Barreiros,

Ion jelly: a tailor-made conducting material for smart electrochemical devices, Chem. Commun. (2008) 5842–5844.
[129] A.F.R. Pimenta, A.C. Baptista, T. Carvalho, P. Brogueira, N.M.T. Lourenco, C.A.M. Afonso, S. Barreiros, P. Vidinha, J.P. Borges, Electrospinning of ion jelly fibers, Mater. Lett. 83 (2012) 161–164.
[130] R. Leones, F. Sentanin, L.C. Rodrigues, R.A.S. Ferreira, I.M. Marrucho, J.M.S.S. Esperanca, A. Pawlicka, L.D. Carlos, M. Manuela Silva, Novel polymer electrolytes based on gelatin and ionic liquids, Opt. Mater. 35 (2012) 187–195.
[131] R. Leones, F. Sentanin, L.C. Rodrigues, I.M. Marrucho, J.M.S.S. Esperanca, A. Pawlicka, M.M. Silva, Investigation of polymer electrolytes based on agar and ionic liquids, Express Polym. Lett. 6 (2012) 1007–1016.
[132] T.J. Trivedi, D.N. Srivastava, R.D. Rogers, A. Kumar, Agarose processing in protic and mixed protic-aprotic ionic liquids: dissolution, regeneration and high conductivity, high strength ionogels, Green Chem. 14 (2012) 2831–2839.
[133] G. Yang, Z. Shi, S. Zang, CN102219917 (A).
[134] M. Deepa, A. Awadhia, S. Bhandari, Electrochemistry of poly(3,4-ethylenedioxythiophene)-polyaniline/Prussian blue electrochromic devices containing an ionic liquid based gel electrolyte film, Phys. Chem. Chem. Phys. 11 (2009) 5674–5685.
[135] R. Sydam, M. Deepa, A.K. Srivastava, Electrochromic device response controlled by an in situ polymerized ionic liquid based gel electrolyte, R. Soc. Chem. Adv. 2 (2012) 9011–9021.
[136] P. Jia, W.A. Yee, J. Xu, C.L. Toh, J. Ma, X. Lu, Thermal stability of ionic liquid-loaded electrospun poly(vinylidene fluoride) membranes and its influences on performance of electrochromic devices, J. Membr. Sci. 376 (2011) 283–289.
[137] J.D. Park, J.S. Oh, B.B. Lee, S.J. Mun, KR20070063772 (A).
[138] S. Ahmad, S.A. Agnihotry, M. Deepa, US2010084600 (A1).
[139] H. Mei, J. Gaojun, CN101412776 (A).
[140] B. Li, CN103186004 (A).
[141] S.A. Edwards, R. Fabretto, C. Hall, C.P. Innis, J.H. Mazurkiewicz, P.J. Murphy, G.G. Wallace, WO2007002989 (A1).
[142] J.D. Park, J.S. Oh, B.B. Lee, S.J. Mun, KR100821833 (B1).
[143] L.L. Adams, A. Agrawal, J.P. Cronin, J.C.L. Tonazzi, WO2010003138 (A1).
[144] S. Ahmad, M. Deepa, Ionogels encompassing ionic liquid with liquid like performance preferable for fast solid state electrochromic devices, Electrochem. Commun. 9 (2007) 1635–1638.
[145] P.M.S. Monk, R.J. Mortimer, D.R. Rosseinsky, Electrochromism and Electrochromic Devices, Cambridge University Press, Cambridge, 2007.
[146] P.M.S. Monk, R.J. Mortimer, D.R. Rosseinsky, Electrochromism: Fundamentals and Applications, VCH, Weinheim, 1995.
[147] C. Pozo-Gonzalo, D. Mecerreyes, J.A. Pomposo, M. Salsamendi, R. Marcilla, H. Grande, R. Vergaz, D. Barrios, J.M. Sanchez-Pena, All-plastic electrochromic devices based on PEDOT as switchable optical attenuator in the near IR, Sol. Energy Mater. Sol. Cells 92 (2008) 101–106.
[148] A. Brazier, G.B. Appetecchi, S. Passerini, A. Surca Vuk, B. Orel, F. Donsanti, F. Decker, Ionic liquids in electrochromic devices, Electrochim. Acta 52 (2007) 4792–4797.
[149] B.J. Warner, A. Agrawal, J.P. Cronin, J.C.L. Tonazzi, T.M. MCCleskey, A.K. Burrel, WO2005026827 (A3).
[150] A. Agrawal, J.P. Cronin, J.C.L. Tonazzi, WO2009099900 (A3).
[151] A. Agrawal, A.K. Burrel, J.P. Cronin, T.M. MCCleskey, J.C.L. Tonazzi, B.J. Warner, WO2004001877 (A3).
[152] C. Pozo-Gonzalo, R. Marcilla, M. Salsamendi, J.A. Pomposo, H.J. Grande, WO2009095507 (A1).
[153] A. Georg, A. Georg, Electrochromic device with a redox electrolyte, Sol. Energy Mater. Sol. Cells 93 (2009) 1329–1337.
[154] A. Llordes, G. Garcia, J. Gazquez, D.J. Milliron, Tunable near-infrared and visible-light transmittance in nanocrystal-in-glass composites, Nature 500 (2013) 323–326.
[155] P. Chandrasekhar, B.J. Zay, T. McQueeney, A. Scara, D. Ross, G.C. Birur, S. Haapanen, L. Kauder, T. Swanson, D. Douglas, Conducting polymer (CP) infrared electrochromics in spacecraft thermal control and military applications, Synth. Met. 135-136 (2003) 23–24.

[156] P. Chandrasekhar, B.J. Zay, G.C. Birur, S. Rawal, E.A. Pierson, L. Kauder, T. Swanson, Large, switchable electrochromism in the visible through far-infrared in conducting polymer devices, Adv. Funct. Mater. 12 (2002) 95–103.

[157] P. Chandrasekhar, B.J. Zay, S. Barbolt, R. Werner, G.C. Birur, A. Paris, Variable emittance electrochemical devices for spacecraft application based on ionic liquids coupled with conducting polymers, ECS Trans. 16 (49) (2008) 165–170.

[158] P. Chandrasekhar, B. J.Zay, S. Barbolt, R. Werner, G.C. Birur, A. Paris, High performance variable emittance devices for spacecraft application based on conducting polymers coupled with ionic liquids, AIP Conf. Proc. 1103 (2009) 101–104.

[159] P. Chandrasekhar, B. J.Zay, S. Barbolt, R. Werner, E. Caldwell, Variable emittance skins for active thermal control in spacecraft based on conducting polymers, ionic liquids and specialized coatings, AIP Conf. Proc. 1208 (2010) 99–104.

[160] P. Verge, P.H. Aubert, F. Vidal, L. Sauques, F. Tran-Van, S. Peralta, D. Teyssie, C. Chevrot, New prospects in the conception of IR electro-tunable devices: the use of conducting semi-interpenetrating polymer network architecture, Chem. Mater. 22 (2010) 4539–4547.

[161] C. Chevrot, D. Teyssie, P. Verge, L. Goujon, F. Tran-Van, F. Vidal, P.H. Aubert, S. Peralta, L. Sauques, Electroactive semi-interpenetrating polymer networks architecture with tunable IR reflectivity, Proc. SPIE 7976 (2011).

[162] J.P. Dudon, J. Crouzet, A. Teissier, P.H. Aubert, F. Vidal, C. Chevrot, S. Remaury, Thermal control. Satellites by polymer-based electroemissive device infrared spectra: component design ground thermal testing, 41st international conference on environmental systems, 2011.

[163] L.J. Goujon, F. Tran-Van, F. Vidal, P.H. Aubert, D. Teyssie, C. Chevrot, IR reflectivity change from electroactive IPN, Mol. Cryst. Liq. Cryst. 554 (2012) 95–102.

[164] A. Teissier, J.P. Dudon, P.H. Aubert, F. Vidal, S. Remaury, J. Crouzet, C. Chevrot, Feasibility of conducting semi-IPN with variable electro-emissivity: a promising way for spacecraft thermal control, Sol. Energy Mater. Sol. Cells 99 (2012) 116–122.

[165] F. Wang, M.E. Itkis, E. Bekyarova, R.C. Haddon, Charge-compensated, semiconducting single-walled carbon nanotube thin film as an electrically configurable optical medium, Nat. Photonics 7 (2013) 459–465.

CHAPTER 11

Outlook

Francisco José Hernández Fernández[1], Antonia Pérez de los Ríos[2]
[1]Department of Chemical and Environmental Engineering, Regional Campus of International Excellence "Campus Mare Nostrum," Technical University of Cartagena, Campus La Muralla, Cartagena, Murcia, Spain
[2]Department of Chemical Engineering, Regional Campus of International Excellence "Campus Mare Nostrum," University of Murcia, Campus de Espinardo, Murcia, Spain

It was our intention in the 10 previous chapters to provide essential information for a deep understanding of the nature and properties of ionic liquids as well as review modern separation applications that have benefited from ionic liquid technology. Application of ionic liquids in chemical processes has blossomed within the past decade. Interest in these compounds is still increasing rapidly. If we look to the future, it is difficult to predict what will be discovered since science is unpredictable, which makes exciting our work as scientists. From this perspective, in this chapter we answer some questions that can summarize the advantages of using ionic liquids in separation technology and the progress and challenges in this field.

1. WHAT ARE THE MAIN ADVANTAGES OF USING IONIC LIQUID IN SEPARATION TECHNOLOGY?

In recent years, novel processes based on ionic liquid (IL) have been proposed as effective methods for the selective separation of different chemical species, such as organic compounds (i.e., amines, alcohols, organic acids, ketones, ethers, and aromatic hydrocarbons) [1–5]; mixed gases [6,7]; and metal ions [8,9]. The main advantage of these media are their near-zero vapor pressure and their good chemical and thermal stabilities: they have a large temperature range in which they are stable, with a negligible vapor pressure. Other interesting properties of ILs in the context of separation processes are their high ion conductivity [10] and high solvent power [11]. Furthermore, all the physicochemical properties of ILs, including their hydrophobicity, density, viscosity, melting point, polarity, and solvent miscibility, can be finely tuned by selecting appropriate combinations of cations and anions so the optimal IL might be designed for each specific separation problem [12]. Although, in many cases, ILs have simply been used to replace organic solvents in separations, they often have led to improved process performance, increasing the selectivity and stability of the process. ILs as molecular solvents are capable of selectively separating molecules based on the functional group. The separation mechanism involves the partition of the specific compounds between the ionic liquids and the contacting phases [13]. Regarding the stability parameter, the use of these new solvents could result in the stabilization of the process because of their

Ionic Liquids in Separation Technology
http://dx.doi.org/10.1016/B978-0-444-63257-9.00011-0
Copyright © 2014 Elsevier B.V.
All rights reserved.

negligible vapor pressure and the possibility of minimizing their solubility in the surrounding phases by adequately selecting the cation and anion [14].

2. HOW CAN YOU SELECT THE MOST SUITABLE IONIC LIQUID FOR A SPECIFIC APPLICATION?

Since the beginning of the use of ILs in separation-related applications, the type of ILs used has changed. At first, "conventional" ILs based on imidazolium cations, pyridinium, pyrrolidinium, or tetraalkylammonium combined with hexafluorophosphate or tetrafluroborate anions were used. However, these pioneering ILs show some drawbacks: the hexafluorophosphate anion is known to be quite unstable during hydrolysis and produces toxic and corrosive *hydrofluoric acid* (HF) or fluorides. Then, more stable anions, such as bis[(trifluoromethyl)sulfonyl]imide started to be used. The last generation of ILs used in separating applications are the called "task-specific" ILs (TSILs). TSILs are ILs that contain specific functionalities to enable well-defined chemical properties. For example, TSILs have been used for the selective liquid—liquid extraction of heavy metals from an aqueous solution [15]. In this case, functionalized imidazolium cations with thioether-, urea-, or thiourea-derivatized side chains were used. These functional groups were capable of complexing chemically heavy metals.

As we commented above, an enormous variety of ILs exists by combining different cations and anions. The wide variety of ILs sometimes can be an advantage and sometimes a drawback. The main problem could be finding an adequate IL for a specific separation problem. For that, quantitative structure—activity relationship (QSAR) models can be used as estimation procedures to identify how these individual structural variables of ionic liquids (e.g., cationic group, alkyl chain length of the cation or anion) may evoke the best properties for a specific separation problem. QSARs have been used to predict the physicochemical and biological properties of molecules, considering that the biological activity of a new or untested chemical can be inferred from the molecular structure or other properties of similar compounds whose activities have already been assessed [16—21]. The use of QSARs could help to design ILs with adequate properties for a specific separation problem, reducing the experimental work and the consumption of time and materials. Of course, this methodology requires previous experiments. Furthermore, the model could be enriched and its results could be improved by the inclusion of addicional experimental data.

3. CAN ILs BE CONSIDERED FULLY GREEN SOLVENTS?

Regarding to the consideration of ILs as green solvents, it should be noted that the most important properties of ILs, in contrast to conventional solvents, is their negligible vapor, which can result in lower emissions and, consequently, in a reduced exposure. During the past 5 years, great interest has been shown in the study of the toxicity of ILs [20,22–27]. Different degrees of toxicity have been reported compared to that of chemicals currently used as solvents in the chemical industry, which could be explained by the enormous variety of ILs. Structural information has been obtained from these studies, allowing for the rational design of safer ILs. For instance, the toxicity of an IL is correlated directly with the length of the alkyl substituent in the cation, whereas the anion has a small effect on this parameter [23,24]. Evaluations of the risk posed by ILs to the environment, compared to traditional industrial solvents, must consider not only toxicity but also a presumably reduced exposure to ILs, which is not only caused by the negligible vapor pressure but also the possible lower bioaccumulation of ILs. Bioaccumulation—the capability to penetrate membranes and accumulate in organisms—could be measured by the octanol water partition coefficient: the higher the octanol–water partition coefficient of a compound, the easier it can cross through organic membranes and accumulate in hydrophobic tissues such as the brain. This parameter has been measured for a wide range of ILs and is significantly lower than that of conventional organic solvents because of the ionic nature of ILs [28,29].

Therefore, the adjustable nature of ILs and the structure–toxicity relationship found would allow the design of more environmentally friendly ILs. Deeper investigations are necessary to evaluate further risk. For that, a broader set of test methods should be applied, including studies focusing on exposition pathways as well as bioaccumulation and degradation processes.

4. WHAT ARE THE MAIN AREAS OF IL APPLICATION IN SEPARATION PROCESSES?

Figure 1 gives a selection of the most relevant separation applications of ILs that are discussed, developed, or already technically applied today (some of them have been described in detail in the previous chapters). All applications can be grouped into three interconnected areas: engineering, analytical, and electrochemical applications.

4.1 How the Ionic Liquids Have Been Used in Separation Processes?

ILs have been used as free molecules, for instance, in liquid–liquid extraction [30]. When the compounds of interest are extracted by the ILs phase it is necessary to recovery them from the IL. The compounds could be recovered by re-extraction with another solvent or by distillation from the ILs if the boiling point of the compounds is low and they do not decompose under high temperature. When these possibilities are not viable, supercritical carbon dioxide ($scCO_2$) could be used. $scCO_2$ has been revealed as an interesting clean alternative to classic organic solvents [31]. $scCO_2$ is especially advantageous because it is cheap, nontoxic, and nonflammable; has a near ambient critical temperature (31.18 °C); and has a moderate critical pressure [32,33]. Furthermore, its solvent power can be tuned

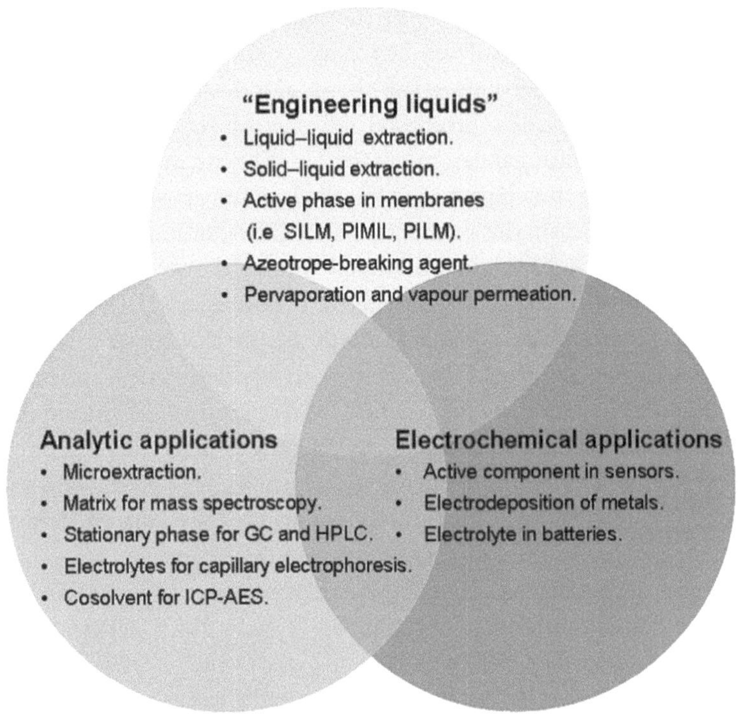

Figure 1 The most relevant separation applications of ionic liquids used in separation technology.

by changing either the pressure or the temperature. For these reasons, $scCO_2$ is frequently designated a "green designer" solvent. It can be used to extract the compounds that are dissolved in the IL phase. The success of the biphasic system formed is based on the high solubility of the $scCO_2$ in the IL phase; the same IL shows low solubility in the $scCO_2$ phase [34,35]. Another aspect to take into account when using liquid–liquid extraction is that the process effluent may contain trace to significant amounts of ILs. The amount of IL released in the process effluent (output) depends on the solubility of the IL in the second phase (water or organic solvent). IL losses can be reduced by selecting a suitable IL: it should possess reduced solubility in the second phase and high extraction capability for the desired compound. If the IL losses are significant, it would be interesting to recover the IL from the effluent for either economic reasons (to be recycled) and/or environmental reasons (possible toxicity of the IL). In this case, it would be necessary to incorporate a separation step downstream, such as distillation or ion exchange, in which case the IL would remain in the residue or in the ion exchange resins, respectively.

Some other applications involve the immobilization of ILs as liquid phase in membranes. One of the main advantages of IL membrane-based separation processes is that they allow continuous operation with a minimal amount of IL (active phase). There are different methodologies for IL immobilization. One of the first methodologies used is the immobilization of the ILs as liquid phase in supported IL membranes (SILMs) by adsorption into a polymeric organic membrane [6,36]. The IL can also be immobilized physically by occlusion in an organic polymer using casting methods (polymer inclusion membrane based on ILs); in this case, the IL acts as an extracting agent and plasticizer at the same time [37,38]. Another recent way to immobilize of ILs is by polymerizing the monomer molecules of ILs (polymeric IL membranes [PILMs]) [39]. Regarding the operational stability of these types of membranes, in general, this parameter increase in the sequence: SILMs < polymer inclusion membranes based on ILs < PILMs. In the case of SILMs, the differential migration of the ILs from the polymeric membrane toward the surrounding phases has been reported and can be attributed to the different degrees of solubility of ILs in contacting phases. In this context, a proper design of an IL comprising cations and anions could reduce its solubility in the surrounding phases; therefore, increasing the stability of the SILM. Furthermore, an important stabilizing effect of the polymeric organic used as support also was reported [40]. As we mentioned earlier, the active phase in ILs membranes is the IL. Therefore, because of their high stability and their high ionic liquid content (100%), PILMs could be a good option to consider.

5. WHICH FACTORS HAVE TO BE CONSIDERED FOR PRACTICAL IMPLEMENTATION OF ILs IN COMMERCIAL SEPARATION PROCESSES?

One of the factors to be considered in commercial separation processes is the price of ILs. On a large scale, the material cost should mainly determine the price of an IL. Therefore, the IL price should be dictated by the price of the cation and anion sources. The cations and anions ordered according to their rough price on an industrial scale occur in the following sequence: imidazolium > piridinium > phosphonium > ammonium (cations) and hexafluoroantimonate > bis[(trifluoromethyl)sulfonil]imide > triflurometanosulfonate > hexafluorophosphate = tetrafluroborate > tetracloroaluminate = acetate > nitrate > metilsulfate > chloride (anions) [41,42]. In this context, more familiar ILs based on the 1-alkyl-3-methylimidazolium cation, coupled with the bis[(trifluoromethyl) sulfonyl]imide ion, can be costly. However, not all ILs are expensive, especially on an industrial scale: some ILs, for example, choline-based and tetraalkylphosphonium or tetraalkylammonium salts, are intrinsically less expensive than others [43]. In general, on a laboratory scale ILs are normally 5–20 times more expensive than molecular solvents, but there are those that are cheaper and those that are much more expensive. Furthermore, the possibility of recycling ILs involves a reduction of the cost to even less than that of molecular solvents. The cost of the technology based on IL also depends on the amount of IL needed. The amount of IL is higher for liquid–liquid extraction; however, for the use of membranes based on ILs, the amount of ILs immobilized is lower. For instance, in the case of SILMs, the amount of IL immobilized is around 2 g/m^2 [44]. Even the more expensive ILs could be used in separation processes if the final products have a high added value. Furthermore, it is worth noting that the price of ILs is shrinking day by day as their use is increasing.

Purity is another aspect to consider when ILs are going to be used in a given process. Different impurities, such as water, halides, breakdown products (i.e., HF), unreacted organic salts, and organics, can be found in ILs [45]. The presence of these impurities should be considered by users since they may influence the solvent properties [46]. Water is a common contaminant; even water-immiscible ILs are hygroscopic and can absorb some water [46]. Therefore, ILs to be used in anhydrous systems should be kept free of water by storing them over a strong desiccant agent such as phosphorus pentoxide. Furthermore, as mentioned earlier, water in ILs containing hydrolyzable groups, such as tetrafluoroborate and hexafluorophosphate anions, may cause partial hydrolysis of the latter anions with the formation of HF. Traces of HF were easily removed in water-immiscible IL by washing first with an aqueous sodium carbonate solution and then with milli-Q water until they reached the pH of the milli-Q water. Small amounts of chloride ion have been found in nonchloride ILs because of the preparation methods using a chloride route. Exchange resins could be used for further purification if a lower halide concentration is required. Organics could also be eliminated by evaporation under reduced pressure in a rotary evaporator because of the near-zero vapor pressure of ILs. Although further purification of commercial ILs could be required depending on the specific application of these solvents, suppliers are continuously providing purer ILs.

Regarding the commercial application of the ILs in separation processes, some key challenges have to be faced: (1) the design of a process of sufficient size for pilot and full-scale implementation; (2) the demonstration of the stability of this process in full-scale processes under real operating conditions; (3) a better understanding of toxicity and the potential environmental impact of ILs to design less toxic and biodegradable ILs, even from removable sources, for industrial applications; and (4) the study of cost-benefit and the economic and life-cycle analyses of the processes that involve ILs.

As evidenced in this book, IL technology could involve important benefits in separation processes. The proposed methodologies to intensify processes could reduce and even eliminate the dependence on volatile organic solvents. The application of ILs—often heralded as the green, high-tech media of the future—in separation has been incessantly progressing during the past 10 years, and no limit to future progress is currently in sight.

REFERENCES

[1] L.C. Branco, J.G. Crespo, C.A.M. Afonso, High selective transport of organic compounds by using supported liquid membranes based on ionic liquids, Angew. Chem. Int. Ed. 41 (2002) 2771–2773.
[2] M. Matsumoto, Y. Inomoto, K. Kondo, Selective separation of aromatic hydrocarbons through supported liquid membranes based on ionic liquids, J. Membr. Sci. 246 (2005) 77–81.

[3] F.J. Hernandez-Fernandez, A.P. de los Rios, M. Rubio, F. Tomas-Alonso, D. Gomez, G. Víllora, A novel application of supported liquid membranes based on ionic liquids to the selective simultaneous separation of the substrates and products of a transesterification reaction, J. Membr. Sci. 293 (2007) 73–80.
[4] P. Scovazzo, A.E. Visser, J.H. Davis Jr., R.D. Rogers, C.A. Koval, D.L. DuBois, R.D. Noble, Supported ionic liquid membranes and facilitated ionic liquid membranes ionic liquids: industrial applications to green chemistry, in: R.D. Rogers, K.R. Seddon (Eds.), ACS Symposium Series 818, Am. Chem. Soc, 2002, p. 68. Washington, DC.
[5] F.J. Hernandez-Fernandez, A.P. de los Rios, F. Tomas-Alonso, D. Gomez, G. Víllora, Kinetic resolution of 1-phenylethanol integrated with separation of substrates and products by a supported ionic liquid membrane, J. Chem. Technol. Biotechnol. 82 (2007) 190–195.
[6] L.A. Neves, N. Nemestothy, V.D. Alves, P. Cserjesi, K. Belafi-Bako, I.M. Coelhoso, Separation of biohydrogen by supported ionic liquid membranes, Desalination 240 (2009) 311–315.
[7] P. Luis, L.A. Neves, C.A.M. Afonso, I.M. Coelhoso, J.G. Crespo, A. Garea, A. Irabien, Facilitated transport of CO_2 and SO_2 through supported ionic liquid membranes (SILMs), Desalination 245 (2009) 485–493.
[8] M. Regel-Rosocka, M. Wisniewski, Selective removal of zinc(II) from spent pickling solutions in the presence of iron ions with phosphonium ionic liquid cyphos IL 101, Hydrometallurgy 110 (2011) 85–90.
[9] F.J. Alguacil, M. Alonso, F.A. Lopez, A. Lopez-Delgado, I. Padilla, H. Tayibi, Pseudo-emulsion based hollow fiber with strip dispersion pertraction of iron(III) using $(PJMTH^+)_2(SO_4^{2-})$ ionic liquid as carrier, Chem. Eng. J. 157 (2010) 272–366.
[10] P. Bonhote, A.P. Dias, N. Papageourgio, K. Kalayanasundaram, M. Gratzel, Hydrophobic, highly conductive ambient-temperature molten salts, Inorg. Chem. 35 (1996) 1168–1178.
[11] T. Welton, Room-temperature ionic liquids. Solvents for synthesis and catalysis, Chem. Rev. 99 (1999) 2071–2083.
[12] T.K. Carlisle, J.E. Bara, C.J. Gabriel, R.D. Noble, D.L. Gin, Interpretation of CO_2 solubility and selectivity in nitrile-functionalized room-temperature ionic liquids using a group contribution approach, Ind. Eng. Chem. Res. 47 (2008) 7005–7012.
[13] A.P. de los Ríos, F.J. Hernández-Fernández, M. Rubio, F. Tomás-Alonso, D. Gómez, G. Víllora, Prediction of the selectivity in the recovery of transesterification reaction products using supported liquid membranes based on ionic liquids, J.Membrane Sci. 307 (2008) 225–232.
[14] L.C. Branco, J.G. Grespo, C.A.M. Afonso, Studies on the selective transport of organic compounds by using ionic liquids as a novel supported liquid membranes, Chem. Eur. J. 8 (2002) 3865–3871.
[15] A.E. Visser, R.P. Swatloski, W.M. Reichert Jr., J.H. Davis, R.D. Rogers, R. Mayton, S. Sheff, A. Wierzbicki, Task-specific ionic liquids for the extraction of metal ions from aqueous solutions, Chem. Commun. 1 (2001) 135–136.
[16] P. Luis, I. Ortiz, R. Aldaco, A. Irabien, A novel group contribution method in the development of a QSAR for predicting the toxicity (*Vibrio fischeri* EC_{50}) of ionic liquids, Ecotoxicol. Environ. Saf. 67 (2007) 423–429.
[17] D.T. Allen, D.R. Shonnard, Green Engineering: Environmentally Conscious Design of Chemical Processes, Prentice-Hall, Upper Saddle River, NJ, 2002. ISBN 0-13-061908-6.
[18] A.A. Frid, E.J. Matthews, Prediction of drug-related cardiac adverse effects in humans-B: use of QSAR programs for early detection of drug-induced cardiac toxicities, Regul. Toxicol. Pharmacol. 56 (2010) 276–289.
[19] F.A. Pasha, M. Muddassar, C. Lee, S. Joo Cho, Mechanism based QSAR studies of N-phenylbenzamides as antimicrobial agents, Environ. Toxicol. Pharmacol. 26 (2008) 128–135.
[20] A. Irabien, A. Garea, P. Luis, Hybrid molecular QSAR model for toxicity estimation: application to ionic liquids, Comput. Aided Chem. Eng. 26 (2009) 63–67.
[21] M. Alvarez-Guerra, A. Irabien, Design of ionic liquids: an ecotoxicity (*Vibrio fischeri*) discrimination approach, Green Chem. 13 (2011) 1507–1516.
[22] R.J. Bernot, M.A. Brueseke, M.A. Evans-White, G.A. Lamberti, Acute and chronic toxicity of imidazolium-based ionic liquids on *Daphnia magna*, Environ. Toxicol. Chem. 24 (2005) 87–92.

[23] A. Romero, A. Santos, J. Tojo, A. Rodriguez, Toxicity and biodegradability of imidazolium ionic liquids, J. Hazard. Mater. 151 (2008) 268–273.
[24] C. Pretti, C. Chiappe, I. Baldetti, S. Brunini, G. Monni, L. Intorre, Acute toxicity of ionic liquids for three fresh water organisms: pseudokirchneriellasubcapitata, *Daphnia magna* and Daniorerio, Ecotoxicol. Environ. Safe 72 (2009) 1170–1176.
[25] J. Ranke, A. Mullera, U. Bottin-Webera, F. Stockb, S. Stoltea, J. Arninga, R. Stormanna, B. Jastorffa, Lipophilicity parameters for ionic liquid cations and their correlation to in vitro cytotoxicity, Ecotoxicol. Environ. Saf. 67 (2007) 430–438.
[26] K.M. Docherty, C.F. Kulpa, Toxicity and antimicrobial activity of imidazolium and pyridinium ionic liquids, Green Chem. 7 (2005) 185–189.
[27] D. Zhao, Y. Liao, Z. Zhang, Review toxicity of ionic liquids, Dental Pract. 35 (2007) 42–48.
[28] L. Ropel, L.S. Belveze, S.N.V.K. Aki, M.A. Stadtherr, J.F. Brennecke, Octanol–water partition coefficients of imidazolium-based ionic liquids, Green Chem. 7 (2005) 83–90.
[29] A.P. de los Rios, F.J. Hernandez-Fernandez, F. Tomas-Alonso, M. Rubio, D. Gomez, G. Víllora, On the importance of the nature of the ionic liquids in the selective simultaneous separation of the substrates and products of a transesterification reaction through supported ionic liquid membranes, J. Membr. Sci. 307 (2008) 233–238.
[30] A.P. de los Ríos, F.J. Hernández-Fernández, F.J. Alguacil, L.J. Lozano, A. Ginesta, I. García-Díaz, S. Sanchez-Segado, F.A. López, C. Godínez, On the use of imidazolium and ammonium-based ionic liquids as green solvents for the selective recovery of Zn(II), Cd(II), Cu(II) and Fe(III) from hydrochloride aqueous solutions, Sep. Purif. Technol. 97 (2012) 150–157.
[31] S.V. Dzyuba, R.A. Bartsch, Recent advances in applications of room-temperature ionic liquid/supercritical CO_2 systems, Angew. Chem. Int. Ed. 42 (2003) 148–150.
[32] T. Dumont, D. Barth, C. Corbier, G. Branlant, Enzymatic reaction kinetic: comparison in an organic solvent and in supercritical carbon dioxide, Biotechnol. Bioeng. 40 (1992) 329–333.
[33] S. Srivastava, G. Madras, J. Modak, Enzymatic synthesis of ethyl palmitate in supercritical carbon dioxide, J. Supercrit. Fluids 27 (2003) 55–64.
[34] L.A. Blanchard, Z. Gu, J.F. Brennecke, High-pressure phase behaviour of ionic liquids/CO_2 systems, J. Phys. Chem. B 105 (2001) 2437–2444.
[35] L.A. Blanchard, J.F. Brennecke, Recovery of organic products from ionic liquids using supercritical carbon dioxide, Ind. Eng. Chem. Res. 40 (2001) 287–292.
[36] F.J. Hernández-Fernández, A.P. de los Ríos, F. Tomás-Alonso, D. Gómez, G. Víllora, Improvement in the separation efficiency of transesterification reaction compounds by the use of supported ionic liquid membranes based on the dicyanamide anion, Desalination 244 (2009) 122–129.
[37] F. Tomás-Alonso, A.M. Rubio, R. Álvarez, J.A. Ortuño, Dynamic potential response and SEM-EDX studies of polymeric inclusion membranes based on ionic liquids, Int. J. Electrochemical Sci. 8 (2013) 4955–4969.
[38] F.J. Hernández-Fernández, A.P. de los Ríos, F. Mateo-Ramírez, S. Galai, S. Sánchez-Segado, M.D. Juárez Ruiz, C. Godínez, L.J. Lozano, Improving the efficiency of microbial fuel cells by using polymer inclusion membranes based on ionic liquids, in: 9th European Congress of Chemical Engineering, The Hague, 2013.
[39] L.C. Tomé, M.A. Aboudzadeh, L.P.N. Rebelo, C.S.R. Freire, D. Mecerreyes, I.M. Marrucho, Polymeric ionic liquids with mixtures of counter-anions: a new straightforward strategy for designing pyrrolidinium-based CO_2 separation membranas, J. Mater. Chem. A 1 (2013) 10403–10411.
[40] A.P. de los Ríos, F.J. Hernández-Fernández, F. Tomás-Alonso, J.M. Palacios, D. Gómez, M. Rubio, G. Víllora, A SEM–EDX study of highly stable supported liquid membranes based on ionic liquids, J. Membr. Sci. 300 (2007) 88–94.
[41] N.V. Plechkova, K.R. Seddon, Applications of ionic liquids in the chemical industry, Chem. Soc. Rev. 37 (2008) 123–150.
[42] P. Wasserscheid, T. Welton, Ionic Liquids in Synthesis, Wiley-VCH Verlag GmbH & Co. KGaA, 2002. ISBN: 3-527-30515-7.
[43] K.J. Fraser, E.I. Izgorodina, M. Forsyth, J.L. Scott, D.R. MacFarlane, Liquids intermediate between “molecular” and “ionic” liquids: liquid ion pairs? Chem. Commun. (2007) 3817–3819.

[44] F.J. Hernández-Fernández, A.P. de los Ríos, F. Tomás-Alonso, D. Gómez, G. Víllora, Improvement in the separation efficiency of transesterification reaction compounds by the use of supported ionic liquid membranes based on the dicyanamide anion, Desalination 244 (2009) 122–129.
[45] P.J. Scammels, J.L. Scott, R.D. Singer, Ionic liquids: the neglected issues, Aust. J. Chem. 58 (2005) 155–169.
[46] K.R. Seddon, A. Stark, M.J. Torres, Influence of chloride, water, and organic solvents on the physical properties of ionic liquids, Pure Appl. Chem. 72 (2000) 2275–2287.

INDEX

Note: Page numbers followed by f indicate figures; t, tables; b, boxes.

G

L

M

N

FSC
www.fsc.org
MIX
Papier | Fördert gute Waldnutzung
FSC® C083411

Kolibri 360 GmbH
Ferdinand-Jühlke-Straße 7
99095 Erfurt, Deutschland
produktsicherheit@kolibri360.de